THE BOOK OF SHELLS

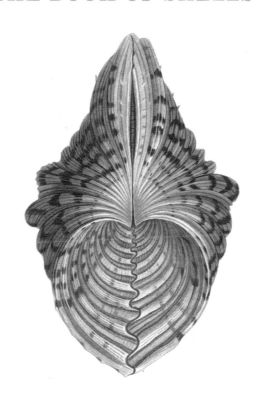

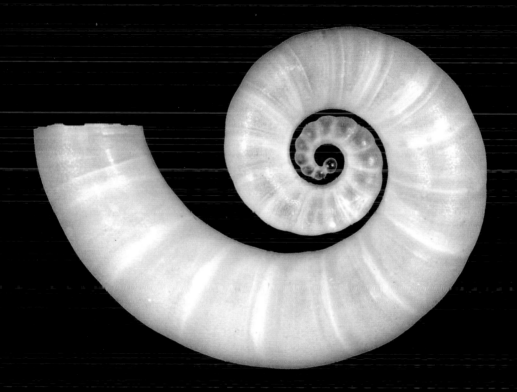

M. G. HARASEWYCH & FABIO MORETZSOHN

THE BOOK OF SHELLS

A LIFE-SIZE GUIDE TO IDENTIFYING AND
CLASSIFYING SIX HUNDRED SEASHELLS

THE UNIVERSITY OF CHICAGO PRESS

Chicago and London

M. G. HARASEWYCH is a curator of the Department of Invertebrate Zoology at the Smithsonian Institution in Washington, D.C., which houses one of the world's largest mollusk collections. He has discovered a variety of new species, written widely for journals and periodicals, and authored *Shells: Jewels from the Sea*.

FABIO MORETZSOHN has a doctorate in zoology and is Assistant Research Scientist and Adjunct Professor of Biology at the Harte Research Institute for Gulf of Mexico Studies, Texas A&M University–Corpus Christi. He is one of the authors of the *Encyclopedia of Texas Seashells*.

The University of Chicago Press, Chicago 60637
The University of Chicago Press, Ltd., London
© The Ivy Press Limited 2010
All rights reserved. Published 2010
Printed in China

18 17 16 15 14 13 12 11 10   1 2 3 4 5

ISBN-13: 978-0-226-31577-5 (cloth)
ISBN-10: 0-226-31577-0 (cloth)

Library of Congress Cataloging-in-Publication Data

Harasewych, M. G.
    The book of shells : a life-size guide to identifying and classifying six hundred seashells / M. G. Harasewych and Fabio Moretzsohn.
        p. cm.
    Includes bibliographical references and index.
    ISBN-13: 978-0-226-31577-5 (cloth : alk. paper)
    ISBN-10: 0-226-31577-0 (cloth : alk. paper) 1. Shells. 2. Mollusks. I. Moretzsohn, Fabio. II. Title.
    QL405.H255 2010
    594.147'7—dc22
                                        2009034321

♾ The paper used in this publication meets the minimum requirements of the American National Standard for Information Sciences—Permanence of Paper for Printed Library Materials, ANSI Z39.48-1992.

This book was conceived, designed, and produced by
Ivy Press
210 High Street, Lewes
East Sussex BN7 2NS
United Kingdom
www.ivy-group.co.uk

Creative Director PETER BRIDGEWATER
Publisher JASON HOOK
Art Director MICHAEL WHITEHEAD
Editorial Director CAROLINE EARLE
Commissioning Editor KATE SHANAHAN
Designer GLYN BRIDGEWATER
Photographs M. G. HARASEWYCH
Additional Research STEVE LUCK, COLIN SALTER
Illustrator CORAL MULA
Map artwork RICHARD PETERS

CONTENTS

Foreword 6

Introduction 8

What is a mollusk? *10*

What is a shell? *14*

Shell collecting *18*

Identifying seashells *22*

The shells *26*
CHITONS *28*
BIVALVES *36*
SCAPHOPODS *168*
GASTROPODS *174*
CEPHALOPODS *630*

Appendices *638*
Glossary 640
Resources 644
Evolutionary classification of the mollusca 646
Index of species by common name 650
Index of species by scientific name 653
Acknowledgments 656

FOREWORD

Shells are the external skeletons of mollusks. Like ancient volumes or tablets, they record the history of the animals that made them. Shells archive every aspect of the animal's life, from its early larval stage through the years, decades or, in some cases, a century or more of life. If fossilized, they may preserve this information for hundreds of millions of years.

If well preserved, the larval shell may tell us whether the animal was brooded or hatched from an egg capsule as a crawling juvenile, or if it spent time in the plankton before metamorphosing into a small version of an adult. All mollusks increase the size of their shells by adding incrementally to their edges; to the margins of shell plates in chitons, to valve edges in bivalves, to apertural margins in scaphopods, gastropods, and cephalopods. Much like tree rings, these sequential layers chronicle the life of the mollusk, sometimes in intricate detail. Some intertidal bivalves, for example, add shell material when the tide is in, but resorb shell when the tide is out, producing a new, recognizable layer with each tidal cycle.

Some shells grow slowly and regularly, others quickly and episodically, producing large sections often demarcated by varices (thickening along the lip of a shell). Most mollusks grow rapidly and in a regular pattern until they reach adulthood, when energy is redirected from growth to reproduction. Some continue to grow in the same general pattern, although much more slowly. Others, such as cowries, have terminal growth, altering the shape of their shell irrevocably in a way that precludes further growth. These adults differ dramatically in appearance from juveniles. They may continue to thicken their shells to become heavier, but not significantly larger.

Many of the most conspicuous attributes of a shell are inherited, and indicate its genealogy. The distinctive shapes of scallops, spider conchs, or chambered nautiluses clearly identify them as members of their respective classes, families, and genera. Other features, such as a flattened limpet shape, may be adaptations to particular habitats that occur independently in many different lineages.

More subtle features of shape and condition provide a wealth of information about the species or even the individual specimen. The presence of large varices and spines indicate that the animals live on hard substrates, while smooth, tapered, elongated shells are characteristic of animals that burrow into sand or mud. Similarly, delicate, frilly spines that remain unbroken reveal a calm, subtidal habitat, while worn or eroded shells are indicators of exposure to waves. Repaired shell breaks or incomplete boreholes bear witness to attacks by predators, while traces of encrusting organisms, boring sponges, and symbionts all add information about the life and times of the animal that produced the shell.

It is second nature to us to admire the delicate shape, color, and beauty of a perfect specimen. Taking the time to "read" each shell as an autobiography of the animal that produced it is often just as rewarding.

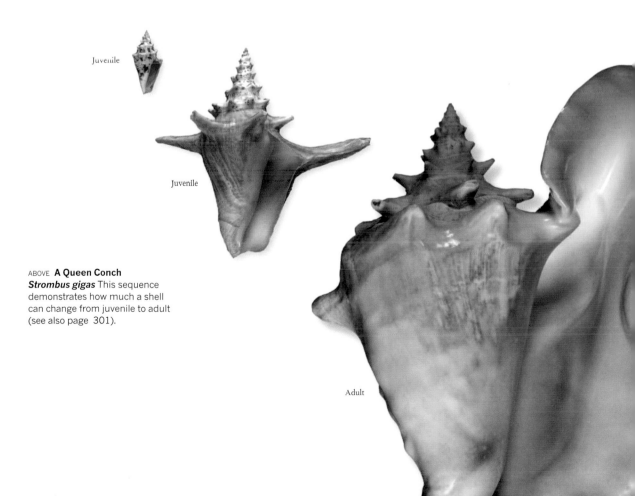

Juvenile

Juvenile

Adult

ABOVE **A Queen Conch**
Strombus gigas This sequence
demonstrates how much a shell
can change from juvenile to adult
(see also page 301).

8

INTRODUCTION

Anyone who has been to the seaside or the shore of a lake or river, or who has walked through the woods or a garden has probably seen and picked up a few shells. Many will have brought these shells home and formed the rudiments of a casual collection on a bookshelf, in a shoe box, or in the garden, without giving the matter further thought. Few, however, will have paused to consider the extraordinary variety of forms into which mollusks mold their shells, each the product of a long evolutionary history and each adapted to a particular habitat.

Although all seashells are made by mollusks, not all mollusks make shells. Of those mollusks that do make shells, the majority live in the seas and oceans of the world, from the tropics to the poles, from above the high tide line, where only wave spray reaches, to the bottoms of ocean trenches. While mollusks originated and diversified in the oceans, a sizeable proportion of species now live on land or in fresh water, the results of numerous independent colonizations of these habitats.

ABOVE **Cellana nigrolineata** Black-lined Limpet (See page 181)

In terms of the number of living species, mollusks are the most diverse animals in the oceans. While the best known and most familiar mollusks tend to be the larger, more conspicuous species, molluscan diversity is dominated by small animals. A recent study of the shelled mollusks from a site in New Caledonia revealed a range in sizes from ⅟₆₄ in (0.4 mm) to 18 in (450 mm), yet the average size was ⅔ in (17 mm). On average, less than 16 percent of the species were larger than 2 in (50 mm), and most are far smaller.

9

When perusing the 600 shells depicted in this book, it is informative to consider that they represent but a fraction of known species of mollusks, and that a proportional sampling of the phylum (a unit by which organisms are classified) would have produced a work dominated by tiny snails. Most major lineages of shelled mollusks living in the sea are represented here, and they are arranged according to current understanding of the branching patterns of their evolutionary history.

BELOW *Pinna rugosa*
Rugose Pen Shell
(See page 66)

While the number of species apportioned to the major classes does reflect their relative diversity, sampling within each class has clearly been skewed toward the larger and more familiar species. Interspersed among these are rare and newly discovered forms, both tiny and large. Many families are not represented at all, for there are far more than 600 families of mollusks, while others are conspicuously over-represented to illustrate the range of sizes and shapes that occur even among relatively closely related species.

Within each family, species are arranged by photographed size, from smallest to largest, without regard to their evolutionary relationships. Each shell is shown at its actual size—shells below ¼ in (5 mm) have been photographed using a scanning electro-microscope (SEM) in order to capture the detail—and supplemented with detail images and nineteenth-century engravings.

WHAT IS A MOLLUSK?

Mollusks are among the oldest and most diverse groups of animals on the planet. Like all taxa, they are defined by their genealogy. That is to say, they have a common ancestor from which all members of the phylum Mollusca, living and extinct, are descended.

EARLY MOLLUSKS

The earliest mollusks were small ($\frac{1}{25}$–$\frac{1}{12}$ in / 1–2 mm), marine, bilaterally symmetrical animals, with an anterior head, a ventral foot, and a posterior mantle cavity that contained paired gills, sensory organs called osphradia, openings of the genital and excretory organs, and the anus. The head contained a mouth with a radula, a ribbonlike feeding structure unique to mollusks that is like a flexible rasp. The foot was an elongated structure used for locomotion, and the visceral mass, situated above the foot, contained the major organ systems, including the heart, kidneys, digestive glands, and gonads. The nervous system consisted of three pairs of ganglia, one for each body region (the head, foot, and viscera). A cuticle covering the body secreted calcareous spicules or scales.

Over the course of geological time, the descendants of this common ancestor diversified and differentiated, giving rise to multiple branches, each with distinctive features and adaptations. Many of the most basal of these branches, the classes within the phylum Mollusca, diverged during the Cambrian period. Some, such as the Gastropoda, Bivalvia, and Cephalopoda, underwent significant anatomical changes, producing combinations of features that enabled rapid exploitation of new environments. Other classes (among them the Polyplacophora,

LEFT **A Trumpet Triton,**
Charonia tritonis.
This gastropod
mollusk is one of
the few predators
of starfish (see
page 381)

Monoplacophora, Scaphopoda) retained their basic anatomical
organization; they persist to the present day, little modified and with
comparatively low diversity. Mollusks are so ancient and diverse that
there are few diagnostic characters that are both unique to Mollusca and
ubiquitous to all its classes.

CHITONS

The chitons (Class Polyplacophora) have elongated, flattened, bilaterally
symmetrical bodies covered by a shell of eight overlapping transverse
plates that are surrounded by a cuticularized girdle (muscular band).
The foot is long and muscular, and flanked on both sides and by a long
mantle cavity that contains multiple pairs of gills (from 6 to 88). The head
is reduced, lacking eyes and tentacles. Light-sensing cells that are unique
to chitons pass through tiny canals in the shell plates. All chitons live in the
ocean, most on rocky bottoms in fairly shallow water where they graze on
algae and sponges.

GASTROVERMS

Gastroverms (Class Monoplacophora) are relatively small ($\frac{1}{36}$–1½ in/
0.7–37 mm), ovate, bilaterally symmetrical mollusks that have a single,
conical, limpetlike shell with eight pairs of serially repeated muscle scars.
They were thought to be extinct, but thirty living species have been
discovered since 1957, nearly all from deepsea habitats (571–21,289 ft/174–
6,489 m), where they inhabit muddy, rocky, or gravelly bottoms. All feed
on organic matter and on small animals in the sediment.

BIVALVES

Bivalves (Class Bivalvia) are the second largest class of mollusks. They have a bilaterally symmetrical body that is completely enclosed in a shell

ABOVE **A Japanese Moon Scallop,** *Amusium japonicum*, a bivalve mollusk, on the ocean bed.

consisting of two valves (left and right) that are connected by an elastic ligament. The head is reduced, and the radula is absent. Most bivalves have a capacious mantle cavity that accommodates large gills. In addition to being a respiratory organs, they filter food particles from the water. Some primitive forms feed directly on the organic matter in fine sediments, a few specialized groups derive nutrition from symbiotic algae or bacteria, while others capture and consume small crustaceans and worms in the deep sea. Most bivalves burrow in sand or mud, some in wood, clay, or coral. Some attach to hard substrates with threadlike strands (byssus), others by cementing one of their valves. Several different groups have adapted to freshwater habitats.

SCAPHOPODS

Scaphopods or tusk shells (Class Scaphopoda) comprise a small group of about 600 living species. They have tall, bilaterally symmetrical bodies completely contained in a long, curved, tapering tubular shell that is open at both ends. Scaphopods lack eyes and gills. They burrow in soft bottoms using a foot that emerges from the larger opening. The smaller opening remains near the surface of the sediment. Scaphopods feed on microscopic organisms in the sediment, which they capture with thin, threadlike tentacles called captacula.

GASTROPODS

Gastropods or snails (Class Gastropoda) comprise the largest class of mollusks. During their larval stage, all gastropods undergo torsion, a process that twists the animal until the formerly posterior mantle cavity is rotated to a position over the head, resulting in an asymmetrical animal with a single coiled shell. Snail shells assume a variety of forms, ranging from microscopic (1/75 in/0.3 mm) to enormous (39 in/1 m). The shell of a snail may be external, internal, or entirely absent. Like bivalves, snails inhabit all marine and freshwater habitats. Unlike any other mollusks, snails developed lungs and have also colonized land environments ranging from forests to mountains to deserts. Snails may be herbivores, carnivores, parasites, filter feeders, detritivores, or even chemoautotrophs.

CEPHALOPODS

The earliest cephalopods (Class Cephalopoda) had external shells, with chambers that were interconnected by a tube that allowed them to become gas-filled and buoyant. During the course of their evolution, the vast majority of cephalopods have lost an external shell. Some, including sepia, cuttlefish, and squid have internal shells that have been reduced to various degrees; octopuses lack any shell at all. Some cephalopod lineages developed the ability to swim by undulating their fins, as well as by jet propulsion.

Cephalopods inhabit all oceans at all depths. Many live in shallow coastal areas, while others are pelagic, spending their lives swimming or drifting through the open ocean at great distances from surface, shore or bottom. Cephalopods range from 1 in (25 mm) to more than 46 ft (14 m) in length, and include both the Giant Squid and the even larger Colossal Squid, the largest known invertebrate. All are predatory, with the head and mouth surrounded by muscular, sucker-bearing tentacles that capture prey, which is then eaten with a parrotlike beak and radular teeth.

LEFT **A Chambered Nautilus, *Nautilus pompilius*** on the ocean bed. This species is one of a very few living cephalopod species with an external shell (see page 632). Nautilus control their buoyancy by regulating the flow of gas into and out of the chambers in their shells.

WHAT IS A SHELL?

As broadly defined, a shell is a hard outer covering that encases certain organisms, usually for the purpose of protecting them from the environment. Many organisms, ranging from microscopic foraminifera to turtles, produce shells using a variety of materials.

HOW A SHELL FORMS

External shells composed of calcium carbonate are secreted by many invertebrate phyla, among them Cnidaria (corals), Arthropoda (crabs and barnacles), Echinodermata (sea urchins), Brachiopoda (lamp shells), and Bryozoa (moss animals), yet the term "shell" or, more specifically "seashell" almost inevitably conjures the image of the calcified external skeleton of a mollusk. These molluscan shells are the subject of this book.

The shell is secreted by the mantle (or pallium), a specialized tissue that is present in every mollusk. One section produces a thin layer of a protein called conchiolin. Other cells secrete a fluid into the narrow space between the animal's tissues and the conchiolin layer. Calcium carbonate crystallizes from the fluid onto the inner surface of the conchiolin, producing a continuously mineralized shell. The shells of all mollusks are secreted outside the animal's tissues. Unlike the bones of vertebrates, shells do not contain cells or DNA.

In all mollusks, shell growth occurs through the addition of new bands of conchiolin along the existing edges of the shell, followed by crystallization of calcium carbonate onto this matrix. Shells can be made thicker by the successive secretion of conchiolin matrix and calcium carbonate to produce additional internal layers.

BIVALVES

The shells of bivalves consist of two separate valves. The tiny larva of a bivalve produces a single, uncalcified, caplike shell, called a pellicle. As the larva grows, it is gradually enveloped by two mantle lobes, each developing a separate center of calcification—the dissoconchs, the parts of the shell produced after the larva metamorphoses, assume the proportions and features of the adult bivalve. Most bivalves are composed of two valves that are mirror images of each other. The shell usually consists of three layers: an outer periostracum, which may be quite thick in some species, and outer and inner shell layers. The outer layer forms surface details such as scales or spines. In some bivalves, the shells have become reduced; in others, they have become incorporated into large, cylindrical tubes.

15

Bivalve External Shell Features

Dorsal

Anterior

Adductor muscle scar

Posterior

Concentric ribs

Ventral

Bivalve Internal Shell Features

Teeth   Hinge plate

Umbo

Ligament

Adductor muscle scar

Adductor muscle scar

Pallial sinus

Denticulate margin

Pallial line

Dorsal Bivalve Features

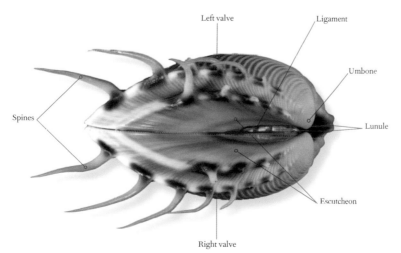

Left valve

Ligament

Umbone

Spines

Lunule

Escutcheon

Right valve

Scaphopod External Shell Features

Apex

Longitudinal ribs

Anterior

Posterior

SCAPHOPODS

The tubular shell of scaphopods originates as a small, caplike shell in the larva. During development, the edges expand to surround the larva and fuse along the opposite side to form a tube. After metamorphosis, growth occurs through the addition of shell to the circular edge of the anterior opening, producing a shell consisting of a periostracum and two to four layers of aragonite, a crystalline form of calcium carbonate. As the shell grows, the length and anterior diameter increase and the inner walls thicken. The posterior opening is maintained at an appropriate diameter by the mantle, which dissolves, constricting portions of the shell.

CHITONS

The shells of chitons are secreted as eight separate plates, and include a head valve, six intermediate valves, and a tail valve. Each valve is composed of four separate layers. The outermost layer is the periostracum; beneath it is the tegmentum; then comes the articulamentum, the thickest and hardest of the layers; the innermost layer is the hypostracum, which is composed of columnar crystals. The valves are held together by muscles and a cuticular girdle that fits between the tegmentum and articulamentum. Depending on the species, the girdle may be covered by proteinaceous hairs, or calcareous spines, granules, or scales.

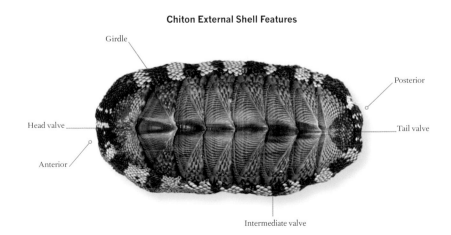

Chiton External Shell Features

Girdle

Posterior

Head valve

Tail valve

Anterior

Intermediate valve

GASTROPODS AND CEPHALOPODS

The shells of gastropods and cephalopods all begin as simple, cap-shaped shells formed by the larvae of these mollusks. Growth occurs by incremental addition to the roughly circular rim to produce a conical shell. In extinct ancestral cephalopods, the caplike larval shells continued to grow into long, narrow, conical tubes. The animals remained near the base of the cone, and periodically sealed off the upper portions of the cone with partitions (septa). The few surviving species of *Nautilus* are the only cephalopods living today that still have external shells. In all other living cephalopods, the shell has become internal, greatly reduced, or is absent.

Gastropods also begin life with caplike larval shells. However, their larvae undergo torsion, a 180-degree twisting of the body that produces anatomical asymmetry. This, in turn, leads to helical coiling of their shells, almost invariably in a right-handed spiral. The shape of the resulting spiral can assume a staggering variety of forms, many of which are adaptations to particular environments. It is not uncommon for similar shell forms to be exhibited by distantly related snails, as a result of convergent evolutionary adaptations to particular habitats. As in the cephalopods, the shell has become reduced, internal, or lost within several lineages of gastropods.

Gastropod External Shell Features

Posterior · Protoconch or larval shell · Axial varix · Suture · Shoulder · Spire · Penultimate whorl · Spiral cords · Parietal shield · Teleoconch · Anal or posterior canal · Tooth at rear of aperture · Shoulder · Teeth inside outer lip · Outer lip or labrum · Aperture · Body whorl · Columella · Columellar teeth · Siphonal canal · Anterior

Apex · Body whorl · Suture · Spiral cord · Anal canal · Aperture · Base · Spiral beaded cord · Umbilicus · Siphonal canal · Peripheral keel

Cephalopod External Shell Features

Parietal wall · Outer lip · Axis · Body whorl

SHELL COLLECTING

Since prehistoric times, humans have acquired shells and accorded them a treasured status. Seashells have been used as tools and currency; they have been incorporated or depicted in ornamental and ritualistic objects; they have held significance for many cultures, even those far from the sea.

HISTORY OF SHELL COLLECTING

The business of gathering shells as specimens goes back to Roman times at least. Collections of shells are among the artifacts found in the ruins of Pompeii. The expansion of the known world in the late Middle Ages led to a fascination in Europe with curiosities brought from distant lands. Merchants and aristocrats assembled collections of rare objects, including shells, that became symbols of their wealth and prestige. Scholars were employed to arrange, organize, and publish on these vast collections, many of which became the foundations of major museums.

TYPES OF SHELL COLLECTION

Most shell collections are general collections that strive to assemble a broad sampling from the enormous variety of shells. Some collectors focus on more specialized collections. Some may be restricted to the varieties of shells that occur in a particular area or habitat. Others confine their interests to a specific group of closely related shells. Cowries, cones, murexes, volutes, olives, and scallops are among the most popular groups for such collectors. Still others prefer to collect micromollusks, specimens with adults that do not exceed ½ inch (10 mm) in length.

New England Neptune

ABOVE **Some collectors combine** interests and collect shells that are illustrated on postage stamps. The type of shell collection is limited only by the imagination of the collector.

FINDING SPECIMENS

Adding specimens to one's collection has its own pleasures. It can involve rewarding personal effort: strolling purposefully on a beach after a storm; searching rocky shorelines at low tide; snorkeling or scuba diving; or other, more specialized means such as rock scrubbing, dredging, or trapping. As well as enabling the collector to select specimens, these activities offer the opportunity to observe shells in their element, and to develop an appreciation for the way each one has become adapted to its environment. Novice collectors should be aware, however, that some regions require permits or a fishing license to collect mollusks, while others either limit or prohibit the collection of living specimens. Collectors should respect the natural habitat, and minimize their impact on the areas in which they collect. After examining the underside of rocks, for example, they should be returned to their original position and orientation.

ESSENTIAL EQUIPMENT

Equipment varies with the type of collecting, and each collector soon develops a personal set of gear optimized for specific habitats. The basics usually include a plastic bucket, a few plastic bags, forceps for retrieving small specimens from crevices, a knife or spatula for dislodging chitons, limpets, or bivalves from the rocks to which they are attached, and perhaps a garden spade and a small sieve for burrowing mollusks. Other items of equipment may include a digital camera and a portable GPS device, but the most important tools are a pencil, a few labels, and a notebook in which to record important details about each specimen collected.

RECORDING DATA

The value of a specimen lies not only in its rarity and perfection, but also in the quality of the collecting information that accompanies it. This should include, at a minimum, the precise location, depth, date, and even time it was collected, as well as ecological information, such as "on rocks exposed at low tide" or "buried in fine sand at edge of eel grass bed." Anyone reading the label should be able to return to the place where the specimen was collected.

PREPARING A SPECIMEN

Unless the collector has a saltwater aquarium, the tissues must be promptly removed from the shell before it is added to a collection. Placing the shell in warm water and bringing it to a boil for a few minutes usually loosens the attachment of the animal to the shell. When the shell is cool enough to touch, the animal can be carefully unwound from the snail or extracted from the bivalve, using forceps or dental tools. Occasionally, a second boiling is required.

Alternatively, the specimens may be frozen by placing the collecting bags in the freezer. After a day or so, the shell is thawed, and the tissues can be carefully removed. Often the specimen has to be frozen and thawed repeatedly to loosen the animal so it can be removed in one piece. Some collectors (and most museums) prefer to have specimens in their natural state. Others choose to remove encrusting organisms and the periostracum by soaking the specimen in diluted bleach, then picking and cleaning with dental tools and toothbrushes.

ABOVE **For specialist** collectors it is vital to label and organize a collection with the precise information about each shell.

ORGANIZING A COLLECTION

In a collection, all the specimens of the same species that were collected together at the same place and time should be grouped together, with a label that includes the detailed collection data. When the specimens are identified, the genus and species should also be added to the label. Many collectors maintain a catalog of their collections, either a handwritten ledger, or more frequently as a spreadsheet on their computers. This allows them to keep track of large and growing collections.

With most collections, there comes a point when specialized storage becomes essential. Most collectors with sizable collections have storage systems similar to those used in museums, namely metal cabinets with multiple shallow drawers filled with rows of paper trays or plastic boxes, each containing one or more specimens together with their label. Collections tend to be arranged according to evolutionary relationships, so that the most closely related specimens are in the same drawer, or close to each other. This makes it easy to find particular species, and also makes the collection useful when comparing a new specimen against identified shells.

Many collectors supplement their collections by exchanging duplicate specimens with other collectors, or by purchasing specimens from shell dealers. There are shell clubs in many cities of the world. Some clubs host annual shell shows, where collectors may display portions of their collections in a variety of categories.

BELOW **The level of** organization in shell collections varies greatly. A large shell collection is usually stored in shallow cabinet drawers to protect specimens from the fading effect of strong sunlight.

21

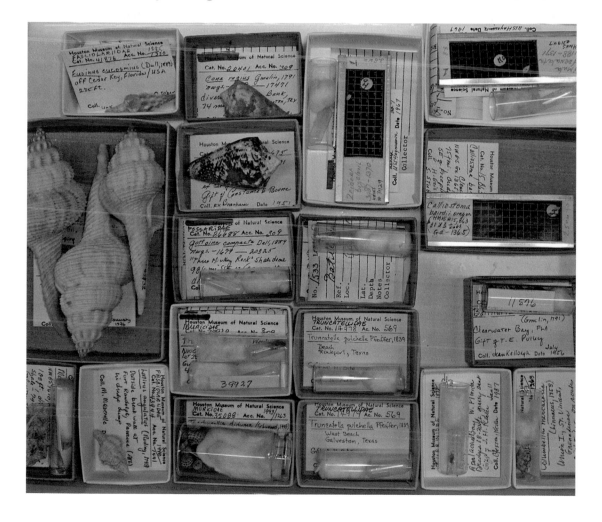

IDENTIFYING SEASHELLS

Identifying a shell may seem a daunting task, especially considering there are roughly 100,000 species of mollusks living today. It is best seen as a process of elimination in which you gradually or rapidly narrow the range of possibilities. Simply knowing that you are dealing with a seashell, for example, immediately rules out all the species of mollusks that do not have a shell, as well as all those that do not live in the sea.

DETERMINING THE CLASS

The first and most basic stage in the process of identification is to determine the class to which the shell belongs. This is done by counting the number of sections that comprise the shell.

If the shell is composed of eight sections, or valves, it is a chiton, of which there are about 1,000 living species. If the shell is composed of two valves, it is a bivalve (about 20,000 species). If the shell consists of a single section (not counting an operculum), then it could be a scaphopod (about 600 species), a cephalopod (six species with an external shell), or a gastropod (more than 50,000 species).

Scaphopods have a long, tapering, tubular shell open at both ends. Cephalopods have shells that are large, planispiral (coiled in a single plane), and subdivided into chambers connected by a siphuncle. Any single-valved shell that has not previously been excluded is likely to be one of the marine snails (gastropods).

23

NARROWING THE OPTIONS

Once the shell has been identified to a class, it needs to be sorted into one of several smaller groups within the class. The identification process continues until the choices are limited to a few species within a single genus, and finally to one of the species within the genus. This is done by comparing the shell directly against known specimens or images until a final identification is achieved.

It should be remembered that this book illustrates fewer than one percent of the known species of seashells. This book will help you to identify many of the more frequently encountered species, but others may require additional references or internet-based research (see Resources on pages 644–645).

SCAPHOPOD SHELL SHAPES

The diversity of scaphopods (tusk shells) is comparatively low, with only some 600 species worldwide. There are far fewer differences among their shells as well.

Living tusk shells are divided into two major groups. The order Dentaliida have shells that taper uniformly, with the widest diameter always at the aperture. The order Gadilida have shells that may be widest at their midpoint, or near the aperture. Each of these groups are further subdivided using such features as the shape of the shell in cross-section, the presence and pattern of longitudinal ribs, as well as the occurrence of longitudinal slits or elongated plugs at the posterior opening.

Tusk shape, widest at aperture

See examples on pages 171, 172, and 173

Tusk shape, widest near the middle of the shell

See example on page 170

GASTROPOD SHELL SHAPES

The gastropods are by far the largest and most diverse group of mollusks. They are subdivided into several major evolutionary branches, including the Patellogastropoda, the Cocculiniformia, the Vetigastropoda, the Neritopsina, the Caenogastropoda, and the Heterobranchia.

Some have very uniform shell shapes; others are extremely diverse and include many vastly different shapes. So, for example, all Patellogastropoda have low, conical, cap-shaped shells, but not all low, conical, cap-shaped shells are Patellogastropods. Cap-shaped shells are widespread among mollusks and occur in each of the major groups of gastropods.

Matching the gastropod specimen to one of the following basic shell shapes will be useful in narrowing down the possibilities in identification—each basic shell shape is referenced to one or more different families of shells that share this basic shape. Each shell may have additional features such as spines that can obscure the basic shape. These additional features are significant at later stages of identification.

An exact match might be found among the species illustrated in this book. However, given the diversity of gastropods, it may be that only a close similarity will be detected, allowing the shell to be ascribed to a family or perhaps a genus, but not to a species. At this point, additional research will be needed for an identification to species. Several useful references and internet web sites are recommended in the Resources section on pages 644–645.

Cap shape	**Ear shape**	**Low conical**	**Tall conical**	**Inverted conical**

See examples on pages 179, 182, 186, 192, 208, 247, 309, 317, 386, 452, and 628	*See examples on pages 200–201, 207, 223, and 357–358*	*See examples on pages 218–219, 222, 227, 231, 271, 273–276, 278–279, 410, and 608*	*See examples on pages 248, 259, 264–268, 307, 387, 392, 429, 599, and 616*	*See examples on pages 540, 586, 589, and 598*

Egg shape	**Spindle shape (fusiform) / Biconical**	**Mace shape (club shape)**	**Spherical**	**Flattened sphere**

See examples on pages 226, 493, 505, and 623	*See examples on pages 186, 293, 299, 343, 375, 377, 438, 444, 474, 550, and 581*	*See examples on pages 372–373, 408, 461, and 472*	*See examples on pages 245, 359, 389, and 470*	*See examples on pages 199, 228, 240, 243, 269, 280, 282, 356, and 612*

Barrel shape	**Pear shape**	**Irregular / Unwound**	**With terminal, flared lip**

See examples on pages 369–371, 455, 562, and 617	*See examples on pages 361–364, 368, and 383*	*See examples on pages 260–261, 285, 322-323, 469, 558, and 629*	*See examples on pages 294, 300, and 303*

BIVALVE SHELL SHAPES

The bivalves are the second most diverse group of mollusks, and are far less variable in the shape of their shells than are the gastropods. The major subdivisions are based in large part on gill anatomy. They include the Protobranchia, the Pteriomorphia, the Paleoheterodonta, and the Heterodonta.

Bivalves that burrow in sand tend to be bilaterally symmetrical, while those that are attached to hard substrates or are free-living tend to be asymmetrical to varying degrees. As with gastropods, some shapes are specific to particular groups, while others occur in multiple groups and may be more indicative of a common habitat than close relationships.

Compare the bivalve to be identified with the basic shell shapes, disregarding spines or other surface sculpture. Additional features such as hinge type, size, color pattern, and surface sculpture will be helpful to distinguish among the families, genera, and species that share a common shell shape. Consult the Resources section (pages 644–645) for additional useful references for identification.

25

Round, discus shape	Triangular, paddle shape	Triangular, ax shape	Fan shape	Irregular/ asymmetrical

See examples on pages 49, 93, 106, 110, 117, 121, 127, 138, 140, and 151	*See examples on pages 51, 54, 104, and 168.*	*See examples on pages 48, 66, 67, 102, 142, and 146*	*See examples on pages 69–90 and 128*	*See examples on pages 43, 57, 58–63, 62, 96–98, 103, 105, 111, and 166–167*

Boat-shape	Heart shape (in posterior view)	Elongated ellipse	Rectangular

See examples on pages 44, 47, 92, 95, 99, 114–115, 125, and 159	*See examples on pages 118–119, 123, and 124*	*See examples on pages 40, 55, 94, 120, 143, 152–153, 155, and 164–165*	*See examples on pages 154 and 156*

CHITON SHELL SHAPE

Chitons are composed of eight separate shell plates surrounded by a tough girdle.

All living chitons can be assigned to one of four groups, the Lepidopleurina, Choriplacina, Ishnochitonina, and Acanthochitonina. These groups are distinguished from each other on the presence or absence of the tegmentum layer and of slitted insertion plates. The specimen has to be disarticulated for these characteristics to be observed. However, the more common species can be identified by the sculptural patterns on the valves as well as by the granules, spicules, scales, or spines present on the girdle.

Shield shape

See examples on pages 33 and 34

CEPHALOPOD SHELL SHAPE

While the diversity of living Cephalopods is considerable, only six primitive species retain an external shell—all of them members of the genus *Nautilus*. Two of the species have an umbilicus on each side of their shells, the other four species do not. There is only a single living species with a coiled internal shell, *Spirula spirula*.

Hundreds of cephalopod species have reduced, internal shells, including dozens of species of cuttlefish, whose shell is the familiar cuttlebone. What appear to be paper-thin, coiled shells are egg cases produced by female *Argonauta*, a relative of the octopus that has no shell.

Helmet shape

See example on page 632

THE SHELLS

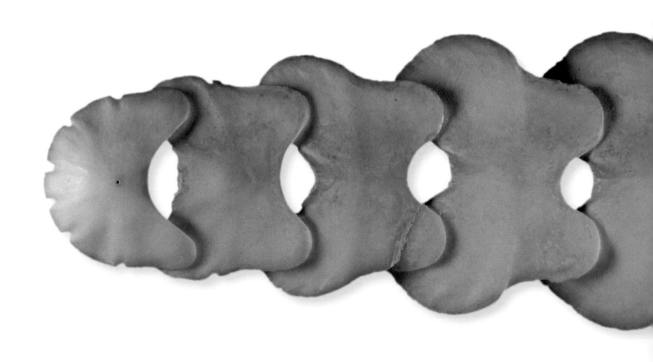

CHITONS

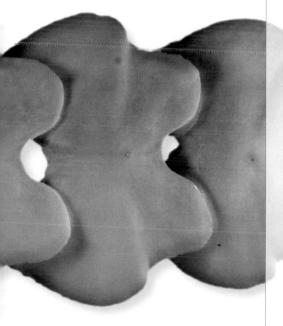

The chitons are primitive mollusks that are covered by a shell composed of eight shell plates or valves held in place by a muscular, cuticle-covered girdle. There are about 1,000 species of chitons living today, ranging in size from ⅛ in (3 mm) to about 16 in (40 cm). All live in the oceans, mostly on rocks and hard bottoms in tropical and temperate regions at intertidal to shallow subtidal depths. Chitons "taste" the substrate using a special organ before scraping the rock surface to feed on algae and encrusting animals. One group of chitons is predatory, capturing small crustaceans and other invertebrates beneath an enlarged region of the girdle.

FAMILY	Ischnochitonidae
SHELL SIZE RANGE	⅝ to 1 in (15 to 25 mm)
DISTRIBUTION	Europe, Mediterranean, and Red Sea
ABUNDANCE	Common
DEPTH	Intertidal to 3,300 ft (1,000 m)
HABITAT	Red calcareous algae
FEEDING HABIT	Grazer and deposit feeder

SHELL SIZE RANGE
⅝ to 1 in
(15 to 25 mm)

PHOTOGRAPHED SHELL
1 in
(25 mm)

30

CALLOCHITON SEPTEMVALVIS
SMOOTH EUROPEAN CHITON
(MONTAGU, 1803)

Callochiton septemvalvis is a typical chiton, and has, like all other chitons, eight valves rather than the seven suggested by its name. Montagu believed that this species was characterized by having only seven valves, but the specimen he used to describe the species was aberrant and missing one valve. *Callochiton septemvalvis* is a common species in shallow waters around most of Europe, from Scandinavia to the Canary Islands, as well as in the Mediterranean and Red Sea. It is found on red coralline algae and other hard substrates. It is a grazer but may also feed on deposits of organic sediments. There are some 200 living species in the family Ischnochitonidae worldwide.

RELATED SPECIES

Ischnochiton wilsoni Sykes, 1896, from South Australia and Victoria, Australia, has a longer and more slender shell that may be pinkish white, with many streaks grayish in the center of the valves, and brown on the lateral areas. *Stenochiton longicymba* (Blainville, 1825), from southern Western Australia to Tasmania, is very elongated in outline, its width being less than one-fifth of its length.

Actual size

The shell of the Smooth European Chiton is medium in size, smooth, broad, and oval in outline. Its valves are moderately elevated and carinated. The body is wide, about half or more the body length. The valves appear smooth to the naked eye, but are composed of diagonal ridges and small granules that are only visible under magnification. The girdle is broad and covered with small spicules. It is orange-red in color, and may have a few white radial bands. The exterior color of the shell valves is brick red to orange, sometimes with traces of green or yellow-orange.

FAMILY	Ischnochitonidae
SHELL SIZE RANGE	1 to 1½ in (25 to 40 mm)
DISTRIBUTION	Western Australia to Tasmania
ABUNDANCE	Common
DEPTH	Shallow subtidal
HABITAT	On sheaths of seagrasses
FEEDING HABIT	Grazer, feeds on seagrasses and algae

SHELL SIZE RANGE
1 to 1½ in
(25 to 40 mm)

PHOTOGRAPHED SHELL
1½ in
(37 mm)

STENOCHITON LONGICYMBA

CLASPING STENOCHITON

(BLAINVILLE, 1825)

31

Stenochiton longicymba has the most elongated shell of the living chitons, with a length-to-width ratio of 7:1. Some fossil chitons had even longer shells, reaching a length-to-width ratio of 32:1. *Stenochiton longicymba* is endemic to Australia, ranging from Western Australia to Tasmania. Its narrow and elongate shape came about as a result of evolutionary adaptation to conditions of life on the blades and root sheaths of *Posidonia australis* and other seagrasses in shallow subtidal depths. It feeds on the seagrass as well as on the algae that grows there.

RELATED SPECIES

Ischnochiton papillosus (C. B. Adams, 1845), from the West Indies and Gulf of Mexico, is a small, shallow-water chiton. It is the most common species in Texas. Its valves are greenish and have fine beads and incised lines. *Nuttallochiton mirandus* (Thiele, 1906), from Antarctica, has a large shell with a wide girdle, and high valves sculptured with strong radial ribs.

The shell of the Clasping Stenochiton is of medium length, elevated, very narrow, and elongate. Its length may reach 6 to 7 times its width, while its height may be more than half the width. The head and tail valves are semi-oval, while the intermediary valves are rectangular. The surface of the valves is smooth, with fine reticulated patterns. The girdle is very narrow, and the width of the valves increases slightly toward the posterior margin. The shell color is brown with creamy white speckles and streaks.

Actual size

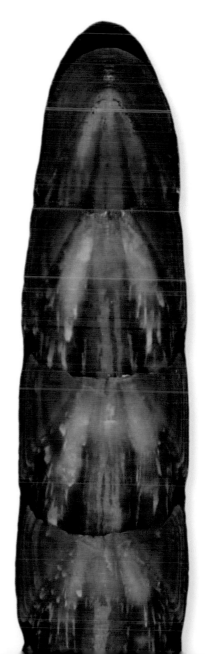

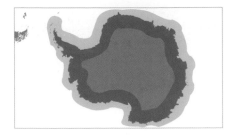

FAMILY	Ischnochitonidae
SHELL SIZE RANGE	1¼ to 4½ in (30 to 120 mm)
DISTRIBUTION	Antarctica
ABUNDANCE	Common
DEPTH	100 to 4,600 ft (30 to 1,400 m)
HABITAT	Hard bottoms
FEEDING HABIT	Grazer, feeds on bryozoans and foraminifera

SHELL SIZE RANGE
1¼ to 4½ in
(30 to 120 mm)

PHOTOGRAPHED SHELL
1½ in
(38 mm)

32

NUTTALLOCHITON MIRANDUS

NUTTALLOCHITON MIRANDUS
(THIELE, 1906)

Nuttallochiton mirandus has a very elevated shell and a wide, leathery girdle surrounding its valves. It is a common circumpolar chiton from Antarctica, ranging from offshore to the deep sea. It is a grazer, feeding primarily on bryozoans and foraminifera. This species of chiton uses its radula to swallow big pieces of bryozoan colonies. In life, the valves overlap with the previous one, unlike the photograph of a preserved specimen shown here. Females and males bend the posterior part of their bodies and release eggs and sperm respectively in the water column. Often found with hexactinellid (glass) sponges.

RELATED SPECIES

Nuttallochiton hyadesi (Rochebrune, 1889), from deep water off Tierra del Fuego, Argentina, and the Weddel Sea, near the Antarctic circle. The valves are similar to *N. mirandus*, but with smaller and more subtle sculpture. *Callochiton septemvalvis* (Montagu, 1803), from European waters in the Atlantic and Mediterranean, has a broad oval shell and may have an orange-red to brick red color.

Actual size

The shell of *Nuttallochiton mirandus* is medium to large in size, elongated oval in shape, and appears saw-toothed in lateral view. Its valves are high, brittle, with an inverted V-shape, and a notch in the center. The sculpture of the valves consists of 8 to 10 strong radial ribs, crossed by fine growth marks. The head valve has stronger radial sculpture than the other valves. The girdle is leathery, wide, and covered with fine and elongate spicules. The shell color is creamy white, sometimes stained in reddish brown.

FAMILY	Mopaliidae
SHELL SIZE RANGE	1⅜ to 3 in (35 to 76 mm)
DISTRIBUTION	Alaska to Baja California
ABUNDANCE	Common
DEPTH	Intertidal
HABITAT	Rocky shore
FEEDING HABIT	Nocturnal grazer

SHELL SIZE RANGE
1⅜ to 3 in
(35 to 76 mm)

PHOTOGRAPHED SHELL
1⅞ in
(47 mm)

MOPALIA LIGNOSA

WOODY CHITON

(GOULD, 1846)

Mopalia lignosa is a common chiton found on intertidal rocky shores from Alaska, U.S.A., to Baja California, Mexico. It prefers to live on the bottoms or sides of large boulders on open coasts, where it feeds on sea lettuce and diatoms, as well as foraminifera and bryozoans. Like some other grazers on rocky shores, these chitons have their radular teeth capped with magnetite to reduce tooth wear. Mopaliids often have a broad, leathery girdle that bears hairs, bristles, or spines, but never scales as in some chitons from other families. There are about 55 species in the family Mopaliidae worldwide, with about 20 living in the northeastern Pacific Ocean.

RELATED SPECIES

Mopalia muscosa (Gould, 1846), also from Alaska to Baja California, has a thick mass of stiff hairs on its girdle. *Katharina tunicata* (Wood, 1815), from Kamchatka, Russia, to the Aleutian Islands and to southern California, is a large chiton, growing to 5 in (130 mm). It has a broad black, leathery girdle that covers most of the dorsum. Only a small, diamond-shaped part of each of the eight plates shows on the dorsum.

The shell of the Woody Chiton is medium-sized and broadly ovate. The thick, leathery girdle is brown (not visible in the photograph), sometimes mottled with green or lighter brown, and has short hairs. In some shells the lines are more irregular and the valves mottled. The sculpture of the valves consists of a strong V-shape ridge, and fine radial lines. The shell color varies from light brown to greenish to dark brown, with prominent lines in light brown or pale green.

Actual size

FAMILY	Chitonidae
SHELL SIZE RANGE	⅜ to 4 in (10 to 100 mm)
DISTRIBUTION	Florida to Venezuela and West Indies
ABUNDANCE	Common
DEPTH	Intertidal to 13 ft (4 m)
HABITAT	Rocky shores
FEEDING HABIT	Nocturnal grazer, feeds on algae

SHELL SIZE RANGE
⅜ to 4 in
(10 to 100 mm)

PHOTOGRAPHED SHELL
2½ in
(64 mm)

34

CHITON TUBERCULATUS
WEST INDIAN CHITON
LINNAEUS, 1758

The shell of the West Indian Chiton is medium in size and ovate, with the girdle covered by scales. The dorsal surface sculpture consists of about 8–9 strong, wavy longitudinal ribs on the triangular areas of each valve, a smooth central strip, and beaded nodules at the terminal valves. The girdle has white and greenish black alternating bands, and the dorsal surface of the valves ranges from grayish to brownish green. The underside of cleaned valves is greenish or bluish white.

Chiton tuberculatus is one of the largest chitons occurring in the Caribbean. It has a handsome shell with a girdle covered by scales that have alternating color bands in white and greenish black. Like most chitons, it is active at night, and feeds by grazing algae on rocks. It has a "homing" behavior, and after short feeding excursions it returns to its original resting place. It may live as long as 12 years. *Chiton tuberculatus* and other chitons occur in high densities locally, and may contribute significantly to the bioerosion of limestone. Chitonids have comblike teeth on the margin of the valves. There are about 100 species in the family Chitonidae worldwide.

RELATED SPECIES

Chiton glaucus Gray, 1828, native to New Zealand and introduced to southern Australia, is the most abundant chiton in New Zealand. It has a dark green and nearly smooth shell. *Acanthopleura granulata* (Gmelin, 1791), from Florida to the southern Caribbean and the Gulf of Mexico, is an abundant chiton. It is about the same size as *C. tuberculatus*, but is more elongated and has short spines on the girdle instead of scales.

Actual size

FAMILY	Acanthochitonidae
SHELL SIZE RANGE	4 to 16 in (100 to 400 mm)
DISTRIBUTION	Hokkaido, Japan, to Aleutian Is.; Alaska to southern California
ABUNDANCE	Locally common
DEPTH	Intertidal to 65 ft (20 m)
HABITAT	Rocky shores
FEEDING HABIT	Nocturnal grazer, feeds on red algae

SHELL SIZE RANGE
4 to 16 in
(100 to 400 mm)

PHOTOGRAPHED SHELL
6¼ in
(160 mm)

CRYPTOCHITON STELLERI

GUMBOOT CHITON

(MIDDENDORF, 1846)

Cryptochiton stelleri is the world's largest chiton, reaching 16 in (400 mm) in length and weighting up to 2 lbs (800 g), although it typically grows to about 6 in (150 mm). It is the only chiton that has all eight plates completely covered by the thick, leathery mantle typical of the group. Its broad foot is yellow or orange. *Cryptochiton stelleri* is a traditional source of food for native peoples, despite its tough meat. It is slow-growing, typically taking some 20 years to reach about 6 in (150 mm) in length; it can live more than 25 years. Because of its slow reproduction and overharvesting, there are concerns about its conservation.

The shell of the Gumboot Chiton is large and thick, with plates that are only loosely interconnected. The shell plates are much smaller than the very large body, and they are completely enclosed within the leathery mantle. The 8 plates, when cleaned and articulated, resemble the vertebrae of mammals. When disarticulated most of the plates have a butterfly shape, and are known as "butterfly" shells that often wash up ashore. The shell plates may be white or robin-egg blue in color.

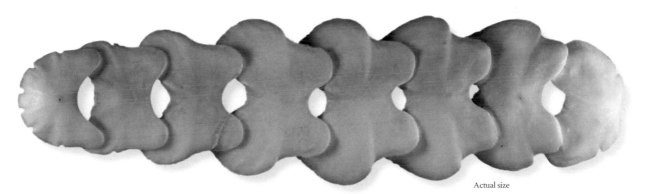

Actual size

RELATED SPECIES

Cryptochiton stelleri is the only species in its genus. Other related chitons include *Acanthochitona pygmaea* (Pilsbry, 1893), from the western central Atlantic. It is a small chiton with very variable coloration, ranging from bright orange to green. The girdle, which is broad and partially covers the shell plates, has tufts of glassy spicules.

BIVALVES

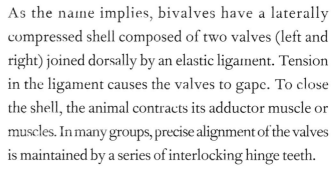

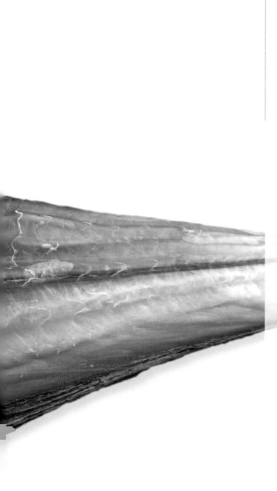

As the name implies, bivalves have a laterally compressed shell composed of two valves (left and right) joined dorsally by an elastic ligament. Tension in the ligament causes the valves to gape. To close the shell, the animal contracts its adductor muscle or muscles. In many groups, precise alignment of the valves is maintained by a series of interlocking hinge teeth.

There are roughly 20,000 living species that inhabit most aquatic habitats. Marine species range from the intertidal zone to the deep sea, from the poles to the tropics. Bivalves have also colonized brackish-water estuaries and freshwater rivers, streams, and lakes.

Bivalves range in size from $\frac{1}{32}$ in (1 mm) to over 40 in (1 m) in length. Most have a capacious mantle cavity that accommodates large gills. In addition to respiration, the gills filter food particles from the water in the majority of bivalves. Some primitive forms feed directly on the organic matter in fine sediments, a few specialized groups derive nutrition from symbiotic algae or bacteria, while others capture and consume small crustaceans and worms in the deep sea. Most bivalves burrow in sand or mud, some in wood, clay or coral. Some attach to hard substrates with threadlike strands of protein (byssus) , others by permanently cementing one of their valves.

FAMILY	Nuculidae
SHELL SIZE RANGE	⅛ to ⅜ in (3 to 10 mm)
DISTRIBUTION	Nova Scotia, Canada to Central America
ABUNDANCE	Abundant
DEPTH	15 to 100 ft (5 to 30 m)
HABITAT	Infaunal in muddy bottoms
FEEDING HABIT	Deposit feeder
BYSSUS	Absent in adults

SHELL SIZE RANGE
⅛ to ⅜ in
(3 to 10 mm)

PHOTOGRAPHED SHELL
¼ in
(6 mm)

NUCULA PROXIMA
ATLANTIC NUT CLAM
SAY, 1822

The shell of the Atlantic Nut Clam is very small, thin but solid, inflated, and obliquely oval. Its valves are the same size and shape; the sculpture is smooth, and inside there are fine radial threads. The hinge has many strong, parallel, triangular teeth, and the ventral margin is finely denticulate. The shell color is whitish gray, and the interior is nacreous white.

Nucula proxima is an abundant microbivalve, a clam that measures less than ⅜ in (10 mm) in maximum size. It is an infaunal bivalve, living buried close to the surface in muddy bottoms, where it forages for organic deposits on which it feeds. The animal has two long labial palps that collect deposits and bring them to the mouth. There are about 160 living species in the family Nuculidae worldwide; they are most common in the deep sea. The family includes some of the smallest known bivalves, but also larger species that can reach about 2 in (50 mm) in length.

RELATED SPECIES

Nucula calcicola Moore, 1977, from the Florida Keys and Gulf of Mexico to Colombia, is one of the smallest known bivalves: its minute shell grows to less than ¹⁄₁₆ in (2 mm) in length. It lives in coral sand. *Acila divaricata* (Hinds, 1843), from Japan to China, is larger and has diverging riblets on its valves.

Actual size

FAMILY	Nuculidae
SHELL SIZE RANGE	¾ to 1¼ in (18 to 30 mm)
DISTRIBUTION	Japan to China
ABUNDANCE	Uncommon
DEPTH	50 to 1,650 ft (15 to 500 m)
HABITAT	Infaunal in muddy bottoms
FEEDING HABIT	Deposit feeder
BYSSUS	Absent in adults

SHELL SIZE RANGE
¾ to 1¼ in
(18 to 31 mm)

PHOTOGRAPHED SHELL
1¼ in
(31 mm)

ACILA DIVARICATA

DIVARICATE NUT CLAM

(HINDS, 1843)

39

Acila divaricata is a variable species of nut clam. A few subspecies have been described on the basis of shell characters, especially shell shape, although all have similar outer shell sculpture. *Acila divaricata* is infaunal in muddy and sandy bottoms, living close to the surface of the sediment. It has a wide bathymetric range, from relatively shallow subtidal to deep levels. Although adults seem to be mostly deposit feeders, juvenile nuculids can feed by filtering particles from the water using their gills. In some species, adults may also be able to filter feed.

RELATED SPECIES

Acila insignis (Gould, 1861), from Japan, has a smaller and more ovate shell, but with sculpture similar to *A. divaricata*. *Nucula proxima* Say, 1822, from Nova Scotia, Canada, to Central America, has a very small shell that is smooth outside and nacreous white inside.

The shell of the Divaricate Nut Clam is small, thick, solid, and ovate. The umbones are prominent and point to the posterior end; the posterior margin has a pointed end. Its main feature is the strong sculpture of diverging ribs on both valves. There is a thick, brown periostracum. Inside, the shell is smooth and nacreous white, with 2 oval and small muscle scars of about the same size.

Actual size

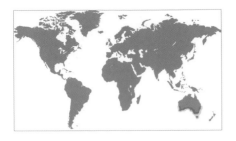

FAMILY	Solemyidae
SHELL SIZE RANGE	1¼ to 2⅜ in (30 to 59 mm)
DISTRIBUTION	South Australia, Tasmania
ABUNDANCE	Common
DEPTH	Subtidal to 33 ft (10 m)
HABITAT	Infaunal in muddy or sandy bottoms
FEEDING HABIT	Modified deposit feeder
BYSSUS	Absent

SHELL SIZE RANGE
1¼ to 2⅜ in
(30 to 59 mm)

PHOTOGRAPHED SHELL
1⅝ in
(43 mm)

40

SOLEMYA AUSTRALIS
AUSTRALIAN AWNING CLAM
LAMARCK, 1818

Solemya australis is a primitive bivalve that lives buried in anaerobic sandy or muddy bottoms with high organic content. It is a modified deposit feeder, and gets its nutrition from symbiotic bacteria living in the gills that oxidize sulfur from the substrate. Some solemyids lack a gut and depend completely on their symbionts. The shell has a high organic content and is fragile, often cracking when dried. Solemyids are adapted for burrowing and live in U- or Y-shaped burrows. There are about 30 living species in the family Solemyidae worldwide, living in all depths and seas, except the polar regions. The family's fossil record traces back to the Devonian Period.

RELATED SPECIES

Solemya velum Say, 1822, which ranges from Nova Scotia to Florida, has a similar, but smaller shell with a yellowish brown periostracum with lighter-colored radial rays. *Solemya togata* (Poli, 1795), from Iceland to Angola and the Mediterranean, is the type species of the genus. Its shell is large for the family and can reach about 3½ in (90 mm) in length.

The shell of the Australian Awning Clam is thin, fragile, and elongated cylindrical in outline. Its diagnostic feature is the glossy, dark brown periostracum with frilly extensions beyond the shell margin. The hinge is toothless and the umbones are located anteriorly. The valves are identical in size and shape. The sculpture consists of broad, flat, oblique radial ribs; inside the shell is smooth, with 2 adductor muscle scars of different sizes. The shell color is dark brown, with white umbones; the interior is gray, with a white ridge extending from the umbo.

Actual size

FAMILY	Nuculanidae
SHELL SIZE RANGE	½ to 1¾ in (12 to 44 mm)
DISTRIBUTION	Western Guatemala to Panama
ABUNDANCE	Uncommon
DEPTH	43 to 240 ft (13 to 73 m)
HABITAT	Sandy and muddy bottoms
FEEDING HABIT	Primarily deposit feeder
BYSSUS	Absent in adults

SHELL SIZE RANGE
½ to 1¾ in
(12 to 44 mm)

PHOTOGRAPHED SHELL
⅝ in
(15 mm)

NUCULANA POLITA

POLISHED NUT CLAM

(SOWERBY I, 1833)

41

Nuculana polita is easily recognized by its large size and sculpture. While most nuculanids have shells with strong concentric ribs, *N. polita* has a mostly smooth shell with a set of oblique incised parallel lines in the posterior half of the shell, and lacks ribs. Nuculanids are infaunal, living partially buried in sand and mud bottoms of high organic content. They are primarily deposit feeders, but can also filter feed. There are between 200 and 250 living species in the family Nuculanidae worldwide, with most species found in deep waters. The first fossil records of nuculanids date from the Devonian Period.

RELATED SPECIES

Propeleda carpenteri (Dall, 1881), which has a wide latitudinal range, from North Carolina south to Argentina, has a small shell with a very long posterior margin. *Adrana suprema* (Pilsbry and Olsson, 1935), from western Mexico to Panama, is one of the largest species in the family, reaching more than 4 in (100 mm) in length. Its shell has both anterior and posterior margins elongated.

Actual size

The shell of the Polished Nut Clam is relatively large for the family, compressed, and elongate sub-elliptical in outline. Its umbones are small, located at about the middle of the hinge, and point posteriorly. The hinge plate is strong and has chevron-shaped teeth. The valves are identical in size and shape, the anterior margin is rounded, and the posterior pointed and elongate. The sculpture is mostly smooth, with fine parallel, incised diagonal lines, and faint concentric growth lines. The shell color is white.

FAMILY	Yoldiidae
SHELL SIZE RANGE	1⅜ to 2¾ in (35 to 70 mm)
DISTRIBUTION	Greenland to North Carolina; Alaska to Puget Sound
ABUNDANCE	Common
DEPTH	60 to 2,500 ft (18 to 760 m)
HABITAT	Infaunal in mud or sand
FEEDING HABIT	Deposit feeder
BYSSUS	Absent

SHELL SIZE RANGE
1⅜ to 2¾ in
(35 to 70 mm)

PHOTOGRAPHED SHELL
2¼ in
(55 mm)

42

YOLDIA THRACIAEFORMIS
BROAD YOLDIA
STORER, 1836

Yoldia thraciaeformis is an infaunal bivalve that lives buried in muddy or sandy bottoms, where it feeds on organic deposits. It can also supplement its diet by filter feeding, pumping water through its siphons by ciliary action of the gills; in doing so, it contributes to bioturbation, the stirring of sediments into the water. In some areas, toxic sediments are stirred back into the water column by yoldiids, affecting other organisms, including commercially important species. There are about 90 living species in the family Yoldiidae worldwide, occurring in both tropical and temperate oceans.

RELATED SPECIES
Portlandia arctica (Gray, 1824), from both sides of the Arctic Ocean, has a small shell with very variable outline, but usually subquadrate and pointed. *Solemya togata* (Poli, 1795), in the related family Solemyidae, from Iceland to Angola, has a thin, cigar-shaped shell with frilly extensions of the periostracum.

Actual size

The shell of the Broad Yoldia is large for the family, compressed, and with an oblong and squarish outline. The posterior end is broadly truncate, and the anterior rounded. Its umbones are prominent and located off-center, closer to the anterior end. The hinge plate is wide, with moderately strong teeth in 2 rows, separated by a large triangular resilifer, a platform for the internal ligament. The sculpture consists of concentric growth lines. The light brown periostracum is dull and polished.

FAMILY	Arcidae
SHELL SIZE RANGE	2 to 4½ in (50 to 120 mm)
DISTRIBUTION	Red Sea to Indo-West Pacific
ABUNDANCE	Common
DEPTH	3 to 165 ft (1 to 50 m)
HABITAT	Fine sand and shell hash bottoms
FEEDING HABIT	Filter feeder
BYSSUS	Present

SHELL SIZE RANGE
2 to 4½ in
(50 to 120 mm)

PHOTOGRAPHED SHELL
2⅛ in
(53 mm)

TRISIDOS TORTUOSA

PROPELLOR ARK

(LINNAEUS, 1758)

43

Trisidos tortuosa has a distinctively twisted shell, which makes it easy to identify. It is a common arcid with a wide distribution, ranging from the Red Sea to the tropical Indo-West Pacific, and from southern Japan to Australia. It is usually found half-buried in muddy to fine sand with abundant shell hash in shallow waters. Other species in the genus *Trisidos* also have twisted shells; *T. tortuosa* is probably the most distorted species. It produces exceptionally long byssus threads that are attached to shell fragments and help anchor the shell in fine sand.

RELATED SPECIES

Trisidos semitorta (Lamarck, 1819), from the Indo-West Pacific, has a less distorted shell. Although it is slightly smaller, its shell is more inflated, and the shell volume may be larger than *T. tortuosa. Arca zebra* (Swainson, 1833), from North Carolina to Brazil, has a long hinge with more than 100 small teeth. It is an edible species, although it reportedly has a bitter taste.

Actual size

The shell of the Propellor Ark is medium-sized, elongated, laterally compressed, and twisted around the long straight hinge. Its sculpture consists of fine radial ribs crossed by even finer concentric growth lines. The umbones are located at about one-third of the hinge length, closer to the anterior end. At about this point, the shell twists some 90 degrees clockwise, so that the posterior end is 90 degrees in relation to the anterior end. The shell color is whitish to yellowish, lighter inside the valves, and the periostracum is brown.

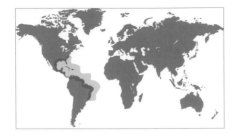

FAMILY	Arcidae
SHELL SIZE RANGE	2 to 4 in (50 to 100 mm)
DISTRIBUTION	North Carolina to Brazil
ABUNDANCE	Common
DEPTH	Intertidal to 460 ft (140 m)
HABITAT	Attached to rocks and coral heads
FEEDING HABIT	Filter feeder
BYSSUS	Present

SHELL SIZE RANGE
2 to 4 in
(50 to 100 mm)

PHOTOGRAPHED SHELL
2⅝ in
(66 mm)

44

ARCA ZEBRA

ATLANTIC TURKEY WING

(SWAINSON, 1833)

The shell of the Atlantic Turkey Wing is medium in size, elongated, and nearly rectangular in outline. It has a long, straight hinge, which can have more than 100 denticles, and prominent umbones; a narrow byssal gape is opposite from the hinge. Its sculpture is dominated by the 24 to 30 irregular radial ribs, crossed by growth ridges. The shell length is about twice the height. The shell color is white with purple-brown irregular stripes; the shell has a white center inside and reddish brown margins.

Arca zebra is a common arcid found in the intertidal zone or shallow waters attached by a byssus to the underside of rocks and coral heads. Its distinctive pattern of purple-brown, zigzag stripes on a white shell resembles the wings of a turkey, hence the popular name. Young turkey wings are usually brightly colored, but coloration fades with increased size. Like other arcids, there are small ocelli on the mantle that are light-sensitive and the animal responds to changes in light intensity and shadows. *Arca zebra* is an important food resource in Venezuela.

RELATED SPECIES

Both *Arca navicularis* Bruguière, 1789, and *Trisidos tortuosa* (Linnaeus, 1758), have a wide distribution from the Red Sea to the Indo-West Pacific. The former has a smaller but similar shell to *A. zebra*. It is collected for food by coastal populations. The latter has a larger but more compressed shell than *A. zebra*, but it is twisted at about mid-length around its long hinge.

Actual size

FAMILY	Arcidae
SHELL SIZE RANGE	1⅜ to 3¼ in (35 to 80 mm)
DISTRIBUTION	Red Sea to Indo-West Pacific
ABUNDANCE	Common
DEPTH	Intertidal to 85 ft (25 m)
HABITAT	Byssally attached to rocks and corals
FEEDING HABIT	Filter feeder
BYSSUS	Present

SHELL SIZE RANGE
1⅜ to 3¼ in
(35 to 80 mm)

PHOTOGRAPHED SHELL
2⅞ in
(74 mm)

BARBATIA AMYGDALUMTOSTUM

ALMOND ARK

(RÖDING, 1798)

45

Barbatia amygdalumtostum is a common arcid that lives in shallow waters attached to crevices and under rocks by a strong byssus. It has a wide distribution, ranging from the Red Sea south to Madagascar, throughout the Indian Ocean, to the west Pacific. Young specimens show two white rays radiating from the white umbones, but the pattern becomes less visible in larger specimens. Like other arcids, its periostracum is thick and hairy. Its valves are of unequal size, and their shape is variable, depending on the space available in the crevices where the animal lives.

The shell of the Almond Ark is medium-sized, moderately compressed, and sub-rectangular in outline. Its umbones are low and positioned closer to the anterior margin; the hinge plate is whitish and bears many small teeth. The ventral and dorsal margins are nearly parallel, and the anterior and posterior margins rounded. The sculpture consists of radial ribs crossed by concentric lines, forming beads. The shell color is red-brown, with a brown, hairy periostracum, and the interior is white, stained with purple-brown.

RELATED SPECIES

Barbatia clathrata (Defrance, 1816), from the Mediterranean and Madeira, has a very small shell, sculpted with radial and concentric ridges, forming heavy knobs. *Anadara grandis* (Broderip and Sowerby I, 1829), from Mexico to Peru, has the largest shell in the family Arcidae, reaching 6 in (150 mm) in length. This bivalve is consumed for food, and has a potential for aquaculture.

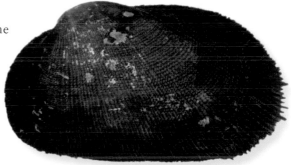

Actual size

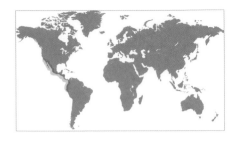

FAMILY	Arcidae
SHELL SIZE RANGE	3 to 6 in (75 to 150 mm)
DISTRIBUTION	Lower California to Peru
ABUNDANCE	Common
DEPTH	Intertidal to shallow subtidal
HABITAT	Mangroves and muddy bottoms
FEEDING HABIT	Filter feeder
BYSSUS	Present

SHELL SIZE RANGE
3 to 6 in
(75 to 150 mm)

PHOTOGRAPHED SHELL
4¾ in
(121 mm)

46

ANADARA GRANDIS
GRAND ARK
(BRODERIP AND SOWERBY I, 1829)

Anadara grandis is one of the largest arcids. The largest recorded specimen is over 6 in (156 mm) in length. *A. grandis* has very heavy, triangular rounded shells. It is edible and has been used for food since prehistoric times; several shell mounds are known in Mexico and Peru. Nowadays it is harvested in large numbers. *Anadara grandis* occurs as Miocene fossils in the Caribbean, but now it lives only in the eastern Pacific. There are about 250 species in the family Arcidae worldwide, with most species living in warm waters. The family has a long fossil history, dating back to the time of dinosaurs.

RELATED SPECIES

Scapharca broughtonii (Schrenck, 1867), from Japan to China, has a medium-large shell that resembles *A. grandis*, but is smaller, more rounded, and with more numerous radial ribs. It is used for food. *Arca zebra* (Swainson, 1833), from North Carolina to Brazil, has a nearly rectangular shell with a long and straight hinge plate with up to 100 denticles. Its shell has purple-brown, zigzag, stripes on a white background.

Actual size

The shell of the Grand Ark is large (for the family), thick, heavy, inflated, and high oblique triangular in outline. Its umbones are large, prominent, and central; the hinge plate is strong, straight and bears about 50 denticles. The valves are of similar size, with the posterior margin longer than the anterior. The sculpture consists of about 26 strong, broad, and flat radial ribs. The shell color is white, but the exterior is covered by a thick, brown periostracum; the interior is porcelaneous white.

FAMILY	Cucullaeidae
SHELL SIZE RANGE	2½ to 4½ in (60 to 120 mm)
DISTRIBUTION	East Africa to Japan and Australia
ABUNDANCE	Common
DEPTH	16 to 825 ft (5 to 250 m)
HABITAT	Sandy and muddy bottoms
FEEDING HABIT	Filter feeder
BYSSUS	Absent in adults

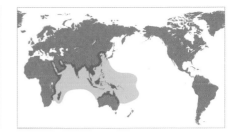

SHELL SIZE RANGE
2½ to 4½ in
(60 to 120 mm)

PHOTOGRAPHED SHELL
4 in
(98 mm)

CUCULLAEA LABIATA

HOODED ARK

(LIGHTFOOT, 1786)

47

Cucullaea labiata is the sole living species in the family Cucullaeidae, a family that has a long fossil history dating back to the Jurassic Period. The family can be separated from the related Arcidae by the structure of the hinge, and the presence of a large projection of the shell that serves as the attachment for the posterior adductor muscle. *Cucullaea labiata* is a large clam that lives on sandy or muddy bottoms, with the anterior end down. Juveniles have a byssus but it is lost in the adults.

RELATED SPECIES

The related family Arcidae has some species, especially those in the genus *Anadara,* that superficially resemble *C. labiata,* including *Anadara subcrenata* (Lischke, 1869), from Japan to China. Its shell is smaller, and shorter, and it has strong, flat ribs, but the straight hinge and strong umbones are reminiscent of the species *C. labiata.*

The shell of the Hooded Ark is large, thin but solid, inflated, and quadrate to triangular in shape. The umbones are tall and prominent, located near the center of the long, straight hinge, which bears small teeth in the center and longer teeth laterally. Its sculpture consists of over 100 fine axial and concentric ribs. A curved projection inside the valves is the attachment site for the posterior adductor muscle. The shell color is purple tan, with a yellowish periostracum; the inner color is purple with white margins.

Actual size

FAMILY	Glycymerididae
SHELL SIZE RANGE	1 to 1½ in (25 to 40 mm)
DISTRIBUTION	Gulf of California to Peru
ABUNDANCE	Common
DEPTH	13 to 80 ft (4 to 24 m)
HABITAT	Soft sediments
FEEDING HABIT	Filter feeder
BYSSUS	Absent in adults

SHELL SIZE RANGE
1 to 1½ in
(25 to 40 mm)

PHOTOGRAPHED SHELL
1½ in
(37 mm)

48

GLYCYMERIS INAEQUALIS
UNEQUAL BITTERSWEET
(SOWERBY II, 1833)

Glycymeris inaequalis is a common clam that ranges from the Gulf of California to Peru. It lives offshore in mud and sandy bottoms. Like other glycymeridids, young shells are byssate, but adults are free-living. Several of the larger species of *Glycymeris* are harvested for food, and the popular name of the shells in the family Glycymerididae, Bittersweet Clams, derives from its taste. There are some 50 living species in the Glycymerididae worldwide, except in polar regions and the deep sea. The family's fossil record traces back to the Cretaceous Period.

RELATED SPECIES

Glycymeris gigantea (Reeve, 1843), which has a limited distribution, from the Gulf of California to Acapulco, Mexico, is one of the largest species in the family, with some shells measuring over 4 in (100 mm) in length. *Glycymeris americana* (Defrance, 1826), from North Carolina to Texas, which is perhaps the largest species, has a wide hinge for the family.

The shell of the Unequal Bittersweet is medium-sized, thick, solid, and obliquely rounded triangular. Its umbones are small and curved posteriorly. The hinge is curved, with teeth in an arch. The valves are nearly identical in size and shape. The valve sculpture consists of about 10 strong radial ribs that are strongest in the center, and smaller radial threads, crossed by weak growth lines. The shell color is a white background with transverse or zigzag brown bands, and the interior is porcelaneous white.

Actual size

FAMILY	Glycymerididae
SHELL SIZE RANGE	½ to 4¼ in (12 to 110 mm)
DISTRIBUTION	North Carolina to Texas
ABUNDANCE	Rare
DEPTH	Intertidal to 160 ft (50 m)
HABITAT	Sandy bottoms
FEEDING HABIT	Filter feeder
BYSSUS	Absent in adults

SHELL SIZE RANGE
½ to 4¼ in
(12 to 110 mm)

PHOTOGRAPHED SHELL
2¼ in
(57 mm)

GLYCYMERIS AMERICANA

GIANT AMERICAN BITTERSWEET

(DEFRANCE, 1826)

Glycymeris americana is one of the largest species of the family Glycymerididae, and the largest in the Americas. It is an uncommon to rare species, and is usually found in moderately shallow water. Its shell is circular and rather compressed, with radial ribs and smaller riblets. Some glycymeridids, such as *G. americana*, are considered inefficient burrowers, and live just below the surface of sandy bottoms. They are believed to be active only at night. Most glycymeridids have circular- to oval-shaped shells, usually ornamented with strong radial ribs, although some species are smooth. The ventral margin is crenulate.

Actual size

RELATED SPECIES

Glycymeris glycymeris (Linnaeus, 1758), which ranges from Norway to the Canary Islands and the Mediterranean, is the type species of the genus and family. Its flesh is considered a delicacy and consumed especially in France. *Glycymeris inaequalis* (Sowerby II, 1833), from the Gulf of California to Peru, is a common and smaller species with an obliquely rounded, triangular shell bearing strong radial ribs.

The shell of the American Bittersweet is large for the family, thick, solid, compressed, and circular. Its umbones are small and located at about the middle of the wide hinge, which bears a curving row of hinge teeth. The valves are identical, slightly wider than tall, and sculptured with moderately weak radial ribs that bear fine riblets. The ventral margin has fine but strong crenulations. The shell color is grayish tan with yellowish brown mottling, and the interior is porcelaneous white.

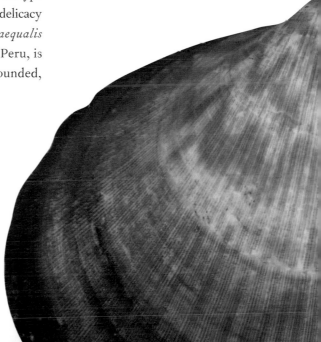

FAMILY	Limopsidae
SHELL SIZE RANGE	¾ to 1¼ in (20 to 33 mm)
DISTRIBUTION	Japonic Province
ABUNDANCE	Uncommon
DEPTH	330 to 2,620 ft (100 to 800 m)
HABITAT	Sandy mud bottoms
FEEDING HABIT	Filter feeder
BYSSUS	Present

SHELL SIZE RANGE
¾ to 1¼ in
(20 to 33 mm)

PHOTOGRAPHED SHELL
1¼ in
(33 mm)

50

LIMOPSIS TAJIMAE
TAJIMA'S LIMOPSIS
SOWERBY III, 1914

Actual size

The shell of the Tajima's Limopsis is
small-medium, thick, solid, compressed, and
obliquely elliptical in outline. Its umbones are
small and centrally located; the hinge plate is
straight, strong, and bears several teeth on each
valve. The valves are roughly identical in size and
shape, with height larger than length. The surface
is nearly smooth, with fine growth lines, but
is obscured by a thick, brown, and hairy
periostracum. The shell itself is white, and
the interior is porcelaneous white.

Limopsis tajimae has a dark brown, hairy periostracum. It lives
on the surface or shallowly buried in sandy mud bottoms in
cold, deep waters off Japan and Taiwan. It has a small byssus
that it attaches to broken shells or pebbles. Because of its weak
byssus and its poor burrowing, it can easily be dislodged from
the sediment. The animal lacks siphons and tentacles, but may
have ocelli on the mantle margin. There are about 25 living
species in the family Limopsidae worldwide, mostly in deep,
cold and temperate waters. The family fossil record extends to
the Cretaceous Period.

RELATED SPECIES

Limopsis cristata Jeffreys, 1876, from Massachusetts to Florida
and Gulf of Mexico, has a tiny shell that looks like a miniature
glycymeridid, nearly circular in outline, and covered with a
thin, light yellowish periostracum. *Limopsis panamensis* Dall,
1902, from Baja California to Panama, has a small, ovate, and
somewhat inflated shell, covered by a hairy periostracum. It lives
in deep waters, and is locally abundant.

FAMILY	Mytilidae
SHELL SIZE RANGE	1 to 2⅜ in (25 to 63 mm)
DISTRIBUTION	Massachusetts to Central America
ABUNDANCE	Common
DEPTH	Intertidal to 2 ft (0.6 m)
HABITAT	Byssally attached to rocks
FEEDING HABIT	Filter feeder
BYSSUS	Present

SHELL SIZE RANGE
1 to 2⅜ in
(25 to 63 mm)

PHOTOGRAPHED SHELL
1⅞ in
(47 mm)

51

ISCHADIUM RECURVUM

HOOKED MUSSEL

(RAFINESQUE, 1820)

Ischadium recurvum is a common epifaunal mytilid that lives on oyster reefs in estuaries, byssally attached to rocks or shells. It can withstand waters with lower salinity than some other mytilids, such as *Brachidontes exustus* (Linnaeus, 1758). Many of the larger and more abundant species of mussels are commercially exploited. Mytilids are usually epifaunal and attached by byssus, but some are borers in coral and limestone. There are between 250 and 400 living species in the family Mytilidae worldwide (expert opinion varies as to the exact number), in all seas and depths, from the intertidal zone to the deep sea. The oldest mytilid fossils are from the Devonian Period.

The shell of the Hooked Mussel is medium-sized, solid, moderately inflated, triangular in outline, and strongly hooked anteriorly. Its umbones are located at the anterior extremity; the hinge plate is narrow, with 3 or 4 small denticles. The valves are about the same size and shape, and have a sculpture of elevated radial lines that branch out toward the posterior margin, and are crossed by concentric growth lines. The shell color ranges from bluish black to chestnut near the margins, and the shell has a purplish interior with a whitish margin.

RELATED SPECIES

Septifer bilocularis (Linnaeus, 1758), from the tropical Indo-Pacific, has a shell similar in size, shape, and sculpture, although less hooked than *Ischadium recurvum*. *Geukensia demissa* (Dillwyn, 1817), from eastern Canada to Florida, resembles *Ischadium recurvum* but has a broader shell that is less anteriorly pointed and ranges from yellow to brown in color.

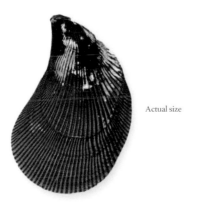

Actual size

FAMILY	Mytilidae
SHELL SIZE RANGE	2 to 6¼ in (50 to 160 mm)
DISTRIBUTION	Worldwide in boreal seas; United States
ABUNDANCE	Abundant
DEPTH	Intertidal to 135 ft (40 m)
HABITAT	Byssally attached to rocks
FEEDING HABIT	Filter feeder
BYSSUS	Present

SHELL SIZE RANGE
2 to 6¼ in
(50 to 160 mm)

PHOTOGRAPHED SHELL
3 in
(76 mm)

52

MYTILUS EDULIS
COMMON BLUE MUSSEL
LINNAEUS, 1758

Actual size

Mytilus edulis, as the name suggests, is an edible mussel that has been collected for food for centuries, especially in Europe. It is still heavily exploited there, where it is harvested from natural populations as well as farmed. It has a wide distribution and grows in all shores with hard substrates in Europe, as well as in boreal waters in the western Atlantic and Pacific. Specimens living in the intertidal zone are smaller than those from deeper waters. *Mytilus edulis* grows in beds as dense as 93 individuals per sq ft (1,000 individuals per m²) with the animals firmly attached by strong byssus to rocks.

RELATED SPECIES

Mytilus californianus Conrad, 1837, which ranges from Alaska to Mexico, is one of the largest species of mytilid; it grows to more than 10 in (250 mm) in length. It is also consumed for food. *Perna viridis* (Linnaeus, 1758), originally from India and Indo-Pacific, is considered an invasive species, and is now widely distributed. It has recently been introduced to Florida, probably via ballast water in ships.

The shell of the Common Blue Mussel is medium-large, solid, and roughly triangular in outline, with a rounded posterior margin. Its umbones are located at the anterior margin, and the hinge line lacks teeth, although there are some small crenulations. The sculpture consists of fine concentric lines. The shell color is brown to nearly black, with a shiny dark periostracum, and the shell has a pearly interior, with a wide, dark purple or blue border.

FAMILY	Mytilidae
SHELL SIZE RANGE	2 to 4½ in (50 to 115 mm)
DISTRIBUTION	Eastern Canada to the Gulf of Mexico
ABUNDANCE	Common
DEPTH	Intertidal to shallow subridal
HABITAT	Salt marshes, near seagrasses
FEEDING HABIT	Filter feeder
BYSSUS	Present

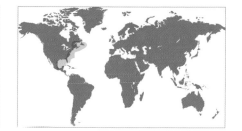

SHELL SIZE RANGE
2 to 4½ in
(50 to 115 mm)

PHOTOGRAPHED SHELL
3 in
(77 mm)

GEUKENSIA DEMISSA

ATLANTIC RIBBED MUSSEL

(DILLWYN, 1817)

Geukensia demissa lives byssally attached to rocks or other mussels and buried in soft sediment in intertidal mud flats, often among the roots of the marsh cordgrass, *Spartina alterniflora*. It seems to thrive in slightly polluted areas. It forms dense beds that can consist of more than 930 individuals per sq ft (10,000 individuals per m²). Shell growth rate increases with increasing depth, although mortality also increases with depth because the mussels become more exposed to predators. Given the mussel's high densities, it is ecologically important. It was unintentionally introduced to California in the 1880s along with *Crassostrea virginica*, which was brought in for aquaculture.

RELATED SPECIES

Geukensia granosissima (Sowerby III, 1914), from Florida to Quintana Roo, Mexico, has a similar shell to *G. demissa*, but with more numerous radial ribs. Some researchers believe it is a subspecies of *G. demissa*. *Ischadium recurvum* (Rafinesque, 1820), from Massachusetts to Central America, has pointed beaks, and a smaller, strongly hooked shell, with bifurcating radial ribs.

The shell of the Atlantic Ribbed Mussel is medium-sized, thin but strong, moderately inflated, elongated, and fan-shaped. Its umbones are low and located near the anterior margin; the hinge plate is narrow and lacks teeth. The valves are about the same size and shape, and the ventral margin is crenulated. The sculpture consists of numerous branching radial ribs, and the interior is glossy. The shell color is variable, ranging from yellowish brown to dark brown, and the interior is white; it is sometimes iridescent.

Actual size

FAMILY	Mytilidae
SHELL SIZE RANGE	2¾ to 8 in (70 to 200 mm)
DISTRIBUTION	Indian Ocean to Southwest Pacific
ABUNDANCE	Locally abundant
DEPTH	Intertidal to 65 ft (20 m)
HABITAT	Byssally attached to rocks
FEEDING HABIT	Filter feeder
BYSSUS	Present

SHELL SIZE RANGE
2¾ to 8 in
(70 to 200 mm)

PHOTOGRAPHED SHELL
3¼ in
(83 mm)

54

PERNA VIRIDIS
GREEN MUSSEL
(LINNAEUS, 1758)

Perna viridis is an edible mussel originally from the coasts of India and widely distributed in the Indo-Pacific Ocean. Because it grows rapidly and has a wide environmental tolerance, it is a successful colonizer and an invasive species. It has been introduced to many places, including Florida, where it is believed to have arrived as larvae in the ballast water of commercial ships. In the West Indies, it is spreading by "island-hopping." It has a potential for cultivation for food, as well as to be used as an indicator species to test for pollution. *Perna viridis* can grow to 8 in (200 mm), although most specimens are less than half of that size.

RELATED SPECIES

Perna perna Linnaeus, 1758, from both sides of the southern Atlantic, is a large mussel that forms dense beds in the intertidal and shallow subtidal. It is heavily exploited commercially. *Mytilus edulis* Linnaeus, 1758, which is found worldwide in boreal waters, is another commercial species. It is particularly popular in Europe, where it is a traditional seafood item.

Actual size

The shell of the Green Mussel is medium-large, moderately thin but solid, inflated, and triangular in outline. Its umbones are pointed and terminal, located anteriorly. The hinge plates narrow, with one small tooth in the right valve and two in the left. The sculpture consists of fine growth lines and faint radial striae. Young shells are green with bluish margins; adults grow brown patches. The interior is iridescent pale bluish green.

FAMILY	Mytilidae
SHELL SIZE RANGE	3 to 5 in (75 to 130 mm)
DISTRIBUTION	Indo-Pacific
ABUNDANCE	Common
DEPTH	Intertidal to 65 ft (20 m)
HABITAT	Bores into limestone and dead coral
FEEDING HABIT	Filter feeder
BYSSUS	Present

SHELL SIZE RANGE
3 to 5 in
(75 to 130 mm)

PHOTOGRAPHED SHELL
4⅝ in
(118 mm)

LITHOPHAGA TERES

CYLINDER DATE MUSSEL

(PHILIPPI, 1846)

Lithophaga teres is one of the largest of the date mussels, a group of mytilids also known as boring bivalves. Some, like *Lithophaga teres*, bore into dead coral or limestone, while others specialize in live corals. The latter needs to keep the aperture of the borehole constantly open to avoid being trapped by the growing coral. Species in the genus *Lithophaga* usually bore with the aid of an acidic secretion of the mantle; some species also use their shells as a file to bore into the rock or coral. The bivalve starts boring as a juvenile, and produces a large borehole with a narrow aperture.

RELATED SPECIES

Botula fusca (Gmelin, 1791), which ranges from North Carolina to Brazil, bores into limestone by mechanical means, using its shell to file the soft rock. Its shell becomes very eroded, and needs constant repairs. *Geukensia demissa* (Dillwyn, 1817), from Florida to Venezuela, seeks refuge from crabs by living in the intertidal zone, sometimes in large clusters.

The shell of the Cylinder Date Mussel is medium-sized, inflated, and elongated cylindrical in outline. The umbones are located near the anterior end; the hinge is about half of the shell length, and toothless. Its sculpture consists of many ribs perpendicular to the anteroposterior axis, in the lower part of the shell, while the upper part is smooth. The shell is covered by a thick brown periostracum; its interior is iridescent.

Actual size

FAMILY	Pteriidae
SHELL SIZE RANGE	3 to 12 in (75 to 300 mm)
DISTRIBUTION	Red Sea to Indo-Pacific
ABUNDANCE	Common
DEPTH	Shallow subtidal to 215 ft (65 m)
HABITAT	Byssally attached to rocks
FEEDING HABIT	Filter-feeder
BYSSUS	Present

SHELL SIZE RANGE
3 to 12 in
(75 to 300 mm)

PHOTOGRAPHED SHELL
6½ in
(170 mm)

56

PINCTADA MARGARITIFERA
PEARL OYSTER
(LINNAEUS, 1758)

Actual size

Pinctada margaritifera is one of the main sources of natural or cultivated pearls. It is widely distributed in the Red Sea and throughout the Indo-Pacific; it has been introduced to Florida. Potentially any mollusk can produce pearls, and many species do, but those produced by *P. margaritifera* are of the highest quality. Its name means "pearl bearer." The color of its nacre and pearls varies widely, from white to gray to shades of yellow, rose, or green; there are also the famous Tahitian black pearls, which are usually dark gray or brown. There are about 60 living species in the family Pteriidae in warm waters worldwide.

RELATED SPECIES

Pinctada longisquamosa (Dunker, 1852), from Florida to Venezuela and the Caribbean, is a small species with an obliquely skewed shell, long scales, and very little nacre inside the valves. *Pteria penguin* (Röding, 1798), from the Indo-Pacific and Red Sea, is a large pteriid cultivated for food and pearls in Thailand and the Philippines.

The shell of the Pearl Oyster is large, thick, and subcircular in outline. The posterior and anterior ears are poorly developed. Its sculpture consists of flat concentric scales that project beyond the shell margin. The hinge is straight and toothless. The outside color is dark brown or green with radial white rays; inside, the shell has a thick and shiny nacreous layer, which varies widely in color from silver to green or dark gray; the shell margin lacks nacre and is dark.

FAMILY	Pteriidae
SHELL SIZE RANGE	4 to 12 in (100 to 300 mm)
DISTRIBUTION	Red Sea to Indo-Pacific
ABUNDANCE	Common
DEPTH	Shallow subtidal to 115 ft (35 m)
HABITAT	Byssally attached to gorgonians and rocks
FEEDING HABIT	Filter-feeder
BYSSUS	Present

SHELL SIZE RANGE
4 to 12 in
(100 to 300 mm)

PHOTOGRAPHED SHELL
7 in
(183 mm)

PTERIA PENGUIN

PENGUIN WING OYSTER

(RÖDING, 1798)

Pteria penguin is the main source of half-pearls (mabé pearls), which grow attached to the inside of the valves. *Pteria penguin* is cultivated in Thailand and the Philippines, and collected throughout its wide geographic range for food and pearls. Cultured pearls are produced by surgically introducing a bead of the desired shape into the mantle; the mollusk then covers it with many thin layers of nacre, and, after several years, it becomes a pearl. The size, color, shape, and luster determine the quality of a pearl. Natural pearls are rare and expensive, especially rounded ones, but cultured ones are more affordable.

RELATED SPECIES

Pteria colymbus (Röding, 1798), from the southeastern U.S.A. to Brazil, is a common pteriid that lives attached to gorgonians (sea fans or whips). Like young *P. penguin*, it has a long wing but much smaller shell. *Pinctada margaritifera* (Linnaeus, 1758), from the Red Sea and Indo-Pacific, is the source of the highest quality pearls, including the famous Tahitian black pearls.

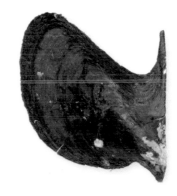

The shell of the Penguin Wing Oyster is large, solid, obliquely ovate, and inflated, with a long wing. The most distinctive feature is the posterior wing, which is very long in young specimens. The rest of the shell grows proportionally larger; adults have a relatively short wing and a tall shell. Its sculpture consists of concentric growth marks. Inside, the valves are covered by brilliant mother-of-pearl, with a wide non-nacreous margin. Outside, a dark and thick periostracum covers the shell.

Actual size

FAMILY	Isognomonidae
SHELL SIZE RANGE	3¼ to 5½ in (80 to 140 mm)
DISTRIBUTION	Indo-Pacific
ABUNDANCE	Common
DEPTH	Intertidal to 33 ft (10 m)
HABITAT	Byssally attached to rocks or mangrove roots
FEEDING HABIT	Filter feeder
BYSSUS	Present

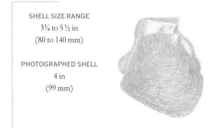

SHELL SIZE RANGE
3¼ to 5½ in
(80 to 140 mm)

PHOTOGRAPHED SHELL
4 in
(99 mm)

ISOGNOMON EPHIPPIUM
SADDLE TREE OYSTER
(LINNAEUS, 1758)

Isognomon ephippium can be separated from other isognomonids by its rounded outline; most species in the family Isognomonidae have more irregularly shaped shells. There is a byssal notch on the right valve, upon which the animal lies; massive byssal threads attach it to mangrove roots and rocks. It is harvested for food and sold in markets in Thailand. A unique feature of isognomonids is the presence of several transverse grooves in the hinge. Most fossil species have much thicker shells than living ones. There are about 20 living species in the family Isognomonidae worldwide, mostly in warm tropical waters.

The shell of the Saddle Tree Oyster is medium-sized, thick, irregularly rounded in outline, and about as wide as it is tall. The hinge is straight and not expanded in wing-like ears, and has about 12 transverse grooves perpendicular to the toothless hinge, each about ¹⁄₃₂ in (1 mm) wide. The sculpture consists of concentric flat scales. The inner surface has nacre, a broad non-nacreous margin, and a single, large muscle scar subcentrally. The shell color is tan to purplish brown.

RELATED SPECIES

Isognomon isognomon (Linnaeus, 1758), from the Indo-Pacific, has a shell with a very variable shape but usually taller than wide; it lies in an erect orientation. *Isognomon bicolor* (C. B. Adams, 1845), from Florida to Brazil, has a small and irregularly ovate shell, which often has two colors (tan and purple), hence its name. It lives attached by a byssus to intertidal to shallow subtidal rocks, sometimes in high densities.

Actual size

FAMILY	Malleidae
SHELL SIZE RANGE	2½ to 4 in (60 to 100 mm)
DISTRIBUTION	Indo-Pacific
ABUNDANCE	Uncommon
DEPTH	17 to 65 ft (5 to 20 m)
HABITAT	Deeply embedded in sponges
FEEDING HABIT	Filter feeder
BYSSUS	Absent in adults

SHELL SIZE RANGE
2½ to 4 in
(60 to 100 mm)

PHOTOGRAPHED SHELL
3⅞ in
(98 mm)

VULSELLA VULSELLA

SPONGE FINGER OYSTER

(LINNAEUS, 1758)

59

Vulsella vulsella is a bivalve that lives deeply embedded in shallow water sponges. It has a byssus when young, but loses it in later life. Juvenile shells are ovoid in outline, but become increasingly elongated dorsoventrally as they grow, hence the name sponge finger oyster. *Vulsella vulsella* is a filter feeder, and filters plankton in the water it pumps in through the inhalant region. Malleids have irregularly shaped shells, even those, like *Malleus albus*, which are free living. There are about 15 living species in the family Malleidae worldwide, in tropical and subtropical waters. The first fossil records of malleids date back to the Jurassic Period.

The shell of the Sponge Finger Oyster is medium-sized, thin, and irregularly shaped. Juveniles have an oblique ovoid shell, but large specimens become dorsoventrally elongated. The hinge is narrow and toothless, with a triangular pit for the ligament. The sculpture and coloration consists of fine concentric growth lines crossed by irregular radial rays that can be discontinuous. Inside, the shell has a nacreous layer, and a non-nacreous margin.

RELATED SPECIES

Vulsella spongiarum Lamarck, 1819, from Australia, also lives associated with sponges. It is often found in high densities, as many as 1,800 specimens per about ½ lb (220 g) of sponge (dry weight). *Malleus albus* Lamarck, 1819, from the Indo-Pacific, is a large species with long extensions of the hinge and an elongated shell, giving it its popular name of Hammer Shell.

Actual size

FAMILY	Malleidae
SHELL SIZE RANGE	6 to 12 in (150 to 300 mm)
DISTRIBUTION	Indo-Pacific
ABUNDANCE	Common
DEPTH	3 to 100 ft (1 to 30 m)
HABITAT	Muddy sand bottoms
FEEDING HABIT	Filter feeder
BYSSUS	Absent in adults

SHELL SIZE RANGE
6 to 12 in
(150 to 300 mm)

PHOTOGRAPHED SHELL
7 in
(180 mm)

60

MALLEUS ALBUS

WHITE HAMMER OYSTER

LAMARCK, 1819

Malleus albus has a very distinctive shell, shaped like a pick ax or a hammer. Like other malleids, its shell is irregularly shaped, in part because of break and repair; the mantle can mend broken parts of the shell rather quickly. Young shells are short and become very long as they grow. *Malleus albus* lives freely on the surface of fine muddy sand bottoms, and loses the byssus during growth. The long projections of the hinge help stabilize the shell on the soft sediment, and prevent the shell from being turned over. It is found in large colonies in some areas.

RELATED SPECIES

Malleus malleus (Linnaeus, 1758), also from the Indo-Pacific, has a shell similar to *M. albus* but darker, and the "handle" part of the hammer may be bent to one side. *Malleus regula* (Forskål, 1775), from the Indo-West Pacific, has a smaller and elongated shell that lacks the extensions of the hinge. It is found in dense beds with *Isognomon isognomon* (Linnaeus, 1758).

Actual size

The shell of the White Hammer Oyster is large, thick, and irregularly shaped like a hammer. Its hinge is long and straight, and both the anterior and posterior extremities of the hinge become elongated. The "handle" of the hammer is the ventral margin, which becomes undulating and extremely developed in adults. The umbones are located at about the mid point of the hinge, near the dorsal margin. The shell color is dirty white outside, and gray or bluish nacreous inside near the ligament.

FAMILY	Ostreidae
SHELL SIZE RANGE	1½ to 3¼ in (40 to 80 mm)
DISTRIBUTION	North Carolina to Brazil
ABUNDANCE	Abundant
DEPTH	Intertidal to 340 ft (104 m)
HABITAT	Attached to gorgonians and rocks
FEEDING HABIT	Filter feeder
BYSSUS	Absent

SHELL SIZE RANGE
1½ to 3¼ in
(40 to 80 mm)

PHOTOGRAPHED SHELL
2¼ in
(54 mm)

LOPHA FRONS

FROND OYSTER

(LINNAEUS, 1758)

Lopha frons is a small oyster that lives attached to gorgonians, rocks, and other hard substrates. Its shell is variable in shape; when growing on gorgonians, it grows into an elongated shape, and the left valve develops projections that hold on to the stem of the sea whips; when growing on rocks, the shell shape is more ovate. Oysters are economically important and a major source of food worldwide, and the commercially species are well studied; however, there is no commercial fishery for *L. frons*. There are some 50 species in the family Ostreidae worldwide, in tropical and temperate waters.

RELATED SPECIES

Lopha cristagalli (Linnaeus, 1758), from the Indo-Pacific, has a very distinctive shell with a zigzag margin and purple color. Like other oysters, it lives attached to hard substrate. *Ostraea conchaphila* Carpenter, 1857, from Alaska to Panama, is the only native oyster in the eastern Pacific; it is currently considered a species of concern because of competition with non-native oysters such as *Crassostrea gigas* (Thunberg, 1793).

Actual size

The shell of the Frond Oyster is small, compressed, solid, irregularly shaped, and usually either elongated or more ovate in outline. Its left valve is cemented to rocks, or has projections, called claspers, that hold on to sea whips. The margin of the shell is crenulated or saw-toothed. The shell is often encrusted with oysters or other organisms. Its color is reddish to dark brown.

FAMILY	Ostreidae
SHELL SIZE RANGE	3 to 8 in (75 to 200 mm)
DISTRIBUTION	Indo-Pacific
ABUNDANCE	Common
DEPTH	17 to 100 ft (5 to 30 m)
HABITAT	Cemented to rocks and corals
FEEDING HABIT	Filter feeder
BYSSUS	Absent

SHELL SIZE RANGE
3 to 8 in
(75 to 200 mm)

PHOTOGRAPHED SHELL
3¼ in
(79 mm)

62

LOPHA CRISTAGALLI
COCK'S COMB OYSTER
(LINNAEUS, 1758)

Lopha cristagalli has a very distinctive shell, with strong angular folds on both valves, and edges with zigzag margins. While most species in the family Ostreidae have white or dull gray shells, *L. cristagalli* is grayish purple. It lives attached by clasping spines to rocks and corals in shallow subtidal waters. The zigzag margin helps to align the valves and to prevent peeling by predators. Because of the folds on both valves, the space inside the shell is small. It is collected for food in the western Pacific, but there is little commercial interest in its meat.

RELATED SPECIES

Crassostraea virginica (Gmelin, 1791), from the western Atlantic, has been introduced into Europe, North America, and elsewhere, and is commercially important. *Crassostrea gigas* (Thunberg, 1793), originally from Japan and Southeast Asia, and introduced in many places, is a large oyster also of commercial importance.

Actual size

The shell of the Cock's Comb Oyster is medium-sized, solid, oval in outline, and with a very angular edge. About 4 to 8 very large, sharp folds radiate from the umbo, and cause the edge to have deep, V-shaped folds, which interlock with the folds on the opposite valve. The surface has fine wormlike pustules, with some spines near the umbo, and small pustules along the inner margin of the edge; the interior of the valves is smooth. The shell color is usually grayish purple on the outside, and tan inside.

FAMILY	Ostreidae
SHELL SIZE RANGE	4 to 12 in (100 to 300 mm)
DISTRIBUTION	Canada to Brazil, Gulf of Mexico
ABUNDANCE	Common
DEPTH	Intertidal to 30 ft (9 m)
HABITAT	Cemented to rocks, other shells, and hard substrates
FEEDING HABIT	Filter feeder
BYSSUS	Absent

SHELL SIZE RANGE
4 to 12 in
(100 to 300 mm)

PHOTOGRAPHED SHELL
12 in
(300 mm)

CRASSOSTRAEA VIRGINICA

EASTERN AMERICAN OYSTER

(GMELIN, 1791)

63

Crassostraea virginica is the most heavily exploited bivalve in the western Atlantic. It has been collected and cultivated for food for centuries. This species of oyster is a key component of estuarine communities because it forms extensive oyster reefs that provide shelter for many species, reduces turbidity, and removes suspended particles from the water. Its filtration rate can be as high as 10 gallons (38 liters) per hour. It cements itself to hard substrates by a small area of the left valve. *Crassostraea virginica* is a prolific species, and females can produce more than 100 million eggs in a single season. Oysters are considered a delicacy. They are widely consumed raw, fried, or boiled, and are industrially canned.

RELATED SPECIES

Crassostrea rhizophorae (Guilding, 1828), from Florida to Uruguay, is an estuarine oyster and is found on the prop roots of red mangrove. It is heavily harvested throughout its range, and some populations are depleted due to overharvesting and pollution. *Crassostrea ariakensis* (Fujita, 1913), originally from Japan, has a large and heavy shell. It is cultivated in Japan and in the eastern Pacific.

Actual size

The shell of the Eastern American Oyster is large, thick, chalky, and irregularly shaped. Its shape is very variable and usually narrow and elongated, like the photo shown, or it can be ovate in outline. The left valve (which is lower, and cemented to substrate) is cupped, and the right valve flattened. The surface of the left valves may have concentric frills or growth lines. The shell color is grayish white, and the interior is glossy white with a purple, rounded scar.

FAMILY	Gryphaeidae
SHELL SIZE RANGE	4 to 12 in (100 to 300 mm)
DISTRIBUTION	Indo-Pacific
ABUNDANCE	Uncommon
DEPTH	Shallow subtidal to 115 ft (35 m)
HABITAT	Attached to rocks and corals
FEEDING HABIT	Filter feeder
BYSSUS	Absent

SHELL SIZE RANGE
4 to 12 in
(100 to 300 mm)

PHOTOGRAPHED SHELL
3¾ in
(96 mm)

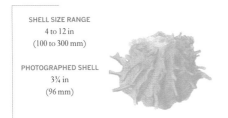

64

HYOTISSA HYOTIS
HONEYCOMB OYSTER
(LINNAEUS, 1758)

The shell of the Honeycomb Oyster is large, thick, heavy, and with an irregular ovate outline. Both left and right valves have radial folds with hollow spines and scales; the ventral margin is wavy and has a roundly zigzag edge. Inside, the shell is smooth to the naked eye, but under magnification it reveals its porous honeycomb structure; a single, large muscle scar is located subcentrally. The shell color ranges from purplish black to light brown on the outside, and it is bluish white inside.

Hyotissa hyotis is a large oyster, which is originally from the tropical Indo-Pacific but has been introduced to Florida. Its shell has a structure that under magnification looks like honeycomb, hence the popular name. Its shell structure and other features distinguish *H. hyotis* from the related Ostreidae. The animal has a dark to black pigmented mantle, and even the gills are dark. *Hyotissa hyotis* cements its shell to rocks, corals, ship wrecks, oil rigs, and other hard substrates. Its shell is usually overgrown with algae and encrusting organisms, providing it with good camouflage. There are five living Gryphaeidae family species worldwide.

RELATED SPECIES

Neopycnodonte cochlear (Poli, 1795), which has a wide distribution in the western Atlantic, Europe, Indo-Pacific, and the Red Sea, has a small and deeply cupped shell. It is the deepest dwelling oyster, with records as deep as 6,900 ft (2,100 m). The related ostreid *Lopha cristagalli* (Linnaeus, 1758), from the Indo-Pacific, has a strong, saw-toothed margin, and a purple shell color.

Actual size

FAMILY	Pinnidae
SHELL SIZE RANGE	1⅜ to 9¼ in (35 to 235 mm)
DISTRIBUTION	Red Sea to Indo-Pacific
ABUNDANCE	Common
DEPTH	Intertidal to 135 ft (40 m)
HABITAT	Rocks and gravel bottoms
FEEDING HABIT	Filter feeder
BYSSUS	Present

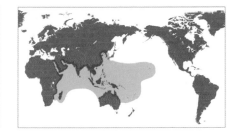

SHELL SIZE RANGE
1⅜ to 9¼ in
(35 to 235 mm)

PHOTOGRAPHED SHELL
5⅛ in
(131 mm)

65

STREPTOPINNA SACCATA

BAGGY PEN SHELL

(LINNAEUS, 1758)

Streptopinna saccata is the most irregularly shaped pen shell, and the single species in the genus *Streptopinna*. Like other pinnids, it lives with its pointed anterior end deeply buried in sand or gravel bottoms, and attached to rocks with byssus threads. The wide posterior end remains out of the substrate. The animal color ranges from white to blue-green, mottled with white and black blotches. There are 22 species in the family Pinnidae living in tropical and subtropical waters worldwide, with most occurring in the Indo-West Pacific.

The shell of the Baggy Pen Shell is medium in size, flexible, fragile, translucent, and triangular. It is irregularly shaped, and its shape depends on its surroundings. The sculpture consists of 5 to 12 smooth or coarse radial ribs, with a triangular area near the anterior end, which is smooth. The posterior end is broad, has a wavy edge, and gapes. Part of the interior of the shell is covered with nacre. The shell color ranges from gray-white to yellow to red-brown.

RELATED SPECIES

Pinna rugosa Sowerby I, 1835, from Baja California to Ecuador and the Galápagos Islands, is a large pen shell that is used for food by native Indians in western Mexico. Its shell has large tubelike spines. *Atrina seminuda* (Lamarck, 1819), which ranges from the southeastern U.S.A. to Argentina, is a common species with a shell that resembles *P. rugosa*, but is smaller.

Actual size

FAMILY	Pinnidae
SHELL SIZE RANGE	4 to 23¼ in (100 to 590 mm)
DISTRIBUTION	Baja California to Ecuador and the Galápagos Islands
ABUNDANCE	Common
DEPTH	Intertidal to shallow subtidal
HABITAT	Muddy bottoms in mangroves
FEEDING HABIT	Filter feeder
BYSSUS	Present

SHELL SIZE RANGE
4 to 23¼ in
(100 to 590 mm)

PHOTOGRAPHED SHELL
6⅜ in
(162 mm)

66

PINNA RUGOSA
RUGOSE PEN SHELL
SOWERBY I, 1835

Pinna rugosa is a large pen shell common in muddy bottoms in quiet bays and mangroves. It has been used traditionally by the Seri Indians from Sonora, western Mexico, as a sustainable food resource. Pinnid shells are poorly calcified and have a high content of organic matter, which make them flexible to the point that the anterior gape can be closed by muscular contraction. Dried shells become brittle and usually crack. Pinnids are called pen shells because of the vague resemblance to old-style writing quills. Another name, razor clams, comes from the sharp posterior edge that projects outside of the sediment.

RELATED SPECIES
Pinna nobilis Linnaeus, 1758, from the Mediterranean, is the largest bivalve in the region and one of the largest in the world. Its long byssus was used to weave textiles from the time of the Romans. *Pinna rudis* Linnaeus, 1758, is another large European species, occurring from the Mediterranean to the Canary Islands. It has a very variable shell.

The shell of the Rugose Pen Shell is large, flexible, fragile, translucent, and triangular elongated. It has about 8 rows of large tubular spines which are largest near the posterior end; in old specimens the spines may be worn down. The anterior end is slender, pointed, and smooth. The posterior end is broad, about half of the shell length in width. The shell color is light brown or tan, and its interior is partly nacreous.

Actual size

FAMILY	Pinnidae
SHELL SIZE RANGE	6 to 12 in (150 to 300 mm)
DISTRIBUTION	North Carolina to Gulf of Mexico and West Indies
ABUNDANCE	Common
DEPTH	Intertidal to 36 ft (11 m)
HABITAT	Sandy mud bottoms
FEEDING HABIT	Filter feeder
BYSSUS	Present

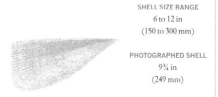

SHELL SIZE RANGE
6 to 12 in
(150 to 300 mm)

PHOTOGRAPHED SHELL
9¾ in
(249 mm)

ATRINA SERRATA
SAW-TOOTHED PEN SHELL
(SOWERBY I, 1825)

67

Atrina serrata is a common pen shell from the southeastern U.S.A., Gulf of Mexico, and the West Indies. It can be recognized by the many rows of small, radial, sharp scales on the exterior of the shell. Pen shells live buried to about half of their length or deeper into the sediment, and are anchored to small rocks by byssus threads. Their shells are fragile and break easily but are also repaired quickly. *Atrina* has a large posterior adductor muscle scar and a large nacreous layer. The related *Pinna* has a small adductor muscle scar located more anteriorly, and a smaller nacreous area that is divided by a sulcus into ventral and dorsal parts.

RELATED SPECIES

Atrina rigida (Lightfoot, 1786), which has a similar distribution and is often washed ashore together with *A. serrata*, has radial rows of large scaled ribs and a dark shell. It is consumed as food in Mexico. *Atrina vexillum* (Born, 1778), from East Africa to Polynesia, has a large, smooth, and broad shell. Its shell is reddish brown to black in color.

The shell of the Saw-toothed Pen Shell
is medium-large, thin, fragile, lightweight, translucent, and triangular. Its sculpture consists of small and sharp scaled ribs arranged in about 30 radial rows. The dorsal hinge is straight and toothless; the ventral margin is rounded, with a squarish posterior edge. Its posterior adductor muscle scar is large and central, and the nacreous layer reaches about three-quarters of the shell length. The shell color is light brown, and the nacreous area iridescent.

Actual size

FAMILY	Pinnidae
SHELL SIZE RANGE	8 to 40 in (200 to 1,000 mm)
DISTRIBUTION	Mediterranean
ABUNDANCE	Common
DEPTH	Low tide to 200 ft (60 m)
HABITAT	Sand, mud, or gravel bottoms
FEEDING HABIT	Filter feeder
BYSSUS	Present

SHELL SIZE RANGE
8 to 40 in
(200 to 1,000 mm)

PHOTOGRAPHED SHELL
16¾ in
(425 mm)

68

PINNA NOBILIS
NOBLE PEN SHELL
LINNAEUS, 1758

Pinna nobilis is one of the largest bivalves, only smaller than *Tridacna gigas* (Linnaeus, 1758) and *Kuphus polythalamia* (Linnaeus, 1767) in length. Like other pinnids, it lives half-buried in the sediment, anchored by its long byssus. Since the time of the Romans until the late 19th century, the byssus was used to produce a golden cloth of high quality; it was known as sea silk. However, due to overharvesting for food, and because of pollution, *P. nobilis* has become endangered. Several species of crustaceans are found in association with pinnids, a fact known since Aristotle's time.

RELATED SPECIES

Pinna bicolor Gmelin, 1791, is a large pen shell with a wide distribution, ranging from East Africa to Hawaii. Its elongated shell may have alternating light and dark brown bans or rays, hence the name. *Atrina serrata* (Sowerby I, 1825), from the southeastern U.S.A. and West Indies, is a common pinnid that has a shell covered with small spines.

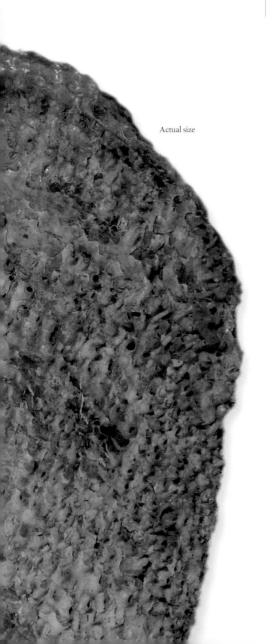

Actual size

The shell of the Noble Pen Shell is very large, thick, elongated, and paddle-shaped. Its sculpture consists of overlapping scales that can be weak or strong; juvenile shells have sharper scales, while old shells tend to have smoother shells. The valves taper sharply toward the pointed anterior umbones, and the posterior end is broad and rounded. The shell color is light brown outside and orange within, and the anterior half is nacreous.

FAMILY	Limidae
SHELL SIZE RANGE	1 to 3 in (25 to 75 mm)
DISTRIBUTION	North Carolina to Brazil
ABUNDANCE	Common
DEPTH	Shallow subtidal to 750 ft (225 m)
HABITAT	Rocky bottoms near reefs
FEEDING HABIT	Filter feeder
BYSSUS	Present

SHELL SIZE RANGE
1 to 3 in
(25 to 75 mm)

PHOTOGRAPHED SHELL
1⅜ in
(37 mm)

LIMA SCABRA

ROUGH LIMA

(BORN, 1778)

69

Lima scabra is one of relatively few bivalves that are capable of swimming. The animal is bright orange and has many long, sticky, and sensitive tentacles. If the animal detects a predator, it detaches its byssus, leaves its nesting place in a rock crevice, and quickly swims away, by rapidly opening and closing its valves, like a scallop. However, unlike a scallop, it swims with its sagittal plane in a vertical orientation, and its many tentacles help with rowing movements. If touched, *L. scabra* detaches some sticky tentacles to delay the predator. There are around 125 living species in the family Limidae worldwide, in tropical and temperate waters.

RELATED SPECIES

Lima lima (Linnaeus, 1758), from the Mediterranean Sea has a spiny shell; (the similar species from the western Atlantic is *Lima caribaea* d'Orbigny, 1853, and not *L. lima*). *Acesta rathbuni* (Bartsch, 1913), an uncommon, deep-water species from the Philippines, is the largest limid. It has a broadly ovate, smooth, and yellow shell.

The shell of the Rough Lima is small, lightweight, and elongated oval in outline. Its valves are nearly identical in size and shape, and have about 22 to 34 scaly radial ribs with flattened spines; these give the shell a rough texture. The margin is crenulated; there is a byssal gape near the hinge. The shell color is white, but a thin, light, brown periostracum covers the shell. Inside, the shell is smooth and porcelaneous white.

Actual size

FAMILY	Limidae
SHELL SIZE RANGE	3½ to 8 in (90 to 200 mm)
DISTRIBUTION	Norway to Azores; Mediterranean
ABUNDANCE	Locally common
DEPTH	135 to 10,500 ft (40 to 3,200 m)
HABITAT	Associated with deep-water corals
FEEDING HABIT	Filter feeder
BYSSUS	Present

SHELL SIZE RANGE
3½ to 8 in
(90 to 200 mm)

PHOTOGRAPHED SHELL
4⅛ in
(103 mm)

70

ACESTA EXCAVATA

EUROPEAN GIANT LIMA

(FABRICIUS, 1779)

The shell of the European Giant Lima is large, thin, translucent in juveniles, and obliquely oval in outline. Its umbones are small, pointed, and turned anteriorly. There is a posterior auricle (wing), and the hinge is toothless, with a triangular pit for the ligament. The valves are about the same size and shape. The sculpture consists of fine radial ribs and fine concentric growth lines, and the interior is shiny. The shell color is light gray with darker gray radial rays, which can be seen through the shell.

Acesta excavata is the largest of the European file shells. It is the largest bivalve associated with (that is, byssally attached to) the *Lophelia pertusa* coral community on the continental shelf. It also lives on steep slopes, in regions where the coral does not grow. In Norway, it occurs in fjords, in depths as shallow as 135 ft (40 m), where it is common, but it occurs much deeper elsewhere. Specimens from the Mediterranean are usually smaller. It is the type species of the genus *Acesta*. The animal is orange, and it has many tentacles. It cannot swim, unlike some other limids.

RELATED SPECIES

Acesta marissinica Yamashita and Habe, 1969, from Japan to China, is one of the largest species in the family. It is an abundant species in China, and has a broad and thin shell. It lives in deep waters. *Lima scabra* (Born, 1778), from North Carolina to Brazil, has a smaller and thin shell, with many radial scaly ribs. It is a good swimmer, and has red, sticky tentacles.

Actual size

FAMILY	Pectinidae
SHELL SIZE RANGE	1 to 2 in (25 to 50 mm)
DISTRIBUTION	California to Western Mexico
ABUNDANCE	Common
DEPTH	Intertidal pools to 820 ft (250 m)
HABITAT	On kelp
FEEDING HABIT	Filter feeder
BYSSUS	Present

SHELL SIZE RANGE
1 to 2 in
(25 to 50 mm)

PHOTOGRAPHED SHELL
1¼ in
(27 mm)

71

LEPTOPECTEN LATIAURATUS

KELP SCALLOP

(CONRAD, 1837)

Leptopecten latiauratus is a common scallop that grows byssally attached to eelgrass, rocks, kelp, and, sometimes pelagic crabs. It is a variable species, and some populations have a more oblique shell. *L. latiauratus* has a wide depth range, and has been recorded as a fossil since the Miocene Period of California (over five million years ago). There are almost 400 living species in the family Pectinidae (true scallops) worldwide, with the highest diversity in the warm waters of the Indo-Pacific, eastern Pacific, and Caribbean.

RELATED SPECIES

Leptopecten bavayi Dautzenberg, 1900, from West Indies to Brazil, has a very small shell, which is fan-shaped and has about 20 sharp riblets. *Cryptopecten speciosum* (Reeve, 1853), which ranges from southern Japan to the New Hebrides, has about 12 broad and scaly ribs and a colorful shell that varies in color.

Actual size

The shell of the Kelp Scallop is small, thin, very lightweight, and slightly convex, with an oblong outline. Its sculpture consists of 12 to 16 broad and undulating radial ribs crossed by fine concentric lines. The ears, which are wide for its shell size, have slightly different sizes. The anterior ear, near the byssal notch, has 6 radial ribs. The shell color is brown-gray to orange, mottled with white or brown chevron marks; the interior of the shell has similar coloration.

FAMILY	Pectinidae
SHELL SIZE RANGE	⅝ to 2⅛ in (15 to 53 mm)
DISTRIBUTION	Florida , Gulf of Mexico to Colombia
ABUNDANCE	Common
DEPTH	3 to 65 ft (1 to 20 m)
HABITAT	Rocks and gravel bottoms
FEEDING HABIT	Filter feeder
BYSSUS	Present

SHELL SIZE RANGE
⅝ to 2⅛ in
(15 to 53 mm)

PHOTOGRAPHED SHELL
1¼ in
(31 mm)

72

CARIBACHLAMYS PELLUCENS
KNOBBY SCALLOP
(LINNAEUS, 1758)

Caribachlamys pellucens was until recently considered an invalid name for this species of scallop, which was known as *C. imbricata* (Gmelin, 1791). A discovery in a museum of a specimen with a label in Linnaeus's hand has validated the name and credited Linnaeus as the proper author of the species. *Caribachlamys pellucens* has 26 minute eyes on the mantle, as well as ten long and many short tentacles.

RELATED SPECIES

Caribachlamys ornata (Lamarck, 1819), from Florida to Brazil and Gulf of Mexico, has a shell with similar shape as *C. pellucens* but with 18 radial ribs and no knobs. *Chlamys squamata* Gmelin, 1791, from the Red Sea and Indo-Pacific, has an elongated shell with scaly ribs. It is edible, but it is not fished commercially.

Actual size

The shell of the Knobby Scallop is small, light, thin, and solid, with an oval fan-shaped outline. Its left valve is flattened, and the right one slightly convex; the ears have very different sizes. The sculpture consists of 8 to 10 radial ribs that have scales on the right valve and knobs on the left valve, arranged in concentric rows. The shell color varies from off-white to rose with reddish rectangular blotches on the left valve; the right valve is a lighter color, and the interior is white with purple and yellow.

FAMILY	Pectinidae
SHELL SIZE RANGE	1 to 2½ in (25 to 60 mm)
DISTRIBUTION	Southern Japan to the New Hebrides
ABUNDANCE	Common
DEPTH	Shallow subtidal to 165 ft (50 m)
HABITAT	Rocks and gravel bottoms
FEEDING HABIT	Filter feeder
BYSSUS	Present

SHELL SIZE RANGE
1 to 2½ in
(25 to 60 mm)

PHOTOGRAPHED SHELL
1⅜ in
(36 mm)

73

GLORIAPALLIUM SPECIOSUM

SPECIOUS SCALLOP

(REEVE, 1853)

Gloriapallium speciosum is a scallop usually found byssally attached to the undersides of dead coral blocks, rocks, and other hard substrates. All species of scallops are epifaunal: they live on the surface of the sediment and are either free living or byssally attached; a few species start out as free living as juveniles but become permanently cemented to hard substrates as adults. Because of its worldwide distribution, and the diversity of shape, sculpture, and colorful patterns, such as those in *G. speciosum*, the family Pectinidae is one of the most popular families among shell collectors.

RELATED SPECIES

Gloriapallium pallium (Linnaeus, 1758), which is an abundant scallop from the Indo-West Pacific, has a similarly shaped shell, but is larger, with about 13 to 15 strong radial ribs without scales. *Pedum spondyloideum* (Gmelin, 1791), from tropical Indo-Pacific, is a coral borer, and lives in crevices in hard corals.

Actual size

The shell of the Specious Scallop is small, solid, thick, lightweight, and colorful, with an ovate outline. The two moderately convex valves are about the same size and shape, and the ears are of unequal size. Both valves have about 12 strong, radial ribs with scales arranged in concentric lines, and the ears have about 4 scaly ribs. The shell color varies from yellow or orange with brown blotches on a white background; the interior is white with the outside color showing through.

FAMILY	Pectinidae
SHELL SIZE RANGE	1 to 3 in (25 to 75 mm)
DISTRIBUTION	Red Sea to Indo-Pacific
ABUNDANCE	Uncommon
DEPTH	3 to 165 ft (1 to 50 m)
HABITAT	Rocky and gravel bottoms
FEEDING HABIT	Filter feeder
BYSSUS	Present

SHELL SIZE RANGE
1 to 3 in
(25 to 75 mm)

PHOTOGRAPHED SHELL
1¾ in
(43 mm)

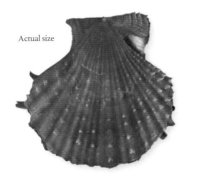

74

CHLAMYS SQUAMATA
SCALY PACIFIC SCALLOP
GMELIN, 1791

Actual size

Chlamys squamata is an edible scallop that is sometimes collected for food in the Philippines. All pectinids are byssate as juvenile, but some become free living as adults. Like all species in the genus *Chlamys* and related genera, *C. squamata* is byssate throughout its life, attaching to rocks and gravel. Pectinids are very diverse and their taxonomy is difficult; small details such as shell microsculpture, the size of the byssal notch on the right valve, and the relative size of the ears are important in identification. The surface usually has radial ribs or fine riblets, and, sometimes, spines or scales.

RELATED SPECIES

Equichlamys bifrons (Lamarck, 1819), from South Australia and Tasmania, is a large scallop harvested commercially. It has 64 small blue eyes on the edge of the mantle. *Caribachlamys pellucens* (Linnaeus, 1758), from Southern Florida to West Indies, has a shell with eight to ten radial ribs, and knobs on the left valve.

The shell of the Scaly Pacific Scallop is small to medium in size, thin, and elongated. The left valve is flatter than the right, and the ears are different sizes. The right valve is more convex, and has a deep byssal notch. Both valves have about 10 to 12 main ribs, 6 or 7 intermediary ones, and projecting scales; the ears also have several ribs. The shell color, both interior and exterior, varies from soft orange, pink, or brown to purple, often with irregular streaks or blotches.

FAMILY	Pectinidae
SHELL SIZE RANGE	1 to 2¼ in (25 to 58 mm)
DISTRIBUTION	Red Sea and tropical Indo-West Pacific
ABUNDANCE	Uncommon
DEPTH	Shallow subtidal to 65 ft (20 m)
HABITAT	Sand and gravel bottoms
FEEDING HABIT	Filter feeder
BYSSUS	Present

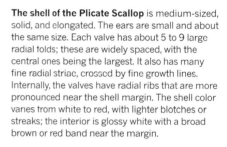

SHELL SIZE RANGE
1 to 2¼ in
(25 to 58 mm)

PHOTOGRAPHED SHELL
2¼ in
(58 mm)

DECATOPECTEN PLICA

PLICATE SCALLOP

(LINNAEUS, 1758)

75

Decatopecten plica is uncommon in most of its wide distribution, but common in the Gulf of Suez. It has an elongated shell, several large radial folds, and a meandering ventral margin. It is found near seagrasses. *Decatopecten plica* is capable of swimming when needed. It is sometimes brought up in crab pots. Because of its relatively small size, it is not fished commercially. Scallops have small but well-developed eyes on the edge of the mantle. Some species have 100 or more eyes, which can be brightly colored. The eyes can sense small changes in light intensity and trigger a flight response if danger is detected.

RELATED SPECIES

Decatopecten striatus (Schumacher, 1817), from Japan to the tropical western Pacific, has a similar shell but differs in the number of radial folds (four to five) and darker shell color. *Lyropecten nodosus* (Linnaeus, 1758) has a large shell with hollow knobs on radial ribs.

The shell of the Plicate Scallop is medium-sized, solid, and elongated. The ears are small and about the same size. Each valve has about 5 to 9 large radial folds; these are widely spaced, with the central ones being the largest. It also has many fine radial striae, crossed by fine growth lines. Internally, the valves have radial ribs that are more pronounced near the shell margin. The shell color varies from white to red, with lighter blotches or streaks; the interior is glossy white with a broad brown or red band near the margin.

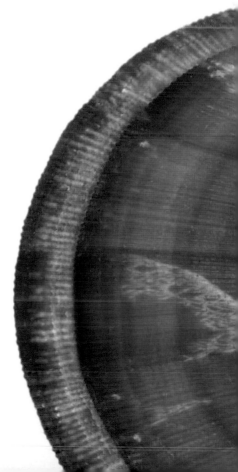

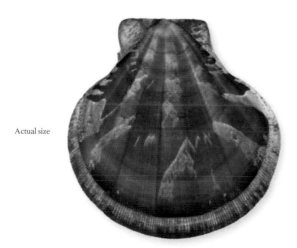

Actual size

FAMILY	Pectinidae
SHELL SIZE RANGE	1¼ to 3 in (30 to 75 mm)
DISTRIBUTION	Eastern U.S.A. and Gulf of Mexico
ABUNDANCE	Uncommon
DEPTH	500 to 1,400 ft (150 to 425 m)
HABITAT	Sand and gravel bottoms
FEEDING HABIT	Filter feeder
BYSSUS	Absent in adults

SHELL SIZE RANGE
1¼ to 3 in
(30 to 75 mm)

PHOTOGRAPHED SHELL
2⅜ in
(59 mm)

76

AEQUIPECTEN GLYPTUS

TRYON'S SCALLOP

(VERRILL, 1882)

Aequipecten glyptus is an edible scallop; because of its relatively small size and the fact that it is not common, it is not fished commercially. It lives in deep waters on sand and gravel bottoms, where it attaches by byssus to rocks and other hard substrates. In byssate scallops, such as *A. glyptus*, there is a series of teeth (the ctenolium) located in the byssal notch in the right valve, which help keep the byssal thread bundles separate, thus increasing the efficiency of attachment. Pectinids range in size from very small to large, with some reaching nearly 12 in (300 mm) in diameter.

RELATED SPECIES

Euvola laurenti (Gmelin, 1791), from the Florida Keys to Venezuela, has a shell that is nearly smooth outside, with fine radial ribs inside. It is a commercial species in Venezuela. *Aequipecten muscosus* (Wood, 1828), from North Carolina to Brazil, is a common, small scallop that is edible but is not fished commercially.

The shell of the Tryon's Scallop is medium in size, thin, flattened, and rounded in outline. The left valve is more flattened than the right, and the ears are about same size. Its sculpture consists of about 17 radial ribs, which become broader and flatter toward the ventral margin, and prickly near the umbones, and fine concentric lines. The left valve has reddish brown or pinkish coloration on the ribs and white in between, while the other valve is lighter; the interior is white.

Actual size

FAMILY	Pectinidae
SHELL SIZE RANGE	2 to 4 in (50 to 100 mm)
DISTRIBUTION	Tropical Indo-Pacific
ABUNDANCE	Common
DEPTH	Shallow subtidal
HABITAT	Attached by byssus in crevices in hard corals
FEEDING HABIT	Filter feeder
BYSSUS	Present

SHELL SIZE RANGE
2 to 4 in
(50 to 100 mm)

PHOTOGRAPHED SHELL
2¾ in
(69 mm)

PEDUM SPONDYLOIDEUM

PEDUM OYSTER

(GMELIN, 1791)

77

Pedum spondyloideum is a coral borer that lives in crevices in hard corals such as several species of *Porites*. As a result of growing in a confined space, its shell is somewhat irregular and flattened. The animal has a beautiful blue-green mantle with many red eyes, which can detect danger; if a shadow is cast over the scallop, it quickly closes its valves. It has been shown that *P. spondyloideum* may help protect its host coral from the coral-eating crown-of-thorns starfish, *Acanthaster planci*, by squirting jets of water, and thus irritating the predator.

The shell of the Pedum Oyster is medium-sized, thin, flattened, subquadrate, weakly inflated, and elongated. As a juvenile it has a rounded outline, but becomes elongated as it grows inside a coral crevice; its adult length can be twice its width. The right valve is broader than the left, and has a deep byssal notch; the left valve has many fine radial riblets. The shell color is off-white, covered by a brown periostracum, and the interior is white, stained with purple.

RELATED SPECIES

Hinnites gigantea (Gray, 1825), from the Aleutian Islands to Mexico, is one of the few scallops that are cemented to hard substrates. It has a very large shell that more resembles that of an oyster than that of a scallop. *Adamussium colbecki* (E. A. Smith, 1902), from Antarctica, is a key species in shallow waters, where it is abundant. It has a very thin, semitranslucent, and flattened shell.

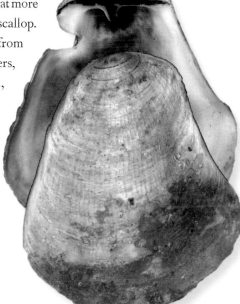

Actual size

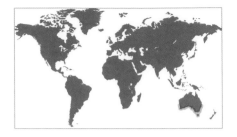

FAMILY	Pectinidae
SHELL SIZE RANGE	2 to 5½ in (50 to 140 mm)
DISTRIBUTION	South Australia and Tasmania
ABUNDANCE	Common
DEPTH	Intertidal to 130 ft (40 m)
HABITAT	Sandy bottoms
FEEDING HABIT	Filter feeder
BYSSUS	Absent in adults

SHELL SIZE RANGE
2 to 5½ in
(50 to 140 mm)

PHOTOGRAPHED SHELL
3¼ in
(80 mm)

78

EQUICHLAMYS BIFRONS
BIFRON'S SCALLOP
(LAMARCK, 1819)

Equichlamys bifrons is a large, free-living scallop that used to be harvested commercially. It has 64 small blue eyes on the mantle edge. It lives in the cold waters of southern Australia, and its growth is seasonal. It grows from late spring until late fall, when the water temperature drops; scallop growth then slows down or even stops, and growth rings are formed. The study of growth rings on the shell allows the scallop's age to be determined. Scallops are filter feeders: they eat phytoplankton, which is pumped into the mantle cavity by ciliary activity of the gills. The water is then filtered by the gills, and the food directed to the mouth.

RELATED SPECIES

Chlamys townsendi (Sowerby III, 1895), from the Red Sea to Pakistan, is one of the largest scallops. It has brownish purple, thick shells. *Decatopecten plica* (Linnaeus, 1758), which has a wide distribution from the Red Sea to the tropical Indo-West Pacific, has an elongated shell with five to nine large radial folds.

Actual size

The shell of the Bifron's Scallop is large and strong, with a rounded outline. Both valves are convex, with the left being more inflated than the right. The ears are about the same size, and the byssal notch is slit-like. There are about 9 strong, rounded radial ribs, with broad interspaces, on the valves; the ears are ridged. The color of the left valve is purple with some white near the umbones, while the right valve has white ribs and pale purple in the interspaces; the interior of the shell is purple.

FAMILY	Pectinidae
SHELL SIZE RANGE	2½ to 4 in (60 to 100 mm)
DISTRIBUTION	Antarctica
ABUNDANCE	Abundant
DEPTH	50 to 16,000 ft (15 to 4,850 m)
HABITAT	Sandy bottoms
FEEDING HABIT	Filter feeder
BYSSUS	Absent in adults

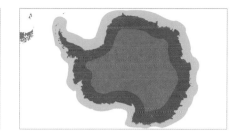

ADAMUSSIUM COLBECKI

COLBECK'S SCALLOP

(E. A. SMITH, 1902)

SHELL SIZE RANGE
2½ to 4 in
(60 to 100 mm)

PHOTOGRAPHED SHELL
3¾ in
(84 mm)

Adamussium colbecki is one of the best-studied mollusks from Antarctica. It is considered a key species in shallow waters; it occurs from shallow subtidal to water as deep as 16,000 ft (4,850 m). Despite several studies, it is difficult to study its year-round growth because of local weather conditions. It is believed that it has a slow growth and metabolism during most of the year, but in the summer its growth rate approaches that of temperate species.

RELATED SPECIES

Aequipecten glyptus (Verrill, 1882), from the southeastern U.S.A. and Gulf of Mexico, is an edible, deepwater scallop that is not fished commercially. It has an ovate shell with about 17 reddish ribs with white interspaces. *Amusium pleuronectes* (Linnaeus, 1758), from the tropical Indo-West Pacific, is a large commercial species that is fished in Taiwan.

Actual size

The shell of the Colbeck's Scallop is large, very thin, and semitranslucent, with a rounded outline. The left valve is flat, and the right one only slightly convex; the ears are small and about the same size. The sculpture consists of about 12 weak costae, more pronounced near the umbones, with very fine concentric lines. The shell color ranges from white to purple, and the interior is iridescent.

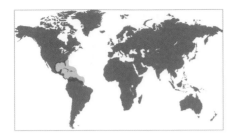

FAMILY	Pectinidae
SHELL SIZE RANGE	2½ to 3½ in (60 to 90 mm)
DISTRIBUTION	Florida to Venezuela
ABUNDANCE	Common
DEPTH	30 to 165 ft (9 to 50 m)
HABITAT	Sandy mud
FEEDING HABIT	Filter feeder
BYSSUS	Absent in adults

SHELL SIZE RANGE
2½ to 3½ in
(60 to 90 mm)

PHOTOGRAPHED SHELL
3⅜ in
(84 mm)

80

EUVOLA LAURENTII
LAURENT'S MOON SCALLOP
(GMELIN, 1791)

Euvola laurentii is the most common scallop commercially trawled in Venezuela, along with *E. papyracea* (Gabb, 1873). Scallops have a single, central adductor muscle that can be rather large in some species. Many of the larger species of scallops, like *E. laurentii*, are edible. The large, round muscle is considered a delicacy and highly prized, and there is an increasing demand for this seafood worldwide. Scallop fisheries are economically important in many countries. Where scallops are consumed fresh, the whole body is eaten, as with oysters, but where it is processed industrially, only the adductor muscle is used.

RELATED SPECIES

Euvola papyracea (Gabb, 1873), from Florida to Brazil, resembles *E. laurentii* but usually has a darker colored shell both interiorly and exteriorly, and is slightly larger. *Amusium obliteratum* (Linnaeus, 1758), from the western Pacific, has a similar but smaller shell with about thirteen ribs.

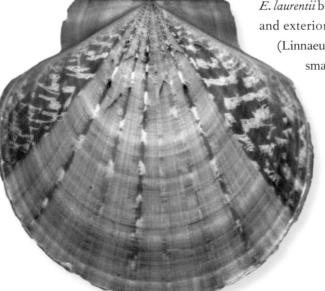

The shell of the Laurent's Moon Scallop is medium-sized, smooth, glossy, and thin, with a rounded outline. Its left valve is nearly flat to slightly concave, and the right is convex; it lies on the convex valve. The ears are nearly equal in size. The sculpture is smooth, and the interior has about 20 sets of paired radial ribs. The left valve color is reddish brown, the right valve lighter in color, and the inside cream.

Actual size

FAMILY	Pectinidae
SHELL SIZE RANGE	3¼ to 5½ in (80 to 140 mm)
DISTRIBUTION	New Caledonia and Australia
ABUNDANCE	Common
DEPTH	Intertidal to 260 ft (80 m)
HABITAT	Sandy bottoms
FEEDING HABIT	Filter feeder
BYSSUS	Present

SHELL SIZE RANGE
3¼ to 5½ in
(80 to 140 mm)

PHOTOGRAPHED SHELL
3⅞ in
(98 mm)

AMUSIUM BALLOTI

BALLOT'S MOON SCALLOP

(BERNARDI, 1861)

Amusium balloti is perhaps the largest species in the genus *Amusium*. It is commercially fished in Australia. Because it is an active swimmer, fishermen need to use otter trawls; in contrast, most scallops are harvested by dredges. Some scallops swim by vigorously clapping the valves: they ingest water when the valves open, and force it through the sides of the hinge when the valves close, which propels the shell forward in the direction of the opening valves. The direction can be changed by directing the water jet with the vellum, a fold of the mantle. Free-living species swim frequently, but those that are byssally attached swim only rarely.

The shell of the Ballot's Moon Scallop is large, thin, heavy, and shiny, with a disk shape. Both valves are similar in size, and nearly smooth, except for very fine concentric and radial striae. Inside, each valve has about 42 to 50 small ribs, often paired. The ears are rather small and are equal in size and shape. The color of the left valve is deep pink, with concentric red-brown bands and some spotting; the right valve is white with some spotting, and the interior is white.

RELATED SPECIES

Amusium japonicum Linnaeus, 1758, from Japan and the western Pacific, has a very similar shell but with about 48 to 54 ribs inside the right valve, and a brick red outer color. It is harvested heavily. *Placopecten magellanicus* (Gmelin, 1791), from Labrador, Canada to North Carolina, U.S.A., is the largest of the American scallops, and is another heavily fished species.

Actual size

FAMILY	Pectinidae
SHELL SIZE RANGE	3¼ to 6 in (80 to 150 mm)
DISTRIBUTION	Mediterranean, Portugal to Angola
ABUNDANCE	Common
DEPTH	80 to 820 ft (25 to 250 m)
HABITAT	Sand, mud and gravel bottoms
FEEDING HABIT	Filter feeder
BYSSUS	Absent in adults

SHELL SIZE RANGE
3¼ to 6 in
(80 to 150 mm)

PHOTOGRAPHED SHELL
4¾ in
(128 mm)

82

PECTEN MAXIMUS JACOBAEUS
ST. JAMES SCALLOP
(LINNAEUS, 1758)

Pecten maximus jacobaeus is famous for being one of the fastest swimmers among scallops; it has been recorded swimming 10 ft (3 m) in just 3 seconds to escape its starfish predators. It is the largest scallop in Europe, where it is commercially fished throughout the Mediterranean, and especially in Italy. It is also known as the Pilgrim Scallop, a reference to the tradition of medieval pilgrims attaching a scallop to their hat or coat when on pilgrimage to Santiago de Compostela, Spain. The pilgrimage was in honor of Saint Jacques (St. James or St. Jacob), hence the species name.

RELATED SPECIES

Pecten maximus maximus (Linnaeus, 1758), from Norway to Madeira and the Canary Islands, has a similar shell and size, but differs in having smoother ridges. *Euvola laurentii* (Gmelin, 1791), from the Florida Keys to Venezuela, has a shell that is nearly smooth outside, with fine radial ribs inside.

Actual size

The shell of the St. James Scallop is large, solid, and broadly fan-shaped. The upper (left) valve is flat and heavy, and the lower (right) is convex and light. Its ears are nearly equal in size, and together comprise about half of the shell width. Each valve has about 14 to 17 broad radial ribs and fine concentric lines. The right valve is white, and the left valve usually brown, white, yellow, or purple.

FAMILY	Pectinidae
SHELL SIZE RANGE	2 to 6½ in (50 to 170 mm)
DISTRIBUTION	Caribbean to Brazil
ABUNDANCE	Common
DEPTH	Intertidal to 500 ft (150 m)
HABITAT	Attached by byssus to rocks
FEEDING HABIT	Filter feeder
BYSSUS	Present

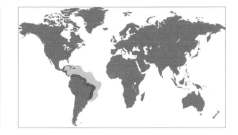

SHELL SIZE RANGE
2 to 6½ in
(50 to 170 mm)

PHOTOGRAPHED SHELL
5½ in
(140 mm)

LYROPECTEN NODOSUS

LION'S PAW

(LINNAEUS, 1758)

83

Lyropecten nodosus is one of the largest scallops in the western Atlantic. It has a very distinctive shell, with several coarse radial ribs bearing hollow knobs; it is usually colored reddish brown but is sometimes bright red, yellow, or orange. Its cultivation for food is still under development in some parts of South America. In culture, *L. nodosus* grows faster at a shallow depth, but has a higher survival rate at deeper depths. This species lives attached to rocks and other hard substrates, including shipwrecks and artificial reefs.

RELATED SPECIES

Nodipecten fragosus (Conrad, 1849), from North Carolina to Florida and Gulf of Mexico, is often confused with the very similar *L. nodosus*, which has a more southern distribution, seven or eight radial ribs, and fewer knobs on the ribs. *Nodipecten magnificus* (Sowerby I, 1835), which is possibly endemic to the Galápagos Islands, has a large and rich-red shell.

The shell of the Lion's Paw is large, thick, heavy, and moderately convex, with a broad fan shape. Each valve has about 7 to 10 large, coarse ribs, which bear hollow knobs arranged in concentric rows; the whole shell is covered by strong radial riblets, and crossed by concentric ridges. The posterior ear is about half the length of the anterior. Inside the valves, the radial ribs show as deep channels. The shell color varies from reddish brown to bright orange or yellow, and the interior is purplish brown.

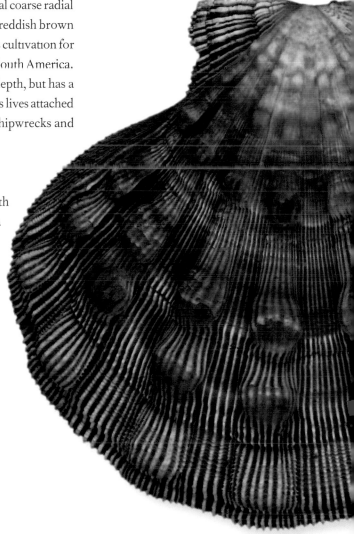

Actual size

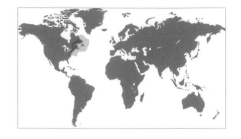

FAMILY	Pectinidae
SHELL SIZE RANGE	4½ to 8 in (120 to 200 mm)
DISTRIBUTION	Labrador, Canada to North Carolina, U.S.A.
ABUNDANCE	Common
DEPTH	6 to 1,250 ft (2 to 380 m)
HABITAT	Sandy bottoms
FEEDING HABIT	Filter feeder
BYSSUS	Absent in adults

SHELL SIZE RANGE
4½ to 8 in
(120 to 200 mm)

PHOTOGRAPHED SHELL
5¾ in
(144 mm)

84

PLACOPECTEN MAGELLANICUS

ATLANTIC DEEPSEA SCALLOP

(GMELIN, 1791)

Placopecten magellanicus is the largest of the American scallops, and one of the most important species fished commercially. It has a wide depth range, and while inshore stocks are becoming depleted from overfishing, offshore wild stocks still appear to be in good shape. Because of increased worldwide demand for scallop meat and an expected shortage in wild-caught scallops, several species are now cultivated, especially in China, Japan, Europe, and Australia. In 1984, Japan was responsible for 94 percent of cultivated scallops worldwide, but by 2004, China had become the top producer, accounting for about 80 percent of world production.

RELATED SPECIES

Pseudoamussium septemradiatum (Müller, 1776), from Norway to West Africa and the Mediterranean, has a small shell with seven broad wavy ribs and a rounded profile. *Pecten maximus jacobaeus* (Linnaeus, 1758), from Portugal to Angola and the Mediterranean, is the largest scallop from Europe, and is fished commercially.

Actual size

The shell of the Atlantic Deepsea Scallop is large, and circular in outline. The two valves are about the same size, and both are slightly convex; the ears are similar in size. Its sculpture is more prominent on the left valve, with numerous radial threads; although the shell looks smooth, minute scales make it feel rough. The left valve is often reddish brown in color, but sometimes lavender or yellow; the right valve is off-white; inside, the valves are cream and shiny.

FAMILY	Propeamussiidae
SHELL SIZE RANGE	⅛ to 3⁄16 in (3 to 4.8 mm)
DISTRIBUTION	Baja California to Ecuador
ABUNDANCE	Uncommon
DEPTH	6 to 1,160 ft (2 to 355 m)
HABITAT	Sandy bottoms
FEEDING HABIT	Filter feeder
BYSSUS	Present

CYCLOPECTEN PERNOMUS

PERNOMUS GLASS SCALLOP

(HERTLEIN, 1935)

SHELL SIZE RANGE
⅛ to 3⁄16 in
(3 to 5 mm)

PHOTOGRAPHED SHELL
¼ in
(5 mm)

85

Cyclopecten pernomus is the smallest of the Panamic glass scallops, and adult shells measure less than 3⁄16 in (5 mm) in length. Its shell is variable in color, usually white peppered with brown to orange maculations. *Cyclopecten pernomus* ranges from shallow subtidal to deep waters. There are about 200 living species in the family Propeamussiidae (glass scallops) worldwide, with most occurring in deep water or polar seas. The family fossil record dates back to the Carboniferous Period.

RELATED SPECIES

Cyclopecten perplexus Soot-Ryen, 1960, from the South Atlantic, is one of the smallest scallops, with its shell reaching about 1⁄16 in (1.5 mm) in length. It broods its larvae in the gills, the first record of brood protection in the scallops. *Propeamussium watsoni* (Smith, 1885) from the western Pacific, has a large, fragile shell with internal radial ribs.

The shell of the Pernomus Glass Scallop is tiny, and fragile, with a rounded outline. The left valve has a sculpture of fine radial striae, while the right valve appears smooth to the naked eye. The left valve is larger than the right valve, and the left ear is slightly larger than the right. The shell color is variable, usually white with irregular brown blotches, with some yellow or orange, but sometimes yellowish white.

Actual size

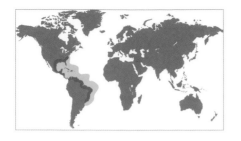

FAMILY	Spondylidae
SHELL SIZE RANGE	3 to 6 in (76 to 150 mm)
DISTRIBUTION	North Carolina to Brazil
ABUNDANCE	Common
DEPTH	Intertidal to 460 ft (140 m)
HABITAT	Cemented to hard substrates
FEEDING HABIT	Filter feeder
BYSSUS	Absent

SHELL SIZE RANGE
3 to 6 in
(76 to 150 mm)

PHOTOGRAPHED SHELL
3⅜ in
(87 mm)

86

SPONDYLUS AMERICANUS
AMERICAN THORNY OYSTER
HERMANN, 1781

Spondylus americanus is a large, spinose, and colorful spondylid that is particularly common on the thousands of offshore oil rigs in the Gulf of Mexico. The world record size for this species is 9½ in (241.5 mm) in length, probably including the spines. It grows cemented by the right valve to rocks, corals, and other hard substrates. The shell is heavily encrusted with sponges, corals, and other marine life, making it well camouflaged.

RELATED SPECIES

Spondylus regius Linnaeus, 1758, from the western Pacific, has sparse but long spines, and many fine hairlike spines in between.

Spondylus linguaefelis Sowerby II, 1847, from the western and central Pacific, is probably the species of spondylid with the most spines, although they are short and thin.

The shell of the American Thorny Oyster is large, heavy, solid, and oval to circular. Its valves have different sizes and shapes, and are decorated with radial ribs and erect spines up to 3 in (75 mm) long. The right valve, permanently cemented to the substrate, is larger than the upper (left) valve. The hinge structure has strong interlocking teeth that form a ball-and-socket type joint. The shell color is variable, from white to yellow and red, and the interior is white with reddish purple near the margins.

Actual size

FAMILY	Spondylidae
SHELL SIZE RANGE	3¼ to 5 in (80 to 130 mm)
DISTRIBUTION	Tropical Western Pacific
ABUNDANCE	Common
DEPTH	16 to 165 ft (5 to 50 m)
HABITAT	Rocks and among coral debris
FEEDING HABIT	Filter feeder
BYSSUS	Absent

SHELL SIZE RANGE
3¼ to 5 in
(80 to 130 mm)

PHOTOGRAPHED SHELL
3½ in
(89 mm)

SPONDYLUS REGIUS

REGAL THORNY OYSTER

LINNAEUS, 1758

Spondylus regius is one of the most striking of the thorny oysters, particularly when it has very long spines. Specimens with the longest spines are found in calm, protected waters. The hinge plate has two large teeth that form a ball-and-socket joint, which is typical of the family. There are about 70 living species in the family Spondylidae worldwide, with most being found in tropical and subtropical waters.

RELATED SPECIES

Spondylus americanus Hermann, 1781, which ranges from North Carolina to Brazil, has a similar shell; it sometimes has long spines, but it more commonly has slightly more closely spaced and shorter spines than *S. regius*. *Spondylus gaederopus* Linnaeus, 1758, from northwest Africa and the Mediterranean, has relatively short, flat spines.

The shell of the Regal Thorny Oyster is large for the family, solid, convex, and subcircular in outline. Its valves are nearly identical in size and shape, and the lower (right) valve has the same sculpture as the top valve. There are 6 main radial ribs, which bear sparse, strong, long, slightly curved spines. Between the main ribs there are 6 or 7 smaller ribs, and in perfect specimens, fine prickly spines. The shell color is reddish brown or rose pink; orange specimens are uncommon. The interior is bluish white.

Actual size

FAMILY	Plicatulidae
SHELL SIZE RANGE	½ to 1½ in (12 to 40 mm)
DISTRIBUTION	North Carolina to Argentina
ABUNDANCE	Common
DEPTH	Intertidal to 400 ft (120 m)
HABITAT	Cemented to hard substrate
FEEDING HABIT	Filter feeder
BYSSUS	Absent

SHELL SIZE RANGE
½ to 1½ in
(12 to 40 mm)

PHOTOGRAPHED SHELL
1¼ in
(31 mm)

PLICATULA GIBBOSA
ATLANTIC KITTEN'S PAW
LAMARCK, 1801

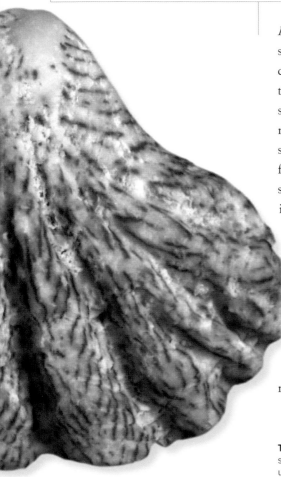

Plicatula gibbosa is an abundant clam that lives cemented by a small area of the right valve to hard substrates such as rocks, corals, or other shells. It usually ranges from the intertidal zone to shallow subtidal, but is also found in deeper waters. Empty shells are washed onto the beach in large numbers. The rich reddish lines found in fresh specimens fade rapidly in beached shells. The large teeth in the hinge plate prevent the valves from opening more than just a slit. There are about ten living species in the family Plicatulidae worldwide, with most found in warm, shallow waters.

RELATED SPECIES

Two plicatulids which have a wide distribution, from the Red Sea to the Indo-West Pacific, are *Plicatula plicata* (Linnaeus, 1767) and *P. australis* Lamarck, 1819. The former has a shell that resembles *P. gibbosa*, with similar sculpture; sometimes it has a more elongate shell, although the shape is quite variable. The latter species has an oval rounded outline and many radial ribs.

The shell of the Atlantic Kitten's Paw is small, solid, compressed, and fan- or tear-shaped. Its umbones are small, central, and terminal; the hinge plate is short and bears large teeth, two in each valve. The right valve is more convex and about the same size as the left valve. The sculpture consists of 5 to 12 broad radial folds, and the interior is smooth. The ventral margin has saw-tooth plications. The shell color is grayish white, with gray or reddish lines, and the interior is white.

Actual size

FAMILY	Anomiidae
SHELL SIZE RANGE	1 to 2¼ in (25 to 56 mm)
DISTRIBUTION	Indo-West Pacific
ABUNDANCE	Common
DEPTH	Intertidal
HABITAT	On mangrove roots
FEEDING HABIT	Filter feeder
BYSSUS	Present

SHELL SIZE RANGE
1 to 2¼ in
(25 to 56 mm)

PHOTOGRAPHED SHELL
1⅜ in
(35 mm)

ENIGNOMIA AENIGMATICA

MANGROVE JINGLE SHELL

(SOWERBY I, 1825)

Enignomia aenigmatica is one of the most curious bivalves. It has an elongate, limpetlike shell, with a hole in the right valve. It is able to crawl around mangrove roots and climb trees with its ribbonlike foot. It is semiterrestrial and can withstand long periods exposed to air. Shells that grow on mangrove leaves are golden in color, while those that live elsewhere are reddish purple (as the one illustrated here). It has a byssus, but can abandon it and move. There are about 15 living species in the family Anomiidae worldwide, with most in temperate waters. The fossil history of the family dates from the Jurassic Period.

RELATED SPECIES

Anomia simplex (d'Orbigny, 1853), which ranges from Eastern Canada to Argentina, has an irregularly circular to oval, translucent, yellowish shell. It is one of the most abundant shells washed on the beaches on the eastern U.S.A. *Placuna placenta* (Linnaeus, 1758), from the tropical Indo-West Pacific, has a large and very flat shell. Young specimens are transparent and often used as a substitute for glass panes.

The shell of the Mangrove Jingle Shell is medium-sized, very thin, fragile, compressed, and elongate to broadly ovate in outline. Its left (upper) valve is limpetlike, convex, and has a narrow slit from the subcentral umbo to the margin of the valve. The right valve is thin, concave, and has a large central hole for the soft byssus, and narrow slit. The sculpture on both valves consists of concentric growth lines. The upper valve is reddish-brown or golden, and the lower valve is translucent silver-white.

Actual size

FAMILY	Placunidae
SHELL SIZE RANGE	4 to 8 in (100 to 200 mm)
DISTRIBUTION	Tropical Indo-West Pacific
ABUNDANCE	Abundant
DEPTH	Intertidal to 330 ft (100 m)
HABITAT	Muddy sand bottoms
FEEDING HABIT	Filter feeder
BYSSUS	Present

SHELL SIZE RANGE
4 to 8 in
(100 to 200 mm)

PHOTOGRAPHED SHELL
4¼ in
(106 mm)

90

PLACUNA PLACENTA
WINDOWPANE OYSTER
(LINNAEUS, 1758)

The shell of the Windowpane Oyster is medium to large, thin, brittle, translucent, and nearly circular in outline. The valves are greatly compressed laterally, with the right valve flat and the left slightly convex. Its surface is mostly smooth, with fine growth lines; inside the shell, a V-shaped ridge radiates from the beak, and the single adductor muscle scar is located subcentrally. Its shell color is silvery white, and nearly colorless in small shells.

Placuna placenta is certainly one of the most flattened bivalves, and it is difficult to imagine how the animal can live within such a narrow space. It is abundant in quiet lagoons, bays, and mangroves, where the animal lies on its right valve and covers itself with a thin layer of mud. It has a long and extensible foot, which is used to clean the mantle cavity but not for locomotion. Its shell is thin and translucent, and has been used for centuries as a substitute for glass panes, especially the young, transparent shells. It is currently used for crafts and lanterns in the Philippines, where it is a major commercial fishery.

RELATED SPECIES

Placuna ephippium (Philipsson, 1788), from the tropical western Pacific, has a similar shell but is more solid and less translucent than *P. placenta*. The related *Anomia simplex* d'Orbigny, 1853, which ranges from Canada to Argentina, has a small, irregular shell that is translucent. It has a byssal hole on the right valve through which the byssus attaches to the substrate.

Actual size

FAMILY	Trigoniidae
SHELL SIZE RANGE	1 to 2 in (25 to 50 mm)
DISTRIBUTION	Endemic to Australia
ABUNDANCE	Common
DEPTH	20 to 260 ft (6 to 80 m)
HABITAT	Infaunal in sand and sandy mud
FEEDING HABIT	Filter feeder
BYSSUS	Present

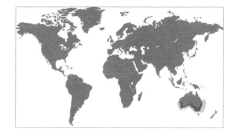

SHELL SIZE RANGE
1 to 2 in
(25 to 50 mm)

PHOTOGRAPHED SHELL
1½ in
(37 mm)

NEOTRIGONIA MARGARITACEA

AUSTRALIAN BROOCH CLAM

(LAMARCK, 1804)

91

Neotrigonia margaritacea is considered a living fossil because it has changed very little in over 65 million years. There are only about six living species of Trigoniidae remaining today, but the family was very diverse during the Mesozoic era (250 to 65.5 million years ago), when the family had a cosmopolitan distribution; the living species are restricted to Australia. The animal of *N. margaritacea* is very active and a fast burrower, despite its inflated shell. It lives half-buried in sand and sandy mud. Its iridescent nacreous shell is used for jewelry.

RELATED SPECIES

The living species of the *Neotrigonia* genus have shells with similar ornamentation. *Neotrigonia lamarcki* (Gray, 1838), from eastern Australia, has a shell similar to *N. margaritacea* but with about 24 radial ribs; the shell color is purplish brown. *Neotrigonia bednalli* Verco, 1907, from southern Australia, has about 26 radial ribs and the shell color varies from white to pink, red or purple.

Actual size

The shell of the Australian Brooch Clam is small, thick, solid, and ovately triangular in outline. There are about 24 strong nodular radial ribs on each valve, crossed by many concentric lines. Inside the valves there are internal grooves near the margin corresponding to the radial ribs. Its hinge teeth are large, V-shaped, grooved, and pearly white to pink, like the inside of the valves. Outside, the valves are covered by a thick, brown periostracum.

FAMILY	Crassatellidae
SHELL SIZE RANGE	1½ to 3¼ in (40 to 80 mm)
DISTRIBUTION	Panama to Venezuela
ABUNDANCE	Rare
DEPTH	95 to 125 ft (29 to 38 m)
HABITAT	Soft bottoms
FEEDING HABIT	Filter feeder
BYSSUS	Absent in adults

SHELL SIZE RANGE
1½ to 3¼ in
(40 to 80 mm)

PHOTOGRAPHED SHELL
3¼ in
(80 mm)

92

EUCRASSATELLA ANTILLARUM

ANTILLEAN CRASSATELLA

(REEVE, 1842)

Eucrassatella antillarum is a large and rare crassatellid that lives offshore on soft bottoms. It ranges from the Pliocene (fossil) to the Recent, and has been found living off Panama and Venezuela. Crassatellids are shallow burrowers and live just below the surface of the soft sediment. Some species brood their young in the mantle cavity. There are about 40 living species in the family Crassatellidae worldwide, with most species in shallow tropical and subtropical waters. The family's fossil record extends back to the Devonian Period.

RELATED SPECIES

Eucrassatella speciosa (A. Adams, 1852), from North Carolina to Colombia, has a smaller, more slender shell with a stronger, concentric sculpture. *Eucrassatella donacina* (Lamarck, 1818), from western and southern Australia, is the largest species in the family Crassatelidae. Its shells have been traditionally used by Australian Aborigines as tools.

The shell of the Antillean Crassatella is medium in size, thick, solid, slightly convex, and rounded subtriangular. Its umbones are large and slightly pointed posteriorly. The valves are identical in size and shape, and they are rostrate (beaklike) posteriorly. The sculpture is mostly smooth, with fine concentric growth lines. The hinge area is heavy and narrow, with a large, triangular socket for the ligament, and 2 to 3 long, diagonal teeth. The shell color is brown, with lighter umbones; it has a chestnut brown interior, and a white band near the ventral margin.

Actual size

FAMILY	Astartidae
SHELL SIZE RANGE	⅜ to 1⅜ in (10 to 35 mm)
DISTRIBUTION	Arctic Seas to West Africa, and Mediterranean
ABUNDANCE	Common
DEPTH	16 to 6,500 ft (5 to 2,000 m) to deep water
HABITAT	Partially buried in mud or gravel bottoms
FEEDING HABIT	Filter feeder
BYSSUS	Absent in adults

SHELL SIZE RANGE
⅜ to 1⅜ in
(10 to 35 mm)

PHOTOGRAPHED SHELL
1⅜ in
(35 mm)

ASTARTE SULCATA

SULCATE ASTARTE
(DA COSTA, 1778)

93

Astarte sulcata is the type species of the genus. It is a common clam in northwestern Europe, around the British Isles, but is also found in the Mediterranean and south to northwestern Africa. It is an infaunal clam, living partially buried in mud, sand, and sandy gravel bottoms. Its shell is often caked with black mud. Although it is found in shallow waters, it is more common offshore; it has been reported from 6,500 ft (2,000 m) deep. There are about 50 living species in the family Astartidae, mostly distributed in cold boreal and Arctic waters. The oldest known astartid fossils date from the Devonian Period.

Actual size

RELATED SPECIES

Astarte borealis (Schumacher, 1817), is a circumboreal species, which in Europe occurs from Norway to the British Isles. It is one of the largest astartids, reaching over 2¼ in (55 mm) in length. It has an extremely heavy shell, covered with a heavy periostracum. *Astarte smithii* Dall, 1886, from Florida and the West Indies to the Gulf of Mexico, is a small astartid from deep waters, with a light brown shell.

The shell of the Sulcate Astarte is small, thick, solid, and broadly oval in outline. Its umbones are prominent, positioned slightly anteriorly, and pointed posteriorly. The hinge plate is thick, with 2 teeth in the right valve and 3 in the left. The sculpture consists of about 20 strong, smooth, concentric ribs. The valves are identical in shape and size, and do not gape. The ventral margin is crenulated. The shell color is chalky white, with a thick, brown periostracum, and a white interior.

FAMILY	Carditidae
SHELL SIZE RANGE	1¼ to 3 in (30 to 75 mm)
DISTRIBUTION	Philippines to Australia
ABUNDANCE	Common
DEPTH	Intertidal to 100 ft (30 m)
HABITAT	Sandy bottoms
FEEDING HABIT	Filter feeder
BYSSUS	Present

SHELL SIZE RANGE
1¼ to 3 in
(30 to 75 mm)

PHOTOGRAPHED SHELL
1¾ in
(45 mm)

94

CARDITA CRASSICOSTA
AUSTRALIAN CARDITA
LAMARCK, 1819

Cardita crassicosta is one of the most heavily scaled species in the family Carditidae. Unlike most carditids, which occur only in certain parts of Australia, *C. crassicosta* has been recorded from all regions around the country. Carditids are usually byssate, and are found nestling in crevices, under rocks, and gravel; there are even some that are shallow burrowers. Most species live in shallow water, but a few are found in the deep sea. Some species, if not all, brood their young into their mantle cavity. There are about 50 species in the family Carditidae worldwide. The family has a long fossil history since the Devonian Period.

RELATED SPECIES

Cardita antiquata (Linnaeus, 1758), from the Mediterranean and neighboring Atlantic, is a small species with an extremely variable color. It has a rounded outline with strong radial ribs. *Beguina semiorbiculata* (Linnaeus, 1758), from the Red Sea to the Western Pacific, is one of the largest species of carditids; it can grow to over 4 in (100 mm) in length. Its shell has weak radial threads.

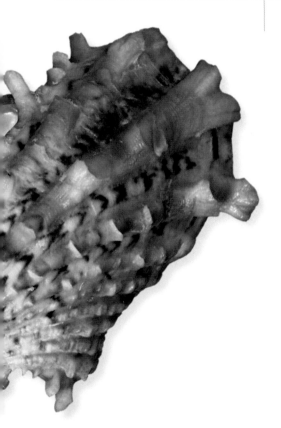

The shell of the Australian Cardita is medium-small in size, solid, scaly, and trapezoidal in outline. Its umbones are small and anterior, and the ventral margin is concave anteriorly. The valves are identical in size and shape, and have 11 to 14 strong radial ribs, wider posteriorly, which bear large, erect scales. The shell color is variable, ranging from pink, orange, yellow, and white to brown, and it has a white interior.

Actual size

FAMILY	Pholadomyidae
SHELL SIZE RANGE	2 to 5 in (75 to 130 mm)
DISTRIBUTION	Caribbean to Colombia
ABUNDANCE	Extremely rare
DEPTH	Shallow intertidal to 80 ft (25 m)
HABITAT	Sandy bottoms
FEEDING HABIT	Deposit feeder
BYSSUS	Absent

SHELL SIZE RANGE
2 to 5 in
(75 to 130 mm)

PHOTOGRAPHED SHELL
3¼ in
(79 mm)

PHOLADOMYA CANDIDA

CARIBBEAN PIDDOCK CLAM

SOWERBY I, 1823

95

Pholadomya candida is among the rarest of bivalves, and is considered to be a living fossil. Several specimens had been collected in the early nineteenth century. It was believed to be extinct until additional shells were collected along the Caribbean coast of Colombia, including a live specimen that was found buried deep in coarse sand in shallow water. *Pholadomya candida* has a pair of large, fused siphons that extend through the gaping valves. The Pholadomyidae are an ancient group of burrowing bivalves that flourished during the Jurassic and Cretaceous Periods, being represented today by only about ten living species.

The shell of the Caribbean Piddock Clam is very thin, fragile, and posteriorly elongated, with a posterior siphonal gape. The umbones are pronounced, rounded, situated anteriorly, and very close together. The margins near the umbones are flared. The hinge is almost smooth, has a short tubercle and pit, and a short external ligament. The shell has strong radial ribs crossed by concentric growth lines. The central 8 or 9 radial ribs are more pronounced, forming a cancellate pattern of beads. The interior of the shell is pitted and nacreous. The color is white both outside and inside.

RELATED SPECIES

All living species are rare, and most occur in deep waters, including *Pholadomya pacifica* Dall, 1907, from the Indo-Pacific; *P. maoria* Dell, 1963, from New Zealand; *P. takashinensis* Nagao, 1943, and *P. levicaudata* Matsukuma, 1989, which are both from Japan; and *Panacca loveni* (Jeffreys, 1881), from Europe.

Actual size

FAMILY	Pandoridae
SHELL SIZE RANGE	½ to 1½ in (13 to 40 mm)
DISTRIBUTION	England to Mediterranean; Canary Islands
ABUNDANCE	Common
DEPTH	Low tide mark to 65 ft (20 m)
HABITAT	Sandy and muddy bottoms
FEEDING HABIT	Filter feeder
BYSSUS	Present

SHELL SIZE RANGE
½ to 1½ in
(13 to 40 mm)

PHOTOGRAPHED SHELL
1¼ in
(31 mm)

96

PANDORA INAEQUIVALVIS
UNEQUAL PANDORA
(LINNAEUS, 1758)

Actual size

Pandora inaequivalvis lives horizontally on the surface of muddy sandy bottoms, or buried shallowly, in shallow waters. Its valves are of markedly different sizes, with the bottom (left) valve larger and more convex than the right, which is nearly flat. There is little space between the valves for the animal to live within. The animal has very short siphons, which are bent upward, to avoid collecting sediment from the bottom. It is the type species of the genus. There are about 25 living species in the family Pandoridae worldwide. Most species are found in the Northern Hemisphere. The family's fossil record extends to the Oligocene Period.

RELATED SPECIES

Pandora ceylanica Sowerby I, 1835, from the Red Sea and Indian Ocean, has a small, fragile, and very compressed shell that resembles *P. inaequivalvis*, but with the posterior end more elongate and pointed. *Pandora arcuata* Sowerby I, 1835, from Lower California to Peru, as the name suggests, has a shell with the posterodorsal margin forming an arc. It is the more common *Pandora* species found in Ecuador and Peru.

The shell of the Unequal Pandora is small, thin, and compressed, with a crescent-shaped outline. Its umbones are small and positioned in the anterior half; the hinge plate is moderately narrow, lacks teeth (but has secondary toothlike ridges), and has an internal ligament. The left valve is larger and more convex than the right, which is almost flat. The sculpture consists of fine concentric lines, and a smooth and glossy interior. The shell color is grayish white, and the interior is white with a pink tint near the posterior margin.

FAMILY	Lyonsiidae
SHELL SIZE RANGE	⅝ to 1¼ in (15 to 31 mm)
DISTRIBUTION	North Carolina to Brazil
ABUNDANCE	Common
DEPTH	18 to 36 ft (5.5 to 11 m)
HABITAT	Grows in sponges
FEEDING HABIT	Filter feeder
BYSSUS	Present

SHELL SIZE RANGE
⅝ to 1¼ in
(15 to 31 mm)

PHOTOGRAPHED SHELL
1¼ in
(31 mm)

ENTODESMA BEANA

PEARLY LYONSIA

(D'ORBIGNY, 1853)

Entodesma beana is a small clam that lives in sponges. Because the available space is sometimes narrow, its shell is irregular and variable in shape. The shell is thin and translucent, and the animal is bright orange, with short, separate siphons. It is a filter feeder, like other lyonsiids. Most lyonsiids live buried vertically in soft sediments, while others are nestlers (such as some species of *Entodesma*), living in rock crevices, or holes in sponges and tunicates. Some species have a thin and long byssus attached to gravel. There are about 45 living species in the family Lyonsiidae worldwide. The family has been known since the Paleocene Period.

RELATED SPECIES

Entodesma navicula (A. Adams and Reeve, 1850), which ranges from the Aleutian Islands to California, is probably the largest species of lyonsiids, reaching 5½ in (141 mm) in length. Its thin shell is fragile when dry. *Lyonsia hyalina* (Conrad, 1831), from Nova Scotia to South Carolina, has a small, fragile shell that has a nacreous interior. The nacre can be seen where the periostracum is eroded, such as around the umbones.

The shell of the Pearly Lyonsia is small-medium, thin, translucent, fragile, and irregularly oval to quadrangular. Its umbones are small and located anteriorly; the hinge plate is narrow and toothless. The valves are unequal in size, with the left one larger than the right, and gape slightly anteriorly and posteriorly. The sculpture consists of irregular concentric lines and weak, raised radial ribs. The shell is translucent pearly.

Actual size

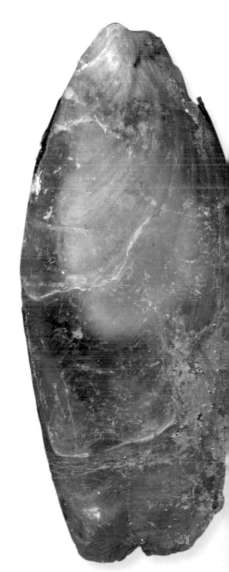

FAMILY	Clavagellidae
SHELL SIZE RANGE	3 to 8 in (75 to 200 mm)
DISTRIBUTION	Indo-West Pacific
ABUNDANCE	Uncommon
DEPTH	135 to 265 ft (40 to 80 m)
HABITAT	Sandy or muddy bottoms
FEEDING HABIT	Filter feeder
BYSSUS	Absent

SHELL SIZE RANGE
3 to 8 in
(75 to 200 mm)

PHOTOGRAPHED SHELL
5⅓ in
(134 mm)

98

BRECHITES PENIS
COMMON WATERING POT
(LINNAEUS, 1758)

Brechites penis is a very distinct and unusual bivalve. Juveniles live freely and have normal shaped shells. When the animal starts to grow, however, it forms a calcareous tube, in which its vestigial shell becomes embedded. The anterior end of the tube is broad and is kept buried in the sediment surface; it has a rounded disk, with many short tubes, and a frilly collar. It is reminiscent of a daisy or the spout of a watering pot. The posterior end of the tube is narrow and open, and is kept just at the surface of the sediment. There are about 15 living species in the family Clavagellidae worldwide.

The shell of the Common Watering Pot is very small, oval, and embedded into a large calcareous tube. Its valves have different sizes, and lack a hinge plate. The anterior end of the calcareous tube has a perforated disk with many short tubes, and a pleated collar. The tube tapers to a narrow and posterior open end, through which the siphons reach the surface. The tube has a sculpture of fine concentric lines. The valves and tube are chalky white.

Actual size

RELATED SPECIES

Brechites giganteus (Sowerby III, 1888), from Japan, is the largest clavagellid, with its calcareous tube reaching over 16 in (400 mm) in length. It lives in gravely sandy bottoms, with part of the tube sticking out of the sediment. *Penicillus philippinensis* (Chenu, 1843), from the tropical western Pacific, has a small shell and forms a calcareous tube similar to *B. penis*, but it often attaches sand grain, gravel, or shell fragments to its tube.

FAMILY	Laternulidae
SHELL SIZE RANGE	1¼ to 2½ in (32 to 63 mm)
DISTRIBUTION	Japan to Hong Kong
ABUNDANCE	Common
DEPTH	Intertidal to shallow subtidal
HABITAT	Muddy bottoms and mangroves
FEEDING HABIT	Filter feeder
BYSSUS	Absent

SHELL SIZE RANGE
1¼ to 2½ in
(32 to 63 mm)

PHOTOGRAPHED SHELL
2½ in
(60 mm)

LATERNULA SPENGLERI
SPENGLER'S LANTERN CLAM
(GMELIN, 1791)

Laternula spengleri is a mud-dwelling clam that lives in the intertidal zone and shallow, subtidal warm waters. It is a slow burrower and lives buried shallowly in the sediment; large specimens may be unable to rebury themselves if displaced. Instead of using a flexible ligament to open the valves, laternulids use their flexible and thin shell. The umbones have a fissure to help in flexing part of the shell. There are about ten living species in the family Laternulidae, all of them distributed in the Indo-West Pacific and one in the Antarctic. The fossil history of the family extends back to the Triassic Period. The relationships of the species in this family are poorly understood.

RELATED SPECIES
Laternula truncata (Lamarck, 1818), from Okinawa to the tropical West Pacific, has a very similar shell to *L. spengleri*, and may represent the same species. *Laternula anatina* (Linnaeus, 1758), from the Indo-West Pacific, is one of the most common species in the family. It has an elongate shell with a rounded posterior margin that is reminiscent of a duck's beak.

The shell of the Spengler's Lantern Clam is medium-sized, very thin, extremely fragile, translucent, and elongate-oval in outline. Its umbones are small, located centrally, and possess a transversal fissure. The hinge is modified, and toothless, with an enlarged spoonlike extension under the umbones. The surface is smooth; it has fine concentric growth lines, and microscopic granules and spines that give it a sandpapery feel. The shell color is gray with a white periostracum, and the interior is off-white and pearly.

Actual size

FAMILY	Periplomatidae
SHELL SIZE RANGE	1½ to 2⅝ in (40 to 65 mm)
DISTRIBUTION	Southern California to Peru
ABUNDANCE	Common
DEPTH	Subtidal to 65 ft (20 m)
HABITAT	Muddy and sandy bottoms
FEEDING HABIT	Filter feeder
BYSSUS	Absent

SHELL SIZE RANGE
1½ to 2⅝ in
(40 to 65 mm)

PHOTOGRAPHED SHELL
2 in
(50 mm)

100

PERIPLOMA PLANIUSCULUM
WESTERN SPOON CLAM
SOWERBY I, 1834

Periploma planiusculum has a spoon-shaped projection of the valve right under the umbones, called a chondrophore, which serves as the attachment site for the internal ligament. It is a common species that lives deeply buried in mud and sand, lying horizontally with the right valve uppermost. The right valve is larger and more convex than the left one. The species name refers to its flattened left valve. There are about 35 living species in the family Periplomatidae worldwide.

RELATED SPECIES

Periploma pentadactylus Pilsbry and Olsson, 1935, from the Pacific coast of Nicaragua to Panama, has a small and bizarre shell with five strong radial ribs that resemble claws. Most peroplomatids do not have radial sculpture. *Periploma margaritaceum* (Lamarck, 1801), which ranges from South Carolina to Brazil, has a small shell with an oblong outline that is truncated posteriorly.

Actual size

The shell of the Western Spoon Clam is medium-sized, thin, fragile, compressed, and quadrate in outline. Its umbones are small, pointed, and face the posterior end, which is shorter than the anterior. The hinge is toothless and narrow, and a long chondrophore extends from under the umbones. The valves are unequal in size, with the right one larger and more convex. The surface is smooth, with fine concentric lines and, sometimes, small pustules. The shell is cream-colored.

FAMILY	Thraciidae
SHELL SIZE RANGE	1½ to 4 in (40 to 100 mm)
DISTRIBUTION	Europe to western Africa; Mediterranean
ABUNDANCE	Uncommon
DEPTH	Intertidal to 200 ft (60 m)
HABITAT	Fine sand, mud and gravel bottoms
FEEDING HABIT	Filter and deposit feeder
BYSSUS	Absent in adults

SHELL SIZE RANGE
1½ to 4 in
(40 to 100 mm)

PHOTOGRAPHED SHELL
1⅞ in
(47 mm)

THRACIA PUBESCENS

PUBESCENT THRACIA

(PULTENEY, 1799)

101

Thracia pubescens is the largest of the European thraciids, and one of the largest in the family. It lives buried in fine sand and mud in shallow water, as well as offshore. It is a slow burrower but is capable of reburying if dislodged. Thraciids usually have a very thin shell. The siphons are separate, and produce two separate mucus canals. There are about 30 living species in the family Thraciidae worldwide, with most distributed in temperate to cold waters.

RELATED SPECIES

Thracia myopsis Möller, 1842, which has circumarctic distribution, and occurs south of British Columbia, has a medium-sized, quadrate oval shell. *Thracia kakumana* Yokoyama, 1927, from the Okhotsk Sea to northern Japan, has a large, thick shell that resembles *T. pubescens* but is broader and shorter.

The shell of the Pubescent Thracia is medium in size, thin, brittle, inflated, and elongate quadrangular in outline. Its umbones are small, rolled in, located at about midline; the hinge is thin and toothless. A triangular pit under the umbones is the attachment site for the ligament. The valves differ in size, with the right valve larger. The sculpture consists of fine, irregular concentric lines, and a finely granulated surface. The shell color is white, often tinged in orange, with a straw-colored periostracum, and the interior is porcelaneous white.

Actual size

FAMILY	Myochamidae
SHELL SIZE RANGE	1¼ to 1½ in (30 to 40 mm)
DISTRIBUTION	New Zealand
ABUNDANCE	Common
DEPTH	Intertidal to 65 ft (20 m)
HABITAT	Sand flats
FEEDING HABIT	Filter feeder
BYSSUS	Absent

SHELL SIZE RANGE
1¼ to 1½ in
(30 to 40 mm)

PHOTOGRAPHED SHELL
1½ in
(37 mm)

102

STRIATE MYADORA

(QUOY AND GAIMARD, 1791)

Myadora striata lives in sandy flats along high-energy beaches at intertidal and shallow, subtidal depths. It is endemic to New Zealand. Its valves are unequal in shape, with a flat left valve and a slightly convex right valve. Shells of the *Myadora* genus are similar to those of *Pandora* (family Pandoridae), with an elongate posterior margin and pointed umbones; however, in the latter, the right valve is the flatter one. Rock lobsters and snappers are among the predators that feed upon *M. striata*. There are about 20 living species of myochamid worldwide, all of which are distributed in the Indo-Pacific. The fossil record of the family extends back to the Miocene Period.

RELATED SPECIES

There are two genera in the family Myochamidae. Species in the genus *Myadora*, such as *M. delicata* Cotton, 1931, a species endemic to South Australia, are free-living. Species in the genus *Myochama*, such as *M. anomioides* Stutchbury, 1830, another species from southern Australia, live attached to other molluscan shells, especially bivalves.

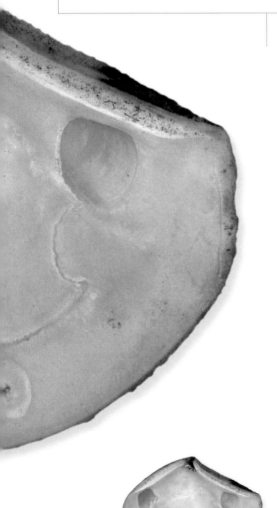

Actual size

The shell of the Striate Myadora is small, thin, compressed, and triangular-ovate in outline. Its umbones are acutely pointed, located about midshell, and pointing posteriorly. The hinge plate is relatively strong. The valves are unequal in size and shape, with a flat left valve and moderately convex right one. The anterodorsal margin is concave, the posterodorsal convex, and the ventral margin rounded. The sculpture consists of fine, irregular concentric lines. The shell color is off-white, and the color of the interior is white.

FAMILY	Verticordiidae
SHELL SIZE RANGE	½ to 1 in (13 to 23 mm)
DISTRIBUTION	Eastern Atlantic
ABUNDANCE	Uncommon
DEPTH	400 to 3,600 ft (120 to 1,100 m)
HABITAT	Soft bottoms
FEEDING HABIT	Filter feeder
BYSSUS	Present

SHELL SIZE RANGE
½ to 1 in
(13 to 23 mm)

PHOTOGRAPHED SHELL
¾ in
(18 mm)

SPINOSIPELLA ACUTICOSTATA

SHARP-RIBBED VERTICORD

(PHILIPPI, 1844)

103

Spinosipella acuticostata is an uncommon, deepwater clam with large, curved umbones. It has been reported by many authors as living in the western Atlantic, but a recent study restricts it to the eastern Atlantic. What used to be called *S. acuticostata* has been described as *S. agnes* Simone and Cunha, 2008, from Brazil. Verticordiids have large, rolled-in umbones, and, usually, strong radial ribs. There are about 50 living species in the family Verticordiidae worldwide; it is well represented in Australia. Most species occur in deep to abyssal waters. The family's fossil record extends to the Paleocene Period.

The shell of the Sharp-Ribbed Verticord is small, thick, solid, inflated, and quadrangular in outline. Its umbones are large, rolled inward, and pointed anteriorly. The hinge plate is weak, and bears 1 rounded tooth in the right valve and a corresponding socket in the left valve. The sculpture consists of 12 to 14 strong and sharp radial ribs, which can be seen near the margins interiorly. The shell color is white to gray, and the interior is porcelaneous white.

RELATED SPECIES

Spinosipella ericia (Hedley, 1911), from Australia, has a shell that is similar to but smaller than *S. acuticostata*, with 18 strong radial ribs that bear minute and fine spines. It is found in deep waters. *Euciroa galathea* (Dell, 1956), from New Zealand and Australia, has a larger, ovate shell that is smooth, with small radial ribs, and a pearly interior.

Actual size

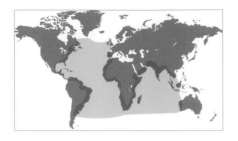

FAMILY	Poromyidae
SHELL SIZE RANGE	½ to ⅝ in (11 to 16 mm)
DISTRIBUTION	Atlantic and Indian Oceans
ABUNDANCE	Rare
DEPTH	6,840 to 12,240 ft (2,084 to 3,730 m)
HABITAT	Deep water
FEEDING HABIT	Carnivore, feeds on polychaetes and other invertebrates
BYSSUS	Present

SHELL SIZE RANGE
½ to ⅝ in
(11 to 16 mm)

PHOTOGRAPHED SHELL
⅝ in
(14 mm)

POROMYA TORNATA
TURNED POROMYA
(JEFFREYS, 1876)

Poromya tornata is a small carnivorous clam that lives in the abyssal zone, and thus is collected infrequently. It has been reported from the Atlantic and Indian oceans. Poromyids usually live in deep waters and have small, thin shells. Like other related clams, poromyids are carnivores and use their enlarged, modified, red inhalant siphon to capture small invertebrates such as polychaetes and crustaceans. The animal also has a crown of sensory tentacles that help it to locate prey. There are about 50 living species in the family Poromyidae worldwide. The fossil record of poromyids extends to the Cretaceous Period.

RELATED SPECIES

Poromya rostrata Rehder, 1943, which ranges from Florida to Uruguay, has a small, fragile shell with a rounded triangular outline, and smaller umbones. It is found in much shallower waters than *P. tornata*. *Poromya perla* Dall, 1908, from the Gulf of Panama (eastern Pacific), has a very similar but even smaller shell than *P. tornata*. It is also found in deep water.

Actual size

The shell of the Turned Poromya is small, thin, fragile, inflated, and rounded triangular in outline. The umbones are very large and curved inward, located at about midshell, and facing anteriorly; the hinge plate is narrow, with 1 tooth in each valve. The valves are about the same size, and have a smooth surface, with only fine concentric growth lines, and microscopic granules arranged in radial lines. The shell color is white, with a brownish yellow periostracum, and a porcelaneous white interior.

FAMILY	Cuspidariidae
SHELL SIZE RANGE	½ to 2 in (12 to 50 mm)
DISTRIBUTION	North Atlantic to Brazil
ABUNDANCE	Uncommon
DEPTH	390 to 9,600 ft (120 to 2,925 m)
HABITAT	Soft bottoms
FEEDING HABIT	Carnivore, feeds on invertebrates
BYSSUS	Absent

SHELL SIZE RANGE
½ to 2 in
(12 to 50 mm)

PHOTOGRAPHED SHELL
2 in
(50 mm)

CUSPIDARIA ROSTRATA

ROSTRATE CUSPIDARIA
(SPENGLER, 1793)

105

Cuspidaria rostrata has a very long, tubelike extension of the posterior margin, called the rostrum. It is a carnivore and feeds on small invertebrates, such as polychaetes, crustaceans, as well as foraminiferans (which are not invertebrates but protists). It lives buried in soft sediments, with the tip of the rostrum near the surface. The siphon tips have sensory tentacles that detect prey; the siphon is capable of extending to twice its relaxed length, and the animal sucks its prey into the mantle cavity. There are about 200 living species in the family Cuspidariidae worldwide, with most in deep and abyssal waters.

Actual size

RELATED SPECIES
Cuspidaria gigantea Prashad, 1932, from northern Western Australia, has a large shell with an even longer rostrum than *C. rostrata*. It lives in depths of over 3,300 ft (1,000 m). Its shell has a sculpture of fine concentric lines. *Cardiomya cleriana* (d'Orbigny, 1846), from Chile, has a small shell with a short and broad rostrum. The shell sculpture consists of strong radial ribs.

The shell of the Rostrate Cuspidaria is medium in size, thin, inflated, and oval in outline with a long, slightly curved, tubelike rostrum. Its umbones are prominent, and the hinge plate is thin, with a single tooth in the right valve. The valves are about the same size and shape. The anterior margin is rounded, and the posterior has a long rostrum that accounts for half of the shell length. The sculpture consists of concentric lines, and the interior is smooth and glossy. The shell color is white with a yellowish periostracum, and the interior is white.

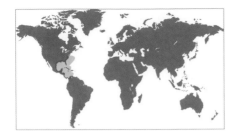

FAMILY	Lucinidae
SHELL SIZE RANGE	½ to 1½ in (12 to 37 mm)
DISTRIBUTION	North Carolina to Central America
ABUNDANCE	Moderately common
DEPTH	Shallow water to 300 ft (90 m)
HABITAT	Soft bottoms
FEEDING HABIT	Filter feeder
BYSSUS	Absent in adults

SHELL SIZE RANGE
½ to 1½ in
(12 to 37 mm)

PHOTOGRAPHED SHELL
1¼ in
(30 mm)

106

DIVARICELLA DENTATA

TOOTHED CROSS-HATCHED LUCINA

(WOOD, 1815)

The shell of the Toothed Cross-Hatched Lucina is medium-sized, thin, inflated, and circular in outline. Its umbones are small, located about centrally; the hinge plate is strong and toothless. The ventral and dorsal margins are dentate. The valves are identical, and do not gape. The sculpture consists of incised, oblique grooves, crossed by 4 to 7 growth stoppages, and the interior is smooth. The shell color is white, and inside the shell is also white with a yellowish tint.

Divaricella dentata has a shell with a distinct sculpture of incised, oblique grooves, and several growth stoppages, which are evident as concentric bands. Like some lucinids, the hinge plate is toothless; the species name refers to the dorsal and ventral internal margins, which are dentate. It is found in shallow water, but also occurs offshore, in soft bottoms. Recent estimates of the diversity in the family Lucinidae provide a figure of as many as 500 living species worldwide, ranging from the intertidal zone to abyssal depths. The family is known since the Silurian Period.

RELATED SPECIES

Divaricella soyoae (Habe, 1951), from Japan, has a shell that is similar in size and shape to *D. dentata*, but has a slightly coarser, oblique sculpture. It lives in moderately deep water. *Codakia tigerina* (Linnaeus, 1758), which ranges from the Red Sea to the Indo-West Pacific, has a large and thick shell with a strong cancellate sculpture. It is fished commercially.

Actual size

FAMILY	Lucinidae
SHELL SIZE RANGE	1 to 2½ in (25 to 60 mm)
DISTRIBUTION	Maryland to Colombia
ABUNDANCE	Common
DEPTH	Intertidal to 10 ft (3 m)
HABITAT	Sandy bottoms
FEEDING HABIT	Filter feeder
BYSSUS	Absent in adults

SHELL SIZE RANGE
1 to 2½ in
(25 to 60 mm)

PHOTOGRAPHED SHELL
1⅝ in
(43 mm)

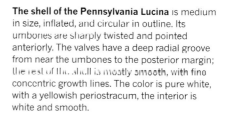

LUCINA PENSYLVANICA

PENNSYLVANIA LUCINA

(LINNAEUS, 1758)

109

Lucina pensylvanica is easily distinguished from other western Atlantic lucines by its circular outline with a strong radial groove that forms a notch on the posteroventral margin. Like other lucinids, *L. pensylvanica* harbors chemosynthetic bacteria in its gills, which allow the animal to live buried deep in hydrogen sulfide-rich sediments. This adaptation allows it to flourish in a habitat that is not suitable for most clams because it is often anoxic. *Lucina pensylvanica* is therefore often the dominant clam species in such habitats.

RELATED SPECIES

Two lucinids that range from North Carolina to Central America are *Lucina leucocyma* (Dall, 1886) and *Divaricella dentata* (Wood, 1815). The former has a small, rounded triangular shell with four large, rounded radial waves, which confer a lobed margin on the shell. The latter has a circular shell with incised, oblique grooves that form chevrons near the anterior. The ventral and dorsal margins are dentate.

The shell of the Pennsylvania Lucina is medium in size, inflated, and circular in outline. Its umbones are sharply twisted and pointed anteriorly. The valves have a deep radial groove from near the umbones to the posterior margin; the rest of the shell is mostly smooth, with fine concentric growth lines. The color is pure white, with a yellowish periostracum, the interior is white and smooth.

Actual size

FAMILY	Lucinidae
SHELL SIZE RANGE	2 to 4 in (50 to 100 mm)
DISTRIBUTION	Eastern Indian Ocean to Western Pacific
ABUNDANCE	Common
DEPTH	Intertidal to 65 ft (20 m)
HABITAT	Infaunal in coralline sand
FEEDING HABIT	Filter feeder
BYSSUS	Absent in adults

SHELL SIZE RANGE
2 to 4 in
(50 to 100 mm)

PHOTOGRAPHED SHELL
2¼ in
(55 mm)

108

FIMBRIA FIMBRIATA
COMMON BASKET LUCINA
(LINNAEUS, 1758)

Actual size

The shell of the Common Basket Lucina is
medium-large, solid, thick, and elongately ovate
in outline. Its umbones are rounded and pointed
anteriorly. The valves are roughly identical in
size and shape, and are not gaping. They are
sculptured with numerous radial ribs crossed
by strong concentric ridges that are lamellose
marginally. The shell color is porcelaneous white,
tinged in pale pink on the dorsal margin; the
interior is off-white, with pinkish margins, and
the hinge is often golden yellow.

Fimbria fimbriata has a large, ovate shell with raised concentric
ridges and radial ribs that form a latticed sculpture. It is found
either completely buried in coralline sand or fully exposed, in
shallow waters. *Fimbria fimbriata* is collected and marketed for
food in southern Japan and the Philippines; its shell is also used
for the shellcraft industry and to make lime. Unlike other lucinids,
F. fimbriata has a short inhalant siphon (typical lucinids lack
one); because of this and other anatomical differences, some
authors believe that the genus *Fimbria* should be segregated in
a separate family.

RELATED SPECIES
Fimbria soverbii (Reeve, 1841), from the southwest Pacific,
has a slightly larger shell with wider and more spaced-out
concentric lamellae than *F. fimbriata*. *Codakia tigerina* (Linnaeus,
1758), from the Red Sea to the Indo-West Pacific, has a large,
subcircular and compressed shell with a cancellate sculpture. It
is consumed as food locally.

FAMILY	Lucinidae
SHELL SIZE RANGE	¾ to 3 in (20 to 75 mm)
DISTRIBUTION	Red Sea to Hawaii
ABUNDANCE	Common
DEPTH	Shallow to 650 ft (200 m)
HABITAT	Muddy bottoms
FEEDING HABIT	Filter feeder
BYSSUS	Absent in adults

SHELL SIZE RANGE
¾ to 3 in
(20 to 75 mm)

PHOTOGRAPHED SHELL
2½ in
(63 mm)

ANODONTIA EDENTULA

TOOTHLESS LUCINA

(LINNAEUS, 1758)

Anodontia edentula has a thin-walled shell with a variable outline and sculpture. This species lives in mud, muddy sands, and mangroves, in shallow waters. It ranges from the Red Sea, throughout the Indo-West Pacific, to Hawaii. Empty shells can be locally abundant in beach drift, as in Midway, Hawaii. Hawaiian shells were named as a separate species, but the difference is mostly the smaller size, and they are thus considered the same species. *Anodontia edentula* is the type species of the genus; *Anodontia* means toothless, as does *edentula*. Both refer to the toothless hinge plate.

RELATED SPECIES

Anodontia alba Link, 1807, from North Carolina to Venezuela, is a common species with a thin, circular, white shell with a yellow interior. It is known as the Buttercup Lucine. Its shell is often used in shellcraft. *Fimbria fimbriata* (Linnaeus, 1758), which ranges from the eastern Indian Ocean to western Pacific, has a thick, ovate shell; its raised concentric lines form a lattice pattern.

The shell of the Toothless Lucina is medium in size, thin, inflated, and subcircular in outline. Its umbones are inflated and turned anteriorly; the hinge plate is narrow and toothless. The valves are identical in size and shape, and do not gape. In juveniles, the sculpture is predominantly radial, but as the shell grows this fades and concentric lines become more visible, including several growth pauses. The shell color is white, both inside and out.

Actual size

FAMILY	Lucinidae
SHELL SIZE RANGE	2½ to 5 in (60 to 130 mm)
DISTRIBUTION	Red Sea to Indo-West Pacific
ABUNDANCE	Common
DEPTH	Intertidal to 65 ft (20 m)
HABITAT	Sandy bottoms
FEEDING HABIT	Filter feeder
BYSSUS	Absent in adults

SHELL SIZE RANGE
2½ to 5 in
(60 to 130 mm)

PHOTOGRAPHED SHELL
2½ in
(64 mm)

CODAKIA TIGERINA
PACIFIC TIGER LUCINE
(LINNAEUS, 1758)

Codakia tigerina is a large and common lucinid with a wide distribution, ranging from the Red Sea to the West Pacific, in shallow warm waters. It lives buried in sandy bottoms, especially near coral reefs. This species of lucinid is collected for food in the Philippines and Tonga; its flesh is also chewed with betel nut, and the shell is used in the shellcraft industry or to make lime. The shell is thick and has a cancellate sculpture that can have more than 100 radial ribs. Note that the species name is frequently misspelled "tigrina."

RELATED SPECIES
Codakia distinguenda (Tryon, 1872), from Baja California to Panama, is the largest living species of Lucinidae, which can reach up to 6 in (150 mm) across. Its shell is very similar to *C. tigerina*. *Fimbria fimbriata* (Linnaeus, 1758), from the eastern Indian Ocean to western Pacific, is another large, edible species. Its shell is ovate and has raised concentric lamellae and radial ribs.

The shell of the Pacific Tiger Lucine is large, heavy, solid, compressed, and subcircular in outline. Its umbones are small, located at the midline of the thick hinge, and facing anteriorly. The sculpture is cancellate, with concentric and radial ridges of about equal strength and spacing; it has a rough feel to the touch. The shell color is cream to white, sometimes tinged with yellow or pink on the umbones; the interior is yellow, with a white margin, and is tinged with pink on the dorsal margin.

Actual size

FAMILY	Ungulinidae
SHELL SIZE RANGE	½ to 1⅛ in (12 to 27 mm)
DISTRIBUTION	Portugal to Senegal, and Mediterranean
ABUNDANCE	Common
DEPTH	Intertidal to 16 ft (5 m)
HABITAT	Rock crevices
FEEDING HABIT	Filter feeder
BYSSUS	Present

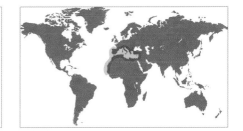

SHELL SIZE RANGE
½ to 1⅛ in
(12 to 27 mm)

PHOTOGRAPHED SHELL
1⅛ in
(27 mm)

UNGULINA CUNEATA

ROSY DIPLODON

(SPENGLER, 1782)

111

Ungulina cuneata is a small, crevice-dwelling bivalve that lives in shallow waters from Portugal to Senegal, and in the Mediterranean. It is most common in the western Mediterranean. Because it depends on the space available in crevices, its shell is variable in shape. Ungulinids usually have a shell with a coarse, concentric surface sculpture, and the median tooth in the hinge is divided in two. *Ungulina cuneata* is the type species of the genus. There are about 50 living species in the family Ungulinidae worldwide, with most species occurring in cold or deep waters.

RELATED SPECIES

Diplodonta rotundata (Montagu, 1803), from the British Isles to Angola, has a circular shell with concentric lines. It ranges from the intertidal zone to 12,600 ft (3,850 m). *Phlyctiderma semiaspera* (Philippi, 1836), from North Carolina to Argentina and the Gulf of Mexico, has a small, chalky, and nearly circular shell. It looks smooth but has microscopic pits in concentric lines.

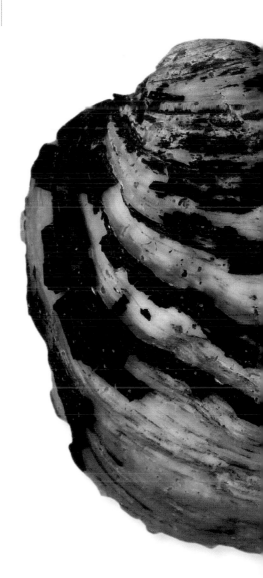

The shell of the Rosy Diplodon is small, thick, and inflated, with an irregular circular outline. Its shell is variable, and the sculpture consists of weak or strong concentric lines. The hinge has 2 cardinal teeth, with the median one split in half. Its shell color is rusty or olive brown, and the interior of the valve is pinkish.

Actual size

FAMILY	Chamidae
SHELL SIZE RANGE	1 to 2 in (25 to 50 mm)
DISTRIBUTION	Quintana Roo, Mexico to Brazil
ABUNDANCE	Uncommon
DEPTH	6 to 250 ft (2 to 76 m)
HABITAT	Broken shell and coarse bottoms
FEEDING HABIT	Filter feeder
BYSSUS	Absent

SHELL SIZE RANGE
1 to 2 in
(25 to 50 mm)

PHOTOGRAPHED SHELL
1¾ in
(46 mm)

112

ARCINELLA ARCINELLA
TRUE SPINY JEWEL BOX
(LINNAEUS, 1767)

The shell of the True Spiny Jewel Box is small to medium in size, thick, solid, and irregularly rounded to subquadrate in outline. Its umbones are low. The diagnostic features are the long, slender spines arranged in about 16 to 35 radial rows, with coarse granulations between ribs. It lives cemented to hard substrates by either left or right valve. The shell color ranges from white to yellowish or pinkish, and the interior is stained with reddish purple.

Arcinella arcinella has one of the spiniest shells in the family Chamidae; a group in which most species have some spines or lamellae. In early life it lives cemented to hard substrates; as an adult it is free-living on shell hash or coarse sand. Chamids somewhat resemble oysters, but differ by having two adductor muscles instead of one. Additionally, unlike oysters, chamids can cement to hard substrates using either the left or right valve, and this can vary within species. The umbones point to the anterior. There are about 70 species in the family Chamidae, with the most diversity found in shallow water in the tropics. The family has a fossil record dating from the Cretaceous Period.

RELATED SPECIES
Arcinella cornuta Conrad, 1866, from North Carolina to Texas, has a very similar shell that is slightly smaller but heavier and broader, with fewer spines. *Chama lazarus* (Linnaeus, 1758), from the Indo-West Pacific, has a large shell with wide, foliose spines. It lives cemented to hard substrates throughout its life.

Actual size

FAMILY	Chamidae
SHELL SIZE RANGE	2 to 5½ in (50 to 140 mm)
DISTRIBUTION	Red Sea to Indo-West Pacific
ABUNDANCE	Common
DEPTH	Intertidal to 100 ft (30 m)
HABITAT	Cemented to rock bottoms
FEEDING HABIT	Filter feeder
BYSSUS	Absent

SHELL SIZE RANGE
2 to 5½ in
(50 to 140 mm)

PHOTOGRAPHED SHELL
2¼ in
(58 mm)

CHAMA LAZARUS

LAZARUS JEWEL BOX

(LINNAEUS, 1758)

113

Chama lazarus probably has the largest spines in the family Chamidae. It is a common species from the Red Sea to the Indo-West Pacific. Its shell has many long, frilly spines and lamellae, on both valves. It lives in shallow, clean waters, and does not tolerate waters with too much suspended particulate or brackish waters. This species is more common along open coasts. It is sometimes collected by coastal peoples for food, but more recently it is mostly collected because of its beautiful shell. It lives permanently cemented to hard substrates by its left valve, which is shallower than the right.

RELATED SPECIES

Chama frondosa Broderip, 1835, which ranges from the Gulf of California to Ecuador and the Galápagos Islands, has a shell with many foliated lamellae. Well-preserved specimens are rare in collections. *Arcinella arcinella* (Linnaeus, 1767), from Quintana Roo, Mexico to Brazil, lives cemented as a juvenile but adults are free-living. It has many long, slender spines on both valves.

The shell of the Lazarus Jewel Box is medium-large, thick, solid, and with an ovate or circular outline. Its umbones are small and located slightly anteriorly; the hinge plate is broad and strong. It is cemented by the left valve, which is shallower and smaller than the right. Both valves are covered with long, wide, foliated spines, arranged in concentric rows. The margin is crenulate. The shell color varies from grayish to yellowish white, tinged with light brown, red, or pink, and the interior is brownish. Younger shells are more brightly colored than larger ones.

Actual size

FAMILY	Hiatellidae
SHELL SIZE RANGE	¾ to 2¾ in (20 to 70 mm)
DISTRIBUTION	North Atlantic to Argentina; eastern Pacific; Antarctic islands
ABUNDANCE	Common
DEPTH	Intertidal to 3,000 ft (900 m)
HABITAT	Byssally attached to crevices in rocks
FEEDING HABIT	Filter feeder
BYSSUS	Present

SHELL SIZE RANGE
¾ to 2¾ in
(20 to 70 mm)

PHOTOGRAPHED SHELL
1¼ in
(30 mm)

114

HIATELLA ARCTICA
ARCTIC SAXICAVE
(LINNAEUS, 1767)

The shell of the Arctic Saxicave is medium in size, solid, irregular, rugose, and oblong-oval in outline. Its umbones are slightly raised, located closer to the anterior end, which is much shorter than the posterior one. The valves are about the same size, but because of its irregular growth, may also be different sizes. The sculpture consists of irregular, concentric growth lines, which are sometimes rugose. The color of the shell is chalky white, and it is covered by a brown periostracum; the interior is white and glossy.

Hiatella arctica is widely distributed, ranging from both sides of the North Atlantic south to Argentina, as well as the eastern Pacific and Antarctic islands. It has a variable shape, and is assumed to be a widely distributed species. However, recent studies on Brazilian specimens suggest that there are at least two different species, based on the spawning times, egg color, radial ribs on the shells, and other features. Shells in the family are often very variable, and the group needs revision. There are about 25 living species recognized in the family Hiatellidae worldwide, with some species living in cold and subantarctic waters.

RELATED SPECIES
Hiatella australis (Lamarck, 1818), from Australia and New Zealand, has an elongate shell with a variable shape and coarse concentric ridges. It also grows in crevices, to which it is attached by byssus. *Panopea glycymeris* (Born, 1778), from Spain to Namibia, West Africa, and the western Mediterranean, is a large clam with long siphons. It has a thick and heavy shell, and burrows deeply in sand and mud.

Actual size

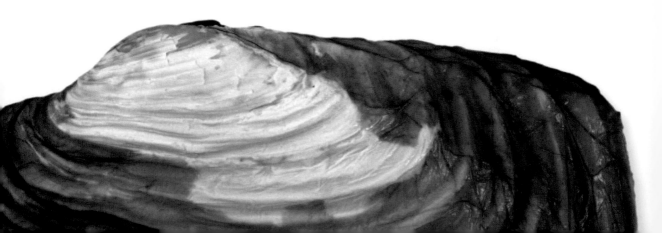

FAMILY	Hiatellidae
SHELL SIZE RANGE	6 to 12 in (150 to 300 mm)
DISTRIBUTION	Spain to Namibia; western Mediterranean
ABUNDANCE	Uncommon
DEPTH	33 to 330 ft (10 to 100 m)
HABITAT	Sand, mud, and gravel bottoms
FEEDING HABIT	Filter feeder
BYSSUS	Present

PANOPEA GLYCYMERIS

EUROPEAN PANOPEA

(BORN, 1778)

SHELL SIZE RANGE
6 to 12 in
(150 to 300 mm)

PHOTOGRAPHED SHELL
9 in
(231 mm)

115

Panopea glycymeris is the largest clam in European waters, and the world's largest burrowing clam. The shell is thick and heavy, and gapes widely on three margins. The animal has very long siphons, reaching about 18 in (450 mm) in length. The siphons cannot be fully withdrawn into the shell. *Panopea glycymeris* lives deeply buried in sand, mud, or gravel bottoms. The animal itself cannot dig to escape a predator; instead it contracts its long siphons, which deters most enemies. Even humans, its main predator, have a hard time dislodging the giant clam from its burrow. It can live to an age of 168 years.

RELATED SPECIES

Panopea abrupta (Conrad, 1849), from Alaska to California, and to southern Japan, has a similar shell, and is one of the largest species in the family. It is known as "geoduck," a Native American name meaning "dig deep." *Hiatella arctica* (Linnaeus, 1767), from North Atlantic to Argentina, the eastern Pacific and the Antarctic, has a much smaller shell that has concentric folds.

Actual size

The shell of the European Panopea is large, thick, heavy, and subquadrate in outline. Its umbones are low and broad, located about centrally, and the hinge has 1 tooth in each valve. The valves are similar in size and shape, and gape widely. The sculpture consists of strong, irregular, concentric growth lines. The shell color of the European Panopea is off-white, with a light-brown periostracum, and the interior is porcelaneous white.

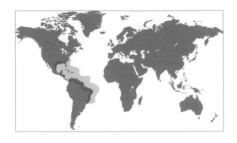

FAMILY	Gastrochaenidae
SHELL SIZE RANGE	¾ to 1¼ in (20 to 30 mm)
DISTRIBUTION	North Carolina to Brazil
ABUNDANCE	Uncommon
DEPTH	33 to 190 ft (10 to 60 m)
HABITAT	Bores into dead coral
FEEDING HABIT	Filter feeder
BYSSUS	Present

SHELL SIZE RANGE
¾ to 1¼ in
(20 to 30 mm)

PHOTOGRAPHED SHELL
1 in
(26 mm)

116

SPENGLERIA ROSTRATA
ATLANTIC SPENGLER CLAM
LINNAEUS, 1758

Actual size

The shell of the Atlantic Spengler Clam is medium-small, thin but strong, and elongated quadrangular, with the posterior end truncated. There is a wide gape in the anteroventral region. The dorsal posterior part of the shell is separated by a strong radial groove, forming an elevated, triangular area with transverse ribbing; this area has a yellow-brown periostracum. The rest of the shell is smooth, with fine growth lines, and it is white in color.

Spengleria rostrata is a borer that is found in limestone, where it grows safe from predators. It is a filter feeder, and its siphons are long and separated distally, with the inhalant and exhalant siphons having separate openings on the surface of the rock, like a "Y." Boring in *S. rostrata* is achieved mechanically by rasping the border of the shell against the substrate, and may be aided by secretions produced by the mantle. Boring organisms contribute to the breakdown of coralline rocks and the formation of fine calcareous sediment. There are some 15 living species in the family Gastrochaenidae distributed worldwide, in shallow tropical and subtropical waters.

RELATED SPECIES
Spengleria mytiloides (Lamarck, 1818) and *Gastrochaena cuneiformis* Spengler, 1783, are two related species from the tropical western Pacific. The former has a shell similar in size and shape to *S. rostrata*. The latter is one of the most common coral borers in Okinawa; it occurs in both living and dead coral, although it is more common in limestone.

FAMILY	Arctictidae
SHELL SIZE RANGE	2⅝ to 5 in (67 to 130 mm)
DISTRIBUTION	Both sides of the North Atlantic
ABUNDANCE	Abundant
DEPTH	45 to 840 ft (15 to 255 m)
HABITAT	Fine to coarse sandy bottoms
FEEDING HABIT	Filter feeder
BYSSUS	Absent

SHELL SIZE RANGE
2⅝ to 5 in
(67 to 130 mm)

PHOTOGRAPHED SHELL
2⅝ in
(67 mm)

ARCTICA ISLANDICA

OCEAN QUAHOG
(LINNAEUS, 1767)

117

Arctica islandica is considered the longest-living mollusk, and one of the longest-living, non-colonial invertebrates: a recent study suggested one specimen to have lived to be between 405 and 410 years of age. The animal grows and matures very slowly; it first reproduces between the ages of 7 and 14 years. The animal has no siphons, and can withstand several days without oxygen. This species is exploited commercially, especially in Europe, where its flesh is used for food, and it is popular as sushi. Its thick and glossy periostracum tends to peel off in dried shells. *Arctica islandica* is the only survivor of a large group of bivalves that appeared in the Triassic Period.

RELATED SPECIES

The shell of the unrelated Northern Quahog (family Veneridae), *Mercenaria mercenaria* (Linnaeus, 1758), resembles *Arctica islandica* in shape and size, but differs in some shell characters, such as the presence of a pallial sinus. The closest-living family is the Trapeziidae, which includes *Trapezium oblongum* (Linnaeus, 1758), from the tropical Indo-Pacific. It has an elongate subquadrate shell.

The shell of the Ocean Quahog is medium-large, thick, heavy, and nearly circular in outline. Its umbones are strong, rounded, and located anteriorly. The hinge line is thick, with 3 cardinal teeth on each valve. The valves are identical in size and shape, and are sculpted with fine concentric growth lines. The inner ventral margin of the valves is smooth. The shell color is off-white, but the valves are covered with a thick brown or black periostracum.

Actual size

FAMILY	Glossidae
SHELL SIZE RANGE	¾ to 1½ in (20 to 40 mm)
DISTRIBUTION	Red Sea to tropical Indo-West Pacific
ABUNDANCE	Common
DEPTH	23 to 230 ft (7 to 70 m)
HABITAT	Sandy and muddy bottoms
FEEDING HABIT	Filter feeder
BYSSUS	Absent in adults

SHELL SIZE RANGE
¾ to 1½ in
(20 to 40 mm)

PHOTOGRAPHED SHELL
1½ in
(38 mm)

118

MEIOCARDIA MOLTKIANA
MOLTKE'S HEART CLAM
(SPENGLER, 1783)

Meiocardia moltkiana is a small clam with a strong keel that runs from the umbones to the posteroventral margin. Its umbones are large, rolled inward, and turned anteriorly, and thus resemble a heart when seen from the anterior margin. The family needs further studies; currently, some species, like *M. moltkiana* are assumed to have wide distributions. The animal has a thin byssus consisting of a few threads, which occurs only in juveniles. There are only about ten living species in the family Glossidae worldwide, with most found in the Indo-Pacific. The family has been known since the Paleocene Period.

RELATED SPECIES

Meiocardia hawaiiana Dall, Bartsch, and Rehder, 1938, from Japan to Hawaii, has a small shell that is subquadrate in outline, and heart-shaped when seen from the anterior margin. It has an angular keel running from the posterior to the ventral margin. *Glossus humanus* (Linnaeus, 1758), which ranges from Norway to Morocco and the Mediterranean, has a large, thin, and inflated shell. It is one of the largest glossids.

Actual size

The shell of the Moltke's Heart Clam is small, thick, solid, inflated, and heart-shaped. Its umbones are large and in-rolled, facing anteriorly; the hinge is strong and bears 2 teeth in each valve. The valves are nearly identical in size and shape, with a strong posteroventral keel, and sharp concentric ribs that end at the keel. The interior is smooth. The shell color is white, sometimes tinged with reddish brown blotches; the interior is white.

FAMILY	Glossidae
SHELL SIZE RANGE	2 to 4½ in (50 to 120 mm)
DISTRIBUTION	Norway to Morocco; Mediterranean
ABUNDANCE	Locally abundant
DEPTH	23 to 820 ft (7 to 250 m)
HABITAT	Sandy and muddy bottoms
FEEDING HABIT	Filter feeder
BYSSUS	Absent in adults

SHELL SIZE RANGE
2 to 4½ in
(50 to 120 mm)

PHOTOGRAPHED SHELL
3⅛ in
(79 mm)

119

GLOSSUS HUMANUS

OXHEART CLAM

(LINNAEUS, 1758)

Glossus humanus is a well-known European clam with coiled umbones that give the shell a resemblance to a mammalian heart. The shape of the umbones easily separates the species from any other European clam. Although it can be locally abundant, it is not easily collected because it occurs offshore. This species is rarely sold for food. It lives shallowly buried in sand or mud, in moderate to deep waters, where it filters particles in the water. The animal lacks siphons; instead, the mantle margin forms inhalant and exhalant regions. The periostracum is thick and it is often eroded near the umbones.

RELATED SPECIES

Meiocardia moltkiana (Spengler, 1783), from the Indo-West Pacific, has a small shell with a sharp dorsal keel and in-rolled umbones. Like other species in the family, it also has a heart-shaped shell. *Meiocardia vulgaris* (Reeve, 1845), from the western Pacific, has an even smaller shell that is very similar to *M. moltkiana*. It lives in sandy bottoms in shallow waters.

The shell of the Oxheart Clam is medium-sized, thin but solid, lightweight, inflated, and globular in outline, resembling a mammalian heart. Its umbones are large and recurved, anteriorly facing; the hinge plate is strong, with 2 teeth in each valve. The valves are about the same size and shape, sculptured externally with fine concentric lines, and with a smooth interior. The shell color is dirty white or fawn, with a thick, red-brown periostracum, and the interior color is white.

Actual size

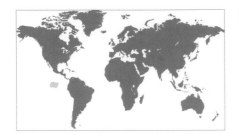

FAMILY	Vesicomyidae
SHELL SIZE RANGE	6 to 10¼ in (150 to 260 mm)
DISTRIBUTION	Off Galápagos Islands, near deepsea hot springs
ABUNDANCE	Locally common
DEPTH	8,050 to 9,050 ft (2,450 to 2,750 m)
HABITAT	Epibenthic on hydrothermal vent systems
FEEDING HABIT	Chemoautotrophic; has symbiotic bacteria
BYSSUS	Absent in adults

SHELL SIZE RANGE
6 to 10¼ in
(150 to 260 mm)

PHOTOGRAPHED SHELL
8 in
(200 mm)

120

CALYPTOGENA MAGNIFICA

MAGNIFICENT CALYPTO CLAM

BOSS AND TURNER, 1980

The shell of the Magnificent Calypto Clam is large, heavy, solid, moderately inflated, and subquadrate elongate in outline. Its umbones are strong but low, and located anteriorly; the hinge is strong, but the teeth are relatively small. The valves are about same size and shape, and gape slightly. The sculpture consists of irregular concentric growth lines. The shell color is chalky white, with a light brown periostracum, and the interior is porcelaneous white.

Before the discovery of the Galápagos Rift in 1977, it was thought that all deepsea organisms depended on food that fell from shallower waters, and, ultimately, on sunlight. *Calyptogena magnifica* was one of the first-known exceptions to this rule. Thanks to symbiotic, sulfur-oxidizing bacteria, it obtains its energy from hydrocarbons that leak from cracks in the ocean bottom. *Calyptogena magnifica*, tubeworms, and other organisms thrive and become giants near undersea hydrothermal vents and cold methane seeps. There are about 30 living species in the family Vesicomyidae, with most in deep water, including some restricted to sulfur-rich hot springs and cold seeps.

RELATED SPECIES

Calyptogena diagonalis Barry and Kochevar, 1999, which is endemic to cold seeps off Costa Rica and Oregon, has a large shell that resembles *C. magnifica*. *Calyptogena soyoae* Okutani, 1957, from Sagami Bay, Japan, is also associated with hydrocarbon seeps. It can reach densities as high as 833 clams per sq yd (1,000 per m²). In contrast with *C. magnifica*, which lives on basaltic rocks, *C. soyoae* is infaunal (buried) in soft sediments.

Actual size

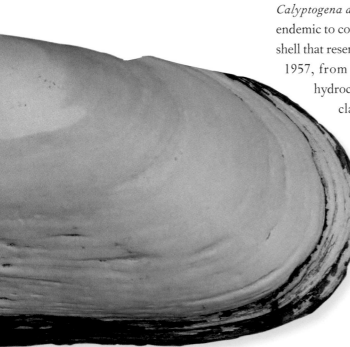

FAMILY	Cardiidae
SHELL SIZE RANGE	¾ to 2½ in (20 to 60 mm)
DISTRIBUTION	Florida to Brazil; Angola to Cape Verde
ABUNDANCE	Uncommon
DEPTH	Intertidal to 65 ft (20 m)
HABITAT	Sandy bottoms and seagrass beds
FEEDING HABIT	Filter feeder
BYSSUS	Absent in adults

SHELL SIZE RANGE
¾ to 2½ in
(20 to 60 mm)

PHOTOGRAPHED SHELL
⅞ in
(23 mm)

PAPYRIDEA SOLENIFORMIS
SPINY PAPER COCKLE
(BRUGUIÈRE, 1789)

121

Papyridea soleniformis is found on both sides of the Atlantic Ocean, from Florida south to Brazil and the Gulf of Mexico, as well as in West Africa, from Angola to Cape Verde. It has a thin and fragile shell that is variable in shape and color. The foot in cardiids is strong and muscular; it is capable of fast burrowing and even of swimming for short distances to escape predators. There are about 250 living species in the family Cardiidae worldwide, with most of the species in shallow waters in temperate and tropical regions. The oldest known fossil identified as a cardiid dates from the Triassic Period.

RELATED SPECIES
Papyridea aspersa (Sowerby I, 1833), from Baja California to Peru, has a very similar shell, in size, shape, and color. It is considered by some as a subspecies of *P. soleniformis*. *Corculum cardissa* (Linnaeus, 1758), from the Red Sea to Indo-West Pacific, has a shell that is very compressed anteroposteriorly, but is greatly expanded laterally, with a strong keel that gives it a heart shape.

The shell of the Spiny Paper Cockle is small to medium in size, thin, fragile, compressed, and subovate in outline. Its umbones are low and located slightly closer to the anterior margin. The valves are about the same size and shape, and gape posteriorly. The sculpture consists of 40 to 48 primary radial ribs, crossed by fine concentric lines; the interior is glossy. The posterior margin is denticulate, and the posterior radial ribs are spinose. The shell color is white or pinkish mottled with reddish brown; the interior is white with the external colors showing.

Actual size

FAMILY	Cardiidae
SHELL SIZE RANGE	¾ to 1½ in (20 to 40 mm)
DISTRIBUTION	Southern California to Ecuador
ABUNDANCE	Common
DEPTH	Intertidal to 330 ft (100 m)
HABITAT	Sandy and gravelly bottoms
FEEDING HABIT	Filter feeder
BYSSUS	Absent

SHELL SIZE RANGE
¾ to 1½ in
(20 to 40 mm)

PHOTOGRAPHED SHELL
1¼ in
(30 mm)

122

AMERICARDIA BIANGULATA
WESTERN STRAWBERRY COCKLE
(BRODERIP AND SOWERBY I, 1829)

Americardia biangulata varies from a small to medium in size and has a squarish outline and many broad and flat radial ribs. It is a common clam in shallow waters on sandy or gravelly bottoms near rocks; it also occurs offshore, in deeper waters. It has been recorded as a Pliocene fossil in Baja California. Cardiids are known as cockles or heart cockles because the combined shape of the articulated valves resembles a heart. Cardiids vary widely in shape and size, from small to the largest bivalves, the giant clams (formerly in family Tridacnidae); most have radial sculpture. The adductor muscles are about equal in size.

RELATED SPECIES

Clinocardium nuttallii (Conrad, 1837), which occurs in the northern Pacific, from Japan to California, is the largest of the North Pacific cockles, reaching over 5½ in (140 mm) in length. It has many strong radial ribs. *Cardium costatum* (Linnaeus, 1758), from western Africa, has a very distinctive, large, wide shell, with sharp, thin, and hollow radial ribs.

The shell of the Western Strawberry Cockle is small-to medium-sized, thick, slightly glossy, and squarish in outline. Its umbones are prominent and convex, central, and turned posteriorly; the hinge plate is thick and bears strong teeth of different sizes in both valves. The valves are similar in size and shape, with a rounded anterior end and truncated posterior end. The sculpture consists of about 26 broad, low radial ribs. The shell color is yellowish with brownish blotches; the interior is white, colored with purple or red markings.

Actual size

FAMILY	Cardiidae
SHELL SIZE RANGE	⅝ to 2¼ in (15 to 55 mm)
DISTRIBUTION	Red Sea to Indo-West Pacific
ABUNDANCE	Abundant
DEPTH	Intertidal to 165 ft (50 m)
HABITAT	Sandy and muddy bottoms
FEEDING HABIT	Filter feeder
BYSSUS	Absent

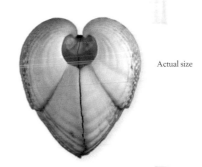

SHELL SIZE RANGE
⅝ to 2¼ in
(15 to 55 mm)

PHOTOGRAPHED SHELL
1½ in
(37 mm)

LUNULICARDIA AURICULA

PACIFIC HALF COCKLE

(NIEBUHR *IN* FORSSKÅL, 1775)

123

Lunulicardia auricula has a shell with a trapezoidal outline and a strong radial keel that is higher than it is wide. It is widely distributed from the Red Sea to the Indo-West Pacific, and is abundant in intertidal sand flats in sheltered bays. It is a species of interest for commercial harvesting in northwestern Australia. The animal lacks siphons, and has a large, sickle-shaped foot that is capable of rapid burrowing, although the animal lives shallowly buried. Its muscular foot can make the shell jump from the substrate and flip itself to escape predators such as starfish.

RELATED SPECIES

Fragum hemicardium (Linnaeus, 1758), from the Red Sea, has a similar shell, with a trapezoidal-triangular outline and a radial keel; however, it is broader than *L. auricula*, and has more radial riblets. *Corculum cardissa* (Linnaeus, 1758), from the Red Sea to Indo-West Pacific, has a shell that is strongly compressed anterodorsally but expanded laterally, with a sharp keel that gives it a heart shape.

Actual size

The shell of the Pacific Half Cockle is medium in size, solid, inflated, and trapezoidal in outline. Its umbones are prominent and sharp, and are located posteriorly; its hinge plate is strong, bent, and thickened under umbones. The area anterior to umbones (lunule) is deeply sunken. The valves are similar in size and shape. The sculpture consists of 18 to 27 radial ribs with rounded nodules, and a sharp radial keel. The shell color is white, both internally and externally, with brownish markings outside.

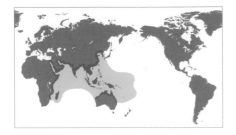

FAMILY	Cardiidae
SHELL SIZE RANGE	1½ to 3¼ in (40 to 80 mm)
DISTRIBUTION	Red Sea to Indo-West Pacific
ABUNDANCE	Abundant
DEPTH	Intertidal to 65 ft (20 m)
HABITAT	Sandy bottoms among reefs
FEEDING HABIT	Filter feeder
BYSSUS	Present

SHELL SIZE RANGE
1½ to 3¼ in
(40 to 80 mm)

PHOTOGRAPHED SHELL
1⅝ in
(41 mm)

124

CORCULUM CARDISSA
TRUE HEART COCKLE
(LINNAEUS, 1758)

Corculum cardissa has a curiously distorted shell: it is strongly compressed anteroposteriorly but expanded laterally, forming a sharp keel that is heart-shaped in outline. One side is concave and the other convex. It is a common species that lives near coral reefs, and can form dense colonies. It rests horizontally with its flatter side down on sandy bottoms in shallow water. The shell is thin, with small translucent "windows" that allow light to reach inside the shell. Like the related giant clams, *C. cardissa* grows symbiotic algae within its mantle and gills, which provide nutrients to the clam. Its shell is collected and used for shellcraft.

RELATED SPECIES

Corculum dionaeum (Broderip and Sowerby I, 1829), from the South Pacific, has a shell that is very similar to *C. cardissa*, but much smaller, reaching about ¾ in (20 mm) in length. It may be a form of *C. cardissa*. *Tridacna gigas* (Linnaeus, 1758), from the eastern Indian Ocean and tropical western Pacific, is the largest living bivalve. It has symbiotic zooxanthellae algae, which produce energy that allows the clam to grow faster.

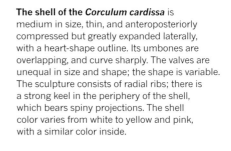

The shell of the *Corculum cardissa* is medium in size, thin, and anteroposteriorly compressed but greatly expanded laterally, with a heart-shape outline. Its umbones are overlapping, and curve sharply. The valves are unequal in size and shape; the shape is variable. The sculpture consists of radial ribs; there is a strong keel in the periphery of the shell, which bears spiny projections. The shell color varies from white to yellow and pink, with a similar color inside.

Actual size

FAMILY	Cardiidae
SHELL SIZE RANGE	4 to 5 in (100 to 125 mm)
DISTRIBUTION	Mauritania to Angola
ABUNDANCE	Uncommon
DEPTH	Offshore, up to 230 ft (70 m)
HABITAT	Sandy bottoms
FEEDING HABIT	Filter feeder
BYSSUS	Absent

SHELL SIZE RANGE
4 to 5 in
(100 to 125 mm)

PHOTOGRAPHED SHELL
4¼ in
(107 mm)

CARDIUM COSTATUM
GREAT RIBBED COCKLE
LINNAEUS, 1758

125

Cardium costatum has a striking shell, easily recognizable because of its strongly keeled radial ribs. It lives along the western coast of Africa, ranging from Mauritania to Angola, buried in soft bottoms in offshore waters. Occasionally, it washes ashore by the thousands after a storm. Internal molds of the shell are abundant in deposits from the Pliocene Epoch near Luanda, Angola, although actual shells are rare.

RELATED SPECIES

Cardium indicum Lamarck, 1819, from the western Mediterranean and northwestern Africa, differs by having scaly projections on the anterior and posterior ribs, a wider gape, and pinkish or fawn coloration; *Trachycardium egmontianum* (Shuttleworth, 1856), from North Carolina to Brazil, is a common species with strong axial ribs with imbricate scales; and *Plagiocardium pseudolima* (Lamarck, 1819), from the Indo-Pacific, is one of the largest cockles, with a heart-shaped profile and many broad, flat axial ribs.

The shell of the Great Ribbed Cockle is large, thin, and inflated, with valves gaping posteriorly. The most distinctive feature of the species is the presence of about 16 to 17 strongly keeled, radial ribs with an acute triangular profile on each valve. Inside the valve, wide, flat grooves correspond to the main radial ribs. The radial ribs are crossed by fine concentric growth lines visible on the outside of the shell. The long hinge is nearly straight, with strong cardinal and lateral teeth. The color is pure white or off-white, with some orange-brown between the ribs.

Actual size

FAMILY	Cardiidae
SHELL SIZE RANGE	2¾ to 6 in (70 to 150 mm)
DISTRIBUTION	Red Sea to Indo-West Pacific
ABUNDANCE	Common
DEPTH	Shallow subtidal to offshore
HABITAT	Muddy sand bottoms
FEEDING HABIT	Filter feeder
BYSSUS	Absent

SHELL SIZE RANGE
2¾ to 6 in
(70 to 150 mm)

PHOTOGRAPHED SHELL
5 in
(130 mm)

126

PLAGIOCARDIUM PSEUDOLIMA
GIANT COCKLE
(LAMARCK, 1819)

Plagiocardium pseudolima is one of the largest species of cockles (besides the giant clams, which have recently been transferred to the family Cardiidae). Although it usually grows to about 6 in (150 mm), the largest known specimen, collected in Mozambique, measures 7 in (181 mm). It is collected for food locally throughout its range from the Red Sea to the Indo-West Pacific. Like other cockles, it is very active and moves about with its strong foot. Cockles are characterized by strong radial ribs and a hinge with teeth curving outward. They have two adductor muscles scars inside the valves.

RELATED SPECIES

Dinocardium robustum (Lightfoot, 1786), from Virginia to Texas and Mexico, is another large cockle that is consumed as food. Its shell is taller than it is wide, and is obliquely ovate. *Papyridea soleniformis* (Bruguière, 1789), from Florida to Brazil, and also from the eastern Atlantic, has a thin, oval elongate shell. It lives near seagrasses in shallow waters.

Actual size

The shell of the Giant Cockle is very large, heavy, inflated, and with a heart-shaped profile. Both valves are mirror images of one another, and have large, rounded umbones. Its sculpture consists of about 36 to 40 flat radial ribs that bear cup-shaped spines about midway toward the ventral margin; concentric growth marks cross the radial ribs. The periostracum has bristles that emerge from each spine. The shell color is pale yellow to reddish brown, with violet concentric bands near the ventral margin; the interior is white.

FAMILY	Cardiidae
SHELL SIZE RANGE	6 to 16 in (150 to 400 mm)
DISTRIBUTION	Tropical Indo-West Pacific
ABUNDANCE	Common
DEPTH	Intertidal to 20 ft (6 m)
HABITAT	Hard bottoms (juveniles); sandy bottoms (adults)
FEEDING HABIT	Has symbiotic algae
BYSSUS	Absent in adults

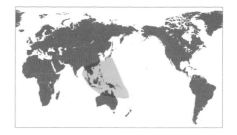

SHELL SIZE RANGE
6 to 16 in
(150 to 400 mm)

PHOTOGRAPHED SHELL
5¾ in
(147 mm)

HIPPOPUS HIPPOPUS
BEAR PAW CLAM
(LINNAEUS, 1758)

127

Hippopus hippopus has the most strongly sculptured and triangular shell among the giant clams. Its shell shape and sculpture, with a few strong and several smaller scaly riblets in between, readily distinguishes *H. hippopus* from other giant clams. Juveniles are byssally attached to hard bottoms, but as they grow the byssus is lost, and adults are free-living on sandy bottoms. *H. hippopus* is collected for food and shellcraft. There are nine living species of giant clams recognized, in two genera, *Hippopus*, with two species, and *Tridacna*, with seven. The latter used to be classified in their own family, Tridacnidae, but recently have been shown to belong in the family Cardiidae.

RELATED SPECIES

Hippopus porcellanus Rosewater, 1982, from the Philippines, Indonesia, and New Guinea, has slightly thinner valves, a semicircular outline, and moderately flat and smooth radial ribs. *Tridacna derasa* (Röding, 1798), from the tropical western Pacific, has one of the smoothest shells; it is one of the deepest-dwelling giant clams, occurring in depths to 115 ft (35 m).

Actual size

The shell of the Bear Paw Clam is large in size, very thick and heavy, and globose and triangular in outline. Its valves are strongly sculptured by about 7 to 12 radial folds and many riblets that bear short scales in between. The hinge line is about half of the shell length; the ventral margin (which is kept upright in life) is long and undulating. Juveniles have a byssal gape but it is nearly closed in adults. The shell color is creamy white with purple and brown blotches, and the interior white.

FAMILY	Cardiidae
SHELL SIZE RANGE	6 to 17¾ in (150 to 450 mm)
DISTRIBUTION	Indo-Pacific except Hawaii
ABUNDANCE	Common locally
DEPTH	Intertidal to 33 ft (10 m)
HABITAT	Coral reefs in somewhat protected localities
FEEDING HABIT	Has symbiotic algae
BYSSUS	Present

SHELL SIZE RANGE
6 to 17¾ in
(150 to 450 mm)

PHOTOGRAPHED SHELL
6⅛ in
(155 mm)

128

TRIDACNA SQUAMOSA
FLUTED GIANT CLAM
LAMARCK, 1819

The shell of the Fluted Giant Clam is very large, thick, and heavy, with a semicircular outline. Its valves change from moderately compressed in juveniles to strongly inflated in adults. The byssal gape is medium-sized, and bordered by 6 to 8 crenulations. The sculpture consists of 5 or 6 broad radial folds with large bladelike concentric scales, which are delicate and easily broken. The shell color is often grayish white but sometimes tinged with orange or yellow; the interior is porcelaneous white.

Tridacna squamosa is the second heaviest shell in the family Cardiidae, with only *Tridacna gigas* (Linnaeus, 1758) being heavier. It lives in shallow, clear waters, and is byssally attached to coral reefs throughout its life. Like other giant clams, it is collected for food and shell trade. Because of overharvesting, the once abundant giant clams were becoming endangered, and in 1983 a worldwide ban on the sale of its meat and shell was passed. Since then, its aquaculture has been developed to restock reefs and to produce meat for food. Giant clams are now becoming popular in the aquarium industry.

RELATED SPECIES
Tridacna gigas (Linnaeus, 1758), from the eastern Indian Ocean and tropical western Pacific, is the largest and heaviest shelled mollusk. Its shell resembles that of *T. squamosa* but it lacks the leaflike scales, and has a more meandering ventral margin. *Tridacna derasa* (Röding, 1798), from the western Pacific, has the second largest shell among the giant clams, although it is lighter than the smaller *T. squamosa*.

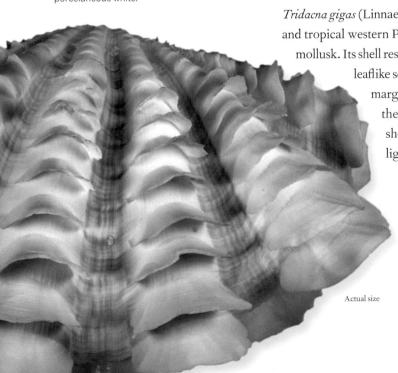

Actual size

FAMILY	Cardiidae
SHELL SIZE RANGE	6½ to 14 in (170 to 350 mm)
DISTRIBUTION	Red Sea to Indo-West Pacific
ABUNDANCE	Common
DEPTH	Shallow subtidal to 65 ft (20 m)
HABITAT	Embedded in corals in shallow water
FEEDING HABIT	Has symbiotic algae
BYSSUS	Present

SHELL SIZE RANGE
6½ to 14 in
(170 to 350 mm)

PHOTOGRAPHED SHELL
6½ in
(171 mm)

129

TRIDACNA MAXIMA

ELONGATE GIANT CLAM

(RÖDING, 1798)

Tridacna maxima excavates shallow depressions in corals and lives partially or fully embedded in the substrate, to which it is byssally attached both as a juvenile and an adult. It has the most elongate shell among the giant clams. Giant clams are primarily filter feeders as young juveniles, but as they grow they develop a symbiotic relationship with an alga (zooxanthellae), which grows within the large mantle of the clam. The optimal conditions for giant clams include clear, shallow, and warm waters; the clam lives with the ventral margin widely open to expose its mantle to sunlight. The algae benefit by getting protection, and the clams benefit by feeding on the nutrients that are produced by the algae.

The shell of the Elongate Giant Clam is large, thick, heavy, and elongately ovate in outline. Its sculpture consists of 6 to 12 broad, convex radial folds, with the central 5 or 6 being much stronger; each bears evenly spaced, low, erect, concentric scales. The hinge line is less than half of the shell length, and the byssal gape is large, with plicae at the edges. The shell color, both exterior and interior, is grayish white, sometimes tinged with pinkish orange or yellow.

RELATED SPECIES

Tridacna crocea Lamarck, 1819, from the eastern Indian Ocean and western Pacific, has a smaller but similar shell, with a more triangular ovate outline. It is also byssally attached throughout its life. *Tridacna squamosa* Lamarck, 1819, which has a wide distribution, from the Red Sea to the Indo-West Pacific, has a very distinctive shell with large bladelike scales.

Actual size

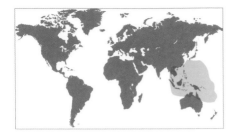

FAMILY	Cardiidae
SHELL SIZE RANGE	12 to 54 in (300 to 1,370 mm)
DISTRIBUTION	Eastern Indian Ocean and tropical western Pacific
ABUNDANCE	Formerly common
DEPTH	6 to 65 ft (2 to 20 m)
HABITAT	Sandy bottoms near coral reefs
FEEDING HABIT	Has symbiotic algae
BYSSUS	Absent in adults

SHELL SIZE RANGE
12 to 54 in
(300 to 1,370 mm)

PHOTOGRAPHED SHELL
30¼ in
(756 mm)

130

TRIDACNA GIGAS
GIANT CLAM
(LINNAEUS, 1758)

Tridacna gigas is the largest and heaviest bivalve that ever lived; it can reach 54 in (1,370 mm) in length, although most specimens are about half of that size. Besides its unrivalled size, its shell is readily recognized by the elongate triangular projections on the ventral margin. *Tridacna gigas* is hermaphrodite, and large specimens can produce more than 100 million eggs in one season. The Pearl of Allah, a famous, non-nacreous and irregular concretion produced by a giant clam, is the largest known pearl, and measures almost 9½ in (240 mm) in diameter. Most *Tridacna* species are now considered vulnerable, and are protected from international trade.

RELATED SPECIES

Tridacna maxima (Röding, 1798), which ranges from the Red Sea to the tropical Indo-West Pacific, lives embedded in the coral substrate. Its mantle is brightly and variably colored. *Hippopus hippopus* (Linnaeus, 1758), from the tropical Indo-West Pacific, has the most triangularly shaped shell among the giant clams; it has several ribbed radial folds.

The shell of the Giant Clam is extremely large, massive, heavy, inflated, and subovate to fan-shaped. Its valves are about the same size and shape, and have 4 to 6 deep radial folds, with weaker radial riblets, which are crossed by concentric growth lines. The ventral margin of the shell has elongate triangular projections. The shell color is off-white, but it is often heavily encrusted with marine debris; the interior is porcelaneous white.

Actual size

FAMILY	Veneridae
SHELL SIZE RANGE	⅛ to ¼ in (3 to 5 mm)
DISTRIBUTION	Nova Scotia to Texas and Bahamas
ABUNDANCE	Locally abundant
DEPTH	Intertidal to 16 ft (5 m)
HABITAT	Bays and estuaries
FEEDING HABIT	Filter feeder
BYSSUS	Present

SHELL SIZE RANGE
⅛ to ¼ in
(3 to 5 mm)

PHOTOGRAPHED SHELL
⅛ in
(3 mm)

GEMMA GEMMA

AMETHYST GEM CLAM

(TOTTEN, 1834)

131

Gemma gemma is one of the smallest of the venerids. It lives in mud flats in bays and estuaries, where it can occur in densities reaching 83,333 clams per sq yd (100,000 per m²). There is a simple byssus to help anchor the shell in soft sediments. It has a wide distribution in the eastern Atlantic, and it has been introduced accidentally, along with oysters, to California and Washington, where it is considered an invasive species. Although not aggressive, it is an opportunist and may displace native clams. The family Veneridae is the most diverse of the extant bivalve families, with more than 800 living species worldwide, in both temperate and tropical waters.

RELATED SPECIES

Parastarte triquetra (Conrad, 1846), which ranges from Florida to Texas, is another very small venerid. Its shell resembles *G. gemma*, but it is higher, and the shell color is tan or brown. *Bassina disjecta* (Perry, 1811), which is endemic to southern Australia, has a shell with wide, lamellose concentric ribs; it is known as the Wedding Cake Venus.

The shell of the Amethyst Gem Clam is very small, thin but strong, and inflated. Shell shape is variable, ranging from broadly oval to triangular in outline, but it is usually about equal in height and width. The umbones are near the center, and the valves are fairly smooth and glossy, with concentric growth lines. The shell color varies from grayish to lavender, with a purplish tinge near the umbones, and a whitish interior.

Actual size

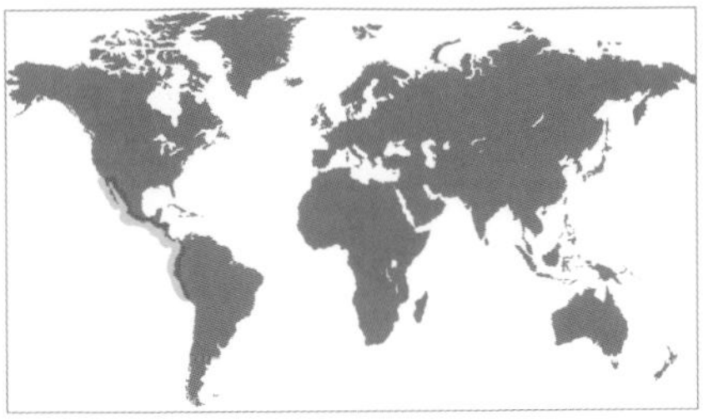

FAMILY	Veneridae
SHELL SIZE RANGE	1¼ to 3¼ in (30 to 80 mm)
DISTRIBUTION	Western Mexico to Peru
ABUNDANCE	Common
DEPTH	Intertidal to 80 ft (25 m)
HABITAT	Infaunal in sandy bottoms
FEEDING HABIT	Filter feeder
BYSSUS	Absent

SHELL SIZE RANGE
1¼ to 3¼ in
(30 to 80 mm)

PHOTOGRAPHED SHELL
1¼ in
(31 mm)

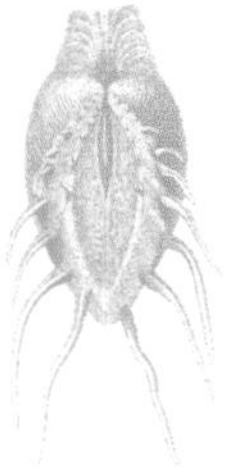

PITAR LUPANARIA

SPINY VENUS

(LESSON, 1830)

Pitar lupanaria is one of the most spectacular clams, with long, erect spines, and a colorful shell. It is one of only a few species in the genus *Pitar* in the large family Veneridae to have spines on the shell, and *P. lupanaria* has the longest ones; most other species in the genus lack spines. The clam lives buried in sand in shallow water, with the spines pointing upward, surrounding its siphons. The spines may protect the siphons from predators. *Pitar lupanaria* is commonly found washed ashore on sandy beaches, especially after a storm.

Actual size

The shell of the Spiny Venus is medium in size, solid, moderately thick and inflated, and ovately triangular in outline. Its valves have wide, erect, concentric ribs that are stronger anteriorly, and end in 2 radial rows of long and short, pointed, and open spines posteriorly. The umbones are prominent and pointed anteriorly. The shell is creamy white to pale pink, tinged with violet and with violet blotches at the base of the spines; the interior is white.

RELATED SPECIES

Pitar dione (Linnaeus, 1758), from Florida to Venezuela, is the Caribbean sister species; it usually has shorter spines than *P. lupanaria*, but the shell is very similar in shape and color. *Periglypta reticulata* (Linnaeus, 1758), which is widely distributed, ranging from the Red Sea to Hawaii, is a large venerid collected locally for food. Its shell is solid and has strong concentric lamellae.

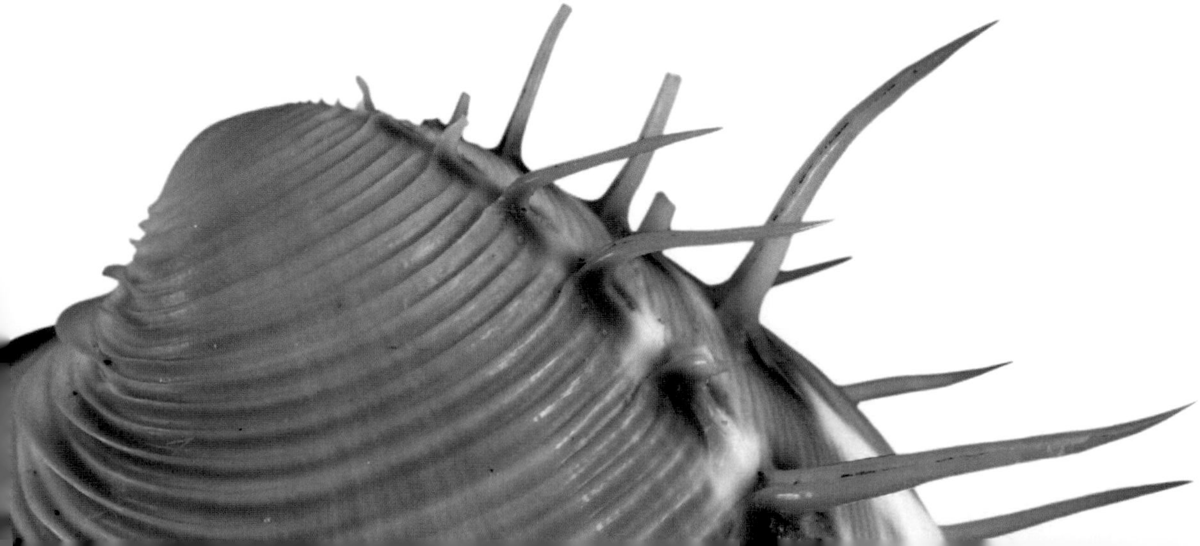

FAMILY	Veneridae
SHELL SIZE RANGE	1¼ to 2½ in (30 to 60 mm)
DISTRIBUTION	Red Sea to Indo-Pacific
ABUNDANCE	Common
DEPTH	Shallow subtidal to 80 ft (25 m)
HABITAT	Sandy bottoms in coral reefs
FEEDING HABIT	Filter feeder
BYSSUS	Absent

SHELL SIZE RANGE
1¼ to 2½ in
(30 to 60 mm)

PHOTOGRAPHED SHELL
1¾ in
(44 mm)

133

LIOCONCHA CASTRENSIS
CAMP PITAR VENUS
(LINNAEUS, 1758)

Lioconcha castrensis has a beautiful shell that is popular among collectors. It is a common species in shallow sandy bottoms in coral reef areas throughout the Indo-Pacific. It is harvested for food in some countries, and its shell used in shellcraft. Like many infaunal bivalves, it inserts its muscular foot into the sand; the foot changes shape and acts like an anchor; the animal then contracts the foot muscles, and pulls its shell into the sand. The Veneridae include species with very diverse shapes and sizes. Many are economically important species.

RELATED SPECIES

Pitar lupanaria (Lesson, 1830), which ranges from Western Mexico to Peru, is easily recognizable by its long spines. *Dosinia discus* (Reeve, 1850), from the western Atlantic, is one of several venerid species with a nearly circular outline and compressed shell.

The shell of the Camp Pitar Venus is heavy, subovate, and wider than it is tall. The prominent, rounded umbones are situated along the anterior half of the shell. The surface of the shell is smooth and shiny, with fine growth lines. Both valves are equal in size and shape, and somewhat inflated. The hinge plate is strong, with three cardinal teeth in each valve (the front lateral tooth being more developed). The internal surface is smooth and glossy white, with a very shallow pallial line. The shell color is creamy white, with large tentlike brown or black markings

Actual size

FAMILY	Veneridae
SHELL SIZE RANGE	1⅜ to 3¼ in (35 to 80 mm)
DISTRIBUTION	Red Sea to tropical Indo-West Pacific
ABUNDANCE	Common
DEPTH	Intertidal to 65 ft (20 m)
HABITAT	Shallow mudflats and sandy bottoms
FEEDING HABIT	Filter feeder
BYSSUS	Absent

SHELL SIZE RANGE
1⅜ to 3¼ in
(35 to 80 mm)

PHOTOGRAPHED SHELL
1⅞ in
(47 mm)

134

TEXTILE VENUS
(GMELIN, 1791)

Paphia textile has a wide natural distribution, occurring from the Red Sea to the tropical Indo-West Pacific. It is also one of the few species that have migrated north into the Mediterranean via the Suez Canal. It can dominate the molluscan fauna in deeper waters in the eastern Mediterranean. In its normal range, it co-occurs with *Paphia undulata* (Born, 1778), with which it is often confused in the literature. The latter is heavily exploited for food, especially in Thailand. *Paphia textile* is an infaunal clam, and burrows shallowly in mudflats and sandy bottoms. This species is often collected for food, along with several other species of *Paphia* and *Tapes*.

RELATED SPECIES

Macrocallista nimbosa (Lightfoot, 1786), from North Carolina to the Gulf of Mexico, is one of the largest species in the family Veneridae, reaching over 7 in (180 mm) in length. Its shell looks like a giant *Paphia textile*. *Petricola pholadiformis* (Lamarck, 1818), from eastern Canada to West Indies, bores into clay and soft rocks. Its shell resembles the pholadid Angel Wing, *Cyrtopleura costata* (Linnaeus, 1758).

The shell of the Textile Venus is medium-sized, strong, moderately inflated, and elongately ovate elliptical in outline. Its umbones are markedly anterior and located about one-third of the shell length from the anterior margin. The valves are identical in size and shape, and do not gape. The posterodorsal margin is nearly straight, and the ventral margin gently curved. The valves are smooth and glossy, and only have shallow concentric growth lines. The shell color is cream to pinkish brown, with light brown, zigzag lines, and the interior is white.

Actual size

FAMILY	Veneridae
SHELL SIZE RANGE	1 to 2¾ in (25 to 70 mm)
DISTRIBUTION	Eastern Canada to West Indies; introduced to eastern Atlantic
ABUNDANCE	Common
DEPTH	Intertidal to 25 ft (8 m)
HABITAT	Bores into mud and soft rocks
FEEDING HABIT	Filter feeder
BYSSUS	Absent in adults

SHELL SIZE RANGE
1 to 2¾ in
(25 to 70 mm)

PHOTOGRAPHED SHELL
2 in
(51 mm)

PETRICOLA PHOLADIFORMIS
FALSE ANGEL WING
(LAMARCK, 1818)

135

Petricola pholadiformis is native to the northwestern Atlantic, where it ranges from eastern Canada to the West Indies and the Gulf of Mexico. It was introduced to the eastern Atlantic, along with oysters, in the late 1800s. It bores into peat, mud, clay, or limestone in intertidal and shallow subtidal depths. Its shell can be irregular as a consequence of the hard substrate into which it bores. Its shell resembles the larger and more attractive Angel Wing, *Cyrtopleura costata* (Linnaeus, 1758), a pholadid. *Petricola pholadiformis* is collected for food or used as bait. Recent studies have shown that shells that had been distinguished as a separate family, the Petricolidae, belong to the family Veneridae.

RELATED SPECIES
Petricola stellae Narchi, 1975, which ranges from Brazil to Uruguay, has a similar shell, but differs by its smaller size, fewer radial ribs, and anatomical details. Some specimens are found in intertidal reefs made by polychaetes (*Phragmatopoma lapidosa*). *Petricola parallela* (Pilsbry and Lowe, 1932), from Baja California to Nicaragua, has the most slender shell in the genus *Petricola*.

Actual size

The shell of the False Angel Wing is medium in size, fragile, inflated, and elongately oval. Its umbones are elevated and located anteriorly; the hinge is narrow and has 2 and 3 teeth in the right and left valves respectively (this feature distinguishes it from pholadids, which lack teeth). The valves have about 60 radial ribs, crossed by concentric growth lines with 10 in the anterior end being much stronger than others. Its shell color is chalky white or off-white, occasionally pinkish.

FAMILY	Veneridae
SHELL SIZE RANGE	2 to 4 in (50 to 100 mm)
DISTRIBUTION	Red Sea to Indo-Pacific
ABUNDANCE	Common
DEPTH	Intertidal to 80 ft (25 m)
HABITAT	Sandy and muddy bottoms near coral reefs
FEEDING HABIT	Filter feeder
BYSSUS	Absent

SHELL SIZE RANGE
2 to 4 in
(50 to 100 mm)

PHOTOGRAPHED SHELL
2⅛ in
(54 mm)

136

PERIGLYPTA RETICULATA
RETICULATE VENUS
(LINNAEUS, 1758)

Periglypta reticulata has a wide distribution, ranging from the Red Sea to the Indo-Pacific. It has a large shell that is variable in color and shape, from nearly circular to quadrate. It is a common species in intertidal to shallow water, and lives infaunally in sandy or muddy bottoms, often near coral reefs. Its valves do not gape, therefore the animal needs to open the valves slightly for the siphons to extend to the surface of the sediment. The mantle edge forms a zipperlike seal to keep the mantle cavity tightly closed, with the opening being the two siphons.

RELATED SPECIES

Periglypta multicostata (Sowerby I, 1835), from lower California to Peru and the Galápagos Islands, is one of the largest species in the family Veneridae; it can reach more than 6 in (150 mm) in length. Its shell has strong concentric ridges. *Mercenaria mercenaria* (Linnaeus, 1758), from eastern Canada to Gulf of Mexico, also has a large shell with concentric ridges. Both species are collected for food.

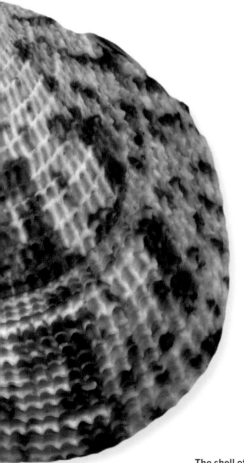

The shell of the Reticulate Venus ranges from medium to large, and is thick, solid, inflated, and suborbicular to quadrate in outline. Its umbones are heavy and anteriorly depressed; the hinge plate is broad and has three strong teeth in each valve. The sculpture of the valves consists of concentric nodulose waving ridges crossed by radial riblets that give a reticulated appearance. The shell color is cream with brown blotches, and the interior is white, with an orange hinge.

Actual size

FAMILY	Veneridae
SHELL SIZE RANGE	1½ to 3 in (40 to 75 mm)
DISTRIBUTION	New South Wales to South Australia
ABUNDANCE	Common
DEPTH	Shallow subtidal to 165 ft (50 m)
HABITAT	Sandy mud banks
FEEDING HABIT	Filter feeder
BYSSUS	Absent in adults

SHELL SIZE RANGE
1½ to 3 in
(40 to 75 mm)

PHOTOGRAPHED SHELL
2½ in
(63 mm)

137

BASSINA DISJECTA

WEDDING CAKE VENUS

(PERRY, 1811)

Bassina disjecta has a beautiful and distinctive shell, rounded and triangular with frilly lamellose concentric ribs. It is usually found in shallow subtidal depths, in tropical and temperate waters, in sandy mud banks. It is a shallow burrower, and lies buried with the posterior margin of the shell close to the surface. The animal has two siphons, one inhalant and one exhalant, which allow it to remain buried in the sediment and circulate water through ciliary pumping. The lamellose ribs help stabilize the shell in the fine sediment.

RELATED SPECIES

Bassina pachyphylla (Jonas, 1839), which has a similar distribution to *B. disjecta*, has a very different shell in shape, color, and sculpture. It has a smooth shell with light and darker brown radial bands. *Pitar lupanaria* Lesson, 1830, from western Mexico to Peru, has a shiny, ovate triangular shell with two radial rows of long spines in the posterior margin, and erect concentric ribs.

The shell of the Wedding Cake Venus is medium in size, thin but solid, compressed, and rounded triangular to elongately ovate in outline. Its most prominent feature is the presence of about 6 to 8 broadly spaced, lamellose concentric ribs that are frilly and turned upward at the edges. The umbones are small and anteriorly facing. The valves are identical in size and shape. The shell color is white, sometimes with pink on the lower side of the lamellae, and it has a white interior.

Actual size

FAMILY	Veneridae
SHELL SIZE RANGE	2½ to 4 in (60 to 100 mm)
DISTRIBUTION	Red Sea to Indo-West Pacific
ABUNDANCE	Common
DEPTH	Intertidal to 65 ft (20 m)
HABITAT	Sandy bottoms
FEEDING HABIT	Filter feeder
BYSSUS	Absent

SHELL SIZE RANGE
2½ to 4 in
(60 to 100 mm)

PHOTOGRAPHED SHELL
2¾ in
(70 mm)

138

CALLISTA ERYCINA
REDDISH CALLISTA
(LINNAEUS, 1758)

Callista erycina is a venerid found in the intertidal zone and shallow waters, buried in sandy bottoms. It has an elegant, glossy shell, with raised, flat concentric ridges, which is decorated with broken brown and tan radial rays. It is collected locally for food but is not a major commercially important species. Venerids lack special adaptations, and in general are anatomically uniform. The animal has a pair of large adductor muscles that are usually similar in size. The valves are usually identical in size and shape, and vary widely in size, 100-fold within the family, from ¹⁄₁₆ in (1.5 mm) to over 6½ in (170 mm).

RELATED SPECIES
Callista chione (Linnaeus, 1758), from the British Isles to northwest Africa and the Mediterranean, has a large shell which is similar but smooth. It is exploited commercially in the Mediterranean. *Dosinia ponderosa* (Gray, 1838), from Baja California to Peru, and the Galápagos Islands, has a ponderous shell, which was collected by Native Americans in prehistoric times for food and to be used as tools.

Actual size

The shell of the Reddish Callista ranges from medium to large in size, and is thick, heavy, inflated, and rounded triangular in outline. Its umbones are large, rounded, and pointed anteriorly; the hinge plate is thick, with 3 teeth on each valve, and has a relatively long, external hinge. The sculpture consists of incised, concentric lines that result in broad, flattened ridges. Internally, the shell is smooth, with 2 large muscle scars. Its color is pale yellow with brown radial rays and lines, and the interior white with orange margins.

FAMILY	Veneridae
SHELL SIZE RANGE	3 to 6 in (75 to 150 mm)
DISTRIBUTION	Eastern Canada to Gulf of Mexico
ABUNDANCE	Common
DEPTH	Intertidal to 50 ft (15 m)
HABITAT	Soft bottoms, often near seagrasses
FEEDING HABIT	Filter feeder
BYSSUS	Absent

SHELL SIZE RANGE
3 to 6 in
(75 to 150 mm)

PHOTOGRAPHED SHELL
4⅛ in
(104 mm)

MERCENARIA MERCENARIA

NORTHERN QUAHOG

(LINNAEUS, 1758)

139

Mercenaria mercenaria is the main clam fished on the U.S.A.'s east coast, and ranks second in value only to oysters. It has been consumed as food for thousands of years by Native Americans, who produced large heaps of shells along the coast. The popular name, quahog, is a Native American word for clam. Native peoples used the shell of *M. mercenaria* to produce beads, called wampum, that were used as money; the purple ones made from near the lip of the shell were particularly valuable. The scientific name, *Mercenaria mercenaria*, derives from the Latin, meaning "wages." It is the state shell of Rhode Island.

RELATED SPECIES

Mercenaria campechiensis (Gmelin, 1791), which ranges from New Jersey to Central America, has a shell similar in sculpture and shape, but with concentric ridges throughout the entire shell, and a larger maximum size. It was once considered a subspecies of *M. mercenaria*, but molecular studies have confirmed it is a separate species.

The shell of the Northern Quahog is large, thick, heavy, inflated, and ovately triangular. Its umbones are large, rounded, and twisted anteriorly. The valves are identical in shape and size, and have closely spaced, coarse, concentric ridges throughout the shell, except in the center of the valves, which are smooth in adult shells. The ventral margin is finely crenulated. The color is dirty gray to off-white, and the interior is white, often tinged in purple near the margin.

Actual size

FAMILY	Veneridae
SHELL SIZE RANGE	3 to 6 in (75 to 150 mm)
DISTRIBUTION	Lower California to Peru, and the Galápagos Islands
ABUNDANCE	Common
DEPTH	10 to 200 ft (3 to 60 m)
HABITAT	Mud and sand flats, seagrasses
FEEDING HABIT	Filter feeder
BYSSUS	Absent in adults

SHELL SIZE RANGE
3 to 6 in
(75 to 150 mm)

PHOTOGRAPHED SHELL
5 in
(129 mm)

140

DOSINIA PONDEROSA

PONDEROUS DOSINIA

(GRAY, 1838)

Dosinia ponderosa is the largest species in the genus *Dosinia*, and one of the largest in the family Veneridae. Its thick and heavy shell was used as tools in prehistoric times by Native Americans, who apparently preferred to break the ventral margin of the shell. However, it was probably more often collected for food.

Dosinia ponderosa lives buried close to the surface on muddy or sandy bottoms, often near seagrasses. It is a common species in shallow water. It is also common in Pliocene and upper Miocene fossil beds in northern Peru and Ecuador.

RELATED SPECIES

Dosinia discus (Reeve, 1850), from Virginia to Texas and the Bahamas, has a similarly shaped but smaller shell. Its sculpture is composed of fine concentric grooves. *Callista erycina* (Linnaeus, 1758), which ranges from the Red Sea to the western Pacific, has a large and thick, ovately triangular shell, with strong concentric growth ridges and radial bands of different colors.

Actual size

The shell of the Ponderous Dosinia is large, thick, heavy, moderately inflated, and subcircular in outline. Its umbones are pointed and sharply turned anteriorly, and located at about the middle of the thick hinge plate. The valves are identical in size and shape, smooth and polished in the central part, and bear fine concentric growth lines. Inside, the surface is smooth, and there are 2 large muscle scars of different sizes and shapes. The shell color is white with a tan periostracum, and the interior is white.

FAMILY	Tellinidae
SHELL SIZE RANGE	⅜ to ⅝ in (8 to 15 mm)
DISTRIBUTION	North Carolina to Brazil
ABUNDANCE	Abundant
DEPTH	15 to 600 ft (5 to 180 m)
HABITAT	Sand, offshore
FEEDING HABIT	Filter feeder
BYSSUS	Absent

SHELL SIZE RANGE
⅜ to ⅝ in
(8 to 15 mm)

PHOTOGRAPHED SHELL
⅖ in
(11 mm)

STRIGILLA PISIFORMIS

PEA STRIGILLA

(LINNAEUS, 1758)

141

Strigilla pisiformis is an abundant small clam that lives offshore but is often cast ashore in great numbers in the Bahamas. It is used in the shellcraft industry, especially in Florida. It has a similar sculpture to that of other species in the genus *Strigilla*. Tellins have two adductor muscles of about the same size, although they can be different shapes. They are fast burrowers and live deeply buried in soft bottoms; they have very long siphons which, when extended, can be more than five times the length of the shell. There are about 350 living species in the family Tellinidae distributed worldwide.

RELATED SPECIES

Strigilla mirabilis (Philippi, 1841), from North Carolina to Brazil, has a very similar shell in size and sculpture, but with a subovate outline and four to six chevrons near the posterior margin. *Phylloda foliacea* (Linnaeus, 1758), from the Indo-West Pacific, has a large, broadly oval shell that is collected for food and used in the shellcraft industry.

The shell of the Pea Strigilla is small, thin, inflated, and oval in outline. Its valves are smooth and sculptured with incised oblique striae, with 2 sets of chevrons near the posterior margin. The hinge line is narrow, with a short ligament. The umbones are pinkish and slightly closer to the anterior end. The shell color is white, and the interior is white tinged with pink in the deepest part of the valves.

Actual size

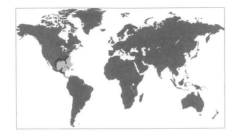

FAMILY	Tellinidae
SHELL SIZE RANGE	1 to 1½ in (25 to 40 mm)
DISTRIBUTION	North Carolina to Central America
ABUNDANCE	Uncommon
DEPTH	Intertidal to 33 ft (10 m)
HABITAT	Sandy bottom
FEEDING HABIT	Suspension feeder
BYSSUS	Absent

SHELL SIZE RANGE
1 to 1½ in
(25 to 40 mm)

PHOTOGRAPHED SHELL
1½ in
(39 mm)

142

TELLIDORA CRISTATA
WHITE CRESTED TELLIN
(RÉCLUZ, 1842)

Tellidora cristata has an unusual shape for a tellin: a triangular shell with a rounded ventral margin, and sawtooth ridges on the anterior and posterior margins, which make this species easily recognized. Unlike many tellins, it buries shallowly in soft sediments; it is found in bays, lagoons, and estuaries, and feeds on suspended particles. It is an uncommon species. Most tellins are ecologically important because of their abundance; many species are consumed by humans or animals. The family's fossil record extends back to the Cretaceous Period.

RELATED SPECIES

Tellidora burneti (Broderip and Sowerby I, 1829), from the Gulf of California to Ecuador, is the Pacific analog. Its shell is slightly larger, but it has a similar shape. *Tellina radiata* Linnaeus, 1758, from South Carolina to Brazil, has a large shell with broad, pink radial rays that resemble a sunrise, hence its popular name, Sunrise Tellin.

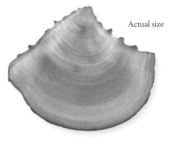

Actual size

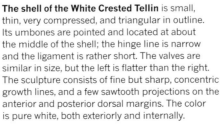

The shell of the White Crested Tellin is small, thin, very compressed, and triangular in outline. Its umbones are pointed and located at about the middle of the shell; the hinge line is narrow and the ligament is rather short. The valves are similar in size, but the left is flatter than the right. The sculpture consists of fine but sharp, concentric growth lines, and a few sawtooth projections on the anterior and posterior dorsal margins. The color is pure white, both exteriorly and internally.

FAMILY	Tellinidae
SHELL SIZE RANGE	2 to 4⅛ in (50 to 105 mm)
DISTRIBUTION	South Carolina to Brazil
ABUNDANCE	Common
DEPTH	Intertidal to 330 ft (100 m)
HABITAT	Infaunal in sandy bottoms
FEEDING HABIT	Suspension feeder
BYSSUS	Absent

SHELL SIZE RANGE
2 to 4⅛ in
(50 to 105 mm)

PHOTOGRAPHED SHELL
2⅞ in
(73 mm)

TELLINA RADIATA

SUNRISE TELLIN

LINNAEUS, 1758

143

Tellina radiata is one of the largest species in the large family Tellinidae. It is moderately common in shallow waters, buried in sand. It has a handsome white shell with pink radial rays that give it a resemblance to a sunrise, hence its popular name. Today the giant Shell Oil Company is known worldwide by its logo of a scallop (based on the European species, *Pecten jacobaeus*), but in 1900, when the company was called Shell Transport and Trading Company, the logo was based on the Sunrise Tellin. The famous red and yellow scallop logo was introduced in 1904, and changed over time to its current form and colors.

RELATED SPECIES

Tellina cumingii Hanley, 1844, from Baja California, Mexico to Colombia, has an elongate shell with the posterior end narrowed. It is sculptured with fine concentric lamellae. *Tellina magna* Spengler, 1798, from North Carolina to the West Indies, is perhaps the largest of the living tellins, and can reach more than 5½ in (140 mm) in length.

The shell of the Sunrise Tellin is large for the family, solid, smooth, and elongated oval in outline. The anterior margin is rounded and the posterior slightly narrowed. Its valves are mostly smooth and highly polished, with fine concentric lines; inside 2 large adductor muscle scars are situated near the hinge line. The shell color is white with pink or rose radial rays, and bright red umbones; the interior is flushed with yellow in the center, with the external colors showing near the ventral margin.

Actual size

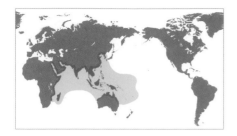

FAMILY	Tellinidae
SHELL SIZE RANGE	3 to 4 in (75 to 100 mm)
DISTRIBUTION	Persian Gulf to Indo-West Pacific
ABUNDANCE	Locally common
DEPTH	Intertidal to 165 ft (50 m)
HABITAT	Sandy mud bottoms
FEEDING HABIT	Deposit and filter feeder
BYSSUS	Absent

SHELL SIZE RANGE
3 to 4 in
(75 to 100 mm)

PHOTOGRAPHED SHELL
3½ in
(87 mm)

144

PHYLLODA FOLIACEA

FOLIATED TELLIN

(LINNAEUS, 1758)

Phylloda foliacea has a large and handsomely colored shell. It is found from the intertidal zone to about 165 ft (50 m), in sandy bottoms. Tellins live deeply buried in the sediment, with their valves in a horizontal orientation. The long siphons are separate and operate independently, with the inhalant siphon being longer than the exhalant. The inhalant siphon sucks up organic debris that is on the surface of the sediment like a vacuum cleaner.

RELATED SPECIES

Two small white tellins from the western Atlantic that have strong crenulations at the dorsal margins are *Phyllodina squamifera* (Deshayes, 1855), which ranges from North Carolina to Brazil, and *Tellidora cristata* (Récluz, 1842), from North Carolina to Central America. The former has a triangular shell with a rounded ventral margin, and the latter has an elongated oval shape.

Actual size

The shell of the Foliated Tellin is large, thin, lightweight, very compressed, and broadly triangular in outline. Its umbones are small and nearly central, and the hinge plate is narrow, with 2 teeth in the right valve and none in the left. The valves gape slightly anteriorly; the anterior is rounded and the posterior truncated. The valves have fine concentric sculpture, the posterior region has oblique ridges, and the posterior margin a few thornlike spines. The shell color is yellowish orange or reddish, and the interior is tinged with pink.

FAMILY	Donacidae
SHELL SIZE RANGE	½ to 1 in (12 to 25 mm)
DISTRIBUTION	Virginia to western Gulf of Mexico
ABUNDANCE	Locally and seasonally abundant
DEPTH	Intertidal to 1 ft (0.3 m)
HABITAT	Sandy beaches
FEEDING HABIT	Filter feeder
BYSSUS	Absent

SHELL SIZE RANGE
½ to 1 in
(12 to 25 mm)

PHOTOGRAPHED SHELL
⅝ in
(17 mm)

DONAX VARIABILIS

COQUINA DONAX

SAY, 1822

Donax variabilis lives on the surf zone of exposed sandy beaches. The small, shallowly buried animal is moved up and down the beach by the turbulence of incoming waves, and rapidly reburies itself as each wave recedes. *Donax variabilis* is extremely diverse in variety in color among individual specimens. Biologists believe that such polychromism may be an antipredatory adaptation, to prevent birds, one of its main predators, from forming a single search image of the species. However, humans have no problem collecting it in large numbers; it is used to make delicious soups. There are about 60 living species in the family Donacidae worldwide.

RELATED SPECIES

Donax serra Röding, 1798, from Namibia to South Africa, has a larger and more wedge-shaped shell, with a more pointed posterior margin. It is used extensively for food and fishing bait in South Africa, where it is very common. *Hecuba scortum* (Linnaeus, 1758), from the Indian Ocean, has a triangular shell with the posterior margin pointed and drawn out.

The shell of the Coquina Donax is small, strong, and elongate wedge-shaped. Its umbones are relatively small, positioned posteriorly; the hinge is strong, and bears 2 teeth in each valve. The valves are identical in size and shape, and sculptured with radial ribs that are strongest posteriorly. The posterior margin is truncate, the anterior rounded, and the ventral margin crenulated internally. The shell color is extremely variable: pure white to yellow, red, purplish, and pink, with or without radial bands; the interior color is also variable.

Actual size

FAMILY	Donacidae
SHELL SIZE RANGE	2 to 3½ in (50 to 90 mm)
DISTRIBUTION	Indo-West Pacific
ABUNDANCE	Common
DEPTH	Intertidal to shallow subtidal
HABITAT	Muddy bays
FEEDING HABIT	Filter feeder
BYSSUS	Absent

SHELL SIZE RANGE
2 to 3½ in
(50 to 90 mm)

PHOTOGRAPHED SHELL
2½ in
(63 mm)

146

HECUBA SCORTUM

LEATHER DONAX

(LINNAEUS, 1758)

The shell of the Leather Donax is medium in size, thick, inflated, and triangular in outline. Its umbones are prominent, located roughly centrally, and turned posteriorly; the hinge is strong, with two teeth in each valve. The valves are about the same size and shape, and the posterior end is drawn out and pointed. The sculpture consists of fine radial lines crossed by irregular concentric lines, forming a lattice pattern centrally, and, sometimes, spiny projections posteriorly. The shell color is grayish white, and the shell is purplish inside.

Hecuba scortum is easily recognized by its triangular shell with a drawn-out and pointed posterior margin, and a sharp, curved keel. It is a large species for the family Donacidae, in which most species are much smaller. It can be very common in parts of its wide distribution in the Indo-West Pacific. It is fished and marketed in India. *Hecuba scortum* lives buried, in a vertical position, right under the surface in muddy bays, in the intertidal zone and shallow waters. Its valves do not gape, but close tightly. Its siphons are relatively short.

RELATED SPECIES

Donax deltoides Lamarck, 1818, from Australia, which has a triangular shell, and is the most common, large bivalve in New South Wales. It is found in coastal middens, which demonstrates its previous importance as food for the Australian Aborigines. *Donax variabilis* Say, 1822, from New Jersey to Central America, and the Gulf of Mexico, lives high on the intertidal zone. It can be abundant, and has a small, wedge-shape shell, which varies widely in coloration.

Actual size

FAMILY	Psammobiidae
SHELL SIZE RANGE	½ to ¾ in (12 to 20 mm)
DISTRIBUTION	Southern Florida to Brazil
ABUNDANCE	Common locally
DEPTH	Intertidal to 3 ft (1 m)
HABITAT	Slopes of sandy beaches
FEEDING HABIT	Filter feeder
BYSSUS	Absent

SHELL SIZE RANGE
½ to ¾ in
(12 to 20 mm)

PHOTOGRAPHED SHELL
½ in
(14 mm)

HETERODONAX BIMACULATUS

SMALL FALSE DONAX

(LINNAEUS, 1758)

147

Heterodonax bimaculatus is a small, colorful clam that is found on sandy beach slopes, where it quickly buries itself to a depth of up to ten times its shell length. It filters water with suspended particles as the waves come in. The species name refers to two large, oblong, crimson spots inside the valves, which can be pale in some shells. Fish are its main predator; they nip the tips of the clam's long siphons. As a defense mechanism, the siphon tips can be autotomized (cast off). There are about 130 living species in the family Psammobiidae worldwide, both in tropical and temperate waters, and range from shallow to deep waters.

The shell of the Small False Donax is small, solid, compact, compressed, and rounded triangular in outline. Its umbones are pointed and located slightly closer to the posterior end, which is truncated. The valves are smooth and have fine concentric growth lines; the interior is glossy, with 2 muscle scars. The shell color is variable, ranging from creamy white to orange or purple, with radial rows of purplish dots or radially rayed. The interior is more vividly colored than the exterior.

RELATED SPECIES

Soletellina diphos (Linnaeus, 1771), from the tropical Indo-West Pacific, is a large clam that lives in muddy bottoms. It is an important commercial species in Taiwan, and is collected for food in other parts of its range. *Asaphis deflorata* (Linnaeus, 1758), from North Carolina to Brazil, is a common species in gravel sand bottoms. Since its meat is considered gritty, it is collected for use as bait, not for food.

Actual size

FAMILY	Psammobiidae
SHELL SIZE RANGE	2 to 4½ in (50 to 120 mm)
DISTRIBUTION	Tropical Indo-West Pacific
ABUNDANCE	Common
DEPTH	Intertidal to 100 ft (30 m)
HABITAT	Infaunal in muddy bottoms
FEEDING HABIT	Deposit and filter feeder
BYSSUS	Absent

SHELL SIZE RANGE
2 to 4½ in
(50 to 120 mm)

PHOTOGRAPHED SHELL
2⅖ in
(61 mm)

148

SOLETELLINA DIPHOS
DIPHOS SANGUIN
(LINNAEUS, 1771)

Soletellina diphos is an important commercial psammobiid species in Taiwan, where it is collected for food. It is considered a delicacy in the Philippines. However, it can accumulate toxins produced by red tide and cause shellfish poisoning. Like other psammobiids, it is an active burrower, and lives buried up to 12 in (30 cm) deep in muddy bottoms, usually in shallow waters. Its valves gape slightly; the strong and compressed foot extends through the gaping anterior margin, and the long siphons through the posterior margin. *Soletellina diphos* is the type species of the genus.

RELATED SPECIES

Gari elongata (Lamarck, 1818) and *Asaphis violascens* (Forsskål, 1775), both of which range from the Red Sea to the tropical West Pacific, are actively collected for their prized meat, as are other common species of psammobiids. However, the flesh of some species, such as *Asaphis deflorata* (Linnaeus, 1758), from Florida to Brazil, have a gritty texture, and therefore have limited human consumption.

Actual size

The shell of the Diphos Sanguin is thin but solid, compressed, and elongate-oval in outline. Its umbones are low, located anteriorly. The anterior margin is rounded, and the posterior narrower and sometimes pointed. The valves are about equal in size, and gape slightly both anteriorly and posteriorly; they lack strong sculpture, and have only fine concentric growth marks. The shell color is dark purple, covered with a brown periostracum with 2 pale rays, and the interior is deep purple.

FAMILY	Psammobiidae
SHELL SIZE RANGE	1¾ to 3 in (45 to 75 mm)
DISTRIBUTION	Red Sea to Indo-Pacific
ABUNDANCE	Common
DEPTH	Intertidal to 65 ft (20 m)
HABITAT	Coarse and fine sandy bottoms
FEEDING HABIT	Suspension feeder
BYSSUS	Absent

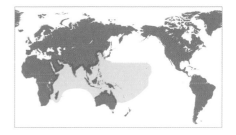

SHELL SIZE RANGE
1¾ to 3 in
(45 to 75 mm)

PHOTOGRAPHED SHELL
2⅝ in
(66 mm)

ASAPHIS VIOLASCENS

PACIFIC ASAPHIS

(FORSSKÅL, 1775)

Asaphis violascens is a deep-burrowing clam that can burrow to a depth of about 8 in (20 cm). It lives in coarse sandy or gravelly bottoms. It is common in shallow tropical waters, ranging from the Red Sea, throughout the Indian Ocean, and to the central Pacific. It is collected for food, marketed locally, and its shell is used in shellcraft. In China, it can reach densities of up to 50 clams per sq yd (60 per m²). Psammobiids are infaunal bivalves, often found in sediment with high organic content. Most species are believed to be deposit feeders, although some are filter feeders.

RELATED SPECIES

Asaphis deflorata (Linnaeus, 1758), which ranges from Florida to Brazil, is very similar and differs by anatomic characters, such as the details of its alimentary tract, as well as thinner radial ribs on the shell. *Heterodonax bimaculatus* (Linnaeus, 1758), from southern Florida to Brazil, is an active burrower that lives on sandy beach slopes in protected bays. It has a small, colorful shell.

The shell of the Pacific Asaphis is medium in size, moderately inflated, thick, and elongately ovate in outline. Its umbones are rounded and situated anteriorly, and the hinge line has 2 cardinal teeth in each valve. The anterior margin is rounded and the posterior is subtruncate. The sculpture consists of numerous strong radial ribs and weaker concentric growth lines. The shell color is white, tinged with purple or orange rays, and the interior is yellow and purple.

Actual size

FAMILY	Semelidae
SHELL SIZE RANGE	1 to 1⅜ in (25 to 34 mm)
DISTRIBUTION	North Carolina to Uruguay
ABUNDANCE	Common
DEPTH	3 to 65 ft (1 to 20 m)
HABITAT	Sandy and muddy bottoms
FEEDING HABIT	Deposit and suspension feeder
BYSSUS	Absent

SHELL SIZE RANGE
1 to 1⅜ in
(25 to 34 mm)

PHOTOGRAPHED SHELL
1¼ in
(32 mm)

150

SEMELE PURPURASCENS
PURPLISH AMERICAN SEMELE
(GMELIN, 1791)

The shell of the Purplish American Semele is thin, slightly inflated, and oval in outline. Its umbones are small, pointed, and slightly posterior; the hinge plate is narrow, with 2 cardinal teeth in each valve, and an internal ligament. The valves are identical in size and shape, with rounded margins. The sculpture is smooth, with fine concentric growth lines. The shell color is gray or cream, with purple and orange markings, and the interior is glossy, with purple, orange, or brown blotches.

Semele purpurascens has a wide longitudinal range, from North Carolina to Uruguay. It is an active burrower and lives deeply buried in sandy and muddy bottoms with high organic content. It is a deposit and suspension feeder, using its long, incurrent siphon to collect surface sediment, as well as to inhale water that it filters with its gills to collect suspended particles. There are about 65 living species in the family Semelidae distributed worldwide, with about half of the species found in the eastern Pacific. The family's fossil record dates to the Eocene Period.

RELATED SPECIES

Semele proficua (Pulteney, 1799), has a similar distribution, but reaches farther south to Argentina. Its shell is about the same size, but it is oval to circular in outline, has a smoother outer surface, and is white. *Semele solida* (Gray, 1828), from the Magellanic province (encompassing Chile and Argentina), has a thick, heavy, and much larger shell with a nearly circular outline. It is also white in color.

Actual size

FAMILY	Semelidae
SHELL SIZE RANGE	1½ to 3¼ in (40 to 80 mm)
DISTRIBUTION	Peru and Chile
ABUNDANCE	Common
DEPTH	Intertidal to 65 ft (20 m)
HABITAT	Sandy bottoms
FEEDING HABIT	Filter feeder
BYSSUS	Absent

SHELL SIZE RANGE
1½ to 3¼ in
(40 to 80 mm)

PHOTOGRAPHED SHELL
2¼ in
(56 mm)

SEMELE SOLIDA

SOLID SEMELE

(GRAY, 1828)

151

Semele solida is a commercially important semelid clam that ranges from Peru to Chile, and is exported to many countries, especially in Asia. It is one of ten species of clam fished commercially in the region. It is sold fresh, frozen, or salted, either as clean flesh or on the half shell. Because of upwelling currents, the coasts of Chile and Peru are among the most productive in the world. *Semele solida* is infaunal, and lives buried in sandy bottoms, ranging from the intertidal zone to shallow subtidal waters. One study in Chile found a mean density of 11 clams per sq yd (13 per m^2), and a life span of 11 years.

RELATED SPECIES

Semele decisa (Conrad, 1837), from California to Baja California, is one of the largest species in the family Semelidae. Its shell is thick and ovate in outline, with one radial ridge and heavy concentric folds. *Semele purpurascens* (Gmelin, 1791), which ranges from North Carolina to Uruguay, has a small, thin shell with a colorful interior, stained in purple with white margins.

The shell of the Solid Semele is medium in size, thick, heavy, compressed, and with a nearly circular outline. Its umbones are small and central, and slightly turned anteriorly; the hinge plate is very strong, with 2 of 4 teeth in each valve thick. The valves are about the same size and shape, with the sculpture consisting of heavy concentric folds, which are somewhat irregular. The shell color is off-white, with a brown periostracum, and the interior is white. The hinge area is tinged with violet.

Actual size

FAMILY	Solecurtidae
SHELL SIZE RANGE	2 to 4 in (50 to 100 mm)
DISTRIBUTION	Massachusetts to Brazil
ABUNDANCE	Common
DEPTH	Intertidal to 33 ft (10 m)
HABITAT	Muddy bottoms
FEEDING HABIT	Deposit and suspension feeder
BYSSUS	Absent

SHELL SIZE RANGE
2 to 4 in
(50 to 100 mm)

PHOTOGRAPHED SHELL
2¾ in
(68 mm)

152

TAGELUS PLEBEIUS
STOUT AMERICAN TAGELUS
LIGHTFOOT, 1786

Tagelus plebeius is an infaunal clam that lives deeply buried in sandy or muddy bottoms with high organic content. It is a common species in salt marshes, mangrove swamps, or mud flats, where it is a deposit and suspension feeder. *Tagelus plebeius* is collected and consumed locally for food. Some species of solecurtids are not consumed by humans but used as bait. Solecurtids live in more or less permanent, vertical burrows. Some species have long siphons that cannot be withdrawn into their shells. There are about 40 living species in the family Solecurtidae in tropical and warm waters worldwide. The family's fossil record dates back to the Cretaceous Period.

RELATED SPECIES

Tagelus californianus (Conrad, 1837), which ranges from California to Mexico, is one of the largest species in the family Solecurtidae, and can reach over 4½ in (120 mm) in length. Its shell is longer and thicker than *T. plebeius*. Its meat is used for bait. *Solecurtus strigilatus* (Linnaeus, 1758), from the Mediterranean and western Africa, is a common species in shallow waters. It is collected for food.

The shell of the Stout American Tagelus is medium-sized, moderately inflated, and elongately oval to nearly rectangular in outline. Its umbones are blunt and low, located at about the middle of the shell length; the hinge has 2 small cardinal teeth. The valves are identical in size and shape, and rounded and widely gaping at both ends. The surface is smooth, with irregular, weak concentric growth lines. The shell color is white or tan, with a heavy brownish yellow or brown periostracum. The interior is white.

Actual size

FAMILY	Solecurtidae
SHELL SIZE RANGE	2 to 4 in (50 to 100 mm)
DISTRIBUTION	Mediterranean to western Africa
ABUNDANCE	Common
DEPTH	Intertidal to 50 ft (15 m)
HABITAT	Sandy and muddy bottoms
FEEDING HABIT	Deposit and suspension feeder
BYSSUS	Absent

SHELL SIZE RANGE
2 to 4 in
(50 to 100 mm)

PHOTOGRAPHED SHELL
3 in
(77 mm)

SOLECURTUS STRIGILATUS
SCRAPER SOLECURTUS
(LINNAEUS, 1758)

153

Solecurtus strigilatus is easily distinguished from other solecurtids in the region by its large size, colorful shell, and oblique sculpture. It lives buried in sand and mud bottoms at depths of about 8 in (200 mm), and its long and thick siphons cannot withdraw completely into the shell. *Solecurtus strigilatus* produces an oblique, Y-shaped burrow, with the shell located in the stem and one siphon in each arm of the "Y." It is an active burrower; it expels water from the mantle cavity to help with the burrowing process during an escape response. The animal is collected for food and sold locally.

RELATED SPECIES

Solecurtus divaricatus (Lischke, 1869), from the Indo-West Pacific, has a shell similar in size and shape, but it has more pronounced sculpture, with radial ribs crossed by concentric growth lines. It is collected for food throughout southeastern Asia. *Tagelus plebeius* Lightfoot, 1786, which ranges from Massachusetts to Brazil, has a more slender, elongate, and smooth shell that is covered with a thick periostracum.

Actual size

The shell of the Scraper Solecurtus is medium in size, thin but solid, inflated, and rectangular. Its umbones are low and located at about the middle of the shell length; the hinge has 2 teeth in the right valve and 1 in the left. The valves are similar in size and shape, and gape at both ends. The sculpture consists of strong and coarse growth lines crossed by oblique, wavy striae. The shell color ranges from tan to pink with 2 white rays, and the interior is white tinged with pink.

FAMILY	Solenidae
SHELL SIZE RANGE	3 to 6½ in (75 to 170 mm)
DISTRIBUTION	Norway to Senegal; Mediterranean
ABUNDANCE	Common
DEPTH	Intertidal to 65 ft (20 m)
HABITAT	Sandy and muddy bottoms
FEEDING HABIT	Filter feeder
BYSSUS	Absent

SHELL SIZE RANGE
3 to 6½ in
(75 to 170 mm)

PHOTOGRAPHED SHELL
3 in
(76 mm)

154

SOLEN MARGINATUS
GROOVED RAZOR CLAM
PULTENEY, 1799

The shell of the Grooved Razor Clam is thin, fragile, and elongate-rectangular in outline. Its umbones are flat and inconspicuous; the hinge plate is narrow, with 1 tooth in each valve. The valves are identical in size and shape, and gape anteriorly. The sculpture consists of fine concentric lines, and a prominent vertical groove just behind the dorsal margin. The shell color is light orange-brown, with a light brown periostracum, and the interior is white that is stained in pink.

Solen marginatus is one of the largest species of solenids. It is a common species that lives buried in clean sand or mud; it is a fast burrower. In the Mediterranean it is collected and sold in local markets for food. One of the methods of collection involves pouring some salt into the burrow, which has a keyhole aperture, causing the animal to emerge on the sand surface. There are about 60 living species in the family Solenidae worldwide, with most living in the intertidal zone and shallow waters, especially in warm, tropical regions. The fossil history of the family dates from the Eocene Period.

RELATED SPECIES
Solen grandis Dunker, 1861, which ranges from Japan to Indonesia, has one of the largest shells in the family, as the name suggests. Its flesh is considered a delicacy. *Solen ceylonensis* Leach, 1814, from the western Indian Ocean, is another large solenid. It has a shell with very straight sides that do not taper, and which end abruptly on both extremities.

Actual size

FAMILY	Pharidae
SHELL SIZE RANGE	1½ to 3¼ in (40 to 80 mm)
DISTRIBUTION	Indo-West Pacific
ABUNDANCE	Common
DEPTH	16 to 115 ft (5 to 35 m)
HABITAT	Sandy and muddy bottoms
FEEDING HABIT	Filter feeder
BYSSUS	Absent

SHELL SIZE RANGE
1½ to 3¼ in
(40 to 80 mm)

PHOTOGRAPHED SHELL
2 in
(50 mm)

SILIQUA RADIATA

SUNSET SILIQUA

(LINNAEUS, 1758)

155

Siliqua radiata has a distinct, strong, whitish radial ridge on the inner side of each valve, running from the umbone to the ventral margin. This ridge corresponds in position with the white radial band on the outer surface of each valve. *Siliqua radiata* is the type species of the genus. Other species in this genus have a similar ridge, although shells in the related genus *Ensis* do not. *Siliqua radiata* is a common species living in shallow subtidal waters in fine sand or mud bottoms. There are about 65 living species in the family Pharidae worldwide. The fossil record of the family extends to the Cretaceous Period.

RELATED SPECIES

Siliqua patula (Dixon, 1789), from Alaska to Russia, is a large pharid with an oval-oblong shell. It is harvested extensively for food, both commercially and recreationally. *Ensis siliqua* (Linnaeus, 1758), from Norway to the Iberian Peninsula, and the Mediterranean, has an elongate and narrow shell.

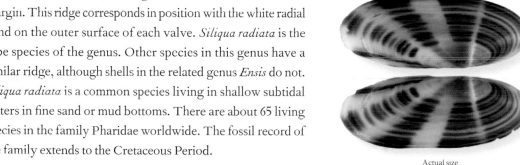

Actual size

The shell of the Sunset Siliqua is medium in size, thin, fragile, polished, very compressed, and elongate-oblong in outline. Its umbones are inconspicuous, situated anteriorly, with a narrow hinge plate and small teeth. The valves are identical in size and shape, and the surface is sculptured with fine concentric lines and faint radial striae. Inside each valve there is a strong radial ridge extending from the umbones to the ventral margin. The shell color is white with 4 broad purple rays; the interior color is similar to the exterior.

FAMILY	Pharidae
SHELL SIZE RANGE	6 to 9 in (150 to 230 mm)
DISTRIBUTION	Norway to Iberian Peninsula; Mediterranean
ABUNDANCE	Common
DEPTH	Intertidal to 230 ft (70 m)
HABITAT	Fine sand and mud
FEEDING HABIT	Filter feeder
BYSSUS	Absent

SHELL SIZE RANGE
6 to 9 in
(150 to 230 mm)

PHOTOGRAPHED SHELL
6⅜ in
(162 mm)

156

ENSIS SILIQUA
GIANT RAZOR SHELL
(LINNAEUS, 1758)

The shell of the Giant Razor Shell is large, thin, fragile, inflated, and elongate-rectangular in outline. Its umbones are inconspicuous, located close to the anterior margin; the hinge is narrow, with small teeth. The sculpture consists of smooth concentric lines, and a diagonal line from the umbones to the posterior ventral edge. The valves are identical in size and shape, and gape at both ends. The shell color is whitish with violet-brown stains, with a yellow-brown periostracum. The shell interior is white and has purple tints.

Ensis siliqua is the largest species in the family Pharidae, with some shells reaching 9 in (230 mm) in length and 1 in (25 mm) in width. The shell is very elongate and narrow, and resembles the old-style razors, hence the popular name of razor or jackknife clams. *Ensis siliqua* lives in deep vertical burrows in fine sandy bottoms, in intertidal flats and offshore. It can bury itself rapidly up to a depth of nearly 20 ft (6 m). The shells in the genus *Ensis* are similar, and muscle scars are useful in identification. *Ensis siliqua* was formerly abundant around Belgium, but now *E. directus* is the dominant species.

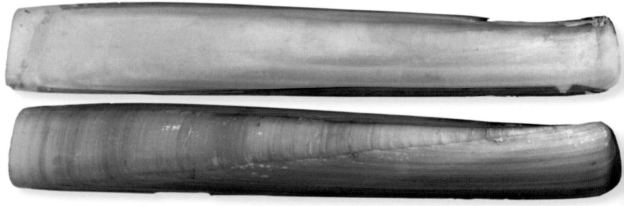

Actual size

RELATED SPECIES
Ensis tropicalis Hertlein and Strong, 1955, from western Panama, has a smaller and more elongate shell than *E. siliqua*. Its length is about ten times the width, and the valves are slightly arcuate. *Siliqua radiata* (Linnaeus, 1758), from the Indo-West Pacific, has a shell that is oval-oblong, with broad, dark violet radial bands, hence its popular name, Sunset Siliqua.

FAMILY	Mactridae
SHELL SIZE RANGE	1½ to 3¼ in (38 to 83 mm)
DISTRIBUTION	New Jersey to Argentina
ABUNDANCE	Common
DEPTH	Intertidal to 36 ft (11 m)
HABITAT	Sandy bottoms
FEEDING HABIT	Filter feeder
BYSSUS	Absent

SHELL SIZE RANGE
1½ to 3¼ in
(38 to 83 mm)

PHOTOGRAPHED SHELL
2¼ in
(70 mm)

RAETA PLICATELLA

CHANNELED DUCK CLAM

(LAMARCK, 1818)

157

Raeta plicatella is a common mactrid species; single, empty valves are commonly washed ashore, but live specimens are rarely found. The hinge plates are the thickest part of the shell, and often the only part that remain of the paper-thin shells found on the beach. *Raeta plicatella* is an infaunal clam that is capable of fast burrowing with its muscular foot. The large, spoon-shaped resilium below the umbones is a characteristic of the mactrids. There are about 150 living species in the family Mactridae worldwide, mostly in shallow waters. The earliest records of the family date back to the Cretaceous Period.

RELATED SPECIES

Anatina anatina (Spengler, 1802), which ranges from North Carolina to Brazil, has a shell that resembles *R. plicatella*, but it is more elongate and smoother. *Spisula solidissima* (Dillwyn, 1817), from Nova Scotia to North Carolina, is a commercially important clam. It is one of the largest bivalves in the western Atlantic, and as the name suggests, has a solid and thick shell.

Actual size

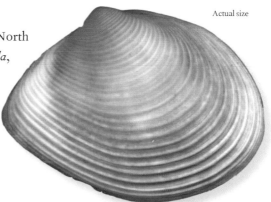

The shell of the Channeled Duck Clam is paper thin, fragile, lightweight, inflated, and broadly oval in outline. Its umbones are small but pointed, and face the anterior margin, which is narrow and pointed. The hinge plate is strong, with 3 teeth in the left valve, 2 in the right, and a spoon-shaped resilium under the umbones. The sculpture consists of smooth, concentric ribs that show as grooves on the inside. The shell color is white.

FAMILY	Mactridae
SHELL SIZE RANGE	1½ to 3¾ in (40 to 95 mm)
DISTRIBUTION	Indian Ocean to Philippines
ABUNDANCE	Common
DEPTH	3 to 65 ft (1 to 20 m)
HABITAT	Sandy bottoms
FEEDING HABIT	Filter feeder
BYSSUS	Absent

SHELL SIZE RANGE
1½ to 3¾ in
(40 to 95 mm)

PHOTOGRAPHED SHELL
3⅛ in
(78 mm)

158

MACTRA VIOLACEA
VIOLET MACTRA
GMELIN, 1791

Mactra violacea is an edible mactrid that is harvested commercially in India and for subsistence elsewhere throughout its range. It lives buried in sandy bottoms in shallow water. It can grow to a relatively large size. Many of the larger mactrids, and those that occur in abundance, are frequently used for food and fished commercially. Their flesh reportedly has a slightly acrid and peppery taste. Their siphons are joined and often enclosed within a periostracal sheath from the posterior margin of the shell; they can be withdrawn into the shell. The animal has a strong foot that is capable of rapid burrowing.

RELATED SPECIES

Tresus capax (Gould, 1850), which ranges from Alaska to California, in shallow water, is the largest mactrid, and the largest intertidal clam in Alaska. Its thick and heavy shell reaches 11 in (280 mm) in length. *Spisula solidissima* (Dillwyn, 1817), from Nova Scotia to North Carolina, is one of the largest clams in the Atlantic Ocean. It can be abundant locally.

Actual size

The shell of the Violet Mactra is medium in size, thin, fragile, glossy, and rounded triangular in outline. Its umbones are prominent, located at about the middle of the shell length, and slightly turned anteriorly. The hinge plate is relatively thick, with a large spoon-shaped depression under the umbones, and with 3 teeth in the left valve and 2 in the right. The valves are about the same size, and the surface is smooth, with fine concentric lines. The shell color ranges from white to purple, and is usually darker near the umbones; the interior is light purple.

FAMILY	Mactridae
SHELL SIZE RANGE	3¼ to 4½ in (80 to 120 mm)
DISTRIBUTION	Vietnam to southern Australia
ABUNDANCE	Common
DEPTH	Intertidal to 50 ft (15 m)
HABITAT	Sand or mud bottoms
FEEDING HABIT	Filter feeder
BYSSUS	Absent

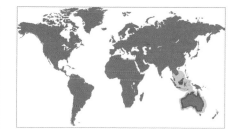

SHELL SIZE RANGE
3¼ to 4½ in
(80 to 120 mm)

PHOTOGRAPHED SHELL
3¼ in
(84 mm)

LUTRARIA RHYNCHAENA

SNOUT OTTER CLAM

JONAS, 1844

159

Lutraria rhynchaena is a well-known mactrid from eastern and southern Australia. It also occurs in Vietnam, where it is abundant and fished commercially. It is now being cultured in Vietnam, where it is an economically important clam. Cultured clams grow to market size in about one year. The flesh is fragrant, rich in proteins, and considered a delicacy. *Lutraria rhynchaena* burrows deeply in sandy or muddy bottoms, below the lowest tidemark. Its long siphons reach the surface, and they are the only sign one can see of the clam while buried in the sediment.

RELATED SPECIES

Lutraria lutraria (Linnaeus, 1758), which ranges from Norway to Morocco and the Mediterranean, has a larger and more oval shell. It is consumed for food where it is common, but its flesh is not considered of high quality. *Raeta plicatella* (Lamarck, 1818), from New Jersey to Argentina, has a paper-thin shell that is commonly found as single valves on the beach.

Actual size

The shell of the Snout Otter Clam is medium in size, thick, solid, inflated, and elongate-oblong in outline. Its umbones are small, and located near anterior margin; the hinge plate is thick, and has a large spoon-shaped socket. The valves are about the same size and shape, and gape widely. The anterior margin is short and angularly rounded; the posterior margin is elongated and rounded; and the dorsal and ventral margins are nearly parallel. The shell color is off-white, and is covered by a brown periostracum; the interior is white porcelaneous.

FAMILY	Mactridae
SHELL SIZE RANGE	4 to 8 in (102 to 200 mm)
DISTRIBUTION	Nova Scotia to North Carolina
ABUNDANCE	Abundant
DEPTH	Intertidal to 420 ft (130 m)
HABITAT	Infaunal in sand
FEEDING HABIT	Filter feeder
BYSSUS	Absent

SHELL SIZE RANGE
4 to 8 in
(102 to 200 mm)

PHOTOGRAPHED SHELL
4½ in
(116 mm)

160

SPISULA SOLIDISSIMA

ATLANTIC SURF CLAM

(DILLWYN, 1817)

The shell of the Atlantic Surf Clam is large, thick, solid, and triangular-ovate in outline. Its umbones are tall and acute, located at about the center of the valves and facing anteriorly; the hinge line is thick and bears a large spoon-shaped cavity, the attachment of the internal ligament. The valves are smooth, with fine concentric growth lines. The shell color is off-white, covered by a thin, yellowish brown periostracum, and the interior is white.

Spisula solidissima is one of the largest bivalves in the western Atlantic; it can reach 8⅞ in (226 mm) in length and 30 years of age. It is an important commercial species, fished primarily off New Jersey and on Georges Bank. After violent storms, the shells are cast ashore; in one event, an estimated 50 million clams were found in a 10-mile (16-km) stretch. Its popular name derives from the clam being found in the surf zone on sandy beaches, although the commercial fishing beds are located offshore. The animal can use its muscular foot to leap in order to escape predation by giant whelks and starfishes.

RELATED SPECIES

Spisula hemphillii (Dall, 1894), from central California to Baja California, is one of a few large mactrids in the eastern Pacific. Its shell resembles *S. solidissima* but can be more elongate anteriorly. *Mactra violacea* Gmelin, 1791, from the tropical Indo-West Pacific, also has a similar, although smaller, shell. It is covered with a light brown periostracum, which is often eroded near the umbones, revealing its lavender shell.

Actual size

FAMILY	Mesodesmatidae
SHELL SIZE RANGE	½ to 2¼ in (13 to 57 mm)
DISTRIBUTION	Newfoundland, Canada to New Jersey
ABUNDANCE	Common
DEPTH	Intertidal to 330 ft (100 m)
HABITAT	Sandy bottoms
FEEDING HABIT	Filter feeder
BYSSUS	Absent

SHELL SIZE RANGE
½ to 2¼ in
(13 to 57 mm)

PHOTOGRAPHED SHELL
1¼ in
(30 mm)

MESODESMA ARCTATUM

ARCTIC WEDGE CLAM

(CONRAD, 1831)

161

Mesodesma arctatum is an infaunal clam that lives on sandy bottoms, and is often found in the intertidal zone of high-energy sandy beaches. Like other bivalves found in such habitat, *M. arctatum* has a large, muscular foot that is capable of rapid burrowing. Its wedge-shaped shell is oriented with the pointed (anterior) side down for easier penetration into the sediment. In Nantucket Island, Pleistocene fossils of *M. arctatum* are found next to living specimens. Some of the larger species of mesodesmatids, in New Zealand and Chile, are commercially fished for food. There are about 40 living species in the family Mesodesmatidae worldwide.

The shell of the Arctic Wedge Clam is medium in size, thick, solid, compressed, smooth, and triangular in outline. Its umbones are prominent, and are located closer to the posterior end; the hinge plate is thick, bearing a spoon-shaped cavity under the umbones, and strong teeth on both valves. The posterior margin is short and forms the base of the wedge; the anterior end is rounded. The valves are about the same size and shape, with fine concentric lines. Inside, the muscle scars are well defined. The shell color is whitish with a yellowish periostracum, and the interior is cream.

RELATED SPECIES

Mesodesma donacium (Lamarck, 1818), from Peru to Chile, has a large, wedge-shaped shell. It is heavily fished for food and bait. *Mesodesma ventricosa* (Gray, 1843), endemic to New Zealand, is probably the largest species of mesodesmatid. It can grow to over 4 in (100 mm) in length. It occurs mostly in the intertidal zone of sandy beaches. Once it was heavily harvested for food, but now it is protected and its collection is prohibited.

Actual size

FAMILY	Myidae
SHELL SIZE RANGE	3 to 6 in (75 to 150 mm)
DISTRIBUTION	Labrador to North Carolina, western Europe. Introduced to Alaska to California
ABUNDANCE	Common
DEPTH	Intertidal to 250 ft (75 m)
HABITAT	Sandy and muddy bottoms
FEEDING HABIT	Filter feeder
BYSSUS	Absent in adults

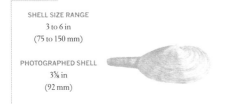

SHELL SIZE RANGE
3 to 6 in
(75 to 150 mm)

PHOTOGRAPHED SHELL
3⅝ in
(92 mm)

162

MYA ARENARIA

SOFT SHELL CLAM

LINNAEUS, 1758

Mya arenaria is a large and edible clam native to both coasts of the northern Atlantic. It was accidentally introduced along with oysters to the Pacific Ocean, from Alaska to California. It lives deeply buried in mud, sand, or gravel, and only the tips of its siphons show on the surface of the sediment. When it is disturbed, it withdraws its long siphons or "neck." It is the third most important clam fishery in the U.S.A. *Mya arenaria* is the main food for large predators such as the walrus and cod. There are about 25 living species in the family Myidae worldwide, with many in the northern hemisphere.

The shell of the Soft-Shell Clam is medium-large, moderately thick, solid, inflated, and ovately elongate in outline. Its umbones are strong and located closer to the posterior margin; the hinge plate has an erect, spoonlike projection below the umbo in the left valve. The right valve is slightly larger than the left; the valves gape widely. The sculpture consists of wrinkled concentric growth lines. The shell color is chalky white, covered with a thin grayish or tan periostracum, and the interior is white.

RELATED SPECIES

Mya truncata Linnaeus, 1758, is circumboreal. As the name suggests, it has a truncated shell, with the posteriorly margin ending rather abruptly. It is a common clam in Greenland and Iceland, where it is considered a delicacy. *Sphaenia fragilis* (H. Adams and A. Adams, 1854), from Georgia to Uruguay, has a small and fragile shell. It lives byssally attached to rock crevices.

Actual size

FAMILY	Corbulidae
SHELL SIZE RANGE	¾ to 1⅛ in (20 to 28 mm)
DISTRIBUTION	Japan to Vietnam and China
ABUNDANCE	Common
DEPTH	Intertidal to 100 ft (30 m)
HABITAT	Muddy bottoms
FEEDING HABIT	Filter feeder
BYSSUS	Present

SHELL SIZE RANGE
¾ to 1⅛ in
(20 to 28 mm)

PHOTOGRAPHED SHELL
1 in
(26 mm)

CORBULA ERYTHRODON

RED-TOOTHED CORBULA

(LAMARCK, 1818)

Corbula erythrodon is a large species of corbulid; most species have smaller shells. It is a common clam ranging from Japan to Vietnam and China, and is found from the intertidal zone to shallow subtidal waters. Like other corbulids, it is an infaunal filter feeder, living buried in mud. It has a thin byssus that is attached to gravel. Its valves close tightly and do not gape. Corbulids tolerate wide variations in conditions such as salinity, oxygen, and pollution. There are about 100 living species in the family Corbulidae worldwide, with most in shallow waters. The family is known from fossils since the Jurassic Period.

RELATED SPECIES

Varicorbula gibba (Olivi, 1792), which ranges from Norway to Angola, and the Mediterranean, is an abundant species that lives from the intertidal zone to deep water. It has a small and variable shell. *Corbula amethystina* (Olsson, 1961), from Panama to Ecuador, has a large, thick shell that resembles *C. erythrodon* in size, but with a more rostrate posterior margin, smaller umbone, and weaker concentric lines.

The shell of the Red-toothed Corbula is medium in size, very thick, heavy, inflated, and triangular-ovate in outline. Its umbones are large, and slightly turned posteriorly. The hinge plate is strong, with a single strong tooth in the right valve. The right valve is larger and more convex than the left. The sculpture consists of widely spaced, strong, concentric lines, and the interior is smooth, with well-marked muscle scars. The shell color is white, with a thin, brown periostracum, and the interior is white, stained in purplish red.

Actual size

FAMILY	Pholadidae
SHELL SIZE RANGE	⅝ to 1½ in (15 to 40 mm)
DISTRIBUTION	Ireland to the Ivory Coast; Mediterranean
ABUNDANCE	Common
DEPTH	Shallow subtidal to 1,000 ft (300 m)
HABITAT	Infaunal in mud, wood, and sandstone
FEEDING HABIT	Filter feeder
BYSSUS	Absent

SHELL SIZE RANGE
⅝ to 1½ in
(15 to 40 mm)

PHOTOGRAPHED SHELL
1⅛ in
(30 mm)

164

PHOLADIDEA LOSCOMBIANA
PAPER PIDDOCK
GOODALL IN TURTON, 1819

Pholadidea loscombiana is a common borer of submerged wood, mud, and sandstone. Juveniles have a small foot that extends through the anteroventral gape; once the animal reaches maturity, the foot becomes vestigial and the gape closes. The siphons are long and fused. The animal has bioluminescent organs in the mantle. Although its shell is fragile, it is used as a file to bore the relatively hard substrate. Because pholadids have an extra plate, the mesoplax, they were originally classified by Linnaeus as "Multivalva" along with barnacles and chitons. The oldest known fossils from the family Pholadidae are from the Carboniferous Period.

RELATED SPECIES

Cyrtopleura costata (Linnaeus, 1758), from Massachusetts to Brazil, has a large, thin, elongate white shell with raised radial ribs bearing fluted scales. Its beautiful shell is reminiscent of an angel's wings, hence its popular name, Angel Wing. *Pholas orientalis* Gmelin, 1791, from the Indo-West Pacific, has a shell similar to *C. costata*, but it is smaller and more elongate.

Actual size

The shell of the Paper Piddock is medium in size, thin, fragile, inflated, and elongately ovate in outline. Its umbones are low, rolled-in, and located anteriorly; the hinge plate is toothless in adults. The valves are similar in size and shape. They gape anteriorly in juveniles; in adults, the anteroventral gape closes with a fragile extension (the callum), and a chitinous tubelike growth develops in the posterior margin. The sculpture consists of concentric scaly ridges divided by a deep, median radial groove. The shell color is off-white.

FAMILY	Pholadidae
SHELL SIZE RANGE	4 to 8 in (100 to 200 mm)
DISTRIBUTION	Massachusetts to Brazil
ABUNDANCE	Common locally
DEPTH	Intertidal to 3 ft (1 m)
HABITAT	Muddy bottoms
FEEDING HABIT	Filter feeder
BYSSUS	Absent

SHELL SIZE RANGE
4 to 8 in
(100 to 200 mm)

PHOTOGRAPHED SHELL
7⅛ in
(183 mm)

CYRTOPLEURA COSTATA

ANGEL WING

(LINNAEUS, 1758)

165

Cyrtopleura costata is certainly the most beautiful and the largest of the Pholadidae. Pholadids are infaunal bivalves that bore into mud, clay, limestone, shale, and wood. They have thin, usually elongate, ribbed, and scaly shells that the animal uses to bore into the substrate, aided by its muscular foot. *Cyrtopleura costata* is a locally common species in shallow subtidal waters, burrowing in soft mud to depths of up to 3 ft (1 m). It is harvested commercially in Mexico and Cuba. Because of its rapid growth, it has potential for aquaculture. There are about 100 species of pholadids, or piddocks, worldwide.

RELATED SPECIES

Pholas campechiensis Gmelin, 1791, has a similar distribution and a shell that resembles *C. costata*, but is smaller and more elongate; *Pholas dactylus* Linnaeus, 1758, from the Mediterranean, also has a less inflated and more elongate shell with less prominent sculpture; *Jouannetia quillingi* Turner, 1955, from South Carolina to the Gulf of Mexico, has a smaller, globular shell, and bores into soft shale.

Actual size

The shell of the Angel Wing is large, thin, fragile, and elongate. The valves are equal in size and shape, inflated, and gape anteriorly. They are sculptured with raised radial ribs that form short spines in concentric rows; internally, the shell is ribbed and pitted, corresponding to the external sculpture. The hinge is toothless, and the margin is reflected, hiding the umbos. As in other pholadids, paired spoon-shaped projections (the apophyses), used for muscle attachment, extend from under the umbos. The apophyses and a third plate, the mesoplax, are usually missing from empty shells. The color is chalky white, occasionally with a pink tint.

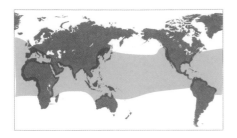

FAMILY	Teredinidae
SHELL SIZE RANGE	¹⁄₁₆ to ½ in (2 to 12 mm)
DISTRIBUTION	Cosmopolitan
ABUNDANCE	Common
DEPTH	Intertidal to 26 ft (8 m)
HABITAT	Bores into wood primarily
FEEDING HABIT	Primarily xylophagous; also filter feeder
BYSSUS	Absent

SHELL SIZE RANGE
¹⁄₁₆ to ½ in
(2 to 12 mm);

PHOTOGRAPHED SHELL
⅛ in
(3 mm)

166

TEREDO NAVALIS
NAVAL SHIPWORM
LINNAEUS, 1758

Teredo navalis is perhaps the bivalve that has caused the most economic damage through the destruction of wooden boats, piers, and other wooden structures into which it bores. Shipworms have a reduced shell, a long, wormlike body, and calcareous structures, called pallets, that plug the entrance of the burrow. The pallets are more important in species identification than the shells, which vary little between species. The animal bores into wood by mechanical abrasion, using its modified shell. There are about 70 living species in the family Teredinidae, most in tropical shallow waters worldwide.

RELATED SPECIES

Kuphus polythalamia (Linnaeus, 1767), from the Indo-Pacific, has a small shell but forms a thick calcareous tube that can grow longer than the shell of a giant clam. Unlike other teredinids, it lives in mud in mangroves. *Bankia carinata* (Gray, 1827), with a cosmopolitan distribution, has very similar shells to *Teredo navalis*, but its pallets are segmented.

The shell of the Common Shipworm is vestigial, minute, thin, inflated, trilobed, and helmetlike in outline. Its valves are identical in size and shape, and gape widely both anteroventrally and posteriorly; there is a deep, right-angled notch. The anterior surface of the valves is finely sculpted with rows of minute teeth, which are used to rasp the wood. There is a long and narrow rib near the umbo. The shell color is white, as are the simple and paddlelike pallets.

 Actual size

FAMILY	Teredinidae
SHELL SIZE RANGE	6 to 60 in (150 to 1,532 mm)
DISTRIBUTION	Indo-Pacific
ABUNDANCE	Common
DEPTH	Shallow subtidal
HABITAT	Muddy bottoms in mangroves
FEEDING HABIT	Filter feeder
BYSSUS	Absent

SHELL SIZE RANGE
6 to 60 in
(150 to 1,532 mm)

PHOTOGRAPHED SHELL
34 in
(863 mm)

KUPHUS POLYTHALAMIA

MUD TUBE CLAM

(LINNAEUS, 1767)

167

Kuphus polythalamia holds the distinction of being the longest shelled mollusk. Although its shell is small in size, the animal produces a very long calcareous tube (see photograph below) that can grow longer than the shell of a Giant Clam, and can reach up to 60 in (1,532 mm) in length. Unlike other teredinids, which bore into waterlogged wood, *K. polythalamia* lives buried in mud in mangroves, where it filters water. Shipworms have reduced shells and long wormlike bodies, and sometimes form calcareous tubes. They are hosts to symbiotic bacteria that allow them to digest cellulose.

RELATED SPECIES

Shells of teredinids are small and difficult to collect and identify, so they are not popular among collectors. *Teredo navalis* Linnaeus, 1758, a common cosmopolitan species, is one of the best known shipworms. *Neoteredo reynei* (Bartsch, 1920) and *Nausitora fusticula* (Jeffreys, 1860) are two species that bore into mangroves in the western Atlantic.

Actual size

The shell of the Mud Tube Clam is relatively small, but the calcareous tube that it produces is heavy and extremely long. The tube is irregular, cylindrical, rounded anteriorly, and slowly tapering toward the posterior (aperture), which terminates in 2 tubes fused together (above), through which the inhalant and exhalant siphons extend. In one 40-in (1,000-mm) long specimen, the anterior of the tube is 4½ in (120 mm) in diameter, tapering to about 1½ in (40 mm) in diameter at the posterior end of the tube. The valves are relatively small, only a few inches or centimeters in length, and fit within the white calcareous tube.

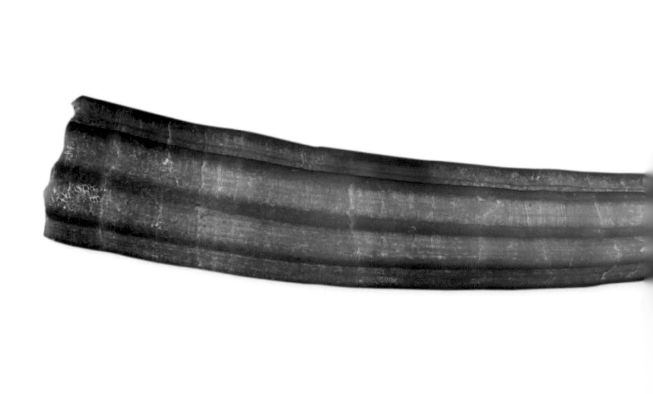

SCAPHOPODS

Scaphopods are mollusks with long, tapering, tubular, curved shells that are open at both ends. The shells resemble elephant tusks, but are coiled in a single plane. They may be polished and smooth, or have longitudinal ribs. Some smaller species are broader in the middle than at the ends. Shells range in size from ⅛ in (3 mm) to 6 in (15 cm). All of the nearly 600 extant species live in the oceans at subtidal to abyssal depths, where they burrow in soft sediments. The larger opening is the anterior end, through which the foot is extended and burrows down into the substrate. The smaller, posterior opening remains near the surface of the sediment. Scaphopods feed by extending multiple, thin tentacles from two lobes above the head throughout the sediment to capture small prey. Food is ground by specialized, crushing radular teeth.

Because of their subtidal, burrowing habitat, scaphopods are rarely encountered living, but shells occasionally wash up on shore.

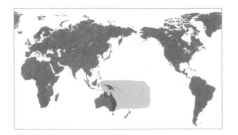

FAMILY	Gadilidae
SHELL SIZE RANGE	5⁄32 to 11⁄64 in (4.0 to 4.3 mm)
DISTRIBUTION	Western South Pacific
ABUNDANCE	Uncommon
DEPTH	35 to 930 ft (10 to 285 m)
HABITAT	Coral sand and mud bottoms
FEEDING HABIT	Micro-omnivore, primarily foraminiferans

SHELL SIZE RANGE
5⁄32 to 11⁄64 in
(4.0 to 4.3 mm)

PHOTOGRAPHED SHELL
5⁄32 in
(4.1 mm)

170

CADULUS SIMILLIMUS
CADULUS SIMILLIMUS
WATSON, 1879

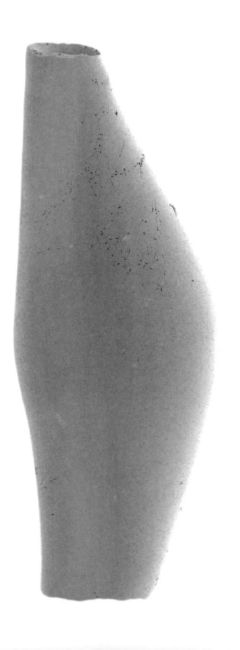

Cadulus simillimus is a very small, gadilid scaphopod from the tropical western Pacific. It ranges from relatively shallow waters to substantial depths, buried in coral sand or muddy bottoms. Like other gadilids, its shell has both ends narrowed, and the middle part of its tubular shell is widest. Most gadilids have small, thin, glossy, and sometimes translucent shells. *Cadulus simillimus* has a particularly short and stout shell, resembling a fat bottle with a slight bend. Some gadilids have notches on the apex, and the aperture may be oblique.

RELATED SPECIES

Gadila mayori (Henderson, 1920), from Florida and Gulf of Mexico, has a very small, translucent shell, which is swollen centrally, with both ends narrowed. It lives offshore in sandy or muddy bottoms. *Polyschides magnus* (Boissevain, 1906), from central Japan to the tropical western Pacific, has a relatively large shell for the family, reaching over 1¼ in (30 mm) in length. Its shell is gently curved, glossy, and the apex has four notches. The aperture is only slightly narrower than the maximum diameter of the shell.

Actual size

The shell of *Cadulus simillimus* is very small, thin, glossy, translucent, narrowed at both ends, very slightly bent, and with a fat bottle shape. Its apex is narrow, circular, and smooth, and without any notches. The aperture is circular, with a thin edge, and is about twice as wide as the apex. There is a slight bulge on the concave side of the shell. The shell surface is smooth, and the shell color is whitish translucent.

FAMILY	Dentaliidae
SHELL SIZE RANGE	2 to 4 in (50 to 100 mm)
DISTRIBUTION	Red Sea to Australia
ABUNDANCE	Uncommon
DEPTH	Intertidal to 133 ft (40 m)
HABITAT	Sandy bottoms
FEEDING HABIT	Micro-omnivore, primarily foraminiferans

SHELL SIZE RANGE
2 to 4 in
(50 to 100 mm)

PHOTOGRAPHED SHELL
3¼ in
(80 mm)

DENTALIUM ELEPHANTINUM
ELEPHANT TUSK
LINNAEUS, 1758

171

Dentalium elephantinum is a large, thick scaphopod that is easily recognized by its dark green color anteriorly, which fades to white posteriorly. It lives buried in sandy bottoms with the posterior (narrow) end sticking out of the surface. *Dentalium elephantinum* is the type species of the genus. Empty shells of scaphopods are often used by hermit crabs and sipunculan worms. Some hermit crabs have become specialized to live exclusively in scaphopod shells, with one cheliped serving as an operculum to plug the shell. There are more than 200 living species in the family Dentaliidae worldwide. The family's fossil record extends to the mid-Triassic Period.

RELATED SPECIES

Dentalium vernedei Sowerby II, 1860, from Japan to the Philippines, is probably the largest scaphopod, reaching nearly 6 in (150 mm) in length. Its shell is slender, tapering, and yellowish in color. *Dentalium aprinum* Linnaeus, 1758, from the Indo-Pacific, somewhat resembles *D. elephantinum*, but is smaller and more slender, and its color is a lighter shade of green. It is a common species of dentaliid.

The shell of the Elephant Tusk is large, thick, solid, heavy, colorful, and slightly curved. Its anterior end is rounded, wide, and about three times as wide as the posterior end. There is a notch in the posterior end. The sculpture consists of about 10 strong, rounded longitudinal ribs that run the entire length of the shell, and fine growth lines. The shell color is dark green near the anterior end, fading to white toward the posterior opening.

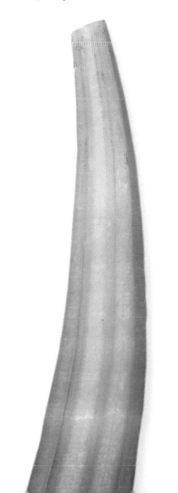

Actual size

FAMILY	Dentaliidae
SHELL SIZE RANGE	2 to 4 in (50 to 100 mm)
DISTRIBUTION	California to Chile
ABUNDANCE	Common
DEPTH	4,800 to 10,800 ft (1,500 to 3,300 m)
HABITAT	Sandy and muddy bottoms
FEEDING HABIT	Micro-omnivore, primarily foraminiferans

SHELL SIZE RANGE
2 to 4 in
(50 to 100 mm)

PHOTOGRAPHED SHELL
3¾ in
(95 mm)

172

FISSIDENTALIUM MEGATHYRIS
COSTATE TUSKSHELL
(DALL, 1890)

The shell of the Costate Tuskshell is large, thick, strong, and broad. Its anterior opening is circular, and over seven times wider than the narrow posterior opening. The apex (posterior end) has a long slit on the convex side. Its shell bends only slightly. The sculpture consists of many longitudinal striae that run the entire length of the shell, which are crossed by fine growth lines. The region around the anterior opening generally forms a smooth, thinner ring around the aperture. The shell is chalky white, with a light brown periostracum.

Fissidentalium megathyris is a large, deepwater scaphopod that ranges from off the coast of California to Chile. Like many scaphopods, it feeds almost exclusively on benthic foraminifera; one specimen studied had 188 foraminifera in its stomach. Due to its large size, it may be a major predator of foraminifera in its region. Scaphopods capture their food with oral tentacles, then pass it to the buccal pouch, where it is crushed by its radular teeth. Other food items that are eaten by scaphopods include ostracods, bivalve spat (juvenile), and eggs. Because most scaphopods are found in deep water, they are not as well studied as shallow water mollusks.

RELATED SPECIES
Laevidentalium lubricatum (Sowerby II, 1860), from Australia, is a common species ranging from the intertidal zone to 3,300 ft (1,000 m). It is a medium-sized species, with a slender, smooth, and slightly bending shell. *Dentalium elephantinum* Linnaeus, 1758, from the Red Sea to Australia, has a large, broad shell with longitudinal rounded ribs. It is dark green anteriorly, fading to white posteriorly.

Actual size

FAMILY	Dentaliidae
SHELL SIZE RANGE	2½ to 3½ in (60 to 90 mm)
DISTRIBUTION	Red Sea to Indo-West Pacific
ABUNDANCE	Uncommon
DEPTH	150 to 500 ft (45 to 155 m)
HABITAT	Sandy bottoms
FEEDING HABIT	Micro-omnivore, primarily foraminiferans

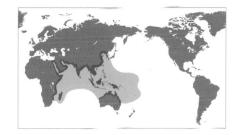

SHELL SIZE RANGE
2½ to 3½ in
(60 to 90 mm)

PHOTOGRAPHED SHELL
3¾ in
(96 mm)

ANTALIS LONGITRORSA

ELONGATE TUSK
(REEVE, 1843)

173

Antalis longitrorsa has a long, slender, and smooth shell. It lives offshore, buried in sandy bottoms. It has a wide distribution, ranging from the Red Sea to the Indo-West Pacific. Like other scaphopods, its shell is a hollow, curved, conical tube that is open at both ends. The anterior end, usually the widest, is the aperture through which its muscular, cylindrical foot extends. The posterior end is the apex; it is often ornamented with one or more notches, or a slit. The apex is cast off periodically to increase the diameter of the opening for respiratory currents.

RELATED SPECIES

Dentalium formosum Adams and Reeve, 1850, from Japan to the Philippines, has a medium-large, relatively short, and stout shell with many low longitudinal ribs. It is one of the most colorful scaphopods, with a maroon to brick-red color. *Dentalium elephantinum* Linnaeus, 1758, from the Red Sea to Australia, has a large, thick shell with about ten rounded longitudinal ribs. It is another of the few colorful scaphopods, with a dark green shell that fades to white posteriorly.

The shell of the Elongate Tusk is medium in size, thin, slender, long, and slightly curved. Its anterior end is circular, relatively narrow, and about three times wider than the posterior end. The shell tapers to a pointed posterior end. The surface is smooth, with very fine growth lines. The shell is translucent and ranges from orange to off-white in color.

Actual size

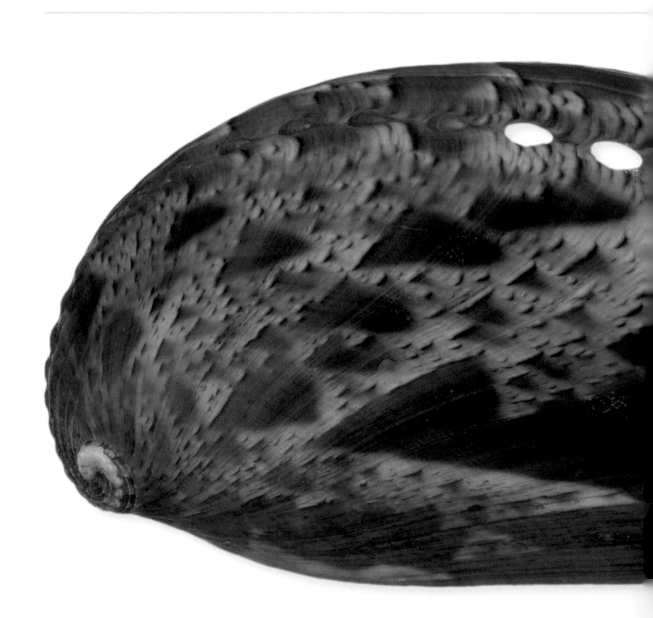

GASTROPODS

With nearly 100,000 living species, the gastropods are by far the largest and most diverse class of mollusks. During their larval stage, all gastropods undergo torsion, a process that produces asymmetry by twisting the animal until the formerly posterior mantle cavity is rotated to a position over the head. Gastropods are characterized by having a single coiled shell that may become limpet-like in some groups, or reduced or lost in others. The shells of most gastropods are asymmetrical and dextrally coiled, although there are examples of mutant individuals as well as species that have shells that are sinistral, with mirror-image asymmetry.

Gastropods occupy all marine habitats from the intertidal to the deepest trenches, from the equator to the poles. Multiple groups have colonized freshwater habitats, others have evolved the ability to breathe air and have radiated to conquer diverse terrestrial habitats, including mountains and deserts.

Snails range in size from $\frac{1}{75}$ in ($\frac{1}{3}$ mm) to 39 in (1 m) in length. Most are mobile, crawling or burrowing through a variety of substrates. Some cement themselves to hard substrates, others are external or internal parasites. Gastropods may be herbivores, carnivores, parasites, filter feeders, detritivores, or rarely even chemoautotrophs.

FAMILY	Patellidae
SHELL SIZE RANGE	¼ to 2¾ in (20 to 70 mm)
DISTRIBUTION	Western Europe and the Mediterranean
ABUNDANCE	Abundant
DEPTH	Intertidal
HABITAT	Rocky shores
FEEDING HABIT	Grazer, feeds on algae
OPERCULUM	Absent

SHELL SIZE RANGE
¼ to 2¾ in
(20 to 70 mm)

PHOTOGRAPHED SHELL
1 in
(25 mm)

176

PATELLA VULGATA
COMMON EUROPEAN LIMPET
LINNAEUS, 1758

Patella vulgata is the type species of the genus *Patella* and of the family Patellidae. It is an abundant limpet, especially around the British and Irish coasts, but is also found in the Mediterranean. It lives on rocks in the intertidal zone. It is more prevalent on rocks exposed to strong waves, than on rocks in calmer waters. *Patella vulgata* living higher up on the rocks usually have taller shells than those in lower positions. Like some other patellids, *P. vulgata* is protandric—it undergoes a sex change, starting out as male and changing to female as it grows larger. It has been used as food by humans for many centuries. There are more than 70 species of Patellidae worldwide.

RELATED SPECIES

Patella ferruginea (Gmelin, 1791), which has many thick, radial ribs, is endemic to the Mediterranean and is the most endangered marine species in European waters. *Patella variabilis* Krauss, 1848, from South Africa has, as the name suggests, a shell that is very variable in size, shape, and coloration; and *P. cochlear* Born, 1778, from southern Africa, has a spoon or teardrop shape.

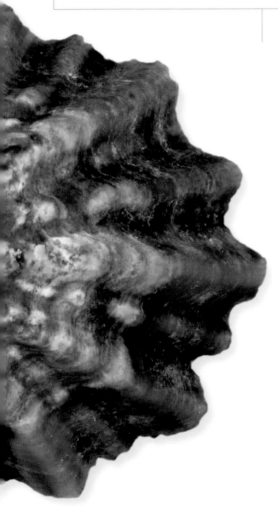

The shell of the Common European Limpet is conic, high or low, broad, and ovate. It has coarse radial ribs, although some shells are nearly smooth, and the margin may be crenulated or smooth. The apex is located centrally or subcentrally, slightly anterior. The shell color is light brown to gray, the interior is pale orange, and the muscle scar is usually lighter.

Actual size

FAMILY	Patellidae
SHELL SIZE RANGE	2 to 4 in (50 to 100 mm)
DISTRIBUTION	South Africa to Mozambique
ABUNDANCE	Common
DEPTH	Intertidal
HABITAT	Rocky shores
FEEDING HABIT	Grazer, feeds on crustose brown algae
OPERCULUM	Absent

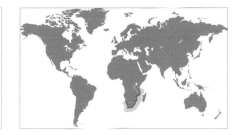

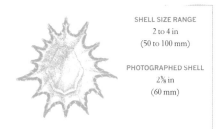

SHELL SIZE RANGE
2 to 4 in
(50 to 100 mm)

PHOTOGRAPHED SHELL
2⅜ in
(60 mm)

PATELLA LONGICOSTA

LONG-RIBBED LIMPET

LAMARCK, 1819

Despite being quite variable in shell shape, *Patella longicosta* is easily recognizable in having the longest radiating ribs among the several star-shaped limpets. The radial ridges add strength to the shell, and also help disperse the force of breaking waves through the shell. *Patella longicosta* is a common limpet on intertidal rocks, occurring mostly in South Africa but also found in Mozambique. Its ecology is well known, and studies have shown *P. longicosta* to be a territorial limpet, that has a mutually beneficial relationship with the crustose brown alga *Ralfsia verrucosa*, maintaining an algal garden around its home scar, an indentation in the rock to which it returns.

RELATED SPECIES

Patella barbara Linnaeus, 1758, from southern Africa, has a smaller, star-shaped shell with more numerous and sharper radial ridges than *P. longicosta*; *P. chapmani* Tenison-Woods, 1876, from southern Australia, also has a smaller, star-shaped shell but with eight axial ridges that are rounded; and *P. cochlear* Born, 1778, from southern Africa, has a spoon or teardrop shape.

The shell of the Long-ribbed Limpet is large, thick, solid, and star-shaped. There are 10 major ridges that project radially as sharp points of a star, as well as a few smaller intermediate ridges. The shell is low and broad. The shell color is dark or light brown, and the shell is often encrusted with algae. The interior is nacreous white, with the central muscle scar brownish.

Actual size

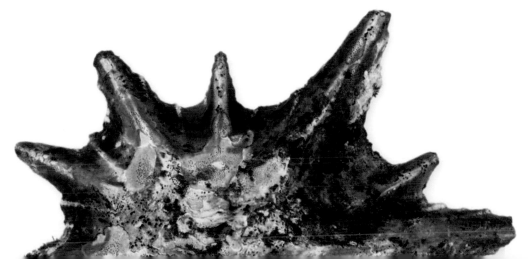

FAMILY	Patellidae
SHELL SIZE RANGE	¼ to 2¾ in (20 to 70 mm)
DISTRIBUTION	Endemic to southern coast of Africa
ABUNDANCE	Abundant
DEPTH	Intertidal
HABITAT	Rocky shores
FEEDING HABIT	Grazer, feeds on encrusting coralline algae
OPERCULUM	Absent

SHELL SIZE RANGE
¼ to 2¾ in
(20 to 70 mm)

PHOTOGRAPHED SHELL
2⅝ in
(65 mm)

178

PATELLA COCHLEAR
SPOON LIMPET
BORN, 1778

The shell of the Spoon Limpet is conic and low, with the anterior part of the shell pinched, giving it its distinctive spoon or teardrop shape, despite the variable and irregular outline. The dorsum has large and small radial grooves and is often eroded and encrusted. Unlike keyhole limpets, it does not have a hole at the apex. The interior of the shell is glossy, light gray to bluish to brown, with a dark gray or black horseshoe-shaped muscle scar.

Patella cochlear has a unique spoon-shaped shell. Because it is one of the most common of the several species in South Africa and easily accessible, its ecology has been well studied. It is a grazer of encrusting coralline algae, and like some other species of limpets, it has a territorial behavior and defends its grazing grounds from other limpets. Limpets have a broad muscular foot that keeps them strongly attached to rocks in high-energy shores, making them some of the most conspicuous mollusks on intertidal rocky shores worldwide. Several limpets are used as food, and some are considered delicacies. Because of their slow growth and overharvesting, many are endangered and their fisheries now regulated.

RELATED SPECIES

The popular name "limpet" refers to several groups of mollusks that have similar shell shapes but have evolved independently. The term "true limpets" refers only to gastropods in the order Patellogastropoda, including those in the family Patellidae, such as *Patella longicosta* Lamarck, 1819, from South Africa, and *P. flexuosa* Quoy and Gaimard, 1834, from the Indo-Pacific. The largest living limpet is *P. mexicana* Broderip and Sowerby I, 1829, from Baja California to Peru—this species is now seriously threatened.

Actual size

FAMILY	Patellidae
SHELL SIZE RANGE	1½ to 4½ in (40 to 120 mm)
DISTRIBUTION	Namibia and South Africa
ABUNDANCE	Common
DEPTH	Shallow subtidal to 23 ft (7 m)
HABITAT	Kelp forest
FEEDING HABIT	Grazer, feeds on epiphytic algae
OPERCULUM	Absent

SHELL SIZE RANGE
1½ to 4½ in
(40 to 120 mm)

PHOTOGRAPHED SHELL
2¾ in
(69 mm)

PATELLA COMPRESSA

KELP LIMPET

LINNAEUS, 1758

Patella compressa is easily recognized by its uniquely elongated shell. Instead of living on rocks, like most limpets, it has adapted to live on the kelp *Ecklonia maxima*, the largest of the marine brown algae. *Patella compressa* feeds on epiphytic algae growing on the surface of the kelp, and it travels up and down the stipe (stem or stalk), protecting its territory from other limpets. Adults live on the stipes of kelp, while juveniles usually aggregate on the fronds, thus avoiding competition. As the shell grows, its base grows concavely to fit around the cylindrical stipe.

RELATED SPECIES

Patella miniata Born, 1778 and *P. granatina* Linnaeus, 1758, both from South Africa, have similar shells with an oval outline and many radial ribs, but the former has a white muscle scar and the latter a dark brown one; and *P. vulgata* Linnaeus, 1758, from western Europe and the Mediterranean, has a coarsely ribbed and variable shell, with some specimens having a high profile.

Actual size

The shell of the Kelp Limpet is large, elongately ovate, and laterally compressed. It can have a moderately high profile, with a subcentral apex that is posteriorly recurved. Juvenile shells are less elongated but become increasingly more elongated as they grow. The sculpture consists of fine radial ribs and faint spiral lirae, with the area near the apex smooth. The exterior color is pale orange-brown to gray, matching the color of the kelp; the interior is grayish white, and near the margin the shell is orange.

FAMILY	Patellidae
SHELL SIZE RANGE	6 to 14 in (150 to 350 mm)
DISTRIBUTION	Baja California to Peru
ABUNDANCE	Formerly common
DEPTH	Intertidal to shallow subtidal
HABITAT	Rocky shores
FEEDING HABIT	Grazer, feeds on algae
OPERCULUM	Absent

SHELL SIZE RANGE
6 to 14 in
(150 to 350 mm)

PHOTOGRAPHED SHELL
12 in
(301 mm)

180

PATELLA MEXICANA
GIANT MEXICAN LIMPET
BRODERIP AND SOWERBY I, 1829

The shell of the Giant Mexican Limpet is large, very thick, heavy, broad, and ovate in shape. Juvenile shells have ribs projecting beyond the margins, but as the shell grows it becomes smoother in outline. Adult shells are usually eroded and chalky white. The shell is slightly narrower at the anterior than the posterior end, and has a low profile. The shell color is white, as is the interior. The muscle scar is light brown and horseshoe-shaped.

Patella mexicana is the largest of all limpets, growing to over 12 in (305 mm) in length, although most shells found in the last 30 years are less than half of that size. It was formerly a common species, but because its flesh is considered a delicacy, it has been overfished and is now endangered. It is protected throughout Mexico. It is apparently extinct in the Gulf of California, and only small shells are currently found in most of its range. It now survives only in hard-to-reach places, where it is protected by high-energy waves. The animal is black with some white markings. This limpet was used to make shell pendants or *oyohualli*, by the Toltecs.

RELATED SPECIES
Patella longicosta Lamarck, 1819, from southern Africa, has a star-shaped shell with long radial ridges; *P. argenvillei* Krauss, 1848, from Namibia to South Africa, has a shell similar to, but smaller than, that of *P. mexicana*, with fine radial ribs; and *P. kermadecensis* Pilsbry, 1894, from the Kermadec Islands, New Zealand, has a large and broad shell.

Actual size

FAMILY	Nacellidae
SHELL SIZE RANGE	1¼ to 3¼ in (30 to 80 mm)
DISTRIBUTION	Japan to Korea and Taiwan
ABUNDANCE	Common
DEPTH	Intertidal
HABITAT	Rocky shores
FEEDING HABIT	Grazer, feeds on algae
OPERCULUM	Absent

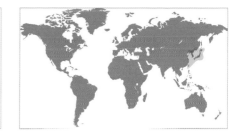

SHELL SIZE RANGE
1¼ to 3¼ in
(30 to 80 mm)

PHOTOGRAPHED SHELL
1¼ in
(31 mm)

CELLANA NIGROLINEATA

BLACK-LINED LIMPET

(REEVE, 1854)

181

Cellana nigrolineata is a common intertidal limpet ranging from Japan to Korea and Taiwan. Its name notwithstanding, the shell often has orange stripes (like the shell depicted here), although it can also have black stripes. The shell is variable in shape and coloration. Like most nacellids, *C. nigrolineata* is edible and harvested for food, and is considered a delicacy. Some nacellids retain their eggs in the mantle cavity and brood their young. There are some 40 living species in the family Nacellidae worldwide, with most in the Indo-Pacific. Although some species live on large seaweeds, the majority are intertidal grazers on rocky shores.

RELATED SPECIES

Cellana mazatlandica (Sowerby I , 1839), which is endemic to the Bonin Islands, Japan, has a large, oval shell with about 40 high, rounded, radial ribs. The shell margin is scalloped, and light brown or orange in color. *Nacella clypeater* (Lesson, 1831), which ranges from Peru to Chile, has an oval to circular shell, with a low conical profile. It has many low radial ribs, and a nacreous interior with a dark center.

Actual size

The shell of the Black-lined Limpet is medium in size, thin, lightweight, and conical in outline. Its apex is relatively high, off-centered, and closest to the anterior margin. The sculpture consists of low radial ribs and fine concentric growth lines; it may be glossy or rough. The shell color varies from a pale yellow background with black radial lines to a gray background with orange radial lines. The interior is nacreous, with an orange-brown center.

FAMILY	Nacellidae
SHELL SIZE RANGE	½ to 1⅞ in (14 to 48 mm)
DISTRIBUTION	Argentina and southern Chile
ABUNDANCE	Common
DEPTH	Intertidal to 330 ft (100 m)
HABITAT	Rocky shores
FEEDING HABIT	Grazer, feeds on algae
OPERCULUM	Absent

SHELL SIZE RANGE
½ to 1⅞ in
(14 to 48 mm)

PHOTOGRAPHED SHELL
1⅝ in
(43 mm)

182

NACELLA MYTILINA

MYTILINE LIMPET

(HELBLING, 1779)

Nacella mytilina is a thin-shelled yet sturdy limpet that lives in the intertidal zone but can also be found in deeper waters. It is a grazer, and scrapes algae off rocky shores with its long radula, the rasping organ of mollusks, which has hundreds of minute teeth. In some species of nacellids, the radular ribbon can be five times longer than the shell length, because of the wear and tear of the radular teeth. Species in the genus *Nacella* are restricted to subantarctic and Antarctic regions, while the other nacellids are usually found in tropical regions. The tip of South America contains the highest diversity of *Nacella* species.

RELATED SPECIES

Nacella fuegiensis (Reeve, 1855), from Tierra del Fuego, the southern tip of South America, between Argentina and Chile, has a medium-sized shell with 40 to 60 radial ribs. Its color is variable, from dark brown to red-brown. *Cellana nigrolineata* (Reeve, 1854), from Japan to Taiwan, has an oval shell with either black or orange radial stripes.

Actual size

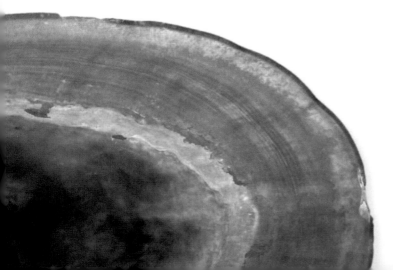

The shell of the Mytiline Limpet is medium in size, thin, translucent, and elongate-oval in outline, with a moderately low height. Its apex is small, curved, and pointed, and located anteriorly. The surface is smooth and glossy, with fine radial riblets crossed by fine concentric growth lines. The shell color is light brown to yellow-brown, and covered by a thin periostracum; the interior of the shell is gray, with a light brown center.

FAMILY	Nacellidae
SHELL SIZE RANGE	1¾ to 3½ in (45 to 90 mm)
DISTRIBUTION	Endemic to Bonin Islands, Japan
ABUNDANCE	Abundant
DEPTH	Intertidal
HABITAT	Rocky shores
FEEDING HABIT	Grazer, feeds on algae
OPERCULUM	Absent

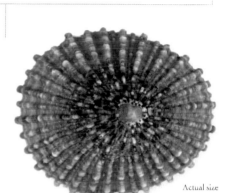

SHELL SIZE RANGE
1¾ to 3½ in
(45 to 90 mm)

PHOTOGRAPHED SHELL
2⅛ in
(53 mm)

CELLANA MAZATLANDICA
BONIN ISLAND LIMPET
(SOWERBY II, 1839)

183

Cellana mazatlandica is a limpet with a very limited distribution; it is restricted to the Bonin Islands (Ogasawara Islands), an archipelago of 30 islands located about 620 miles (1,000 km) directly south of Tokyo, and halfway between Tokyo and Guam. It is locally abundant, living on intertidal rocks. Field studies suggest its life span is about three to four years. In 1966, an attempted introduction into Guam to develop its fishery failed, but transplant experiments within the Bonin Islands have been successful. The species name was given in error for Mazatlán, in western Mexico. A better name, *boninensis*, was proposed by Pilsbry in 1891, but *mazatlandica* has priority.

RELATED SPECIES

Cellana testudinaria (Linnaeus, 1758), from the tropical Indo-West Pacific, has a semitranslucent shell with weak sculpture. *Cellana sandwicensis* Pease, 1861, which is endemic to Hawaii and known as "opihi," resembles *C. mazatlandica* in having strong radial ribs, but it is smaller and more elongate. Because it is considered a delicacy, the opihi fishery is regulated in order to protect the species from being overharvested.

Actual size

The shell of the Bonin Island Limpet is medium in size, relatively thin, conic, and ovate. The apex is acute and located slightly off-center, closer to the anterior end. The sculpture consists of about 40 strong and scaly radial ridges that alternate with intermediary ones, and concentric growth lines that coincide with concentric bands of varying colors. The margin of the shell is scalloped. The shell color is usually dull light brown to orange, and the interior silvery with a light brown margin and a brown muscle scar.

FAMILY	Nacellidae
SHELL SIZE RANGE	1⅜ to 4 in (35 to 100 mm)
DISTRIBUTION	Tropical Indo-West Pacific
ABUNDANCE	Abundant
DEPTH	Intertidal and shallow subtidal
HABITAT	Rocky shores
FEEDING HABIT	Grazer, feeds on algae
OPERCULUM	Absent

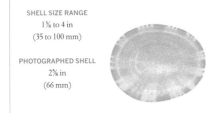

SHELL SIZE RANGE
1⅜ to 4 in
(35 to 100 mm)

PHOTOGRAPHED SHELL
2⅜ in
(66 mm)

184

CELLANA TESTUDINARIA

COMMON TURTLE LIMPET

(LINNAEUS, 1758)

The shell of the Common Turtle Limpet is large, semitranslucent, low, conical, and ovately rounded. The apex is on the midline of the shell, about a third of the distance from the anterior end. The sculpture is weak, with many low, rounded radial riblets and concentric growth marks. The shell color is greenish brown with dark brown radial rays, forming zigzag or chevron marks. The interior is bluish silver with the muscle scar whitish gray to yellow-brown.

Cellana testudinaria is a common limpet from the tropical southwestern Pacific. It lives on volcanic rocks in the intertidal zone or just below the tide line, on exposed shores. The radula has few but robust and elongate teeth, which are tipped with iron oxides to increase resistance to abrasion when scraping algae from rocks. In some species of *Cellana,* the radular ribbon can be very long, up to five times the length of the shell, the highest radular-length-to-shell-length ratio among the gastropods. Some nacellids brood their young in a brood pouch under the shell; the young crawl away when the eggs hatch. The Common Turtle Limpet is used as food by humans throughout its range.

RELATED SPECIES

Cellana solida (Blainville, 1825), from southern Australia, has a scalloped margin and strong, rounded radial ribs; and *C. mazatlandica* (Sowerby II, 1839), from the Indo-Pacific, has a broad shell that is similar to *C. solida*, but with a light brown central scar and shell margin.

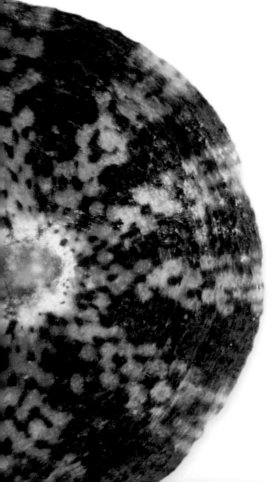

Actual size

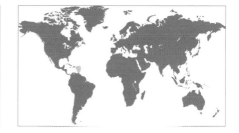

FAMILY	Acmaeidae
SHELL SIZE RANGE	⅝ to ⅞ in (15 to 21 mm)
DISTRIBUTION	Lesser Antilles, West Indies
ABUNDANCE	Uncommon
DEPTH	1,365 to 3,500 ft (415 to 1,050 m)
HABITAT	Coarse sand and broken shells
FEEDING HABIT	Grazer
OPERCULUM	Absent

SHELL SIZE RANGE
⅝ to ⅞ in
(15 to 21 mm)

PHOTOGRAPHED SHELL
⅞ in
(21 mm)

PECTINODONTA ARCUATA

ARCUATE PECTINODONT

DALL, 1882

185

Pectinodonta arcuata is a small, deepwater limpet from St. Thomas and St. Lucia, in the West Indies. It has been found on bottoms ranging from coarse sand, shell hash, and lava sand, to mud. The animal is blind, and has fewer teeth per row in the radula than other acmaeids. *Pectinodonta arcuata* is the type species of the genus *Pectinodonta*. A few other species are recognized in the genus, with the highest diversity found in New Zealand and Australia. Many species previously included in the family Acmaeidae have been reclassified to the family Lottiidae. Therefore, there are now only about ten living species in the family Acmaeidae.

RELATED SPECIES

Acmaea subrotundata Carpenter, 1865, from the Pacific side of Nicaragua to Panama, has a small, nearly circular, brown shell, with the apex located at the center or slightly off-centered. *Acmaea mitra* (Rathke, 1833), which ranges from Alaska to Baja California, has a small, high, conical, white shell. Its surface may be smooth or may have concentric growth marks.

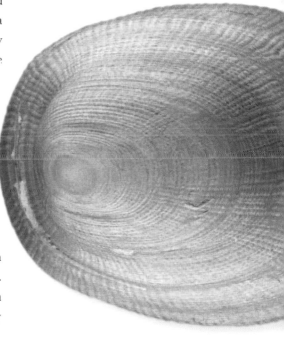

Actual size

The shell of the Arcuate Pectinodont is small, thin but solid, and elongately oval in outline, with a high apex. Its apex is rounded and polished, facing the anterior margin; the sides of the slope are convex. The sculpture consists of a large number of radial riblets crossed by concentric growth lines, forming a cancellate pattern. The margin of the shell can be smooth or slightly denticulate. The shell color is off-white, and the interior is white, with a beige margin and center.

FAMILY	Acmaeidae
SHELL SIZE RANGE	¾ to 1½ in (19 to 38 mm)
DISTRIBUTION	Alaska to Baja California
ABUNDANCE	Common
DEPTH	Low intertidal to 100 ft (30 m)
HABITAT	Rocky bottoms
FEEDING HABIT	Grazer
OPERCULUM	Absent

SHELL SIZE RANGE
¾ to 1½ in
(19 to 38 mm)

PHOTOGRAPHED SHELL
1⅛ in
(29 mm)

186

ACMAEA MITRA
WHITE-CAP LIMPET
(RATHKE, 1833)

Actual size

Acmaea mitra has a taller shell than most acmaeids; its shell height can reach about 80 percent of the shell length. It ranges from Alaska to Baja California, and lives in cold water from the lower intertidal zone to offshore. It grazes on algae that grow on rocky bottoms. Empty shells are often washed ashore. Acmaeids are distributed worldwide, ranging from the lower intertidal zone to the deep ocean. Most acmaeids live on rocky bottoms, but some species in the genus *Pectinodonta* are found on wood and vestimentiferan (deepsea worms) tubes. Acmaeids are more closely related to the deepwater limpets in the family Lepetidae than the Lottiidae, which dwell in shallower water.

RELATED SPECIES

Bathyacmaea nipponica Okutani, Tsuchida, and Fujikura, 1992, from Sagami Bay, Japan, lives in deep water, ranging from 3,600 to 3,950 ft (1,100 to 1,200 m). It has a small shell with a relatively high apex, an oval outline, and many broad radial ribs. *Pectinodonta arcuata* Dall, 1882, is a deepwater limpet that lives in the Lesser Antilles. It has a thin shell with an elongate-oval outline, a high apex, and convex sides.

The shell of the White-cap Limpet is medium in size, thick, oval in outline, and high-conical in profile. Its apex is pointed, located near the center, and closest to the anterior margin. The anterior side of the slope is slightly convex, and the posterior straight or slightly concave. The sculpture consists of concentric growth lines, but they are often covered by growths of coralline algae. The margin of the shell is thin and sharp. The shell color is chalky white, and the interior is gray.

FAMILY	Lepetidae
SHELL SIZE RANGE	½ to 1 in (11 to 25 mm)
DISTRIBUTION	North Japan and Bering Sea to Washington State
ABUNDANCE	Abundant
DEPTH	Shallow subtidal to 480 ft (145 m)
HABITAT	Rocks in muddy areas
FEEDING HABIT	Deposit feeder
OPERCULUM	Absent

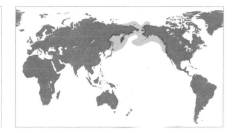

PHOTOGRAPHED SHELL
⅞ in
(21 mm)

CRYPTOBRANCHIA CONCENTRICA

RINGED BLIND LIMPET

(MIDDENDORFF, 1851)

187

Cryptobranchia concentrica is one of the largest and most common species in the family Lepetidae. It lives in cold waters of the northern Pacific and Arctic oceans, ranging from shallow subtidal to deeper offshore waters. As the popular name suggests, the animal is blind. It also lacks gills as well as the chemosensory organ (the osphradium) that is used by other gastropods to locate food. Unlike most limpets, it is believed to feed on deposits rather than graze on algae. There are only a few living species, perhaps less than 20, in the family Lepetidae. Most species live in cold waters worldwide, usually in the deep sea.

RELATED SPECIES

Limalepeta lima Dall, 1918, which ranges from the Kurile Islands to Hokkaido, northern Japan, has a large shell for the family, reaching about 1¼ in (30 mm) in length. It has an oval outline and a low, conical shell. *Lepeta fulva* (Müller, 1776), from Scandinavia to the Azores, has a very small shell that is often red or orange in color. It ranges from shallow subtidal waters to the deep sea.

Actual size

The shell of the Ringed Blind Limpet is small, thin but strong, oval in outline, and has a relatively low apex that is rounded and may or may not be pointed. It is directed anteriorly, and located closer to the anterior margin. The surface consists of concentric growth lines, which can be fine or frilly, and fine radial lines. The shell margin is smooth and thin. The shell color is off-white, the interior is porcelaneous white.

FAMILY	Lottiidae
SHELL SIZE RANGE	⅜ to 1½ in (10 to 38 mm)
DISTRIBUTION	Southern Alaska to Baja California
ABUNDANCE	Abundant
DEPTH	Intertidal to shallow subtidal
HABITAT	Stalks or holdfasts of seaweeds
FEEDING HABIT	Grazer, feeds on algae
OPERCULUM	Absent

SHELL SIZE RANGE
⅜ to 1½ in
(10 to 38 mm)

PHOTOGRAPHED SHELL
½ in
(13 mm)

188

LOTTIA INSESSA
SEAWEED LIMPET
(HINDS, 1842)

Lottia insessa is an abundant limpet that lives only on the stalks and holdfasts of the feather boa kelp, *Egregia menziesii*. It grazes on other algae growing on the kelp, as well as the kelp itself. It ranges from southern Alaska to Baja California, but it is uncommon north of Oregon. This species experiences high mortality in the winter. Even the largest specimens are generally less than a year old. There are about 100 living species in the family Lottiidae worldwide, with the highest diversity on the west coast of North America. Most lottiids are grazers on algae growing on the rocky intertidal zone.

RELATED SPECIES

Patelloida saccharina (Linnaeus, 1758), from the Indo-West Pacific, has a very variable shell, with projecting radial ribs that give it a star-shaped outline. *Lottia gigantea* (Sowerby I, 1834), which ranges from Washington State to Baja California, is the largest species in the family Lottiidae. It is a key species in its habitat, and if removed, the diversity of limpets and algae will eventually decrease.

Actual size

The shell of the Seaweed Limpet is medium in size, relatively thick, somewhat glossy, and varying in outline from oval to elongate-oval. Its apex is rounded and located closer to the anterior margin. The shell height can be as much as three-quarters of the shell length. The sculpture consists of fine radial lines crossed by fine concentric growth lines, and is mostly smooth. The shell color is light to reddish brown, and the interior is similar but lighter near the margin, and darker at the center.

FAMILY	Lottiidae
SHELL SIZE RANGE	⅝ to 2 in (15 to 50 mm)
DISTRIBUTION	Tropical Indo-West Pacific
ABUNDANCE	Abundant
DEPTH	Intertidal to 20 ft (6 m)
HABITAT	Rocky shores
FEEDING HABIT	Grazer
OPERCULUM	Absent

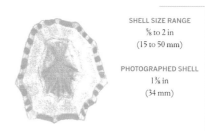

SHELL SIZE RANGE
⅝ to 2 in
(15 to 50 mm)

PHOTOGRAPHED SHELL
1⅜ in
(34 mm)

PATELLOIDA SACCHARINA
PACIFIC SUGAR LIMPET
(LINNAEUS, 1758)

Patelloida saccharina is an abundant limpet that lives in intertidal to shallow subtidal depths along rocky shores, especially along exposed coasts. It ranges from Sri Lanka to Melanesia, and from Japan to Australia. Its shell is star shaped, and is used for shellcraft in the Philippines. Lottiids and other related limpets are considered among the most primitive gastropods, based on their anatomy. However, not all gastropods with a limpetlike shell are closely related. This shape is evolved independently in several groups in response to a similar habitat. *Cheilea equestris* from the family Hipponicidae and *Concholepas concholepas* from the family Muricidae are but two examples.

RELATED SPECIES

Patelloida conulus Dunker, 1871, from Korea, has a small shell with an oval outline, and a very elevated spire that can be almost as high as the shell is long. *Lottia lindbergi* Sasaki and Okutani, 1994, from Japan and Korea, has a similar size as *P. saccharina*, but with an oval shell with a pointed apex and weak riblets in juveniles that grow smoother with growth.

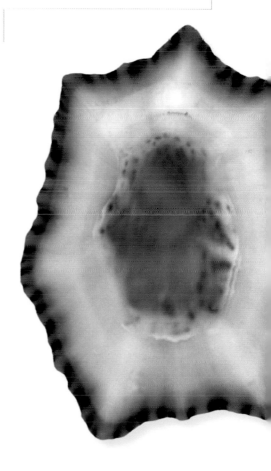

Actual size

The shell of the Pacific Sugar Limpet is medium in size, solid, opaque, and shaped like a webbed foot. Its apex is relatively high, located near the center of the shell, and often eroded. The sculpture consists of 7 to 9 strong, raised, radial ribs that project beyond the margin, and smaller intermediary ribs. The shell color is variable, ranging from dark to light colored, with or without decorations between major ribs, and the inner surface is white with a dark margin.

FAMILY	Lottiidae
SHELL SIZE RANGE	1½ to 4½ in (38 to 121 mm)
DISTRIBUTION	Washington State to Baja California
ABUNDANCE	Common
DEPTH	High to middle intertidal
HABITAT	Rocky shores
FEEDING HABIT	Grazer
OPERCULUM	Absent

SHELL SIZE RANGE
1½ to 4½ in
(38 to 121 mm)

PHOTOGRAPHED SHELL
3¼ in
(82 mm)

190

LOTTIA GIGANTEA
GIANT OWL LIMPET
(SOWERBY 1, 1834)

The shell of the Giant Owl Limpet is large, light, ovate, and low in profile. The apex is located at about one-eighth of the shell length from the anterior end. The exterior of the shell is often eroded, but the maculated pattern can usually be seen near the margins. The interior central area is light brown; the muscle scar is bluish white and horseshoe-shaped, opened anteriorly. The interior margin has a brown band.

Because of its comparatively small genome size, *Lottia gigantea* is being actively studied and will likely be one of the first mollusks to have its entire genome sequenced. It has a relatively small genome (about 500 million base pairs) compared with other mollusks (average about 1.8 billion base pairs). It is a territorial animal, aggressively protecting about 155 sq in (1,000 cm²) of rock, where it grows a "garden" of algae near its home scar. *Lottia gigantea* is a key species in its habitat, and when removed the densities of other limpets temporarily increase until they deplete the algal film. Most then perish.

RELATED SPECIES

Tectura scutum (Rathke, 1833), from Alaska to Baja California, has a subcentral apex, a maculated shell, and a low profile; *Scurria scurra* (Lesson, 1830), from Peru to the Falkland Islands, has a small, smooth shell with a high profile; and *Patelloida alticostata* Angas, 1865, from southern Australia, has a small shell with about 20 rounded radial ribs.

Actual size

FAMILY	Cocculinidae
SHELL SIZE RANGE	⅛ to ¼ in (3 to 5 mm)
DISTRIBUTION	Puerto Rico Trench
ABUNDANCE	Rare
DEPTH	17,000 to 28,200 ft (5,200 to 8,600 m)
HABITAT	Sunken wood
FEEDING HABIT	Grazer
OPERCULUM	Absent

SHELL SIZE RANGE
⅛ to ¼ in
(3 to 5 mm)

PHOTOGRAPHED SHELL
¼ in
(5 mm)

MACLEANIELLA MOSKALEVI

MOSKALEV'S MACLEANIELLA

LEAL AND HARASEWYCH, 1999

191

Macleaniella moskalevi is the deepest-dwelling known mollusk.
It has been collected at depths of 28,200 ft (8,600 m), close to the
deepest part of the Puerto Rico Trench, the deepest region in
the Atlantic Ocean. The sides of the Puerto Rico Trench drop
off very steeply. Wood and decaying plant matter from the
surrounding islands sink into the trench regularly. *Macleaniella
moskalevi* has been collected on fragments of wood that float out
to sea, then become waterlogged and sink. There are about 50
living species in the family Cocculinidae worldwide; many live
in the deep sea, including some in the abyssal and hadal zones,
but others occur in shallower waters.

Actual size

RELATED SPECIES

Fedikovella caymanensis (Moskalev, 1976), from the
Cayman Trench, between Cayman Island and Jamaica,
is the next deepest-dwelling known cocculinid. It has
a smaller shell with a cancellate sculpture, and lives at a
slightly shallower depth. *Cocculina japonica* Dall, 1907, from
off Japan, has a larger shell with an elongate-oval outline. The
apex is small and located centrally; the sculpture has fine
radial punctuation and fine concentric growth lines.

The shell of the Moskalev's Macleaniella is very
small, thin, arched, elevated, and elongately oval
in outline. Its apex is recurved, and positioned
posteriorly, below the highest point, which is
equivalent to about half of the shell length in
height. The surface is apparently smooth, but
examination by electron microscope reveals a
fine cancellate sculpture formed by low radial
threads and fine concentric growth marks. The
aperture is broad, the margin smooth, and there
is a septum inside the aperture.

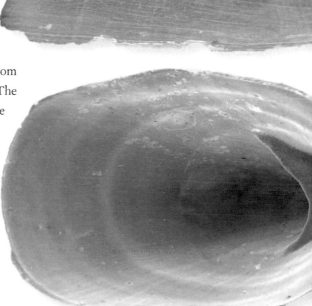

FAMILY	Addissoniidae
SHELL SIZE RANGE	⅜ to ¾ in (8 to 20 mm)
DISTRIBUTION	Mediterranean to the Azores; northeast U.S.A.
ABUNDANCE	Uncommon
DEPTH	230 to 6,000 ft (70 to 1,830 m)
HABITAT	Sandy bottoms
FEEDING HABIT	Detritivore
OPERCULUM	Absent

SHELL SIZE RANGE
⅜ to ¾ in
(8 to 20 mm)

PHOTOGRAPHED SHELL
¾ in
(19 mm)

192

ADDISONIA EXCENTRICA
PARADOXICAL BLIND LIMPET
(TIBERI, 1857)

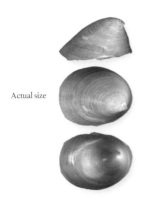

Actual size

Addisonia excentrica is a small, deepwater, limpetlike snail. It has a broad distribution, ranging from the Mediterranean to the Azores and in the northeastern U.S.A. It has a moderately high, conical shell with a thin periostracum. *Addisonia excentrica* has a large, asymmetrical gill, a relatively simple radula, and no eyes. It lives on and within hatched egg cases of sharks and skates, known as mermaid's purses, and feeds on organic remains. If you find these egg cases, check inside for this uncommon gastropod. There are only four living species in the family Addisoniidae worldwide.

The shell of the Paradoxical Blind Limpet is small, thin, fragile, and relatively high. It is asymmetric and limpetlike. The apex is pointed, curved, and located posteriorly, below the highest part of the shell. The larval shell breaks off during early development. The sculpture consists of fine, concentric growth lines, and the surface is smooth inside. The aperture is broadly oval, and the margin smooth. The shell is white, both inside and out.

RELATED SPECIES

Addisonia brophyi McLean, 1985, from the eastern Pacific, has a much smaller but similar shell. *Addisonia enodis* Simone, 1996, from southeastern Brazil, has been found at a depth of 260 ft (80 m) on sandy bottoms. It has a shell similar to *A. excentrica*, but with a more rounded apex and lower shell.

FAMILY	Caymanabyssiidae
SHELL SIZE RANGE	up to ⅛ in (3 mm)
DISTRIBUTION	Cayman Trench, Caribbean
ABUNDANCE	Rare
DEPTH	Hadal zone, from 22,110 to 22,300 ft (6,740 to 6,800 m)
HABITAT	Sediments with animal and plant debris
FEEDING HABIT	Herbivore, feeds on plant debris
OPERCULUM	Absent

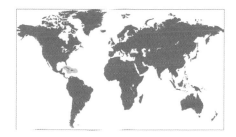

SHELL SIZE RANGE
up to ⅛ in
(3 mm)

PHOTOGRAPHED SHELL
⅛ in
(3 mm)

CAYMANABYSSIA SPINA

CAYMANABYSSIA SPINA

MOSKALEV, 1976

193

Caymanabyssia spina is one of the deepest-dwelling mollusks. It lives in depths that are greater than any mountains in North America and Europe are high. Only a handful of other mollusks are known to live in deeper waters. *Caymanabyssia spina* is a very small, limpetlike gastropod that has a shell covered by a thick organic layer, the periostracum. Because of the great depth and pressure, the seawater would dissolve the calcareous shell if it were exposed. The animal lacks eyes, since there is no light in its habitat; instead, it uses short tentacles positioned near the mouth to locate its food. It feeds on bacterial films on the surface of the wood (originating on land) from the surrounding shallower regions.

RELATED SPECIES

Amphiplica plutonica Leal and Harasewych, 1999, also from abyssal depths in the Cayman Trench, is a larger member of the family Caymanabyssiidae, but still small, reaching less than ½ in (13 mm) in length. *Copulabyssia riosi* Leal and Simone, 2000, from southern Brazil, is another limpetlike gastropod from the abyssal plain, at depths of 4,300 ft (1,320 m).

The shell of *Caymanabyssia spina* is very small, thin, and limpet-shaped, with an ovate outline. Its shell profile is low, and the larval shell is rounded and smooth, closer to the posterior margin. The rest of the shell is covered by short spines arranged in concentric ovals. The periostracum is thick, and has one of the highest organic protein contents among gastropods. The inside of the shell is smooth. The shell color is yellowish white, and the interior is white.

Actual size

FAMILY	Pseudococculinidae
SHELL SIZE RANGE	up to ⅛ in (3 mm)
DISTRIBUTION	Bahamas
ABUNDANCE	Rare
DEPTH	Bathyal zone, 1,700 ft (520 m)
HABITAT	Mud and debris bottoms
FEEDING HABIT	Herbivore, feeds on plant debris
OPERCULUM	Absent

SHELL SIZE RANGE
up to ⅛ in
(3 mm)

PHOTOGRAPHED SHELL
⅛ in
(3 mm)

194

NOTOCRATER YOUNGI
YOUNG'S FALSE COCCULINA
MCLEAN AND HARASEWYCH, 1995

Notocrater youngi is a very small, limpet-like gastropod from deep waters off the Bahamas. Like other deepwater limpets, because it lives in darkness it lacks eyes; instead, it depends on other sensory organs, such as cephalic tentacles covered with minute cilia. It feeds on plant debris that comes from shallower waters, such as algal holdfasts, or decaying wood from nearby shores. *Notocrater youngi* lives above the carbonate compensation depth, so its thin calcareous shell will not dissolve in seawater. Therefore, its shell has only a thin periostracum. The shells of mollusks that live below the carbonate compensation depth would dissolve in that environment, and require the protection of an intact periostracum.

RELATED SPECIES

Notocrater houbricki McLean and Harasewych, 1995, is also from off the Bahamas, but from a shallower depth (1,350 ft or 410 m). It is slightly smaller than *N. youngi*, and has the apex more centrally located. *Kaiparapelta askewi* McLean and Harasewych, 1995, from off Charleston, South Carolina, also has a very small, but broader shell. The shell and periostracum are thin, and its sculpture consists of weak growth lines.

The shell of Young's False Cocculina is very small, thin, ovate in outline, and moderately high. The larval shell is smooth to the naked eye, and located closer to posterior margin. The sculpture near the apex consists of weak concentric ribs and fine radial striae; it changes to beaded concentric lines nearer the margin of the shell. The shell edge is thin and sharp. The muscle scar is not well marked inside the shell. The shell color is white, and the periostracum is thin.

Actual size

FAMILY	Peltospiridae
SHELL SIZE RANGE	less than ⅛ to ¼ in (2 to 5 mm)
DISTRIBUTION	Endemic to hydrothermal vents, East Pacific Rise
ABUNDANCE	Rare
DEPTH	8,645 ft (2,635 m)
HABITAT	Restricted to hydrothermal vents
FEEDING HABIT	Detritivore
OPERCULUM	Corneous, rounded, multispiral

SHELL SIZE RANGE
less than ⅛ to ¼ in
(2 to 5 mm)

PHOTOGRAPHED SHELL
less than ⅛ in
(2 mm)

PACHYDERMIA LAEVIS

SMOOTH PACHYDERM SHELL

WARÉN AND BOUCHET, 1989

195

Pachydermia laevis is a tiny gastropod that is endemic to hydrothermal vents in the deep sea along the East Pacific Rise. It is associated with the tubes of the polychaete worm, *Alvinella pompejana*, which lives in extreme depth and temperatures, near the hot vents. Peltospirids have separate sexes, but the males lack a penis; the gametes are released into the water and find their way to the female. Fertilization occurs in the ovary. Currently there are about 15 living species in the family Peltospiridae. All are restricted to hydrothermal vents in the Pacific Ocean.

RELATED SPECIES

Depressigyra planispira, *Solitigyra reticulata*, and *Lirapex humata*, all described by Warén and Bouchet, 1989, are also restricted to hydrothermal vents in the East Pacific Rise. There are likely more species to be discovered, since the deep sea is still poorly known, and only a small portion of it has been sampled.

The shell of the Smooth Pachyderm Shell is very small, thin, fragile, and coiled, with a relatively tall spire. Its spire and body whorls are coiled and rounded. However, the last quarter whorl is more loosely wound, and does touch the previous whorls. The sculpture is smooth to the naked eye, but shows fine growth lines under the scanning electron microscope. The aperture is circular, and the lip smooth. The shell color is white, and the aperture is not nacreous.

Actual size

FAMILY	Pleurotomariidae
SHELL SIZE RANGE	1 ¾ to 2 ½ in (45 to 66 mm)
DISTRIBUTION	Southern Gulf of Mexico to West Indies
ABUNDANCE	Rare
DEPTH	420 to 1,800 ft (130 to 550 m)
HABITAT	Rocky habitats in deep water
FEEDING HABIT	Carnivore, feeds on sponges and soft corals
OPERCULUM	Corneous, circular

SHELL SIZE RANGE
1 ¾ to 2 ½ in
(45 to 66 mm)

PHOTOGRAPHED SHELL
2 ⅛ in
(52 mm)

196

PEROTROCHUS QUOYANUS
QUOY'S SLIT SHELL
(FISCHER AND BERNARDI, 1856)

Perotrochus quoyanus is a rare, deepwater gastropod that feeds on sponges and soft corals. Extant species of Pleurotomariidae are considered living fossils: they look like their extinct relatives and have changed little in millions of years. The Quoy's Slit Shell was the first of the living pleurotomariids to be discovered; before that, they were known only from fossils from the age of the dinosaurs. The family is characterized by the presence of a slit along the midline of the aperture, which is used as the exhalant channel to expel water and wastes. As the shell grows, the back of the slit is filled to form a selenizone or slit band, that is visible on the earlier whorls.

RELATED SPECIES

Entemnotrochus adansonianus (Crosse and Fisher, 1861), from the Gulf of Mexico, Bermuda to Brazil, has a long slit. *Entemnotrochus rumphii* (Schepman, 1879), from Japan to the Philippines, has the largest shell in the family. The Scissurellidae, a family of tiny gastropods that also have shells with shallow slits, are distantly related to the Pleurotomariidae.

Actual size

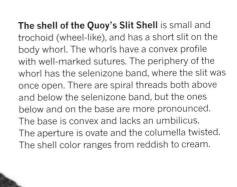

The shell of the Quoy's Slit Shell is small and trochoid (wheel-like), and has a short slit on the body whorl. The whorls have a convex profile with well-marked sutures. The periphery of the whorl has the selenizone band, where the slit was once open. There are spiral threads both above and below the selenizone band, but the ones below and on the base are more pronounced. The base is convex and lacks an umbilicus. The aperture is ovate and the columella twisted. The shell color ranges from reddish to cream.

FAMILY	Pleurotomariidae
SHELL SIZE RANGE	4 to 10 in (100 to 250 mm)
DISTRIBUTION	Southern Japan to Philippines
ABUNDANCE	Rare
DEPTH	Deep water, to 1,000 ft (300 m)
HABITAT	Rocky bottoms
FEEDING HABIT	Carnivore, feeds on sponges
OPERCULUM	Corneous, multispiral, and large

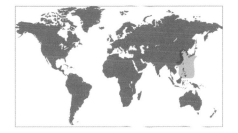

SHELL SIZE RANGE
6 to 10 in
(150 to 250 mm)

PHOTOGRAPHED SHELL
4½ in
(114 mm)

ENTEMNOTROCHUS RUMPHII

RUMPHIUS' SLIT SHELL

(SCHEPMAN, 1879)

Entemnotrochus rumphii is the largest species of slit shell. Like other slit shells, it lives in deep water and feeds on sponges. Slit shells all have corneous opercula; some species, such as *Mikadotrochus hirasei*, have a small operculum, but all species of *Entemnotrochus* have a large operculum that plugs the aperture. The ancestors of slit shells are among the oldest gastropods, having appeared some 500 million years ago. Pleurotomariids were once more diverse, but only about 30 recent species are currently known. Because of their deepwater habitat, most species are considered rare, but recent data suggests that some are common in their habitat.

RELATED SPECIES

Perotrochus quoyanus (Fischer and Bernardi, 1856), from the southern Gulf of Mexico to the West Indies, has a small shell with a short slit. *Mikadotrochus hirasei* Pilsbry, 1903, from Japan and the Philippines, has a medium-sized shell with a tall spire and a wide but shallow slit; sometimes albino shells are found.

The shell of the Rumphius' Slit Shell is large and heavy yet fragile. It has a tall spire and a narrow and very long slit. The suture is well impressed. The spire whorls are slightly convex and the narrow selenizone divides the whorls in about half. Its sculpture comprises fine spiral lines and oblique axial lines that coincide with reddish streaks against a creamy white background. The aperture is large and nacreous, and the umbilicus is wide and deep.

Actual size

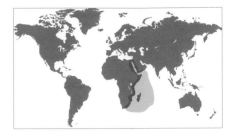

FAMILY	Scissurellidae
SHELL SIZE RANGE	less than ⅛ in (1 to 2 mm)
DISTRIBUTION	Red Sea, western Indian Ocean
ABUNDANCE	Common
DEPTH	Intertidal to 10 ft (3 m)
HABITAT	Sandy bottoms and on algae
FEEDING HABIT	Detritivore
OPERCULUM	Corneous, round and multispiral

SHELL SIZE RANGE
less than ⅛ in
(1 to 2 mm)

PHOTOGRAPHED SHELL
less than ⅛ in
(1 mm)

198

SCISSURELLA ROTA

ROTA SCISSURELLE

YARON, 1983

Scissurella rota is a common micromollusk from the Red Sea and East Africa. It is found in the intertidal zone and shallow subtidal waters, on sandy bottoms and on algae, and feeds on fine detritus. Scissurellids are easily recognized by the presence of a slit at the shoulder, and look like miniature slit shells (pleurotomariids), to which they are related. Scissurellid shells are usually white, sometimes translucent, and have fine cancellate or axial sculpture. There are at least 170 living species in the family Scissurellidae worldwide, but many species have yet to be named. Scissurellids occur from the intertidal zone to the deep sea in all oceans.

RELATED SPECIES

Scissurella coronata Watson, 1885, from Japan to Australia and Fiji, has a small, low shell with strong diagonal axial, elevated ribs that are crossed by fine spiral lines. Its slit is narrow and short. *Anatoma crispata* (Fleming, 1828), has a wide distribution, from the Mediterranean to the Azores, the West Indies, California, and Japan. It has a globose shell with a moderately high spire. Its slit is narrow and about one-quarter of a whorl long.

Actual size

The shell of Rota Scissurelle is very small, thin, fragile, globose, and coiled with a flattened spire. Its spire is small, the suture well marked, and the body whorl large. The sculpture consists of 16 radial ribs crossed by about 15 axial ribs. The slit is bordered by an elevated margin, and about one-quarter of a whorl long; the selenizone has raised crescents. The aperture is wide, the outer lip thin, the columella smooth, and the umbilicus wide and deep. The shell is white and semitranslucent.

FAMILY	Scissurellidae
SHELL SIZE RANGE	up to ⅛ in (1 to 4 mm)
DISTRIBUTION	Mediterranean, North Atlantic to West Indies, and Japan
ABUNDANCE	Common
DEPTH	50 to 2,000 ft (15 to 600 m)
HABITAT	Sandy bottoms and on algae
FEEDING HABIT	Fine detritus browser
OPERCULUM	Corneous, rounded, multispiral

SHELL SIZE RANGE
up to ⅛ in
(1 to 4 mm)

PHOTOGRAPHED SHELL
⅛ in
(4 mm)

ANATOMA CRISPATA

CRISPATE SCISSURELLE

(FLEMING, 1828)

199

Anatoma crispata is the type species of the genus *Anatoma*. It is often found on sandy and shelly bottoms, and on algae. It has a wide distribution, occurring in the Mediterranean, North Atlantic to West Indies, Japan, and Alaska to Baja California. Like the pleurotomariids, the slit in scissurellids is used to expel wastes and gametes. Most scissurellids are considered micromollusks, although a few species grow to be larger than ⅖ in (10 mm). Some species of scissurellid aggregate by the thousands for mass spawnings at night; they are attracted by light and are capable of sustained swimming.

RELATED SPECIES

Anatoma parageia Geiger and Sasaki, 2009, from Sagami Bay, Japan, is a rare representative of the genus *Anatoma* from shallow waters. Its shell is very small and low, with a relatively wide but short slit at the periphery of the whorls. *Scissurella rota* Yaron, 1983, which ranges from the Red Sea to the western Indian Ocean, has a small shell with 16 radial ribs crossed by 15 axial ribs.

The shell of Crispate Scissurelle is very small, thin, fragile, globose, and conic in outline. Its spire is moderately tall, the suture well marked, and the whorls are convex. The open slit at the periphery of the whorls is narrow, about a quarter of a whorl long, and bordered by elevated margins. The sculpture consists of fine radial ribs crossed by spiral cords. The aperture is wide, the outer lip thin, and the columella smooth, with a folded columellar margin. The umbilicus is narrow and deep. The shell color is white.

Actual size

FAMILY	Haliotidae
SHELL SIZE RANGE	¼ to 1⅛ in (7 to 28 mm)
DISTRIBUTION	Western Pacific
ABUNDANCE	Uncommon
DEPTH	Intertidal to 133 ft (40 m)
HABITAT	Coral reefs
FEEDING HABIT	Herbivore, grazes on algae
OPERCULUM	Absent

SHELL SIZE RANGE
¼ to 1⅛ in
(7 to 28 mm)

PHOTOGRAPHED SHELL
½ in
(14 mm)

200

HALIOTIS JACNENSIS
JACNA ABALONE
REEVE, 1846

Haliotis jacnensis is one of the smallest species of abalone. It lives in the tropical western Pacific, ranging from Japan to Indonesia and west to Polynesia. It is found from the intertidal zone to offshore, on hard bottoms and near coral reefs. Like all abalones, it has one spiral row of holes, or tremata; in *Haliotis jacnensis*, the holes are usually raised. Its coloration is extremely variable, ranging from brown to brightly colored red or yellow shells. The sculpture is also variable; some shells have strong, scaly radial ribs and axial growth lines, while in others the sculpture is less pronounced.

RELATED SPECIES

Haliotis pourtalesii Dall, 1881, from the Gulf of Mexico to northern South America, is an uncommon abalone usually found in deep water. It has a small shell with an orange coloration. It is rarely collected alive. *Haliotis asinina* Linnaeus, 1758, from the tropical Indo-West Pacific, has a thin, smooth, and elongate shell with the highest shell length-to-width ratio in the family.

The shell of the Jacna Abalone is small for the family, thin, low, and elongate-oval in outline. Its spire is low and close to the posterior margin, and the suture is not well marked. The sculpture usually consists of spiral rows of scaly ribs, and axial growth marks, but it can also be smoother. The internal surface is smooth and nacreous. The dorsal holes are elevated, and the last 3 or 4 are open. The shell coloration is very variable: shells can be pale brown or they may be brightly colored in red or yellow.

Actual size

FAMILY	Haliotidae
SHELL SIZE RANGE	1⅛ to 3¼ in (27 to 80 mm)
DISTRIBUTION	Red Sea, Indo-West Pacific
ABUNDANCE	Common
DEPTH	Intertidal
HABITAT	Rocky bottoms
FEEDING HABIT	Herbivore, grazes on algae
OPERCULUM	Absent

SHELL SIZE RANGE
1⅛ to 3¼
(27 to 80 mm)

PHOTOGRAPHED SHELL
1¾ in
(44 mm)

HALIOTIS VARIA

VARIABLE ABALONE

LINNAEUS, 1758

Haliotis varia is a common and widespread abalone, ranging from the Red Sea to the tropical Indo-West Pacific. It lives in the intertidal zone to shallow subtidal depths, on rocky bottoms and coral reefs, and under stones. As the name suggests, it is extremely variable in outer sculpture and coloration. Because of its wide distribution and variability, it has received many names; Linnaeus was the first to describe it, therefore *varia* has priority. The animal is collected for food and its shell is often used in shellcraft. The tremata are slightly elevated, and the last four or five holes are open.

RELATED SPECIES

Haliotis rubra Leach, 1814, which is endemic to Australia, ranging from New South Wales southward to Tasmania, is a large abalone harvested commercially. Its shell is often red-brown and ovate, with large or small axial dorsal folds. *Haliotis fulgens* Philippi, 1845, from Oregon to Baja California, is one of the largest species in the family Haliotidae. It was overfished, and now its fishery is strictly regulated; *H. fulgens* is farmed for both its meat and its shell.

The shell of the Variable Abalone is medium in size, rather inflated, and elongate-oval in outline. Its spire is low and located posteriorly, and the shoulder of the spiral whorls is pronounced. The holes are oval, slightly elevated, and the last 4 or 5 are open. The sculpture is variable, and consists of irregular radial folds crossed by low spiral ribs of varying sizes. The interior surface is smooth and silvery nacreous. The shell color is brown or greenish, and there are sometimes cream patches.

Actual size

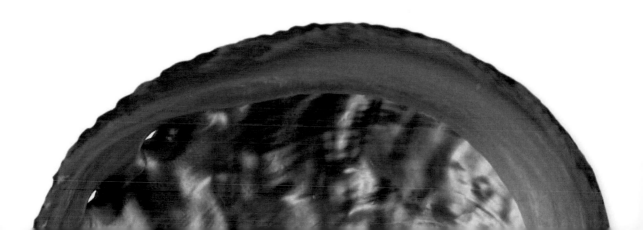

FAMILY	Haliotidae
SHELL SIZE RANGE	2 to 2½ in (48 to 60 mm)
DISTRIBUTION	Endemic to southern Australia
ABUNDANCE	Common
DEPTH	Subtidal to 100 ft (30 m)
HABITAT	Rocky bottoms
FEEDING HABIT	Herbivore, grazes on algae
OPERCULUM	Absent

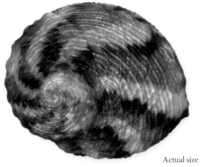

SHELL SIZE RANGE
2 to 2½ in
(48 to 60 mm)

PHOTOGRAPHED SHELL
2 in
(49 mm)

202

HALIOTIS CYCLOBATES

WHIRLING ABALONE

PÉRON, 1816

Haliotis cyclobates has an easily identified shell, with a relatively high spire for the family, and a nearly circular outline. Its outer surface is sculptured with oblique radial folds and spiral ribs, and is decorated with curved, radiating cream rays. It is endemic to southern Australia, ranging from the southern area of Western Australia to Victoria. It is found on rocky bottoms or attached to large shells, from the lower intertidal zone to offshore. Male and female abalones release large amounts of gametes into the water, and fertilization is external. The foot is large and muscular, and the mantle has several small tentacles, in addition to two cephalic tentacles.

RELATED SPECIES

There are several abalones endemic to Australia, including *Haliotis roei* Gray, 1862, from southern Australia, which has a larger, red or red-brown shell, with a high-quality, mother-of-pearl interior. It is fished commercially in Western Australia for its meat and shell. *Haliotis scalaris* Leach, 1814, from Western to South Australia, is one of the most beautiful and recognizable abalones, with a complex sculpture dominated by a thick radial rib.

Actual size

The shell of the Whirling Abalone is medium in size, convex, and nearly circular in outline, with an elevated spire. Its spire is high for the family, with 3 rounded whorls, located posteriorly. Its sculpture consists of many beaded spiral ribs, crossed by weak, obliquely radial folds; the inner surface is smooth and nacreous. The holes are oval, slightly elevated, and the last 5 or 6 are open. The shell color is brown and green, with oblique, radiating cream rays.

FAMILY	Haliotidae
SHELL SIZE RANGE	2¾ to 4 in (70 to 100 mm)
DISTRIBUTION	Endemic to Western Australia
ABUNDANCE	Uncommon
DEPTH	Shallow subtidal to 65 ft (20 m)
HABITAT	Rocky bottoms
FEEDING HABIT	Herbivore, grazes on algae
OPERCULUM	Absent

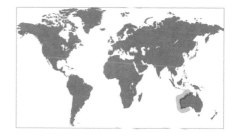

SHELL SIZE RANGE
2¾ to 4 in
(70 to 100 mm)

PHOTOGRAPHED SHELL
2⅝ in
(66 mm)

203

HALIOTIS ELEGANS

ELEGANT ABALONE

PHILIPPI, 1874

Haliotis elegans lives offshore on rocky bottoms, often hiding under stones and in crevices in reefs, so it is difficult to find alive. It is endemic to Western Australia. The shell somewhat resembles the widespread *H. asinina*, because of its very elongate shape. However, while *H. asinina* has a smooth shell, *H. elegans* is sculptured with heavy spiral ribs that are close together; inside, the spiral ribs show as grooves. The foot is wide and very strong, and is capable of clamping tightly to rocks with great tenacity.

RELATED SPECIES

Haliotis tuberculata Linnaeus, 1758, from the Mediterranean and western Africa, has a medium-large shell covered with spiral ribs crossed by radial growth lines. It has been fished for centuries for food and for its high-quality mother-of-pearl. *Haliotis varia* Linnaeus, 1758, from the Red Sea to the Indo-West Pacific, has an oval shell with extremely variable sculpture and coloration. This species is collected for food and the shell used for shellcraft.

The shell of the Elegant Abalone is medium in size, strong, and elongate-oval in outline. Its spire is low and small, and located very close to the posterior margin. The sculpture consists of heavy, raised, and slightly meandering spiral ribs crossed by growth lines; the inner surface is smooth and nacreous, and shows the spiral ribs as grooves. The holes are elongate-oval and the last 8 or 9 are open in juveniles; the number of open holes varies in adults. The shell color is red-brown or orange, with radiating cream or reddish rays.

Actual size

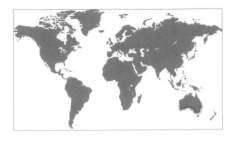

FAMILY	Haliotidae
SHELL SIZE RANGE	2½ to 4 in (60 to 100 mm)
DISTRIBUTION	Endemic from Western to South Australia
ABUNDANCE	Common
DEPTH	Intertidal to shallow subtidal
HABITAT	Rocky bottoms
FEEDING HABIT	Herbivore, grazes on algae
OPERCULUM	Absent

SHELL SIZE RANGE
2½ to 4 in
(60 to 100 mm)

PHOTOGRAPHED SHELL
2⅝ in
(66 mm)

204

HALIOTIS SCALARIS

STAIRCASE ABALONE

LEACH, 1814

Actual size

The shell of the Staircase Abalone is medium in size, thin, and oval in outline. Its spire is moderately elevated, located closer to the posterior margin, and has about 3 whorls. The tremata are elevated, and the last 4 to 6 holes are open. The sculpture is dominated by a strong, central spiral fold and oblique radial lamellae that run from this fold to the previous whorl. There are also other weaker radial and spiral elements; the main ones show in the interior, which is smooth and nacreous. The shell color is orange-red, with curved, radiating cream rays.

Haliotis scalaris is one of the most distinctive and beautiful of the Australian abalones. Its outer sculpture is complex and the main feature is a central spiral fold, with radial lamellae running from the central fold to the previous whorl. It is endemic to Australia, ranging from Western Australia to South Australia. It is found from the intertidal to subtidal zones, under rocks; it is common but never abundant. Like other abalones, it has a depressed and loosely coiled shell.

RELATED SPECIES

Haliotis gigantea Gmelin, 1791, from the Japonic province, has a large red to brown shell, with a relatively weak sculpture of radial and spiral lines. The interior is lined with a highly nacreous mother-of-pearl. *Haliotis cyclobates* Péron, 1816, which is endemic to southern Australia, has a nearly circular shell with a high spire. It has a brown and green outer color and oblique, radial cream rays.

FAMILY	Haliotidae
SHELL SIZE RANGE	2½ to 4½ in (60 to 120 mm)
DISTRIBUTION	Indo-West Pacific
ABUNDANCE	Abundant
DEPTH	Intertidal to 33 ft (10 m)
HABITAT	Rocky
FEEDING HABIT	Herbivore, grazes on algae
OPERCULUM	Absent

SHELL SIZE RANGE
2½ to 4½ in
(60 to 120 mm)

PHOTOGRAPHED SHELL
3 in
(77 mm)

HALIOTIS ASININA

DONKEY'S EAR ABALONE

LINNAEUS, 1758

205

Haliotis asinina is a common species in the tropical Indo-West Pacific. Abalones are herbivores, grazing on algae on rocky substrates. They have a large muscular foot, and can have short bursts of "rapid" movement when necessary to escape predators. The shell has a row of holes, called tremata, near the periphery of the whorl, which are used to exhale water. As the shell grows, the oldest holes are filled, and new ones are formed near the edge of the shell. Where large species occur, abalones are used for food. Their nacreous shells are used in jewelry.

RELATED SPECIES

Haliotis varia Linnaeus, 1758, from the Indo-West Pacific, is a small species with a pearly white interior; *H. clathrata* Reeve, 1846, from East Africa to Samoa, has a small shell of variable coloration; and *H. diversicolor* Reeve, 1846, from Japan to Indonesia, is another small species.

Actual size

The shell of the Donkey's Ear Abalone is thin and elliptical, with the apex near the edge of the shell. It is easily recognized by its elongated shell, which has the highest shell length-to-width ratio in the family. There are 6 or 7 open, ovate holes. The exterior is smooth, with axial growth lines that are crossed by low spiral ridges. The coloration is variable, usually olive green with blotches of brown and cream. The interior is nacreous and white, with a greenish tint.

FAMILY	Haliotidae
SHELL SIZE RANGE	3 to 6 in (75 to 150 mm)
DISTRIBUTION	Oregon to Baja California
ABUNDANCE	Formerly abundant
DEPTH	Intertidal to 16 ft (5 m)
HABITAT	Rocky bottoms
FEEDING HABIT	Herbivore, grazes on algae
OPERCULUM	Absent

SHELL SIZE RANGE
3 to 6 in
(75 to 150 mm)

PHOTOGRAPHED SHELL
4⅜ in
(117 mm)

206

HALIOTIS CRACHERODII

BLACK ABALONE

LEACH, 1814

Actual size

Haliotis cracherodii is an edible, medium-large abalone ranging from Oregon to Baja California. Once it was the most abundant mollusk in the intertidal zone; however, because of overfishing, its population has declined dramatically in recent decades. The withering disease, which affects the foot and results in poor adhesion to rocks, also contributed to its decline; it is estimated that about 20 percent of the population died from this disease between 1985 and 1992. It is currently listed as critically endangered, and its fishery regulated in the U.S.A. *Haliotis cracherodii* has a smooth and dark shell, with the color ranging from dark blue or green to black.

RELATED SPECIES

Haliotis kamtschatkana Jonas, 1845, from Alaska to California and Japan, has a medium to large shell with a wrinkled outer surface and relatively large holes. Overharvesting has led to the closure of its fishery in Canada. *Haliotis fulgens* Philippi, 1845, from Oregon to Baja California, is one of the largest abalones. It is still fished but its harvest is strictly regulated. Its shell is the source of the beautiful mother-of-pearl used in jewelry.

The shell of the Black Abalone is medium-large, thick, smooth, and with an oval outline. Its spire is low and small, and located very close to the posterior margin. The sculpture is nearly smooth, and consists of concentric growth lines; the interior is nacreous, and shows growth marks and more irregularities than most abalones. The holes are oval, slightly elevated, and the last 5 to 7 are open. The shell color ranges from dark blue or green to black, and the interior is nacreous silvery or golden.

FAMILY	Haliotidae
SHELL SIZE RANGE	6 to 10 in (150 to 250 mm)
DISTRIBUTION	Oregon to Baja California
ABUNDANCE	Common
DEPTH	Intertidal to 65 ft (20 m)
HABITAT	Rocky shores
FEEDING HABIT	Herbivore, grazes on algae
OPERCULUM	Absent

SHELL SIZE RANGE
6 to 10 in
(150 to 250 mm)

PHOTOGRAPHED SHELL
8 in
(205 mm)

HALIOTIS FULGENS

GREEN ABALONE

PHILIPPI, 1845

207

Haliotis fulgens is one of the largest species of abalones. Once common in intertidal and shallow depths along the Pacific Northwest, intensive harvesting caused this large abalone to become scarce. Its important fisheries have become strictly regulated. It is now farmed both for its meat, which is considered a delicacy, and for its shell, which is a source of mother-of-pearl for jewelry and decoration.

Actual size

RELATED SPECIES

Haliotis brazieri Angas, 1869, from southern Australia, and *H. queketi* E. A. Smith, 1910, from South Africa, are from temperate waters; *H. planata* Sowerby II, 1882, and *H. fatui* Geiger, 1999, both from the Indo-West Pacific, are examples of tropical species, which usually have smaller shells than those from colder waters. Australasia is particularly rich in abalone species, with several endemic to the region.

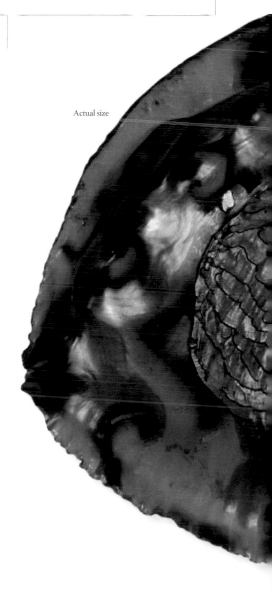

The shell of the Green Abalone is large, thick, and heavy, with an ovate profile, flattened spire, an eccentric nucleus, and a rapidly expanding whorl. It has a spiral row of slightly raised holes at the periphery of the whorl, with the last 5 to 7 holes open. The surface is sculptured by 30 to 40 spiral lirae and radial growth lines. The interior is iridescent, with blue and green tints. A large muscle scar is situated at the center of the aperture. The surface is dull reddish brown, but it is often heavily encrusted with other organisms.

FAMILY	Fissurellidae
SHELL SIZE RANGE	¼ to 1 in (6 to 24 mm)
DISTRIBUTION	Galápagos, Georgia to southern Brazil, Portugal to the Azores
ABUNDANCE	Uncommon
DEPTH	33 to 3,840 ft (10 to 1,170 m)
HABITAT	Rocky bottoms
FEEDING HABIT	Herbivore, grazes on algae
OPERCULUM	Absent

SHELL SIZE RANGE
¼ to 1 in
(6 to 24 mm)

PHOTOGRAPHED SHELL
1 in
(24 mm)

EMARGINULA TUBERCULOSA

TUBERCULATE EMARGINULA

LIBASSI, 1859

The shell of the Tuberculate Emarginula is small, cancellate, strongly curved posteriorly, and oval in outline, with a high apex. Its apex is small, and located below the highest point of the shell. The anterior slope is convex and the posterior slope concave. The sculpture consists of about 26 strong radial ribs, with numerous intermediary ones, crossed by many concentric lines; these form a cancellated pattern, with beads at the intersections and depressed pits between ribs. The margin is crenulated, and the anterior slit is short. The lateral view in the photograph below shows the growth scar. The shell color is off-white.

Emarginula tuberculosa is a small fissurellid with a short slit at the anterior margin. It occurs from shallow subtidal depths to the deep sea, on hard bottoms such as rocks and corals. It has a wide distribution, occurring on both sides of the Atlantic Ocean, and in the Galápagos Islands. Most fissurellids have a dorsal foramen (opening); others, like *E. tuberculosa*, have a slit or notch at the anterior margin; relatively few fissurellids lack both a foramen and slit. There are several hundred living species in the family Fissurellidae worldwide. The earliest fissurellids date from the Triassic Period.

RELATED SPECIES

Emarginula peasei Thiele, 1915, from the Arabian Gulf and Arabian Sea, has a small, flattened shell with a narrow and short slit anteriorly. *Scutus antipodes* (Montfort, 1810), from Australia and New Zealand, has an unusual shell for the keyhole limpet family in that it lacks a "keyhole" or slit. Instead, it has an elongate-oval to rectangular shell that is nearly flat. It looks more like the valve of a bivalve than a limpet.

Actual size

FAMILY	Fissurellidae
SHELL SIZE RANGE	⅝ to 1¾ in (16 to 45 mm)
DISTRIBUTION	Florida to Brazil
ABUNDANCE	Common
DEPTH	Intertidal to 10 ft (3 m)
HABITAT	Rocky shores
FEEDING HABIT	Herbivore, grazes on algae
OPERCULUM	Absent

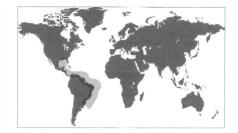

SHELL SIZE RANGE
⅝ to 1¾ in
(16 to 45 mm)

PHOTOGRAPHED SHELL
¼ in
(31 mm)

DIODORA LISTERI
LISTER'S KEYHOLE LIMPET
(D'ORBIGNY, 1847)

209

Keyhole limpets are characterized by a cap-shaped shell
with either an anterior groove, or, more commonly, a dorsal
foramen—the "keyhole." The shell interior is porcelaneous and
has a horseshoe-shaped muscle scar, which is open at the anterior
end. Keyhole limpets occur worldwide, mostly in warm waters
but also in temperate regions, on rocky substrates, usually at
shallow subtidal depths, where they feed on algae. Like other
keyhole limpets, *Diodora listeri* is consumed locally as food,
but there is no commercial fishery for this species.

RELATED SPECIES
Fissurella picta (Gmelin, 1791), from Ecuador to
Argentina, has a large shell rayed in dark and light
colors; *F. aperta* Sowerby I, 1825, an endemic South
African species, has a large hole and elongated shell;
Megathura crenulata (Sowerby I, 1825), from California
to Baja California, is one of the largest keyhole limpets,
growing to 5 in (132 mm), with the dorsal foramen spanning
about one-sixth of the shell length.

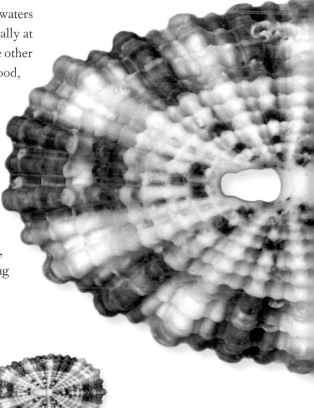

The shell of the Lister's Keyhole Limpet is
conical, thick, moderately elevated, and elliptical.
Its sculpture has strong radial ribs alternating
with smaller ones, with fine riblets between
the ribs (there are 3 radial rib sizes). They are
crossed by strong concentric bands, forming a
cancellate pattern with rounded or scaly nodules.
The keyhole-shaped orifice at the top of the shell
is slightly closer to the anterior end. The exterior
color is cream to brownish with darker radial
bands, and the interior is white to greenish.

Actual size

FAMILY	Fissurellidae
SHELL SIZE RANGE	⅝ to 2 in (15 to 50 mm)
DISTRIBUTION	Northern Mediterranean
ABUNDANCE	Common
DEPTH	Intertidal to shallow subtidal
HABITAT	Rocky shores
FEEDING HABIT	Grazer
OPERCULUM	Absent

SHELL SIZE RANGE
⅝ to 2 in
(15 to 50 mm)

PHOTOGRAPHED SHELL
1¼ in
(44 mm)

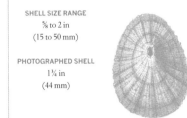

210

DIODORA ITALICA

ITALIAN KEYHOLE LIMPET

(DEFRANCE, 1820)

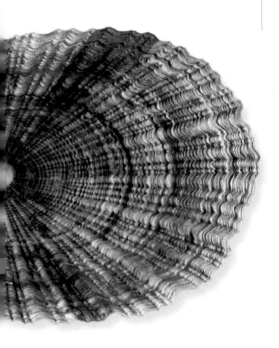

Diodora italica is a common keyhole limpet from the northern Mediterranean. It hides under rocks in the intertidal zone or shallow subtidal depths. It varies widely in sculpture, size, and coloration; as a consequence it has received many names. It is possibly a complex of similar species; a molecular study of the group would shed some light onto its identity and relationships. The animal is orange-yellow, with two short cephalic tentacles. The radula has many teeth per row, with thousands of small teeth in total.

RELATED SPECIES

Diodora patagonica d'Orbigny, 1847, from Chile to southern Argentina, is a common subtidal species. Its shell is oval, thick, and relatively high, with a small foramen and many radial ribs, which are crossed by concentric growth marks. *Emarginula tuberculosa* Libassi, 1859 is widely distributed, occurring in the Galápagos, the western Atlantic, and Europe. It has a small shell with a curved apex and a slit at the anterior margin.

The shell of the Italian Keyhole Limpet is medium-sized, thick, solid, and oval in outline. Its apex is moderately elevated, with the keyhole-shaped foramen located at the highest point, closer to the anterior margin. The posterior margin is wider than the anterior one, and the entire margin is crenulated. Its sculpture consists of elevated radial ribs crossed by concentric growth lines, giving the surface a cancellate appearance. The interior is smooth, with a callus around the foramen. The shell color is cream or gray, with greenish brown radial rays, and a white interior.

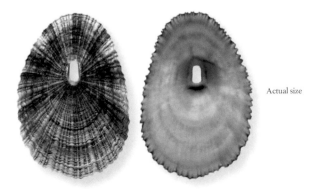

Actual size

FAMILY	Fissurellidae
SHELL SIZE RANGE	2½ to 5¼ in (60 to 132 mm)
DISTRIBUTION	California to Baja California
ABUNDANCE	Common
DEPTH	Intertidal to shallow subtidal
HABITAT	Rocky shores
FEEDING HABIT	Grazer, feeds on algae and tunicates
OPERCULUM	Absent

SHELL SIZE RANGE
2½ to 5¼ in
(60 to 132 mm)

PHOTOGRAPHED SHELL
3½ in
(92 mm)

MEGATHURA CRENULATA

GREAT KEYHOLE LIMPET

(SOWERBY I, 1825)

211

Megathura crenulata is one of the larger species of keyhole limpet, although not the largest. The genus name comes from the Greek word for "large door," in reference to the shell's large, oval keyhole. It is a common limpet, usually found crawling on shallow subtidal rocks in kelp forests, where it feeds on algae and colonial tunicates. *Megathura crenulata* is unique among limpets because its black or gray mantle covers most or all of the shell, showing only the "keyhole." Its hemolymph ("blood") has promising biomedical applications, therefore several companies are currently investing in its aquaculture.

Actual size

RELATED SPECIES

Diodora listeri (d'Orbigny, 1842), from Florida to Brazil, has a small shell with a cancellate pattern of axial and concentric ribs; *Fissurella picta* (Gmelin, 1791), from Ecuador to Argentina, is a large keyhole limpet with a small and narrow keyhole and regular radial stripes; and *Fissurellidea megatrema* (d'Orbigny, 1841), from Brazil, is a medium-sized limpet with an exceptionally large foramen, which can be as large as half of the shell length.

The shell of the Great Keyhole Limpet is large, low, thick, and elliptical in profile. The dorsal foramen, the keyhole, is large, oval, and subcentral. Its sculpture comprises fine, regularly spaced radial ribs and concentric lines. The margin of the shell is irregular and bears small denticles. The dorsal color ranges from reddish brown to gray, and the interior is porcelaneous white. The keyhole margin is white.

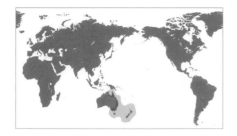

FAMILY	Fissurellidae
SHELL SIZE RANGE	1 to 5 in (25 to 125 mm)
DISTRIBUTION	Eastern Australia and New Zealand
ABUNDANCE	Common
DEPTH	Intertidal to shallow subtidal
HABITAT	Rocky bottoms
FEEDING HABIT	Grazer
OPERCULUM	Absent

SHELL SIZE RANGE
1 to 5 in
(25 to 125 mm)

PHOTOGRAPHED SHELL
4⅝ in
(119 mm)

212

SCUTUS ANTIPODES
SHORT SHIELD LIMPET
(MONTFORT, 1810)

Scutus antipodes is one of the most remarkable species in the family Fissurellidae. Its shell lacks a dorsal foramen (as in the *Fissurella* genus) or a marginal slit (as in *Emarginula*), and instead is a solid, imperforated, shield-shaped shell. The animal is black, large, and fleshy, and partially to completely covers the shell. The animal has two large and long cephalic tentacles, and a tough body. It is a common fissurellid in the intertidal zone and shallow subtidal depths, where it hides under rocks and boulders. It ranges from Queensland to Tasmania and New Zealand. It is also known in Australia as the Elephant Snail.

RELATED SPECIES

Scutus sinensis (Blainville, 1825), from Japan to Thailand, resembles *S. antipodes*, but is smaller and there is a weak anal notch on the posterior margin. It is found under rocks in the intertidal zone. *Diodora italica* (Defrance, 1820), from the northern Mediterranean, has a typical fissurellid shell, with a conical shape and a keyhole foramen at the highest point of the shell. The shell is light colored, with greenish brown radial rays.

Actual size

The shell of the Short Shield Limpet is medium to large in size, thick, strong, smooth, depressed, and elongate-oval to rectangular in outline. Its apex is low, small, and located at about a quarter of the shell length from the anterior margin. The surface is smooth except for concentric growth lines; the interior is entirely smooth. The shell margin is smooth and thick. The color is white or off-white, and the interior is white.

FAMILY	Turbinidae
SHELL SIZE RANGE	1¼ to 3¼ in (30 to 80 mm)
DISTRIBUTION	West Pacific
ABUNDANCE	Uncommon
DEPTH	165 to 650 ft (50 to 200 m)
HABITAT	Rubble bottom
FEEDING HABIT	Herbivore, feeds on algae
OPERCULUM	Calcareous, thick, ovate

SHELL SIZE RANGE
1¼ to 3¼ in
(30 to 80 mm)

PHOTOGRAPHED SHELL
2¼ in
(56 mm)

BOLMA GIRGYLLA

GIRGYLLA STAR SHELL

(REEVE, 1861)

213

Bolma girgylla is the most spectacular species in its genus: its shell can be brightly colored and have foliated spines that may grow to half of the shell diameter. The spines are hollow and very fragile; perfect specimens with intact spines are rarely collected. *Bolma girgylla* has recently been listed as a rare species in the Philippines, and is, therefore, prohibited for exportation. Its long spines indicate that it lives in deep, quiet waters. Like other turbinids, it is a herbivore, grazing on algae. The operculum is thick and calcareous.

RELATED SPECIES

Bolma guttata (Adams, 1863), also from the West Pacific, has a smaller shell that resembles *B. girgylla* in shape but with a row of shorter hollow spines on the lower part of body whorl. *Angaria delphinus melanacantha* (Reeve, 1842), which ranges from the Philippines to Vietnam, has a flattened spire and a row of long and curved spines along the shoulder.

The shell of the Girgylla Star Shell is light and conic, with a tall spire. The whorls have spiral rows of beaded ridges on the upper part, and stronger spiral, beaded ridges on the base. There are two rows of long, projecting, foliated spines, the upper row being more developed. The aperture is elliptical, and the columella smooth and strong. The shell color can be pale yellow to bright orange, greenish, or brown. The aperture may be stained in yellow or orange.

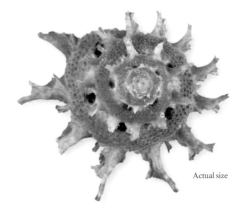

Actual size

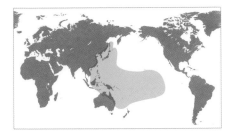

FAMILY	Turbinidae
SHELL SIZE RANGE	1⅜ to 2¾ in (35 to 70 mm)
DISTRIBUTION	West Pacific
ABUNDANCE	Locally abundant
DEPTH	Deep water
HABITAT	Coral reefs and hard bottoms
FEEDING HABIT	Herbivore, grazes on algae
OPERCULUM	Corneous, concentric, circular

SHELL SIZE RANGE
1⅜ to 2¾ in
(35 to 70 mm)

PHOTOGRAPHED SHELL
2¾ in
(69 mm)

214

ANGARIA DELPHINUS MELANACANTHA
IMPERIAL DELPHINULA
REEVE, 1842

Angaria delphinus melanacantha is a common to abundant species living in deeper water near coral reefs. This and many other gastropods are usually collected using tangle nets in the Philippines. Its shell has many spines that vary in development and can grow quite long in some specimens, much longer in calmer waters than in areas with stronger currents. The shells are usually encrusted with algae, corals, and other organisms, and it takes careful work to clean them in order to reveal the perfect shells seen in collections. The operculum is rounded, multispiral, and corneous, unlike those of most turbinids, which have a calcareous operculum.

RELATED SPECIES

Angaria sphaerula (Kiener, 1839), from the West Pacific, resembles the Imperial Delphinula, but its spines can be longer and the spire slightly higher; *A. vicdani* Kosuge, 1980, which is also from the Philippines, has even longer spines that can equal the diameter of the shell in length. *Angaria tyria* (Reeve, 1842), from the southwestern Pacific, has spines that are either short or lacking, as well as a broad spiral band.

The shell of the Imperial Delphinula is thick, depressed, very spinose, and has a short spire. The large body whorl dominates the shell. The shoulder has long spines that curve upward and inward, while the rest of the shell is covered with several spiral rows of spines. The umbilicus is deep and spinose. The shell color is grayish purple to brown, and the aperture is round, nacreous, and white.

Actual size

FAMILY	Turbinidae
SHELL SIZE RANGE	2¾ to 5 in (70 to 125 mm)
DISTRIBUTION	Japonic region
ABUNDANCE	Common
DEPTH	330 to 1,650 ft (100 to 500 m)
HABITAT	Sandy bottom
FEEDING HABIT	Herbivore, grazes on algae
OPERCULUM	Calcareous, thick

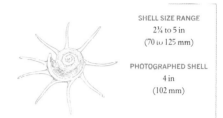

SHELL SIZE RANGE
2¾ to 5 in
(70 to 125 mm)

PHOTOGRAPHED SHELL
4 in
(102 mm)

215

GUILDFORDIA YOKA

YOKA STAR TURBAN

JOUSSEAUME, 1888

Guildfordia yoka is one of the most distinctive seashells, with a flattened shell that bears extremely long spines radiating outward. The spines are produced at the periphery of the shell, thus increasing its effective size. They also make it more difficult for predators to turn its shell over to expose the aperture. *Guildfordia yoka* lives in calm, deep waters, down to 1,650 ft (500 m). Two new species resembling *G. yoka* have recently been described from deep waters off the Philippines.

RELATED SPECIES

Guildfordia triumphans (Philippi, 1841), from Japan to northeastern Australia, has a very similar shell to *G. yoka*, but it is smaller and has shorter and sometimes more numerous spines. *Lithopoma undosa* (Wood, 1828), which ranges from California to Baja California, has a tall, large, top-shaped shell with an undulating ridge along the periphery of the whorls.

Actual size

The shell of the Yoka Star Turban is medium in size, flattened, and disk-shaped, with 7 to 9 long radiating spines. The spire is short, and has a shallow suture. The dorsal surface is ornamented with spiral beaded rows, while the ventral surface has fine, oblique axial lines, and a strong callus around the umbilicus. The extremely long radiating spines can be straight or curved, and they may be as long as the diameter of the shell. The dorsal shell color is iridescent bronze, and the base is cream.

FAMILY	Turbinidae
SHELL SIZE RANGE	1¼ to 4 in (30 to 100 mm)
DISTRIBUTION	Red Sea to Indo-West Pacific
ABUNDANCE	Common
DEPTH	Intertidal to 133 ft (40 m)
HABITAT	Coral reefs and rocky shores
FEEDING HABIT	Herbivore
OPERCULUM	Calcareous, thick, bluish green

SHELL SIZE RANGE
1¼ to 4 in
(30 to 100 mm)

PHOTOGRAPHED SHELL
2 in
(53 mm)

216

TURBO PETHOLATUS
TAPESTRY TURBAN
LINNAEUS, 1758

Turbo petholatus is one of the most beautiful turbinids, with a colorful shell decorated in a tapestry-like pattern. It has a thick, circular operculum stained in green in the center, giving it another popular name, Cat's Eye. It lives from the intertidal zone to shallow depths near coral reefs and hard bottoms, in calm waters. *Turbo petholatus* is collected for its meat and shell; the operculum is used in jewelry. There are more than 200 living species in the family, with most found in tropical and subtropical regions. The earliest turbinid fossil is from the Cretaceous Period.

RELATED SPECIES

The two largest species of turban shells are *Turbo jourdani* Kiener, 1839, from Australia, and *Turbo marmoratus* Linnaeus, 1758, from East Africa to the central Pacific. The former has a smooth shell that resembles *T. petholatus*, but is larger, with a more pointed apex. The latter has a very large and heavy shell, with three heavy spiral ribs, and a rough surface.

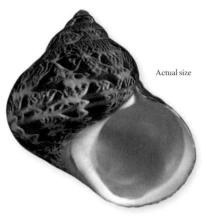

Actual size

The shell of the Tapestry Turban is medium in size, heavy, thick, polished, and turbinate with a moderately high spire. It has rounded, convex whorls and the suture is incised. The surface is smooth and polished. Its aperture is oval, the outer lip thick with a sharp edge, and the columella is smooth. The shell is usually red, orange, or brown, decorated with brown spiral bands with light-colored, axial zigzag or chevron lines. The operculum is smooth, shiny, and bluish green.

FAMILY	Turbinidae
SHELL SIZE RANGE	2¾ to 4½ in (70 to 115 mm)
DISTRIBUTION	New Zealand
ABUNDANCE	Uncommon
DEPTH	Deep water, up to 660 ft (200 m)
HABITAT	Rocky bottoms
FEEDING HABIT	Herbivore, grazes on algae
OPERCULUM	Calcareous, paucispiral, ovate

SHELL SIZE RANGE
2¾ to 4½ in
(70 to 115 mm)

PHOTOGRAPHED SHELL
4½ in
(115 mm)

ASTRAEA HELIOTROPIUM
SUNBURST STAR SHELL
(MARTYN, 1784)

The first specimen of *Astraea heliotropium* was found during one of Captain Cook's voyages to New Zealand. Reportedly, the snail was found clinging to the HMS *Endeavour*'s anchor chain. It was brought back to England, where it was described and soon became a favorite among collectors. The shell is usually overgrown with coral and coralline algae, making it difficult to clean. It is a common species in deep water but has a restricted distribution, endemic to New Zealand, therefore it is uncommon in collections.

RELATED SPECIES

Astralium phoebium (Röding, 1798) from Florida to Brazil, has a similar but smaller shell with more numerous spines; *Bolma girgylla* (Reeve, 1861), from the tropical West Pacific, has a taller shell with foliated spines and no umbilicus; and *Guildfordia yoka* Jousseaume, 1888, from Japan to the Philippines, has a flat shell with seven to nine long, radiating spines.

Actual size

The shell of the Sunburst Star Shell is large, thick, and has a moderately tall spire for the family. The whorls are rounded, and the base is flattened, with a row of large, fluted, triangular scales at the periphery of the whorls. The shell is covered with spiral rows of tubercles, with those on the base being scaly. The umbilicus is deep and wide. The shell color is grayish white and the aperture is pearly white. The thick calcareous operculum is white on the outer side, and brown where attached to the columellar muscle.

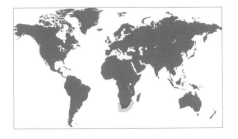

FAMILY	Turbinidae
SHELL SIZE RANGE	1½ to 4½ in (40 to 120 mm)
DISTRIBUTION	Endemic to South Africa
ABUNDANCE	Abundant
DEPTH	Shallow subtidal
HABITAT	Rocky shores
FEEDING HABIT	Herbivore
OPERCULUM	Calcareous, thick, circular, pustulose

SHELL SIZE RANGE
1½ to 4½ in
(40 to 120 mm)

PHOTOGRAPHED SHELL
3⅝ in
(92 mm)

218

TURBO SARMATICUS
SOUTH AFRICAN TURBAN
LINNAEUS, 1758

Turbo sarmaticus is endemic to South Africa, where it is the largest turbinid. It is locally known as "Alikreukel" or giant periwinkle (true periwinkles, littorinids, are unrelated and much smaller). It is edible and considered a delicacy; the tough, muscular animal is cooked, minced, and fried. Its overharvesting led to regulation, and currently only a few can be collected per person with a permit. The animal is green, spotted, and marbled with white and dark green, and the sole of the foot is yellow-orange. The shell is thick and heavy, the operculum is calcareous and circular, and the outer surface has large pustules.

RELATED SPECIES

Turbo reevei Philippi, 1847, from the Philippines to Indonesia, has a medium-sized, glossy shell. Its coloration is very variable, and can range from pale fawn to bright yellow, green, or reddish brown. *Turbo jourdani* Kiener, 1839, which is endemic to Australia, is one of the largest species in the family Turbinidae. It has a relatively high spire, and a circular, smooth, white operculum.

Actual size

The shell of the South African Turban is large, thick, heavy, rough, inflated, and turban-shaped. Its spire is moderately short, with a rounded apex and an incised suture. The whorls are convex and rounded; the surface is rough, with spiral rows of tubercles at the shoulder, fine spiral lines, and axial growth lines. The aperture is large and oval, the outer lip is thick with a sharp edge, and the columella is smooth. The color varies from pale orange to brown; when eroded, a beautiful mother-of-pearl shows through. The aperture is white.

FAMILY	Turbinidae
SHELL SIZE RANGE	3 to 9 in (75 to 230 mm)
DISTRIBUTION	Endemic to southern and western Australia
ABUNDANCE	Uncommon
DEPTH	Intertidal to shallow subtidal
HABITAT	Tidepools and among leafy brown algae
FEEDING HABIT	Herbivore
OPERCULUM	Calcareous, circular, thick

SHELL SIZE RANGE
3 to 9 in
(75 to 230 mm)

PHOTOGRAPHED SHELL
8½ in
(217 mm)

TURBO JOURDANI

JOURDAN'S TURBAN

KIENER, 1839

219

Turbo jourdani is one of the largest species in the family Turbinidae, and the largest turbinid in Australia. It is an uncommon species, and lives from the intertidal zone to shallow subtidal. It is found among leafy brown algae, and sometimes in tidepools. *Turbo jourdani* is sought after for its flesh and large shell, which is prized by shell collectors. The animal is red-brown. There is a form from southern Australia, *Turbo verconis* Iredale, 1937, which differs slightly in having a large, oval, and calcareous operculum, and a white exterior.

RELATED SPECIES

Turbo petholatus Linnaeus, 1758, from the Red Sea to the Indo-West Pacific, has a turban-shaped shell that is glossy and richly decorated with spiral bands and axial lines. It resembles *T. jourdani* but is smaller, with a lower spire, and is more brightly colored. *Turbo cornutus* Lightfoot, 1786, from Japan to China and the Philippines, has a thick turbinate shell with two spiral rows of thick spines, or with five spiral ribs.

Actual size

The shell of the Jourdan's Turban is very large, thick, heavy, polished, and turban-shaped. Its tall spire has a pointed apex, well-marked suture, and rounded, convex whorls. The body whorl is large, with a wide, oval aperture, relatively thin outer lip, and a smooth columella; it lacks an umbilicus. Its surface is smooth, with fine growth marks, and sometimes, low spiral ridges. The shell color is a deep red-brown, and the interior is a porcelaneous white. The large, oval operculum is white outside, and has a thin, brown periostracum inside.

FAMILY	Phasianellidae
SHELL SIZE RANGE	1 to 2 in (25 to 50 mm)
DISTRIBUTION	Endemic to Australia
ABUNDANCE	Common
DEPTH	Intertidal to 33 ft (10 m)
HABITAT	On kelp and under rocks
FEEDING HABIT	Herbivore
OPERCULUM	Calcareous, oval, white

SHELL SIZE RANGE
1 to 2 in
(25 to 50 mm)

PHOTOGRAPHED SHELL
2 in
(49 mm)

PHASIANELLA VENTRICOSA
SWOLLEN PHEASANT
SWAINSON, 1822

Phasianella ventricosa is a common species of pheasant shell with extremely variable color patterns, which include spiral bands, diagonal bands, and axial streaks. It is endemic to Australia. Juveniles live under rocks on reefs, and adults live on kelp, from the intertidal zone to shallow subtidal waters. It is often found in large numbers washed up on the beach, particularly in southern New South Wales. There are some 20 living species in the family Phasianellidae worldwide.

RELATED SPECIES

Phasianella australis (Gmelin, 1791), from Southern Australia and Tasmania, is the largest species in the family, growing to about 4 in (100 mm). *Tricolia speciosa* (Mühfeld, 1824) resembles a miniature *P. australis*, but slightly more elongate.

Actual size

The shell of the Swollen Pheasant is medium in size, strong, smooth, glossy, and cylindrical globose in outline. Its spire is moderately high, with a well-marked suture, and rounded, convex whorls. Its surface is smooth, glossy, and lacks a periostracum. The shell color is extremely variable; usually the background is light, with spiral or axial bands, irregular lines or streaks of pink, cream, reddish-brown, white, or brown.

FAMILY	Trochidae
SHELL SIZE RANGE	¼ to ¾ in (5 to 21 mm)
DISTRIBUTION	Indo-West Pacific
ABUNDANCE	Abundant
DEPTH	Intertidal to 16 ft (5 m)
HABITAT	Sandy mud bottoms
FEEDING HABIT	Herbivore
OPERCULUM	Corneous, circular, multispiral

SHELL SIZE RANGE
¼ to ¾ in
(5 to 21 mm)

PHOTOGRAPHED SHELL
⅝ in
(17 mm)

UMBONIUM VESTIARIUM

COMMON BUTTON TOP

(LINNAEUS, 1758)

Umbonium vestiarium is a small trochid that is abundant in
shallow waters in the Indo-West Pacific. It has a rounded and
flattened shell that is smooth and glossy, and varies widely in
color patterns. Despite its small size, it is edible and collected for
food and used in soups. Its shell is used extensively in shellcraft,
to make shell curtains and shell necklaces. There are several
hundred living species in the family Trochidae worldwide,
especially on hard substrates in shallow tropical and
subtropical waters. The oldest known trochid is from
the Triassic Period.

RELATED SPECIES

Umbonium giganteum (Lesson, 1831), from Japan,
looks like a larger version of *U. vestiarium*, but
has a slightly more conical shape and less colorful
patterns. *Stomatella planulata* (Lamarck, 1816), from
Japan to the southwest Pacific, has a small shell with
a greatly expanding body whorl, so resembles some
abalones in shape, but it lacks the row of holes on the
periphery of the whorl.

Actual size

The shell of the Common Button Top is small, thin, flattened, smooth, and
circular. Its spire is very low, convex, with a fine incised suture that is often
marked by a spiral band. Its surface is glossy, and the body whorl has
rounded sides. The aperture is subtriangular, the outer lip thin, and the
columella smooth. The shell usually has a gray, yellowish, brown, or pinkish
background, with spiral bands, flammules, or streaks.

FAMILY	Trochidae
SHELL SIZE RANGE	½ to ⅞ in (13 to 22 mm)
DISTRIBUTION	Tanzania to South Africa
ABUNDANCE	Common
DEPTH	Intertidal to 6 ft (2 m)
HABITAT	Under rocks
FEEDING HABIT	Herbivore
OPERCULUM	Corneous, thin, and flexible

SHELL SIZE RANGE
½ to ⅞ in
(13 to 22 mm)

PHOTOGRAPHED SHELL
¾ in
(20 mm)

222

CLANCULUS PUNICEUS
STRAWBERRY TOP
(PHILIPPI, 1846)

Clanculus puniceus is a distinctive, attractive, small top shell that resembles a strawberry in color and texture. It is a common species in the intertidal zone and shallow subtidal waters; it is found under rocky rubble, sometimes in large numbers. *Clanculus puniceus* is among the relatively few gastropod shells known to fluoresce or glow under ultraviolet light, as a result of pigments incorporated in the shell.

RELATED SPECIES

Clanculus pharaonius (Linnaeus, 1758), from the Red Sea and Indian Ocean, has a shell similar to *C. puniceus*, but slightly larger, and with more spiral rows of alternate, single black and white beads. *Trochus niloticus* Linnaeus, 1767, from the Indo-Pacific, has a large, heavy, top-shaped shell with a white background and reddish brown flammules.

Actual size

The shell of the Strawberry Top is small, thick, strong, glossy, and shaped like a top. Its spire is moderately high, with convex whorls and well-marked suture. Its sculpture consists of several spiral rows of rounded beads, and a deep umbilicus surrounded by teeth in adults. The aperture is small, protected with projections from the columella and outer lip; the outer lip is lined with black teeth. The shell color is deep or bright red, with spiral rows of beads, in which every fourth bead is black. The aperture is white.

FAMILY	Trochidae
SHELL SIZE RANGE	⅝ to 1¼ in (15 to 30 mm)
DISTRIBUTION	Japan to the Philippines
ABUNDANCE	Common
DEPTH	Intertidal to 33 ft (10 m)
HABITAT	Under rocks on coral reefs
FEEDING HABIT	Herbivore
OPERCULUM	Absent

SHELL SIZE RANGE
⅝ to 1¼ in
(15 to 30 mm)

PHOTOGRAPHED SHELL
1 in
(25 mm)

STOMATELLA PLANULATA

FLATTENED STOMATELLA

(LAMARCK, 1816)

223

The body whorl of *Stomatella planulata* is greatly expanded, and the shell resembles an abalone except that it lacks the spiral row of holes near the periphery of the shell. This convergence in shell morphology is correlated with a convergence in habitat; it lives from the intertidal zone to shallow subtidal depths, hiding under rocks. Its apex and nuclear whorls are very small and close to the posterior margin. It lacks an operculum, and its aperture is nacreous. When disturbed, it can detach part of the foot, in the same manner that a gecko autotomizes its tail.

RELATED SPECIES

Stomatellina sanguinea (A. Adams, 1850), from Japan to the Philippines, has a small shell with rounded whorls, a moderately high spire, and a pointed apex. It can be brightly colored, ranging from red to yellowish. *Clanculus puniceus* (Philippi, 1846), from Tanzania to South Africa, has a striking deep red color and beaded spiral cords. There are four spiral cords with one black granule for every three red ones.

The shell of the Flattened Stomatella is small, thin, flattened, elongate, and ear-shaped. Its spire is nearly flat, the apex very small, near the posterior margin, and the body whorl greatly expanded. The surface is smooth, with fine growth lines and faint spiral grooves; the interior is nacreous with fine spiral lines. The aperture is very large, and the outer lip and columella are smooth. The shell coloration is variable, usually greenish or brownish, with spiral lines or irregular feathered markings. The color can be seen through the nacreous interior.

Actual size

FAMILY	Trochidae
SHELL SIZE RANGE	¾ to 1¼ in (20 to 30 mm)
DISTRIBUTION	Florida to West Indies; northern Gulf of Mexico
ABUNDANCE	Rare
DEPTH	2,000 to 3,500 ft (230 to 1,060 m)
HABITAT	Soft bottoms
FEEDING HABIT	Grazer
OPERCULUM	Corneous, multispiral

SHELL SIZE RANGE
¾ to 1¼ in
(20 to 30 mm)

PHOTOGRAPHED SHELL
1¼ in
(30 mm)

224

GAZA FISCHERI

FISCHER'S GAZA

DALL, 1889

Gaza fischeri is a beautiful gastropod that lives in deep water from off Florida to the West Indies, and the northern Gulf of Mexico. Fresh specimens have an iridescent sheen ranging from pale green to gold and pinkish. It is rare in collections, although it may be common in its habitat as is its sibling species, *Gaza superba*. Some species of the genus *Gaza* have been observed to swim by flapping their broad foot to escape predators. There are currently seven species recognized in the genus, and all occur in deep waters.

RELATED SPECIES

Gaza superba Dall, 1881, with the same distribution, and a shell similar to *G. fischeri*, has a slightly larger shell, more flattened over the suture; the protoconch is lost in adults, and the umbilicus is partially covered. *Tectus niloticus* Linnaeus, 1767, from the tropical Indo-Pacific, is the largest and heaviest of all top shells. Its shell was once used to produce buttons, and its meat for food.

Actual size

The shell of the Fischer's Gaza is small, thin, lightweight, and top-shaped with rounded whorls. The spire is short, and the apex often missing, leaving a tiny circular perforation that is continuous with the umbilicus. The shell surface is smooth, with only fine growth marks and fine spiral lines. The lip is reflected and thickened, and there is a broad callus that completely covers the wide and deep umbilicus in adults. The shell color is a pearly, pale lime green with a purple or gold tint.

FAMILY	Trochidae
SHELL SIZE RANGE	1⅜ to 2 in (36 to 50 mm)
DISTRIBUTION	Bering Sea to Chile
ABUNDANCE	Uncommon
DEPTH	1,150 to 7,000 ft (350 to 2,140 m)
HABITAT	Sandy and muddy bottoms
FEEDING HABIT	Surface-deposit feeder
OPERCULUM	Corneous, circular, thin

SHELL SIZE RANGE
1⅜ to 2 in
(36 to 50 mm)

PHOTOGRAPHED SHELL
1⅜ in
(36 mm)

BATHYBEMBIX BAIRDII
BAIRD'S BATHYBEMBIX
(DALL, 1889)

225

Bathybembix bairdii has a very wide latitudinal range, occurring from Alaska to Chile, in deep water. It has been trawled from sandy and muddy bottoms. It is a surface-deposit feeder, and consumes *Macrocystis pyrifera*, as well as foraminifera and other microorganisms, including some that live in the plankton but sink to the bottom when dead. Deepsea gastropods can feed at high rates, when more food becomes available. There are about nine living species in the genus *Bathybembix* worldwide, all living in deep waters.

RELATED SPECIES

Bathybembix crumpii (Pilsbry, 1893), from Japan to the East China Sea, has a similar shell to *B. bairdii*. It differs by being slightly broader and smaller, and is sculptured with sharper nodules. *Gaza fischeri* Dall, 1889, from Florida to the West Indies, and northern Gulf of Mexico, is another deepwater trochid with a thin-walled shell. Unlike *B. bairdii*, its shell is smooth.

The shell of the Baird's Bathybembix is medium in size, thin, lightweight, and taller than wider. Its spire is moderately high, and often eroded. The sculpture consists of 3 spiral rows of blunt tubercles above the incised suture and several spiral rows of small beads below the suture, as well as of fine growth lines. The aperture is large, with a thin, circular operculum; the outer lip is smooth, as is the columella. The shell lacks an umbilicus. The shell color is white, covered with a thin yellowish or pale orange periostracum; the aperture is white or yellowish.

Actual size

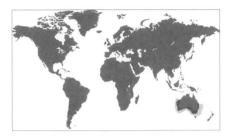

FAMILY	Trochidae
SHELL SIZE RANGE	¾ to 1½ in (20 to 40 mm)
DISTRIBUTION	New South Wales to Western Australia
ABUNDANCE	Common
DEPTH	Intertidal to 10 ft (3 m)
HABITAT	On seaweeds
FEEDING HABIT	Herbivore
OPERCULUM	Corneous, thin

SHELL SIZE RANGE
¾ to 1½ in
(20 to 40 mm)

PHOTOGRAPHED SHELL
1⅜ in
(37 mm)

226

PHASIANOTROCHUS EXIMIUS

GREEN JEWEL TOP

(PERRY, 1811)

Phasianotrochus eximius is a common trochid that lives on seaweed in tide pools and shallow water, along the open coast as well as in semi-protected areas, on seaweed. It ranges from New South Wales to Western Australia. This species varies greatly in color and pattern. It looks like a miniature of the larger phasianellid *Phasianella australis*, from southern Australia. The animal of *Phasianotrochus eximius* has a few sensory tentacles on the sides of the foot, in addition to the pair of cephalic tentacles.

RELATED SPECIES

Phasianotrochus bellulus (Philippi, 1845), from southern Australia, has a smaller and broader shell, with two teeth in the columella. Its shell is also very variable in color and patterns. *Cittarium pica* (Linnaeus, 1758), from the West Indies and Bahamas, has a top-shaped shell with black and white bands.

Actual size

The shell of the Green Jewel Top is medium-sized, smooth, glossy, elongate, and conical. Its spire is tall, with flattened convex whorls, incised suture, and a small, pointed apex. The surface is mostly smooth, with 4 spiral grooves on the upper whorls and 10 on the body whorl. The aperture is oval, the outer lip smooth and thin or sometimes thick; the columella has 1 tooth in mature specimens. The shell color is variable, ranging from mostly solid green, rose, gray, or brown color, to patterned with irregular lines. The aperture is highly iridescent.

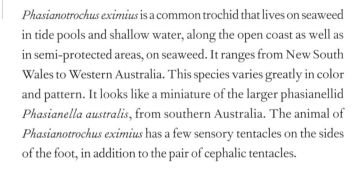

FAMILY	Trochidae
SHELL SIZE RANGE	1½ to 2½ in (38 to 64 mm)
DISTRIBUTION	Southern California to Baja California
ABUNDANCE	Uncommon
DEPTH	20 to 100 ft (6 to 30 m)
HABITAT	Rocky bottoms
FEEDING HABIT	Herbivore
OPERCULUM	Corneous, thin, circular

SHELL SIZE RANGE
1½ to 2½ in
(38 to 64 mm)

PHOTOGRAPHED SHELL
1¾ in
(43 mm)

TEGULA REGINA

QUEEN TEGULA

STEARNS, 1892

227

Tegula regina is an uncommon trochid that ranges from California, including Santa Catalina Island, to Baja California, Mexico. There are about 40 living species in the genus *Tegula*; about half live in the tropical and subtropical waters of the eastern Pacific and Caribbean, while the rest live in the temperate waters of the Pacific coasts of North and South America, and East Asia. Several groups of closely related species coexist in the same area, and are not divided by broad biogeographical or thermal barriers, suggesting that they have experienced radiation along single coastlines.

The shell of the Queen Tegula is medium-sized, thick, and top-shaped. Its spire is moderately high, with flat sides and a projecting, crenulated suture. Its sculpture consists of many diagonal axial cords, and arched lamellae on the base. The aperture is small and oblique, the outer lip crenulated, and the columella has one fold. The shell color is usually dark gray or black, but can have an orange band (as shown here); the base is black or gray, and the aperture iridescent, with a yellow tint.

Actual size

RELATED SPECIES

Tegula funebralis (Adams, 1855), from Vancouver, Canada, to Baja California, Mexico, is a common species in intertidal rocky pools. Its shell is dark gray to black, and the spire often eroded, showing its orange-brown apex. *Norrisia norrisii* (Sowerby II, 1838), which shares the same distribution as *T. regina*, has a larger shell, with a low spire, an umbilicus, and smooth surface.

FAMILY	Trochidae
SHELL SIZE RANGE	1¼ to 2¾ in (30 to 67 mm)
DISTRIBUTION	Gulf of California to western Mexico
ABUNDANCE	Moderately common
DEPTH	Offshore
HABITAT	Kelp beds
FEEDING HABIT	Herbivore
OPERCULUM	Corneous, circular

SHELL SIZE RANGE
1¼ to 2¾ in
(30 to 67 mm)

PHOTOGRAPHED SHELL
1¾ in
(44 mm)

228

NORRISIA NORRISII

NORRIS'S TOP

(SOWERBY II, 1838)

Norrisia norrisii has an unusual shell form among the Trochidae, with a smooth shell with rounded whorls. It has been classified as a turbinid, but anatomical studies confirm it belongs in the family Trochidae. The corneous operculum has a distinctive tufted spiral. The animal is bright red.

RELATED SPECIES

Tegula euryomphala (Jonas, 1844), from Peru and Chile, has a smaller shell with a generally smooth surface, a deep umbilicus, a taller spire, and gray color. *Tegula fasciata* (Born, 1778), which ranges from Florida to Brazil, has a small, smooth, and globose shell. The shell color can be bright or dark.

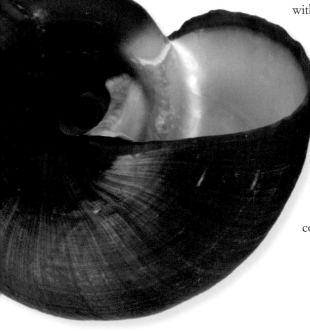

Actual size

The shell of the Norris's Top is swollen, with a very low spire. Its base is concave with a long, narrow umbilicus and the parietal rim is stained turquoise. There is a very large circular aperture with a shallow posterior canal. The interior is iridescently nacreous, and the shell is tan to chestnut, smooth except for growth lines.

FAMILY	Trochidae
SHELL SIZE RANGE	1 to 5⅜ in (25 to 136 mm)
DISTRIBUTION	Caribbean
ABUNDANCE	Common
DEPTH	Intertidal to 23 ft (7 m)
HABITAT	Rocks at open sea
FEEDING HABIT	Herbivore
OPERCULUM	Corneous, circular

SHELL SIZE RANGE
1 to 5⅜ in
(25 to 136 mm)

PHOTOGRAPHED SHELL
2¼ in
(58 mm)

CITTARIUM PICA
WEST INDIAN TOP
(LINNAEUS, 1758)

Cittarium pica is a common edible species of the West Indies, although it is rather tough and needs good culinary preparation, usually in soup. It lives in large groups on and under rocks, usually unsheltered and facing open seas. Because of its striking black and white markings it is also known to collectors as the Magpie Shell.

RELATED SPECIES

Oxystele sinensis (Gmelin, 1791) enjoys a similarly rocky habitat in southern African waters. It is nearly half the size of *Cittarium pica* but has the same proportions and rough surface. The shell is dark steely gray, usually with a white apex. Its umbilical area is white with a rose-tinted edge, and the operculum is yellow.

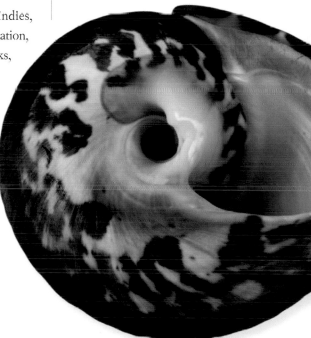

Actual size

The shell of the West Indian Top is heavy with a rough, uneven surface that is a dirty white with blotchy green-black axial bands. Its spire is fairly short with a moderately impressed suture and sloping shoulders. It has a gaping circular aperture from which the callused white columella reaches an open umbilicus. The interior is nacreous, closed by a green-brown multispiral operculum.

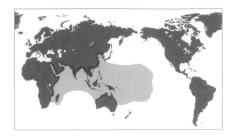

FAMILY	Trochidae
SHELL SIZE RANGE	2 to 6 in (50 to 150 mm)
DISTRIBUTION	Indo-Pacific
ABUNDANCE	Abundant
DEPTH	Subtidal to 60 ft (20 m)
HABITAT	On or near coral reefs
FEEDING HABIT	Herbivore
OPERCULUM	Corneous, circular

SHELL SIZE RANGE
2 to 6 in
(50 to 150 mm)

PHOTOGRAPHED SHELL
3⅞ in
(97 mm)

230

TECTUS NILOTICUS
COMMERCIAL TOP
(LINNAEUS, 1767)

The common name of *Tectus niloticus* refers to its widespread use as a source of mother-of-pearl for buttons and jewelry. It is the largest of the top shells and has the added attraction to commercial fishermen of being edible; the large foot of the animal is boiled and smoked. It is currently fished in modest quantities, mostly for the decorative trade. It is a striking shell, popular with both collectors and interior designers.

RELATED SPECIES

Tectus conus (Gmelin, 1791) is a smaller top shell with a broader range, extending into the Red Sea. Also prized for its mother-of-pearl, *T. conus* has a rounded lip and bottom, and a more tapering spire. It has a spiral sculpture of fine beads or nodules, and the less well-defined axial bands are dusky pink-red.

Actual size

The shell of the Commercial Top is equilateral in profile and smooth apart from the protoconch, which has hollow axial tubercles on the younger whorls. Mature specimens have a broadly bulging final whorl. The shell is white, with a pattern of broad axial bands of dark red-brown. The bottom is concave, and it lacks an umbilicus. The very open downward sloping aperture has a thin outer lip and a labial ridge on the columella.

FAMILY	Trochidae
SHELL SIZE RANGE	1⅝ to 4⅛ in (40 to 105 mm)
DISTRIBUTION	Red Sea, northwestern Indian Ocean
ABUNDANCE	Common
DEPTH	Subtidal to 30 ft (10 m)
HABITAT	In or near coral reefs
FEEDING HABIT	Herbivore
OPERCULUM	Corneous, circular

SHELL SIZE RANGE
1⅝ to 4⅛ in
(4 to 105 mm)

PHOTOGRAPHED SHELL
4 in
(103 mm)

TECTUS DENTATUS

DENTATE TOP

(FORSKÅL *IN* NIEBUHR, 1775)

231

All top shells are herbivorous (feeding mostly on seaweed), although some also eat sponges and bryozoans. *Tectus dentatus* forages for algae in the inner and mid ranges of coral reefs, with a relatively restricted distribution in the Red Sea and the northwestern corner of the Indian Ocean. It is, however, locally abundant. The animal has a large, muscular foot and a well-developed head. In profile, the shell itself looks like a fir tree covered in snow.

RELATED SPECIES

Tectus pyramis (Born, 1778) is of a similar size, but with a wider distribution in the Indo-Pacific. Its spire is shorter and the nodules are less well formed, particularly on the mature lower whorls (which are mottled brown and green), where they may occur as clusters of much smaller warts. A rarer unstained variant, *T. pyramis noduliferus* (Lamarck, 1822) is closer to *T. dentatus* in form and sculpture.

Actual size

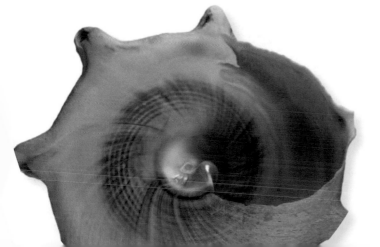

The shell of the Dentate Top is tall with a rough, dirty white surface. On the whorls, above and overlapping the finely incised suture, is a spiral band of evenly spaced, large, extended nodules. The bottom of the shell is flat and smooth with a blue-green stained band on the base.

FAMILY	Calliostomatidae
SHELL SIZE RANGE	1¼ to 1½ in (30 to 39 mm)
DISTRIBUTION	North Carolina to Florida Keys
ABUNDANCE	Uncommon
DEPTH	390 to 1,200 ft (120 to 365 m)
HABITAT	Hard substrates
FEEDING HABIT	Carnivore, feeds on sponges and tunicates
OPERCULUM	Rounded, corneous, multispiral

SHELL SIZE RANGE
1¼ to 1½ in
(30 to 39 mm)

PHOTOGRAPHED SHELL
1½ in
(38 mm)

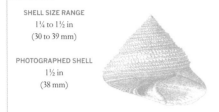

232

CALLIOSTOMA SAYANUM
SAY'S TOP
DALL, 1889

Calliostoma sayanum has a beautiful shell sculpted with beaded spiral rows. It is an uncommon snail from deep water. Calliostomatids usually live on hard substrates, normally found near their prey, which includes sponges, gorgonians, and tunicates. They were formerly included as a subfamily in the related Trochidae, but have recently been recognized as a separate family. There are more than 200 living species in the family Calliostomatidae worldwide.

RELATED SPECIES

Calliostoma zizyphinum (Linnaeus, 1758), from western Europe to the Canary Islands and the Mediterranean, has a very variable shell. *Maurea tigris* (Martyn, 1784), from New Zealand, has the largest shell in the family. Its shell has brown flames on a beige background.

Actual size

The shell of the Say's Top is small to medium, solid, umbilicate, and top shaped. Its spire is pointed, with an angle of about 90 degrees, and the suture moderately well marked. The whorls are usually flat sided, with a rounded periphery. The sculpture consists of about 8 to 10 finely beaded spiral cords on the sides and about 15 on the base. The umbilicus is narrow, deep, and surrounded by a strongly beaded cord. The aperture is subquadrate, the outer lip smooth, and columella slightly thickened. The shell color is golden brown with a red band at the whorl periphery, and a white aperture.

FAMILY	Calliostomatidae
SHELL SIZE RANGE	2 to 4 in (50 to 100 mm)
DISTRIBUTION	Endemic to New Zealand
ABUNDANCE	Moderately common
DEPTH	Intertidal to 700 ft (210 m)
HABITAT	Among boulders
FEEDING HABIT	Carnivore
OPERCULUM	Corneous, circular, yellowish

SHELL SIZE RANGE
2 to 4 in
(50–100 mm)

PHOTOGRAPHED SHELL
2¼ in
(57 mm)

MAUREA TIGRIS
TIGER MAUREA
(MARTYN, 1784)

233

Maurea tigris is the largest of the maureas and the calliostomatids. The family Calliostomatidae has several genera, including the genus Maurea, which is the second most diverse, after the genus *Calliostoma*. Like many species of Maurea, *M. tigris* is endemic to New Zealand, and has been collected throughout the main islands. It ranges from the intertidal zone to deep water, on rocky substrates. *Maurea tigris* is also known as a fossil from Pleistocene deposits.

RELATED SPECIES

Maurea punctulata (Martyn, 1784), another form endemic or restricted to New Zealand, has a blunter spire, more rounded shoulders, and broader, beaded spiral cords broken into beading by growth lines. These raised beads are cream, against a background of streaky pale to mid tan. The siphonal angle in the aperture is less pronounced.

Actual size

The shell of the Tiger Maurea is cream with broad, axial orange-brown zigzags. Its spire is tall, steepening toward the sharp apex. All whorls are covered with very fine spiral beading, and are slightly convex; the body whorl is very swollen toward the base. The aperture is very open and round, with a marked angle at the siphonal notch.

FAMILY	Calliostomatidae
SHELL SIZE RANGE	1¼ to 2⅜ in (30 to 60 mm)
DISTRIBUTION	Northern and central New Zealand
ABUNDANCE	Locally abundant
DEPTH	Intertidal to 300 ft (90 m)
HABITAT	Sandy beaches
FEEDING HABIT	Carnivore
OPERCULUM	Corneous, circular

SHELL SIZE RANGE
1¼ to 2⅜ in
(30–60 mm)

PHOTOGRAPHED SHELL
2¼ in
(58 mm)

234

MAUREA SELECTA
SELECT MAUREA
DILLWYN, 1817

Maurea selecta is one of the largest species in the genus, and like some calliostomatids, it is endemic to New Zealand. It is abundant especially along the Wellington west coast. The shell is top-shaped, with thin walls, and is elegantly ornamented with fine spiral cords bearing small nodules. It is found from the intertidal zone on sandy beaches, to offshore. Shells from deeper water tend to be larger, heavier, and have a taller spire than those from shallow water.

RELATED SPECIES

Maurea pellucida (Valenciennes, 1846) is found only in northern New Zealand. It has a similar sculpture of nodules on the body that, however, also extend across the base. There may also be larger splashes of mid tan around the shoulders, especially on the body whorl.

The shell of the Select Maurea is equilateral in profile with a steep, sharp apex. The spiral whorls are indistinct, the suture only impressed above the body whorl. Fine, well-spaced spiral bands of very small off-white and tan nodules cover the cream to tan shell. The base is paler, with similarly fine but continuous spiral cords. The exterior decoration can be seen through the large, flattened aperture, which has a very thin lip and a callused columella.

Actual size

FAMILY	Seguenziidae
SHELL SIZE RANGE	⅛ to ⅕ in (4 to 5 mm)
DISTRIBUTION	Western Atlantic, from North Carolina to Brazil
ABUNDANCE	Uncommon
DEPTH	325 to 4,050 ft (100 to 1,235 m)
HABITAT	Fine sand and mud bottoms
FEEDING HABIT	Detritivore
OPERCULUM	Circular, thin, concave

SHELL SIZE RANGE
⅛ to ⅕ in
(4 to 5 mm)

PHOTOGRAPHED SHELL
⅛ in
(4 mm)

CARENZIA TRISPINOSA

THREE-ROWED CARENZIA

(WATSON, 1879)

235

Like other members of the family Seguenziidae, *Carenzia trispinosa* is a small species that inhabits fine muddy bottoms at bathyal depths. These animals digest the organic component of the sediments they consume. Despite their small size, many members of the family have very elaborate shell sculpture, often with multiple carinae and complex indentations and expansions of the outer lip.

RELATED SPECIES

Carenzia carinata (Jeffreys, 1877), inhabits comparable depths throughout the Atlantic, from southern Brazil, to the Azores and Canary Islands. It has a shorter, broader, smoother shell, and lacks nodes. *Thelyssa callisto* Bayer, 1971, from the abyssal plain off the Bahamas, is nearly twice the size of *C. trispinosa*, and has a smooth, less stepped shell, with a nearly flat base and a rhomboidal aperture. *Seguenzia lineata* Watson, 1879, another related species that ranges from Mexico to Brazil, has a much taller shell that lacks an umbilicus but has an outer lip with a complex outline.

The Three-rowed Carenzia has a small, conical shell with a rounded base and a broad, deep umbilicus. The spire is tall and weakly stepped. A roughly rectangular aperture forms a narrow carina along the shell periphery, a broader, knobby cord near the suture, and a weaker band midway between these features. The columella has a complex fold. The shell appears white, as the pearly, nacreous inner layer is visible through the thin, unpigmented outer shell layer.

Actual size

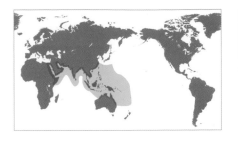

FAMILY	Neritopsidae
SHELL SIZE RANGE	½ to 1⅜ in (13 to 35 mm)
DISTRIBUTION	Red Sea to Okinawa and New Caledonia
ABUNDANCE	Uncommon
DEPTH	Intertidal to 133 ft (40 m)
HABITAT	Caves and other cryptic habitats
FEEDING HABIT	Unknown
OPERCULUM	Calcareous, thick, trapezoidal, white

SHELL SIZE RANGE
½ to 1⅜ in
(13 to 35 mm)

PHOTOGRAPHED SHELL
1¼ in
(30 mm)

236

NERITOPSIS RADULA

RADULA NERITE

(LINNAEUS, 1758)

The shell of the Radula Nerite is small, thick, and globose, with a depressed to relatively high spire. The shell sculpture consists of spiral beaded cords crossed by growth lines, giving it a rough texture that is reminiscent of a radula, hence the name. Its aperture is oval-rounded and the outer lip crenulated; the inner lip has a squarish U-shaped notch in the mid columella. The operculum is calcareous, thick, and trapezoidal, with a squarish peg that fits the notch in the columella. The shell color is off-white, cream, or light brown. The aperture is white and glossy.

Neritopsis radula is considered a living fossil. It is a representative of a group that has a long fossil history, dating to the Middle Devonian Period, some 350 million years ago. About 100 fossil species are recognized. *Neritopsis radula* has been known for more than 250 years, and was considered the sole living species until 1973, when a second species was discovered in Cuba. More recently, several additional living species were discovered from the Red Sea and French Polynesia. *Neritopsis radula* is common in underwater caves and other cryptic habitats, but its biology remains poorly known.

Actual size

RELATED SPECIES

Neritopsis atlantica Sarasúa, 1973, is a rare species from Cuba and Trinidad; it has a small shell that resembles *N. radula*. *Neritopsis richeri* Lozouet, 2009, was recently described from French Polynesia. It differs by having a shell with weaker beads but more numerous spiral cords.

FAMILY	Neritidae
SHELL SIZE RANGE	⅛ to ⅜ in (3 to 10 mm)
DISTRIBUTION	Florida to Brazil, western Spain to northwestern Africa, Mediterranean, Red Sea
ABUNDANCE	Common
DEPTH	Intertidal to 10 ft (3 m)
HABITAT	Grass and seaweed beds
FEEDING HABIT	Herbivore
OPERCULUM	Calcareous, semicircular

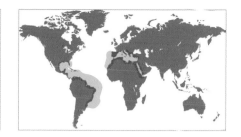

SHELL SIZE RANGE
⅛ to ⅜ in
(3 to 10 mm)

PHOTOGRAPHED SHELL
¼ in
(7 mm)

SMARAGDIA VIRIDIS

EMERALD NERITE

(LINNAEUS, 1758)

237

The nerites are a very large family of small species, of which *Smaragdia viridis* is one of the smallest. Its range is wide, lining both the eastern and western subtropical coasts of the North and South Atlantic. Nerites have colonized almost every kind of aquatic environment from deep sea to freshwater lake, and some species even live in trees. Part of their success may be due to the very close-fitting operculum, behind which they can store water for survival in times and places of relative dryness.

RELATED SPECIES

Theodoxus oualaniensis (Lesson, 1831) is a species that inhabits the shallow grass beds of the Indo-Pacific. It is highly polished, and usually olive green or off-white, with greatly varied markings, among them: broad, cream spiral bands edged in black, blue-black reticulate markings, axial zigzags, or more often a combination of some or all of these.

Actual size

The shell of the Emerald Nerite is smooth and globular, with an almost completely sunken spire. It is bright yellow-green, decorated with several rows of patchy white axial bars, sometimes edged in purple-black. Specimens from deeper water may be white with only the dark edge-marks. There are 7 to 9 teeth on the columella, and a parietal shield of white-green.

FAMILY	Neritidae
SHELL SIZE RANGE	⅛ to ⅖ in (5 to 10 mm)
DISTRIBUTION	Southeast U.S.A. to the Caribbean and Bermuda
ABUNDANCE	Abundant
DEPTH	Intertidal to 3 ft (1 m)
HABITAT	Intertidal rocky shores and tide pools
FEEDING HABIT	Herbivore, grazes on algae
OPERCULUM	Calcareous, paucispiral, with an internal peg

SHELL SIZE RANGE
⅛ to ⅖ in
(5 to 10 mm)

PHOTOGRAPHED SHELL
½ in
(12 mm)

238

ZEBRA NERITE

(LINNAEUS, 1767)

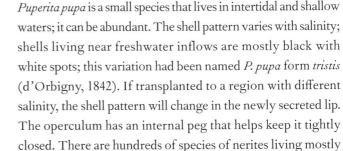

Actual size

Puperita pupa is a small species that lives in intertidal and shallow waters; it can be abundant. The shell pattern varies with salinity; shells living near freshwater inflows are mostly black with white spots; this variation had been named *P. pupa* form *tristis* (d'Orbigny, 1842). If transplanted to a region with different salinity, the shell pattern will change in the newly secreted lip. The operculum has an internal peg that helps keep it tightly closed. There are hundreds of species of nerites living mostly in the tropics worldwide, with some in brackish and fresh water.

RELATED SPECIES

Several related species have patterns similar to *Puperita pupa*, among them *Neritina zebra* (Bruguière, 1792), from Honduras to Brazil, which has an orange or reddish shell and black, diagonal zigzag lines; *N. virginea* (Linnaeus, 1758), from Florida and the Caribbean to Brazil, with a very polychromic shell; and *N. turrita* (Gmelin, 1791), from the western Pacific, which has a yellow shell with broad, irregular, diagonal black bands.

The shell of the Zebra Nerite is small, globular, thick, and solid. The spire is low, often eroded; the body whorl is large, rounded, and smooth, with very fine axial or spiral lines. Its aperture is the typical half-moon shape found in most nerites, closed by a calcareous operculum of the same shape. The columella is straight, with 4 denticles and a callused parietal shield. The shell color is white with irregular, diagonal, zigzag black stripes forming beautiful patterns; the aperture is yellow to orange. Like other nerites, no two shells have the same pattern.

FAMILY	Neritidae
SHELL SIZE RANGE	¾ to 1¼ in (20 to 30 mm)
DISTRIBUTION	Southern and western Pacific
ABUNDANCE	Abundant
DEPTH	Intertidal
HABITAT	Mangroves
FEEDING HABIT	Herbivore
OPERCULUM	Calcareous, semicircular

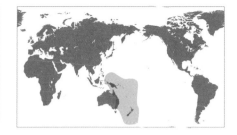

SHELL SIZE RANGE
¾ to 1¼ in
(20 to 30 mm)

PHOTOGRAPHED SHELL
1 in
(24 mm)

NERITODRYAS CORNEA

HORNY NERITE

(LINNAEUS, 1758)

With the exception of a few species, such as *Neritina communis* and *N. pelaronta*, nerites are rather overlooked by collectors. It's a surprising omission; there can be few families that exhibit such a variety of color and pattern on a fairly uniform shell shape, not just between species but even within. Their diversity of decoration is almost matched by their evolutionary ability to adapt to all manner of water conditions from salt to fresh, including, in the case of *N. cornea*, the brackish environment of the mangroves.

Actual size

RELATED SPECIES

Neritina violacea (Gmelin, 1791), which also inhabits the southern and western Pacific, has adjusted to a different level of salinity, and lives near mangroves rather than in them. It is white with moderately fine violet to purple axial zigzags. The aperture (with a much extended parietal shield) is deep orange-brown.

The shell of the Horny Nerite is globular with a very low spire. It is black, with 2 cream to tan spiral bands of short diagonal dashes on the body. The same dashes are less organized on the spire and bottom of the shell. The aperture opens slightly downward; the lip and columella are very white, and the interior shows the exterior pattern.

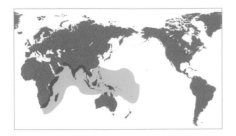

FAMILY	Neritidae
SHELL SIZE RANGE	¾ to 1⅜ in (17 to 35 mm)
DISTRIBUTION	Eastern Africa to central Pacific
ABUNDANCE	Common
DEPTH	Intertidal
HABITAT	Rocks
FEEDING HABIT	Herbivore
OPERCULUM	Calcareous, semicircular

SHELL SIZE RANGE
¾ to 1⅜ in
(17 to 35 mm)

PHOTOGRAPHED SHELL
1 in
(27 mm)

240

NERITA COSTATA
RIBBED NERITE
GMELIN, 1791

Nerita costata is one of the less showy species in the Neritidae family, many of whose members are distinguished by dazzling variations in both color and pattern. But the muted rhythmic sculpture of its exterior does enclose a classical neritid interior: the very thick lip and columella, the semicircular mouth, and its two rows of teeth. Like all Neritidae, it is vegetarian, feeding on the algae that thrive on rocks heated by the sun in warm, shallow, intertidal waters.

RELATED SPECIES

Nerita exuvia (Linnaeus, 1758) is another ribbed nerite living on intertidal rocks, near mangroves in the southwestern Pacific. Its ribs are dark brown, separated by spiral channels of cream or tan. There are more, and narrower, labial teeth than in *N. costata*; those on the columella are less prominent, but the parietal wall has a row of tiny pustules.

Actual size

The shell of the Ribbed Nerite is globular, with only the apex of the sunken spire showing above the body whorl. It is dark brown, with bold spiral ribs patterned only by paler, axial growth lines. The lip has 7 or 8 teeth, the columella 3 or 4; both are thick and white, and the latter is callused and thinly extended across the whorl and apex.

FAMILY	Neritidae
SHELL SIZE RANGE	¾ to 1½ in (20 to 38 mm)
DISTRIBUTION	Western Mexico to Peru, not Galápagos Islands
ABUNDANCE	Uncommon
DEPTH	Intertidal, shallow
HABITAT	Rocks in small river mouths
FEEDING HABIT	Herbivore
OPERCULUM	Calcareous, semicircular

SHELL SIZE RANGE
¾ to 1½ in
(20 to 38 mm)

PHOTOGRAPHED SHELL
1¼ in
(32 mm)

NERITINA LATISSIMA

WIDEST NERITINA

(BRODERIP, 1833)

241

Neritinas vary from other nerites in having a thin, smooth outer lip and several fine folds rather than teeth on the columella. Their most distinctive feature is the small armlike addition to the operculum, which is used to grip onto surfaces. *Neritina latissima* lives on rocks in large colonies in the tidal estuaries of smaller rivers and streams, and it displays the neritid capacity to endure both absence of water (low tide) and varying salinities (from saline to fresh).

RELATED SPECIES

Neritina auriculatum (Lamarck, 1816) has a parietal shield even more greatly expanded in all directions, hence its name the Eared Nerite. The "ears" (above and below the aperture) and outer lip are lilac gray, the columella white, and the rest of the shell olive brown with fine spiral grooves.

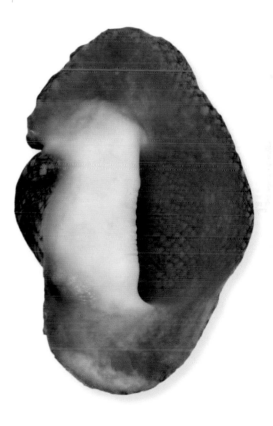

Actual size

The shell of the Widest Neritina is globular with a sunken spire. It is cream to tan with a mauve to olive brown reticulate pattern. The apex is white. However, the body features are overshadowed by the greatly expanded lip and columella, which extend above and below the body whorl. The lip is pale to mid-lilac gray, and the parietal shield white to yellow. There are numerous fine columellar pleats.

FAMILY	Neritidae
SHELL SIZE RANGE	¾ to 2 in (20 to 49 mm)
DISTRIBUTION	Southeastern Florida to Venezuela, Caribbean
ABUNDANCE	Abundant
DEPTH	Intertidal to high tide
HABITAT	Rocks facing open sea
FEEDING HABIT	Herbivore
OPERCULUM	Calcareous, semicircular

SHELL SIZE RANGE
¾ to 2 in
(20 to 49 mm)

PHOTOGRAPHED SHELL
1¼ in
(32 mm)

242

NERITA PELORONTA
BLEEDING-TOOTH NERITE
LINNAEUS, 1758

With its gory splash of color and gap-toothed columella, *Nerita peloronta* is a good visual joke, the puffin of the Neritidae. The joke tends to overshadow the rather beautiful surface decoration of the whorl. The animal can survive above the high tide line thanks to its tight-fitting operculum, which traps and retains moisture inside the shell. The operculum is colorful: externally part red, part blue-green, and internally a deep orange.

The shell of the Bleeding-tooth Nerite is globular, with a low spire. The shallow cream spiral cords have bands of dusky red and gray-black spots, in axial zigzags. There are fine dental pleats on the lip, and 4 prominent teeth on the white columella. The interior is yellow, and the white parietal shield extends thinly over the whorl.

Actual size

RELATED SPECIES

Nerita versicolor Gmelin, 1791, shares the same distribution and habitat. It has an all-white dentate columella, and two prominent teeth inside the outer lip. The spiral cords are more pronounced, and the dots of black and red on them are too few to combine in significant axial bands as they do on *N. peloronta*.

FAMILY	Neritidae
SHELL SIZE RANGE	½ to 1⅝ in (13 to 40 mm)
DISTRIBUTION	Red Sea to Indo-West Pacific
ABUNDANCE	Abundant
DEPTH	Intertidal
HABITAT	Rocks near sand
FEEDING HABIT	Herbivore
OPERCULUM	Calcareous, semicircular

SHELL SIZE RANGE
½ to 1⅝ in
(13 to 40 mm)

PHOTOGRAPHED SHELL
1⅜ in
(36 mm)

NERITA POLITA

POLISHED NERITE

LINNAEUS, 1758

243

The Polished Nerite comes in a bewildering range of color and pattern combinations. No single specimen can be typical, and it is no surprise that researchers have sought to define several subspecies. *Nerita polita antiquata* in northern Australia is one of the most colorful; it is distinguished by a continuous band of yellow-orange that spans the columella and outer lip, within a white margin. Like all nerites, *N. polita* lives in large groups and feeds on algae, which flourishes in tropical intertidal waters.

The shell of the Polished Nerite is globular, with an almost completely sunken spire. It is smooth and usually cream, white, or pale green, with spiral bands of orange, white, cream, or red that may be plain, marbled, or axially lined. The columella is weakly pleated and the lip has very fine lirations (elevations). There may be a pale yellow tint to the margin of the aperture.

RELATED SPECIES

Neritina communis (Quoy and Gaimard, 1832), is one of the few widely collected nerites, brighter and bolder in its infinite variations. Broad spiral bands may be pink, red, black, yellow, or cream, separated by diagonally or axially striped black ones. It is confined to the southwestern Pacific.

Actual size

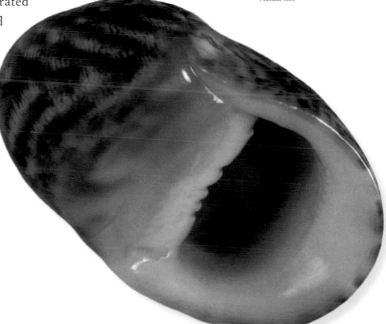

FAMILY	Neritidae
SHELL SIZE RANGE	¾ to 1 in (20 to 50 mm)
DISTRIBUTION	Eastern Africa to western Pacific
ABUNDANCE	Common
DEPTH	Shoreline
HABITAT	High rocks
FEEDING HABIT	Herbivore
OPERCULUM	Calcareous, semicircular

SHELL SIZE RANGE
¾ to 1 in
(20 to 50 mm)

PHOTOGRAPHED SHELL
1½ in
(36 mm)

244

NERITA TEXTILIS

TEXTILE NERITE

GMELIN, 1791

Thick walls and a tight-closing operculum enable many species of Neritidae to withstand apparently harsh habitats. They are able, by these characteristic features, to trap water inside the shell, thereby surviving for extended periods out of water, or even—as with the Textile Nerite—colonizing rocky surfaces above the high tide line that are only splashed by wave spray and never submerged. Perhaps not being subject to the erosive influence of waves has allowed *Nerita textilis* to evolve as the most sculptural of nerites.

RELATED SPECIES

Nerita undata Linnaeus, 1758, has a persistent small spire, and flatter spiral cords than *N. textilis*. Bands of black or olive dashes on the cords may more or less combine to form axial flames against the cream to tan background. It is a smaller shell, living on intertidal rocks of the Indo-Pacific.

The shell of the Textile Nerite is highly sculptural, with a flat spire. Thick spiral ribs are broken by growth lines that give them the appearance of twisted cords. They are white with frequent elongated black dots, rarely contiguous with those on adjacent cords and therefore looking like interwoven black ribbon. The aperture is dentate, and white or pale yellow; the columella is postulate and tinted pale orange at the base of the parietal wall.

Actual size

FAMILY	Neritidae
SHELL SIZE RANGE	⅜ to 1 in (14 to 51 mm)
DISTRIBUTION	Gulf of California to Ecuador, Galápagos Islands
ABUNDANCE	Common
DEPTH	Intertidal to supertidal
HABITAT	Rocks
FEEDING HABIT	Herbivore
OPERCULUM	Calcareous, semicircular

SHELL SIZE RANGE
⅜ to 1 in
(14 to 51 mm)

PHOTOGRAPHED SHELL
1⅝ in
(40 mm)

NERITA SCABRICOSTA
ROUGH-RIBBED NERITE
LAMARCK, 1822

245

Nerita scabricosta is one of the largest nerites, and one of several that have managed to colonize the regularly exposed high tide zone. It is most common toward the north of its distribution; further south it is replaced by a subspecies, *N. s. ornata*, the Ornate Nerite, which has smoother, shallower, more regular ribs. It is sometimes ascribed to the subgenus *Ritena*, whose chief feature is the broad uneven folds on the columellar callus.

The shell of the Rough-ribbed Nerite is globular with a short spire, the body whorl much swollen behind the aperture. It has a sculpture of uneven dark gray to black spiral ribs with occasional elongated spots, streaky orange on the body but white on the spire. The aperture is white. There is a pleated dental ridge on the thickened lip, 4 deep folds on the columella, and irregular pleating on the parietal shield.

RELATED SPECIES
Nerita funiculata (Menke, 1851) occurs in the same range of distribution. It is a small shell, with a mid-gray body of fine, relatively widely spaced, spiral ribs and a fairly flat, pale gray spire. Its lip is finely dentate, and there are irregular, elongated pustules on the parietal callus.

Actual size

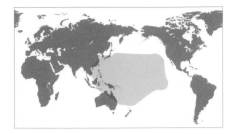

FAMILY	Neritidae
SHELL SIZE RANGE	¾ to 1¾ in (20 to 42 mm)
DISTRIBUTION	Western and central Pacific
ABUNDANCE	Uncommon
DEPTH	Intertidal
HABITAT	Rocks
FEEDING HABIT	Herbivore
OPERCULUM	Calcareous, semicircular

SHELL SIZE RANGE
¾ to 1¾ in
(20 to 42 mm)

PHOTOGRAPHED SHELL
1⅝ in
(41 mm)

246

NERITA MAXIMA
MAXIMUM NERITE
GMELIN, 1791

Nerita maxima feeds at night, when it can be seen at its most active on overhanging rocks, clefts, and sheltered corners of reefs and other intertidal formations. It is edible and, as the name suggests, large; it frequently forms part of the harvest of vakacakau or reef gleaning in the central Pacific. This is the highly skilled and knowledgeable removal (either for commercial gain or for self-sufficiency) of useful marine life from the intertidal zone—a task usually undertaken by women.

RELATED SPECIES

Nerita plicata Linnaeus, 1758, from the Indo-Pacific has more clearly defined spiral cords. This species may bear dark gray patches, as rough axial stripes, but is more often plain off-white to cream. The labial teeth are formed by fewer and deeper folds, and the columellar dental pleats are elongate across the parietal shield.

The shell of the Maximum Nerite is globular, with an engorged downward-sloping aperture. It has a very low spire with a white apex; the rest of the shell is white or off-white, with blurred dark gray patches. These are defined sometimes by very fine spiral grooves, sometimes by axial growth lines, giving a checked appearance. The aperture is white, stained pale apricot on the finely dentate lip and square-toothed columella.

Actual size

FAMILY	Phenacolepadidae
SHELL SIZE RANGE	½ to 1 in (13 to 25 mm)
DISTRIBUTION	Red Sea to India and Sri Lanka
ABUNDANCE	Locally common
DEPTH	Under intertidal rocks and boulders
HABITAT	Rocks in sand and mud bottoms
FEEDING HABIT	Detritivore
OPERCULUM	Vestigial

SHELL SIZE RANGE
½ to 1 in
(13 to 25 mm)

PHOTOGRAPHED SHELL
⅞ in
(21 mm)

PLESIOTHYREUS CYTHERAE
VENUS SUGAR LIMPET
(LESSON, 1831)

247

Although closely related to nerites, species in the family Phenacolepadidae have secondarily acquired a limpet shape. Most live attached to rocks partially buried in mud in shallow waters of tropical and subtropical regions. One group is restricted to deepsea hydrothermal vents. The animals are detritivores. They are unusual among gastropods in having erythrocytes (red blood cells) rather than the bluish, respiratory protein hemocyanin within the blood.

RELATED SPECIES

Plesiothyreus galathea (Lamarck, 1819), from the tropical Indo-Pacific has a smaller, thinner, more elongated shell, with more numerous and finer radial cords that are beaded. *Phenacolepas asperulata* (A. Adams, 1858), from the Indian Ocean, also has a much smaller and more narrowly ovate shell with much finer, beaded radial ribs. *Cinnalepeta pulchella* (Lischke, 1871), which ranges from Japan to Vietnam, has a smaller, orange-colored shell in which the apex extends beyond the posterior margin of the aperture.

The Venus Sugar Limpet has a large, broad, bilaterally symmetrical, cap-shaped shell with a strongly convex anterior and a straight to slightly concave posterior slope. The sculpture consists of numerous broad cords radiating from the apex. As the shell grows, the distance between adjacent cords increases and new cords are formed between them. The aperture is broadly oval, with a horseshoe-shaped muscle scar that is open anteriorly. The shell color is white.

Actual size

FAMILY	Abyssochrysidae
SHELL SIZE RANGE	1¼ to 2 in (30 to 50 mm)
DISTRIBUTION	South Africa; Philippines
ABUNDANCE	Rare
DEPTH	1,640 to 9,200 ft (500 to 2,800 m)
HABITAT	Muddy bottom
FEEDING HABIT	Deposit feeder
OPERCULUM	Corneous, thin, soft

SHELL SIZE RANGE
1¼ to 2 in
(30 to 50 mm)

PHOTOGRAPHED SHELL
1⅜ in
(33 mm)

248

ABYSSOCHRYSOS MELANIOIDES

MELANIOID ABYSSAL SNAIL

TOMLIN, 1927

Abyssochrysos melanioides is a representative of a small family of gastropods that live in very deep waters. Because of its depth, it is rarely collected, but it may be common in its muddy bottom habitat. It is known from off South Africa and the Philippines, but further exploration of the deep sea may reveal it in other places in the Indian Ocean. Since it lives beyond the penetration of sunlight, the animal is blind. There are only six known living species in the family Abyssochrysidae worldwide; all live in deep water. It appears to be related to the family Pseudozygopleuridae, which flourished 200 million years ago.

RELATED SPECIES

Abyssochrysos melvilli (Schepman, 1909), from South Africa, Philippines and Indonesia, has a slightly smaller and more slender shell. The periostracum is olive-green. *Abyssochrysos brasilianum* Bouchet, 1991, from southeastern Brazil, has a shell similar to *A. melvilli*, but smaller, with fewer whorls.

The shell of the Melanioid Abyssal Snail is medium-sized, thin, elongate, and conical in outline. Its spire is very high and accounts for about 80 percent of the shell height, with about 14 whorls; the suture is well marked. The sculpture consists of about 12 to 14 axial ribs with nodules at the bottom of each axial rib. The outer lip is thin and sharp, the columella smooth, and the aperture oval. The shell color is white, covered with an olive brown periostracum, which confers a golden color to the shell. The aperture is white.

Actual size

FAMILY	Cerithiidae
SHELL SIZE RANGE	⅛ to ¼ in (3 to 6 mm)
DISTRIBUTION	Hawaii to French Polynesia
ABUNDANCE	Abundant
DEPTH	Intertidal
HABITAT	Sand and mudflats
FEEDING HABIT	Herbivore
OPERCULUM	Corneous, paucispiral

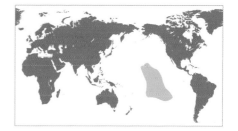

SHELL SIZE RANGE
⅛ to ¼ in
(3 to 6 mm)

PHOTOGRAPHED SHELL
⅛ in
(4 mm)

ITTIBITTIUM PARCUM

POOR ITTIBITTIUM

(GOULD, 1861)

249

The genus *Ittibittium* was established to encompass a group of minute ceriths that share a characteristic protoconch shape as well as several distinctive anatomical features. This species deposits large eggs, each in a separate capsule, with many capsules stacked in a short, gelatinous tube. The larvae develop within their capsules and hatch as crawling juvenile snails. Like all Cerithiidae, they feed on the algae and other organic scraps found in the rich environment of the mudflats, where they live in large groups.

RELATED SPECIES

Ittibittium parcum is the largest of the species of that genus. *Ittibittium nipponkaiense* (Habe and Masuda, 1990) occurs in Japanese waters, while *I. turriculum* (Usticke, 1969) is from the Virgin Islands of the western Atlantic.

The shell of the Poor Ittibittium is shiny and tiny, with a very tall spire with a deeply impressed suture. The whorls have moderately rounded spiral cords of varying width and faint axial ribs, sometimes nodulose on the shoulders—on the younger whorls the cord on the shoulder is slightly raised. The shell is off-white to cream, with chestnut axial streaks. The ovate aperture has thin lips and a deep siphonal notch.

Actual size

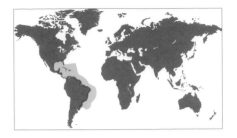

FAMILY	Cerithiidae
SHELL SIZE RANGE	⅝ to 1⅜ in (15 to 36 mm)
DISTRIBUTION	Southeastern Florida to Brazil
ABUNDANCE	Common
DEPTH	Intertidal to 150 ft (50 m)
HABITAT	Sand and mudflats
FEEDING HABIT	Herbivore
OPERCULUM	Corneous, paucispiral

SHELL SIZE RANGE
⅝ to 1⅜ in
(15 to 36 mm)

PHOTOGRAPHED SHELL
1 in
(24 mm)

250

CERITHIUM LITTERATUM
STOCKY CERITH
(BORN, 1778)

There are several hundred species of Cerithiidae, occurring throughout the world's tropical and temperate seas, although there are noticeably fewer in European waters. They are all vegetarian, scavenging for algae on muddy or sandy bottoms. They usually have a well-developed siphonal notch to accommodate their long siphons. Most ceriths have elongated shells, with varying degrees of surface sculpture. The name *litteratum*, means "marked with letters," and refers to the spiral color pattern, which looks like rows of letters on a document.

RELATED SPECIES

Cerithium muscarum Say, 1832, is a more slender cerith from shallow water in the northern reaches of the same region. The spiral dots are more widely spaced, and there are three or four spiral rows of nodules on each whorl, which tend to link axially to form ridges.

The shell of the Stocky Cerith is bulbous with a tall, convexly tapering spire. It is white to cream, with uneven spiral bands of dark brown, broken irregularly by axial growth lines. There is a spiral row of low to moderate nodules just below the suture. The ovate aperture has deep anterior and posterior canals, a short, thin parietal wall, and a pleated outer lip.

Actual size

FAMILY	Cerithiidae
SHELL SIZE RANGE	1⅜ to 2⅛ in (35 to 53 mm)
DISTRIBUTION	Bahamas, northern Cuba
ABUNDANCE	Rare
DEPTH	Intertidal to 3 ft (1 m)
HABITAT	Sand and mudflats
FEEDING HABIT	Herbivore
OPERCULUM	Corneous, paucispiral

SHELL SIZE RANGE
1⅜ to 2⅛ in
(35 to 53 mm)

PHOTOGRAPHED SHELL
1¼ in
(32 mm)

FASTIGIELLA CARINATA

CARINATE FALSE CERITH

REEVE, 1848

251

Fastigiella carinata is the only living example of its genus, although there are fossil records from the Eocene Period of possible related species. The shell itself has been reclassified several times since its original discovery. Only in the 1980s, when a live specimen was collected for the first time, was Reeve's original generic attribution confirmed. It remains extremely rare, with known specimens numbered only in the low hundreds.

RELATED SPECIES

The genus *Fastigiella* is thought to be related to the Indo-West Pacific genus, *Pseudovertagus*. *Pseudovertagus aluco* (Linnaeus, 1758) is the type species, with a very tall spire of sharply stepped whorls accentuated by a spiral of nodules on each shoulder. It is cream with axial dark speckling, especially along the anterior part of the shell.

The shell of the Carinate False Cerith is porcelain-white and conical with a tall spire, the whorls marked only by a finely inscribed suture. The whole shell is spirally sculpted with deep channels and high raised cords, most prominent immediately below the suture. The cord immediately above the suture is flattened. The ovate aperture has a deep anterior and narrow posterior canal, and a chestnut-stained fold in the middle of the columella.

Actual size

FAMILY	Cerithiidae
SHELL SIZE RANGE	1 to 2 in (25 to 50 mm)
DISTRIBUTION	Indo-West Pacific
ABUNDANCE	Uncommon
DEPTH	Subtidal
HABITAT	Sand near mangroves
FEEDING HABIT	Herbivore
OPERCULUM	Corneous, paucispiral

SHELL SIZE RANGE
1 to 2 in
(25 to 50 mm)

PHOTOGRAPHED SHELL
1½ in
(37 mm)

252

CERITHIUM CITRINUM
YELLOW CERITH
SOWERBY II, 1855

All Cerithiidae feed on algae and the detritus of decayed plants, which naturally they find near the shores where those plants grew. In this shallow habitat they are among the commonest families of gastropods. The greatest concentration of cerithiid species is to be found in the Indo-Pacific zone, and many of them, including *Cerithium citrinum*, have adapted to life in or near mangroves, where their organic diet is amply catered for.

RELATED SPECIES

Cerithium novaehollandiae Sowerby II, 1855, is confined to northern Australia, but has a similar sculpture. The suture is slightly deeper, the axial ribs less pronounced, the whitish whorls stained brown over the lower half. It has a narrower aperture than *C. citrinum*, with a shorter but recurved siphonal canal, and the spiral cords have less impact on the profile of the outer lip.

The shell of the Yellow Cerith is conical and the tall spire has a moderate suture. The slightly convex whorls, exaggerated by deep axial ribs, carry close, fine, uneven spiral cords. The body whorl and a flaring pleated outer lip are lemon yellow, younger whorls are paler with some white callusing. The nearly round aperture has a long anterior canal. Its columella and fasciole are white, the latter splashed with dark red.

Actual size

FAMILY	Cerithiidae
SHELL SIZE RANGE	1 to 2 in (27 to 52 mm)
DISTRIBUTION	Eastern Indonesia and Papua New Guinea
ABUNDANCE	Rare in collections
DEPTH	Intertidal to 65 ft (20 m)
HABITAT	Muddy or fine sandy bottoms
FEEDING HABIT	Grazer
OPERCULUM	Corneous, with terminal nucleus

SHELL SIZE RANGE
1 to 2 in
(27 to 52 mm)

PHOTOGRAPHED SHELL
1¾ in
(46 mm)

CLAVOCERITHIUM TAENIATUM

RIBBON CERITH

(QUOY AND GAIMARD, 1834)

Clavocerithium taeniatum has a narrow distribution and is apparently restricted to shallow and offshore waters between eastern Indonesia and Papua New Guinea. Although it may be locally common, it is rare in museum collections. Its shell is variable in sculpture and color: some specimens, such as the one in the illustration, have bright colored spiral bands near the suture, while others may have paler colors. The radula is tiny.

RELATED SPECIES

Two species from the Indo-West Pacific, *Rhinoclavis vertagus* (Linnaeus, 1758) and *Rhinoclavis fasciata* (Bruguière, 1792), superficially resemble *C. taeniatum*. The former differs by having a smaller aperture, and a more slender, larger shell. The latter has an even larger and more slender shell; its shell color pattern is variable, but usually has spiral bands or blotches.

The shell of the Ribbon Cerith is medium-sized, solid, stout, and fusiform with a high spire. Most of the spire whorls have spiral cords and longitudinal riblets that become stronger toward the middle whorls; the last 2 whorls are mostly smooth. The aperture is oval, and the siphonal canal bent backward. The outer lip is thickened, and the columella concave and smooth. The shell color is yellowish white with a pink or tan broad band near the suture; the outer lip and columella are stained in orange-brown, and the aperture is white.

Actual size

FAMILY	Cerithiidae
SHELL SIZE RANGE	1⅜ to 3¾ in (35 to 95 mm)
DISTRIBUTION	Red Sea to Indo–West Pacific
ABUNDANCE	Abundant
DEPTH	Subtidal to 60 ft (18 m)
HABITAT	Fine sand near reefs
FEEDING HABIT	Herbivore
OPERCULUM	Corneous, paucispiral

SHELL SIZE RANGE
1⅜ to 3¾ in
(35 to 95 mm)

PHOTOGRAPHED SHELL
2½ in
(65 mm)

254

RHINOCLAVIS FASCIATA

STRIPED CERITH

(BRUGUIÈRE, 1792)

The Striped Cerith is a visually striking shell, long popular with collectors and blessed with many common names, including Banded Creeper, Banded Vertagus, Punctate or White Cerith, and (most appropriate at its western extent) Pharaoh's Horn. It has a diet of algae that it finds in the sands around offshore reefs. Like many ceriths it can vary enormously in the degree to which it exhibits its decorative features—width and color of banding, for example, and depth of sculpture.

RELATED SPECIES

Rhinoclavis aspera (Linnaeus, 1758) has a similar variety of markings on a shorter, stouter, more convex spire. The axial ribs hinted at on *R. fasciata* are fully developed here although they do not align on adjacent whorls. The siphonal canal is not quite as steeply recurved.

The shell of the Striped Cerith has a very tall, slightly convex spire, usually of 13 or 14 whorls. Below a moderately impressed suture there may be faint vestigial axial ribs, particularly on the upper whorls, between short, axial brown streaks. Its white to cream surface may have several fine or broad spiral bands of mid to dark brown. The lips are white and thickened with 1 mid-columellar fold, and the anterior canal is very steeply recurved.

Actual size

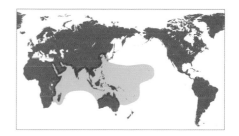

FAMILY	Cerithiidae
SHELL SIZE RANGE	2½ to 6 in (60 to 150 mm)
DISTRIBUTION	Indo Pacific
ABUNDANCE	Abundant
DEPTH	Intertidal to shallow subtidal
HABITAT	Sand, rubble, and reef flats
FEEDING HABIT	Grazer, feeds on microalgal detritus
OPERCULUM	Corneous, ovate, with few whorls

SHELL SIZE RANGE
2½ to 6 in
(60 to 150 mm)

PHOTOGRAPHED SHELL
4 in
(98 mm)

CERITHIUM NODULOSUM

GIANT KNOBBED CERITH

BRUGUIÈRE, 1792

255

Cerithium nodulosum is the largest species in the genus *Cerithium* and one of the largest living cerithiids. It is widespread throughout the Indo-West Pacific, abundant in shallow water on sand, rubble, and reef flats, near the outer edge of reefs. While the identification of most ceriths is difficult because many species are variable and may have similar shells, *C. nodulosum* is easily recognized because of its large size and knobbed sculpture. It is collected for food and the shell trade. Females lay egg masses with a thick axial base that is attached to the substrate, and long filaments with eggs that contain an estimated 66,000 eggs.

RELATED SPECIES

Cerithium erythraeonense Lamarck, 1822, from the Red Sea to Madagascar, is a closely related species, and is sometimes considered a subspecies of *C. nodulosum*. *Cerithium erythraeonense* has a similar but smaller and more slender shell. *Cerithium citrinum* Sowerby II, 1855, from East Africa to the west Pacific, has a slender, medium-sized, pale yellow shell with a long and angled siphonal canal.

The shell of the Giant Knobbed Cerith is large for the family, thick, solid, elongate, and heavily sculptured. Its spire is tall, the suture is well marked, and the spire whorls are strongly angulated at the periphery. Each whorl has a single spiral row of strong tubercles and other weaker spiral ribs. The body whorl and aperture are large, and the outer lip thickened and flared and crenulated in adults. The shell color is dirty white with gray-brown blotches. The aperture is white.

Actual size

FAMILY	Batillariidae
SHELL SIZE RANGE	½ to 1⅝ in (12 to 40 mm)
DISTRIBUTION	Western Mexico to Chile
ABUNDANCE	Abundant
DEPTH	Subtidal to 90 ft (27 m)
HABITAT	Under rocks in estuaries
FEEDING HABIT	Herbivore
OPERCULUM	Corneous, paucispiral

SHELL SIZE RANGE
½ to 1⅝ in
(12 to 40 mm)

PHOTOGRAPHED SHELL
1⅜ in
(33 mm)

256

RHINOCORYNE HUMBOLDTI
RHINO CERITH
(VALENCIENNES, 1832)

Like other members of the family Batillariidae, *Rhinocoryne humboldti* inhabits estuarine mud flats and mangroves along the temperate and tropical eastern Pacific. These animals have the ability to withstand wide variations in temperature and salinity, as well as long periods of starvation and desiccation. *Rhinocoryne humboldti* is atypical of the family in having a sharply shouldered shell with pronounced axial sculpture and a relatively long siphonal canal. It has a very wide distribution, from the tropical sandbars of Central America to the far more temperate waters around Isla de Chiloé in southern Chile.

RELATED SPECIES

Cerithium lifuensis (Melvill and Standen, 1895) is locally common in the central Pacific. It has a more slender shell, which is hazelnut brown with a spiral band of large, off-white nodules on the shoulders and two intermediate spirals of smaller protrusions, all ending in notches on the outer lip.

Actual size

The shell of the Rhino Cerith is slender with a tall, flat spire (the apex is sometimes damaged on beached specimens). It is chestnut brown, often with fine white axial and spiral striations. A single spiral row of large nodules fills each whorl, marking the shoulders and finishing at a large axially pleated notch on the outer lip. The aperture is ovate and white, the interior black, with a deep, recurving anterior canal.

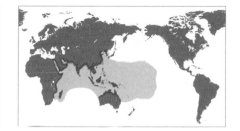

FAMILY	Dialidae
SHELL SIZE RANGE	less than ⅛ to ¼ in (2 to 7 mm)
DISTRIBUTION	Indo-Pacific
ABUNDANCE	Abundant
DEPTH	Intertidal
HABITAT	On algae and coral rubble
FEEDING HABIT	Herbivore, feeds on red and brown algae
OPERCULUM	Corneous, oval

SHELL SIZE RANGE
less than ⅛ to ¼ in
(2 to 7 mm)

PHOTOGRAPHED SHELL
less than ⅛ in
(3 mm)

DIALA ALBUGO

WHITE SPOTTED DIALA

(WATSON, 1886)

257

Diala albugo is a small gastropod that lives on soft sediments, on algae and coral rubble bottoms in the intertidal zone. It is distributed in the tropical Indo-West Pacific. Until 1992, the taxonomy of the family Dialidae was not well established, and many unrelated but similar species were included in this family. A recent revision of the family Dialidae recognized eight living species, all in the genus *Diala*. These species tend to be small, variable, and restricted to the Indo-West Pacific. Dialid shells are small, usually about ⅒–¼ in (3–7 mm) in length, with a tall spire, and with spiral sculpture only.

RELATED SPECIES

Diala varia A. Adams, 1860, which is native to the Red Sea and Indo-Pacific, has been introduced to the eastern Mediterranean via the Suez Canal. It has a shell that is similar in shape and size to *D. albugo*, but with more flattened whorl sides. *Diala flammea* (Pease, 1868), from the tropical Indo-West Pacific, is the dominant micromollusk on soft sediments in some tropical lagoons. It can occur in densities of 50 per teaspoon of sand.

The shell of White Spotted Diala is very small, thin, fragile, glossy, elongate, and conical. Its spire is tall, with about 7 slightly convex whorls, a smooth apex, and an incised suture. The sculpture of the shell consists of microscopic spiral lines, and no varices. The aperture is oval, without a siphonal canal; the outer lip is thin and smooth, as is the columella; there is no umbilicus. The shell color is cream, with orange-brown interrupted spiral stripes; the interior has similar color to the outside.

Actual size

FAMILY	Turritellidae
SHELL SIZE RANGE	¾ to 1½ in (18 to 40 mm)
DISTRIBUTION	Endemic to Japan
ABUNDANCE	Rare
DEPTH	2,300 to 3,600 ft (700 to 1,100 m)
HABITAT	Sandy mud bottoms
FEEDING HABIT	Filter feeder
OPERCULUM	Corneous, circular, multispiral

SHELL SIZE RANGE
¾ to 1½ in
(18 to 40 mm)

PHOTOGRAPHED SHELL
1½ in
(40 mm)

258

ORECTOSPIRA TECTIFORMIS
PAGODA CERITH
(WATSON, 1880)

Orectospira tectiformis is a rare, deepwater gastropod that is endemic to Japan. It lives on sandy mud bottoms. Its shell is pagoda-like, and is so unique that scientists have puzzled over its family placement. As a result, its classification has changed a few times, mostly based on shell features. The radula suggests it belongs in the family Turritellidae, but some researchers classify it in a separate family, Orectospiridae.

RELATED SPECIES

There are only a few known species in the genus *Orectospira*, including *Orectospira shikoensis* (Yokoyama, 1928), also endemic to Japan. Its shell is similar but smaller and narrower. *Turritella terebra* (Linnaeus, 1758), from the Indo-West Pacific is abundant, and one of the largest species in the family.

The shell of the Pagoda Cerith is medium-sized, thin, conical, and pagoda-like. Its spire is tall, with many whorls and an impressed suture; the apex is often missing. The surface of the shell is mostly smooth, with fine axial growth lines, and 1 spiral row of small nodules at the periphery of the whorls right above the suture. Each whorl overhangs the following whorl slightly. The aperture is squarish, the outer lip thin, and the columella smooth and reflected. There is a narrow umbilicus at the base. The shell color, inside and out, is white or off-white.

Actual size

FAMILY	Turritellidae
SHELL SIZE RANGE	2½ to 6½ in (60 to 170 mm)
DISTRIBUTION	Indo-West Pacific
ABUNDANCE	Abundant
DEPTH	Shallow subtidal to 100 ft (30 m)
HABITAT	Sandy, muddy bottoms
FEEDING HABIT	Suspension feeder
OPERCULUM	Corneous, circular

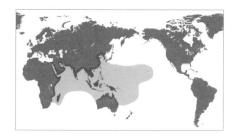

SHELL SIZE RANGE
2½ to 6½ in
(60 to 170 mm)

PHOTOGRAPHED SHELL
5½ in
(141 mm)

TURRITELLA TEREBRA

GREAT SCREW SHELL

(LINNAEUS, 1758)

Turritella terebra is variously known as the Great, Common, or Tower Screw Shell or the Screw Turritella. The Great Screw Shell is abundant and the largest member of the screw shell family, all of which are offshore herbivores. Despite its strikingly tall, perfectly formed spire, the species is not particularly popular with collectors, most likely due to its uniform brown coloring, which is devoid of any markings or patterns. Despite being generally abundant, the species is listed as "vulnerable" in Singapore due to land reclamation.

RELATED SPECIES

Turritella duplicata (Linnaeus, 1758), found in the Indian Ocean, shares a tall, regularly formed spire and neatly rounded aperture, but is stockier and shorter, and the whorls feature two distinctive spiral ridges. *Turritella bicingulata* (Lamarck, 1822), from the Canary and Cape Verde islands to West Africa, has fewer, more rounded axial ridges and a flamelike pattern that runs from the suture to the first ridge of each whorl.

The shell of the Great Screw Shell is notable for its long, very sharp spire, which in adults can be made up of around 30 whorls. Separated by deep sutures, each whorl has 6 clearly defined spiral ridges, with smaller ridges in between. The almost perfectly round aperture is bounded by a thin columella and sharp outer lip. The color ranges from pale to dark brown.

Actual size

FAMILY	Siliquariidae
SHELL SIZE RANGE	1½ to 6 in (40 to 150 mm)
DISTRIBUTION	North Carolina to northern Brazil
ABUNDANCE	Common
DEPTH	80 to 2,400 ft (25 to 730 m)
HABITAT	Embedded in sponges
FEEDING HABIT	Filter feeder
OPERCULUM	Corneous, conical

SHELL SIZE RANGE
1½ to 6 in
(40 to 150 mm)

PHOTOGRAPHED SHELL
3 in
(75 mm)

260

TENAGODUS SQUAMATUS

SLIT WORM SNAIL

(BLAINVILLE, 1827)

The shell of the Slit Worm Snail is medium-sized, thin, fragile, and irregularly coiled. Its spire starts as a conical shell, but is often missing. The whorls are loosely coiled or not coiled at all, with a long, continuous slit that can be smooth or constricted in places. The surface can be smooth or have spiral ridges with scales. The aperture is rounded and the outer lip simple and thin. The slit opens more widely anteriorly. The shell color is off-white, stained a pale orange-brown along the slit.

Tenagodus squamatus has an irregularly coiled shell in which the whorls become completely detached. It lives embedded in sponges, and, like the sponge, it is a filter-feeder. Because the sponge supports the weight of the shell, some functional constraints on shell morphology are relaxed, and the result is a very irregular shell. However, the snail needs to keep up with the growth of the sponge to keep the shell aperture open to the outside. There are about 20 living species in the family Siliquariidae worldwide. Some, but not all, species have a continuous slit on the shell, like *T. squamatus*.

Actual size

RELATED SPECIES

Tenagodus ponderosus (Mörch, 1861), from the Indo-Pacific, has the largest shell in the family, and can reach more than 16 in (400 mm) in length. The first few whorls are regularly, but somewhat loosely, coiled, and the last part of the shell uncoils. *Tenagodus modestus* (Dall, 1881), which ranges from west Florida and the Gulf of Mexico to the Caribbean and Brazil, has a more modestly sized shell, with a slit that consists of a row of oval holes.

FAMILY	Siliquariidae
SHELL SIZE RANGE	1½ to 6 in (40 to 150 mm)
DISTRIBUTION	West Florida to Brazil
ABUNDANCE	Uncommon
DEPTH	120 to 4,830 ft (35 to 1,470 m)
HABITAT	Embedded in sponges
FEEDING HABIT	Filter feeder
OPERCULUM	Corneous, conical

SHELL SIZE RANGE
1½ to 6 in
(40 to 150 mm)

PHOTOGRAPHED SHELL
3⅝ in
(93 mm)

TENAGODUS MODESTUS
MODEST WORM SNAIL
(DALL, 1881)

Tenagodus modestus lives offshore or in deep water, embedded in sponges. It is similar in shape and size to *T. squamatus*, but less commonly collected. The first few whorls grow like those of a normal coiled gastropod, and resemble a turritellid with a high spire; however, the apex is often missing. The operculum is conical, and has long bristles. The animal has short cephalic tentacles, with eyes at the base, a short foot, and a mantle with a long slit corresponding to the slit in the shell.

The shell of the Modest Worm Snail is medium in size, thin, fragile, smooth, and irregularly coiled. The first few whorls of the spire are coiled and may resemble a turritellid shell, but the following whorls are rounded and loosely coiled or completely uncoiled. The surface of the shell is smooth, with fine growth marks. Its slit consists of a series of small oval holes. The aperture is rounded, and the outer lip may be thickened or thin. The shell color ranges from white to pale orange.

Actual size

RELATED SPECIES

Tenagodus anguina (Linnaeus, 1758), from the western Pacific, has a smaller shell with spiral ridges that bear short, erect spines. The slit consists of a series of oval holes. *Tenagodus squamatus* (Blainville, 1821), from North Carolina to northern Brazil, has a shell of similar size, but it can be smooth or have spiral rows of scaly ridges. It can be distinguished from *T. modestus* by its continuous slit.

FAMILY	Planaxidae
SHELL SIZE RANGE	⅜ to ⅝ in (10 to 17 mm)
DISTRIBUTION	Southeastern Florida to Venezuela, Bermuda
ABUNDANCE	Abundant
DEPTH	Intertidal to 10 ft (3 m)
HABITAT	Rocks
FEEDING HABIT	Herbivore
OPERCULUM	Corneous, oval

SHELL SIZE RANGE
⅜ to ⅝ in
(10 to 17 mm)

PHOTOGRAPHED SHELL
⅝ in
(17 mm)

262

PLANAXIS NUCLEUS
BLACK ATLANTIC PLANAXIS
(BRUGUIÈRE, 1789)

Planaxidae is a fairly large family of tropical water snails, resembling Littorinidae but distinguished from that family by having significant apertural canals. On an anatomical level, male littorinids have a penis, unlike their planaxid counterparts. Planaxid eggs are hatched in a pouch within the head, the newborn sometimes released as planktonic larvae, sometimes kept in the oviduct until they are able to crawl. There are six genera, and species occur in fresh as well as saline environments.

RELATED SPECIES
Planaxis lineatus (da Costa, 1778), the Dwarf Atlantic Planaxis, is about half the size of *P. nucleus*. It is more widely distributed, often found as far south as Brazil. It has a relatively taller spire, and spiral grooves across the whole body whorl.

Actual size

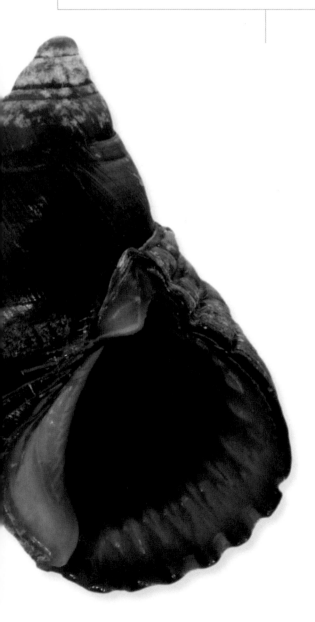

The shell of the Black Atlantic Planaxis is bulbous with a moderate spire, and may be beige to purple-brown, with a pale orange columella and dark interior. Spiral whorls are convex, with a fine spiral groove just below the suture. There are 3 deep grooves below the suture on the body and more across the bottom. The outer lip contains elongated dental ridges, and the aperture has distinct narrow anterior and posterior canals.

FAMILY	Planaxidae
SHELL SIZE RANGE	½ to 1⅜ in (13 to 35 mm)
DISTRIBUTION	Indo-West Pacific
ABUNDANCE	Abundant
DEPTH	Intertidal to shallow subtidal
HABITAT	Rocks
FEEDING HABIT	Herbivore
OPERCULUM	Corneous, thin

SHELL SIZE RANGE
½ to 1⅜ in
(13 to 35 mm)

PHOTOGRAPHED SHELL
1 in
(27 mm)

PLANAXIS SULCATUS

RIBBED PLANAXIS

(BORN, 1778)

263

Planaxis sulcatus has adapted to a sheltered rocky environment at or below low water, where there is little wave disturbance to disrupt the microalgae on which it feeds. When uncovered by the retreating tide, *P. sulcatus* can be seen huddled in groups in dips or cracks in the rock. In life, planaxids have a tan to orange-brown periostracum of a rough, fibrous texture. All have relatively sharp spires, and no umbilicus.

RELATED SPECIES

Planaxis labiosa (Adams, 1853), the Dwarf Pacific Planaxis, is less than half the size of *P. sulcatus*, with a tall spire. It is smooth and shiny; there are faint spiral grooves low on the body but no overall spiral ribs, and it has an attractive and variable pattern of frequent, thin to fine spiral bands in tan to dusky red.

Actual size

The shell of the Ribbed Planaxis is bulbous with a moderately tall spire, and covered in narrow, even, spiral cords. It is white with dark brown dashes on the cords, which may combine as axial zigzags or even completely cover the body whorl. The aperture is white with an orange margin, and has deep narrow siphonal and anal notches. There are long, deep grooves inside the thick outer lip.

FAMILY	Potamididae
SHELL SIZE RANGE	¾ to 2 in (20 to 50 mm)
DISTRIBUTION	Indo-West Pacific
ABUNDANCE	Abundant
DEPTH	Intertidal
HABITAT	Mangroves and mud flats
FEEDING HABIT	Detritivore, feeds on organic detritus
OPERCULUM	Corneous, circular

SHELL SIZE RANGE
¾ to 2 in
(20 to 50 mm)

PHOTOGRAPHED SHELL
1⅜ in
(35 mm)

CERITHIDEA CINGULATA
GIRDLED HORN SHELL
(GMELIN, 1791)

This small horn snail, like other members of the family, is locally very abundant, with population densities reaching up to 416 individuals per sq yd (500 per m²). As well as mangrove swamps and mud flats, Girdled Horn Shells also thrive in brackish or hypersaline fishponds where they feed on diatoms, bacteria, and other organic detritus. In some areas, notably the Philippines, the Girdled Horn Shell is considered a pest, its sheer abundance having a detrimental effect on milkfish aquaculture.

RELATED SPECIES

Terebralia sulcata (Born, 1778), also Indo-West Pacific, is more uniformly gray to gray-brown—lacking pale nodules of *Cerithidea cingulata*—with notably more rounded body whorls. *Cerithidea obtusa* (Lamarck, 1822), again Indo-West Pacific, also has more rounded body whorls; it lacks spiral grooves and therefore the nodular appearance of *C. cingulata*.

The shell of the Girdled Horn Shell features flattened body whorls with distinct axial ridges that are crossed by 2 deep, darker-colored spiral grooves creating 3 rows of flattened, off-white nodules. The elongated shape of the aperture is created by a greatly expanded lip at either end. In general, individuals vary greatly in color from gray to light brown, with 2 or 3 lighter lines per whorl.

Actual size

FAMILY	Potamididae
SHELL SIZE RANGE	1 to 2½ in (25 to 65 mm)
DISTRIBUTION	Indo-West Pacific
ABUNDANCE	Abundant
DEPTH	Intertidal
HABITAT	Estuary mud flats
FEEDING HABIT	Detritivore, feeds on organic detritus
OPERCULUM	Corneous, multispiral, circular

SHELL SIZE RANGE
1 to 2½ in
(25 to 65 mm)

PHOTOGRAPHED SHELL
1⅞ in
(48 mm)

TEREBRALIA SULCATA

SULCATE SWAMP CERITH

(BORN, 1778)

265

Terebralia sulcata is a small and abundant potamidid that ranges from Madagascar to Melanesia. It lives on the stems and roots of mangrove trees. Its shell has a flared outer lip, which the snail uses to press against the substrate to withstand desiccation and predators. Despite its small size, it is used extensively for food and lime material in the Philippines. Because some species are abundant, the family plays an important ecological role. There are more than 100 species of Potamididae worldwide, with the highest diversity in the tropical Indo-Pacific.

RELATED SPECIES

Terebralia palustris (Linnaeus, 1767), from East Africa to the western Pacific, has a similar but much larger and more elongated shell than *T. sulcata*, and is also used as food. *Cerithidea pliculosa* (Menke, 1829), from the Gulf of Mexico and the Caribbean, is a small potamidid that lives in salt marshes.

The shell of the Sulcate Swamp Cerith is small, thick and heavy, and elongated fusiform, with a high spire and flared outer lip. There are many spire whorls, and the suture is deeply incised. Its sculpture of the whorls consists of 4 or 5 spiral cords with axial ridges, which result in a pattern of squarish nodules. The body whorl has beaded spiral cords; the outer lip is thick and flared, and the columella glazed. The shell color is light or dark brown and the aperture is cream.

FAMILY	Potamididae
SHELL SIZE RANGE	2 to 4½ in (48 to 120 mm)
DISTRIBUTION	Indo-West Pacific
ABUNDANCE	Abundant
DEPTH	Intertidal
HABITAT	Mangroves and mud flats
FEEDING HABIT	Detritivore, feeds on organic detritus
OPERCULUM	Corneous, multispiral, circular

SHELL SIZE RANGE
2 to 4½ in
(48 to 120 mm)

PHOTOGRAPHED SHELL
3¼ in
(82 mm)

266

TELESCOPIUM TELESCOPIUM

TELESCOPE SNAIL

(LINNAEUS, 1758)

Actual size

Telescopium telescopium is an abundant snail found in the high intertidal zone in mangroves and intertidal mud flats, where it feeds on organic detritus. It may be seen in aggregations of many thousands at times. Because *T. telescopium* is an amphibious snail, it can stay out of the water for long periods, but during low tides, it clusters together with other snails and becomes inactive. Like some other potamidids, it has a third eye on the mantle, which is capable of sensing light, in addition to the pair of eyes on the cephalic tentacles. It is used as food in Southeast Asia.

RELATED SPECIES

Terebralia palustris (Linnaeus, 1767), from East Africa to the western Pacific, is the largest species in the family Potamididae. Its shell is also conical but it lacks the strong spiral striation of *Telescopium telescopium*. *Tympanotonus radula* (Linnaeus, 1758), from western Africa and Cape Verde, has a shell with a high spire and heavy triangular spines on a spiral row.

The shell of the Telescope Snail is medium-sized, thick, heavy, and conical with a tall spire. Its spire has many whorls, and the suture is weak. The sculpture of the spire consists of 4 strong, flat spiral cords of unequal size alternating with deep spiral grooves. Its base is flat and the body whorl has a rounded periphery. The aperture is relatively small and obliquely quadrangular, and the columella strongly twisted. The shell color is dark brown or black, sometimes with a light brown band, and the aperture is purplish.

FAMILY	Potamididae
SHELL SIZE RANGE	1½ to 7½ in (40 to 190 mm)
DISTRIBUTION	East Africa to west Pacific
ABUNDANCE	Abundant
DEPTH	Intertidal
HABITAT	Mangroves and mud flats
FEEDING HABIT	Detritivore when juvenile; herbivore when adult
OPERCULUM	Corneous, multispiral, circular

SHELL SIZE RANGE
1½ to 7½ in
(40 to 190 mm)

PHOTOGRAPHED SHELL
4½ in
(121 mm)

TEREBRALIA PALUSTRIS

MUD CREEPER
(LINNAEUS, 1767)

267

Terebralia palustris is the largest species in the family Potamididae. Like other horn snails, it lives in the intertidal zone among mangroves, but the juveniles and adults prefer different areas of the mangrove. The radula in *T. palustris* changes in microstructure from the juvenile to the adult, when it switches its diet from fine detritus to a diet that consists of fallen mangrove leaves and fruits. *Terebralia palustris* is an abundant and conspicuous mangrove dweller. It is extensively collected for food throughout its geographic range.

RELATED SPECIES

Terebralia sulcata (Bruguière, 1792), from the Indo-West Pacific, is also a mangrove-dwelling potamidid, although it prefers firm sandy-mud sediments. It lacks the third eye in the mantle, and has a smaller shell than *T. palustris*. *Terebralia semistriata* (Mörch, 1852), from northern Australia, resembles *T. palustris* but is also much smaller, and has a reflected outer lip.

The shell of the Mud Creeper is large, thick, heavy, and conical, with a tall spire. Its spire has many flat-sided whorls. Its sculpture consists of 4 equal-sized, flat spiral ribs, and strong axial ridges that become obsolete on later whorls. The aperture is ovate and grooved, and the outer lip flared and crenulated. The anterior siphonal canal is short, and the columella has a strong callus. The shell color is mostly uniform dark brown, although the spire may be lighter and eroded.

Actual size

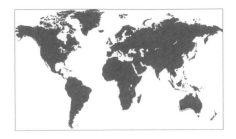

FAMILY	Diastomatidae
SHELL SIZE RANGE	1¼ to 2 in (30 to 50 mm)
DISTRIBUTION	Endemic to south coast of Western Australia
ABUNDANCE	Uncommon
DEPTH	3 to 17 ft (1 to 5 m)
HABITAT	Sandy bottoms and sea grass beds
FEEDING HABIT	Microalgae and detritus feeder
OPERCULUM	Corneous, oval

SHELL SIZE RANGE
1¼ to 2 in
(30 to 50 mm)

PHOTOGRAPHED SHELL
2 in
(50 mm)

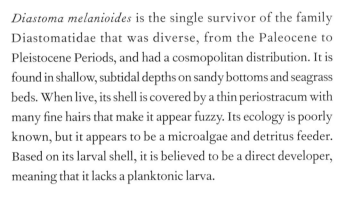

268

DIASTOMA MELANIOIDES
MELANIOID DIASTOMA
(REEVE, 1849)

The shell of the Melanioid Diastoma is medium in size, textured, and turreted. Its spire is tall, with an impressed suture and a pointed apex, and the spire whorl sides are flat to only slightly convex. The sculpture consists of axial folds in the early whorls, crossed by several spiral cords. The axial folds become obsolete toward the body whorl. The aperture is semicircular, the outer lip thin, and the columella has one fold in the middle. The shell color is white or cream, with orange-brown spots or diagonal stripes, and the interior is white.

Diastoma melanioides is the single survivor of the family Diastomatidae that was diverse, from the Paleocene to Pleistocene Periods, and had a cosmopolitan distribution. It is found in shallow, subtidal depths on sandy bottoms and seagrass beds. When live, its shell is covered by a thin periostracum with many fine hairs that make it appear fuzzy. Its ecology is poorly known, but it appears to be a microalgae and detritus feeder. Based on its larval shell, it is believed to be a direct developer, meaning that it lacks a planktonic larva.

RELATED SPECIES

Several gastropods with a superficial resemblance have been erroneously classified in the family Diastomatidae, but now are placed in closely related families, including the Litiopidae, Dialidae, Scaliolidae, Cerithiidae, and others.

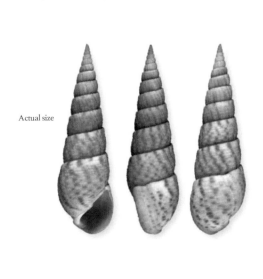

Actual size

FAMILY	Modulidae
SHELL SIZE RANGE	⅝ to 1¼ in (15 to 30 mm)
DISTRIBUTION	East Africa to the Philippines
ABUNDANCE	Common
DEPTH	Intertidal
HABITAT	Weedy, sandy bottoms
FEEDING HABIT	Herbivore
OPERCULUM	Corneous, thin, circular

SHELL SIZE RANGE
⅝ to 1¼ in
(15 to 30 mm)

PHOTOGRAPHED SHELL
1⅛ in
(28 mm)

MODULUS TECTUM

TECTUM MODULUS

(GMELIN, 1791)

269

Modulus tectum is the largest member of the family Modulidae, which has only a single genus *Modulus*, with less than two dozen species. All of them have a top-shaped shell and feature a small tooth on the base of the columella. Found in warm, shallow water, notably seagrass estuaries, the Tectum Modulus—also known as Covered Modulus and Knobby Snail—feeds on microscopic algae.

RELATED SPECIES

Modulus modulus (Linnaeus, 1758), from southeast United States to Brazil, and Bermuda has a taller, more pointed spire, and is generally smaller. *Modulus disculus* (Philippi, 1846), from Gulf of California to Panama is also generally smaller and usually features a wavy-edged outer lip.

Actual size

The shell of the Tectum Modulus features a flattened spire and pronounced axial ridges. Below the sharply cornered shoulder, the body whorl opens rapidly to a large aperture, which is usually uniformly pale cream to white inside. The smooth columella features a prominent tooth at the base. The shell color varies from cream to pale yellow, with light brown to dark gray blotches.

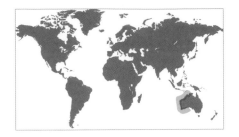

FAMILY	Campanilidae
SHELL SIZE RANGE	3¼ to 8½ in (80 to 215 mm)
DISTRIBUTION	Western Australia
ABUNDANCE	Locally common
DEPTH	3 to 33 ft (1 to 10 m)
HABITAT	Sandy bottoms
FEEDING HABIT	Grazer
OPERCULUM	Corneous, with subcentral nucleus

SHELL SIZE RANGE
3¼ to 8½ in
(80 to 215 mm)

PHOTOGRAPHED SHELL
3½ in
(88 mm)

CAMPANILE SYMBOLICUM

BELL CLAPPER

IREDALE, 1917

The shell of the Bell Clapper is large, thick, heavy, and turret-shaped. The spire is high, the suture incised and sinuous, and the sides of the whorls are flat or slightly concave. Its sculpture is often eroded away, but it consists of one spiral row of blunt tubercles near the suture, and weak spiral and axial lines. The aperture is relatively small and rounded; the outer lip is smooth, and reflected in adults. The siphonal canal is short and twisted. The shell color is chalky white.

Campanile symbolicum is the single living species in the family Campanilidae. The family has a long fossil history that goes back to the Cretaceous Period, and had at least 700 species. Some fossil species were huge and reached more than 3 ft (1 m) in length; they are among the largest gastropods to have ever existed. It has been hypothesized that ecological competition with the Strombidae may have caused the almost total extinction of the Campanilidae. *Campanile symbolicum* has a large, chalky shell that is often eroded and has scars from the attachment of hipponicid gastropods, giving the appearance of a fossil.

RELATED SPECIES

The closest living relatives of *Campanile symbolicum* are in the family Plesiotrochidae, a group of small gastropods with shells less than 1 in (24 mm). One of the species in the family is *Plesiotrochus penetricinctus* (Cotton, 1932), from Australia, which has a small, pagoda-shaped shell.

Actual size

FAMILY	Littorinidae
SHELL SIZE RANGE	⅛ to ¼ in (3 to 5 mm)
DISTRIBUTION	Indo-Pacific
ABUNDANCE	Locally abundant
DEPTH	Supratidal to shallow subtidal
HABITAT	Rocky shores and algal mats
FEEDING HABIT	Herbivore
OPERCULUM	Corneous, circular, multispiral

SHELL SIZE RANGE
⅛ to ¼ in
(3 to 5 mm)

PHOTOGRAPHED SHELL
⅙ in
(4 mm)

PEASIELLA TANTILLA

TRIFLE PEASIELLA

(GOULD, 1849)

Peasiella tantilla is a small representative of the littorinids or periwinkles. Its abundant in some parts of its wide Indo-Pacific distribution, such as in Hawaii, where it occurs on exposed intertidal rocky shores, both above the waterline as well as in tide pools, crevices, and benches. It can also be found on coralline algae in shallow subtidal waters. It has a colorful shell, ranging from yellow to reddish brown. There are about 200 living species in the family Littorinidae worldwide. Most species occur on intertidal rocky shores or above the tide line. The oldest known fossil littorinid dates from the Upper Paleocene Period.

The shell of the Trifle Peasiella is very small, depressed, and conical. Its spire is moderately tall, with a pointed apex, and an impressed suture. Its sculpture is dominated by spiral grooves and a strong ridge on the periphery of the whorls, making the whorls spirally keeled in adults. The aperture is ovate, the outer lip angular, and the columella smooth. The umbilicus is narrow and deep. The shell color ranges from yellow to reddish brown, speckled with white, or with thin brown lines. The interior color is similar to the exterior.

RELATED SPECIES

Peasiella conoidalis (Pease, 1868), from the tropical Indo-West Pacific, has a small conical shell with tubercles on the periphery. It looks more like a miniature trochid than a periwinkle.
Cenchritis muricatus (Linnaeus, 1758), which ranges from Florida to the West Indies and northern South America, has a robust shell sculptured with many spiral rows of tubercles. It lives above the waterline, on rocky shores.

Actual size

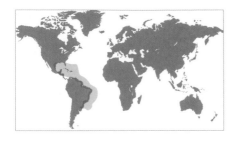

FAMILY	Littorinidae
SHELL SIZE RANGE	⅜ to ⅞ in (10 to 23 mm)
DISTRIBUTION	Southeastern Florida to Brazil
ABUNDANCE	Abundant
DEPTH	Intertidal
HABITAT	Rocks
FEEDING HABIT	Herbivore
OPERCULUM	Corneous

SHELL SIZE RANGE
⅜ to ⅞ in
(10 to 23 mm)

PHOTOGRAPHED SHELL
⅝in
(15 mm)

272

ZIGZAG PERIWINKLE
(GMELIN, 1791)

The Zigzag Periwinkle is a native of the Caribbean, but has since the late twentieth century become established on the western coast of Panama. The sexes of the family Littorinidae are separate, and in the case of *Echinolittorina ziczac* the female lays her eggs individually, each in a floating capsule. Other species of the family lay their eggs in the water or in jelly masses, or hatch them inside an oviduct or pouch within the body of the animal.

RELATED SPECIES

Echinolittorina lineolata had been considered merely a juvenile form of *E. ziczac*, but is now recognized as a separate species. It has a slightly smaller shell, which is very similar in appearance except that it has one less whorl. The columella is a more solid expanse of dark red-brown.

Actual size

The shell of the Zigzag Periwinkle is white and bulbous with a moderately tall spire of 5 or 6 convex whorls with faint spiral grooves. A pale to very dark brown spiral band sits just above the suture, giving rise to diagonal red-brown stripes; these recur on the body whorl, becoming zigzags. The interior is white with very broad, dark lateral bands partially obscuring the exterior pattern. The outer lip is thin, the columella is thickened and pale red.

FAMILY	Littorinidae
SHELL SIZE RANGE	½ to ⅞ in (13 to 22 mm)
DISTRIBUTION	Puget Sound to northern Alaska and northern Japan
ABUNDANCE	Abundant
DEPTH	Intertidal
HABITAT	Rocky shores
FEEDING HABIT	Herbivore
OPERCULUM	Corneous, rounded, multispiral

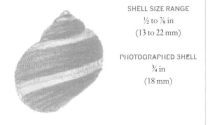

SHELL SIZE RANGE
½ to ⅞ in
(13 to 22 mm)

PHOTOGRAPHED SHELL
¾ in
(18 mm)

LITTORINA SITKANA

SITKA PERIWINKLE

PHILIPPI, 1846

Littorina sitkana is an abundant small periwinkle with strong spiral ridges. It ranges from Puget Sound to northern Alaska, and to northern Japan. Like other periwinkles, it lives in sheltered areas of the rocky intertidal zone, especially the upper intertidal zone. It feeds on diatoms and other algae, by scraping the surface of rocks with its radula. It is estimated that in areas with high densities, these periwinkles can erode about ½ in (10 mm) of intertidal rock per 16 years.

The shell of the Sitka Periwinkle is small, solid, and globose. Its spire is moderately high, with a pointed apex, convex whorls, and an impressed suture. The shell is almost as wide as it is high. The outer sculpture is dominated by about 12 strong spiral ribs. The aperture is oval, the outer lip sharp, and the columella smooth. There are no siphonal or anal canals. The shell color ranges from dirty white to rusty brown, with the columella white. Sometimes there are white spiral bands.

RELATED SPECIES

Echinolittorina placida Reid, 2009, is a recently described periwinkle from the Gulf of Mexico. The Gulf has few natural rocky outcrops, but in the past 100 years, the construction of seawalls made possible a 2,800-mile (4,500-km) expansion of its distribution. *Echinolittorina placida* originated in the southwestern Gulf of Mexico, and now reaches as far north as North Carolina.

Actual size

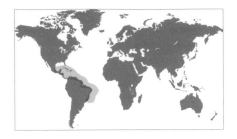

FAMILY	Littorinidae
SHELL SIZE RANGE	½ to 1¼ in (13 to 30 mm)
DISTRIBUTION	Southern Florida, West Indies, northern South America
ABUNDANCE	Abundant
DEPTH	Supertidal
HABITAT	Rocks
FEEDING HABIT	Herbivore
OPERCULUM	Corneous

SHELL SIZE RANGE
½ to 1¼ in
(13 to 30 mm)

PHOTOGRAPHED SHELL
⅞ in
(22 mm)

274

CENCHRITIS MURICATUS
BEADED PERIWINKLE
(LINNAEUS, 1758)

Periwinkles are among the most visible and familiar of shells, because they occur in colonies in areas above the waterline. They are able to do this thanks to a very efficient, tight-fitting operculum that helps them to retain moisture between tides. The Beaded Periwinkle inhabits a supertidal zone of rocks and even trees, up to 30 ft (10 m) above the high-tide mark. It lays its eggs in water, in individual, lozenge-shaped floating capsules.

RELATED SPECIES

Nodilittorina tuberculata (Menkle, 1828) is a smaller littorinid of similarly bold nodular sculpture. It is about half the height of *Cenchritis muricatus*, with fewer spiral rows and less frequent nodes on them; the nodes tend to align vertically within whorls, like beaded axial ribs. They are quite sharp, hence the popular name, the Common Prickly Winkle.

The shell of the Beaded Periwinkle is bulbous, with a swollen body whorl. It is decorated with moderately spaced nodules in regular spiral rows—about 10 on the body, 5 on the spiral whorls. The nodes do not align vertically. There is a wide, round aperture, with a thin outer lip, and the interior is chestnut to dark red. The exterior is off-white, with a broad, faint gray-brown spiral band on each shoulder.

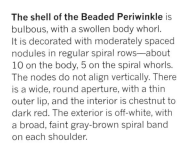

Actual size

FAMILY	Littorinidae
SHELL SIZE RANGE	⅝ to 1⅜ in (15 to 35 mm)
DISTRIBUTION	Indo-Pacific
ABUNDANCE	Abundant
DEPTH	Intertidal
HABITAT	Mangroves
FEEDING HABIT	Herbivore
OPERCULUM	Corneous

SHELL SIZE RANGE
⅝ to 1⅜ in
(15 to 35 mm)

PHOTOGRAPHED SHELL
1 in
(26 mm)

275

LITTORINA SCABRA

MANGROVE PERIWINKLE

(LINNAEUS, 1758)

Periwinkles have successfully colonized shores all around the Pacific rim. Their range extends from Chile and Australia north through the tropics to subarctic Siberia and Alaska. Many scrape algae from the surface of reefs and rocks in and around the water's edge; many others, including *Littorina scabra*, scavenge for rotting plant matter in the brackish waters in and near mangroves.

RELATED SPECIES

A narrower variation with a sharper spire, *Littorina scabra angulifera* (Lamarck, 1822), lives in the mangroves of the tropical western Atlantic. It is identical in decoration, although it lacks the raised spiral cord on its body whorl. The aperture is markedly less open at the posterior, the outer lip pulled a little tighter to the columella there.

Actual size

The shell of the Mangrove Periwinkle is bulbous with a moderately tall spire. It has convex whorls separated by a deeply adpressed suture. The shell color is cream to beige, with fine spiral cords of chestnut to gray dashes that in places coalesce into flames or axial zigzags; one cord at the body shoulder is raised. There is an internal white band beyond the lip that forms an obtuse anterior angle with the columella.

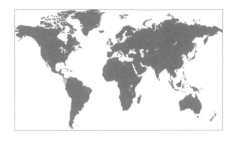

FAMILY	Littorinidae
SHELL SIZE RANGE	¾ to 1¼ in (20 to 32 mm)
DISTRIBUTION	Costa Rica to Colombia
ABUNDANCE	Moderately common
DEPTH	Intertidal
HABITAT	Mangroves
FEEDING HABIT	Herbivore
OPERCULUM	Corneous

SHELL SIZE RANGE
¾ to 1¼ in
(20 to 30 mm)

PHOTOGRAPHED SHELL
1¼ in
(30 mm)

276

LITTORINA ZEBRA

ZEBRA PERIWINKLE

DONOVAN, 1825

Littorina ʒebra is considered by collectors as one of the most attractive species in the genus, with an unusually colorful shell. Like many periwinkles, it lives on the roots and stems of mangroves. It is endemic to a rather narrow region, the western shores of Central America, which suggests a low tolerance to variations in habitat and temperature.

RELATED SPECIES
Littorina modesta (Philippi, 1846) occurs over a greater range of western America. It is smaller, plain pale cream, with well-rounded whorls and spiral cords, a nearly round yellow aperture, and a thin, unextended outer lip.

Actual size

The shell of the Zebra Periwinkle is a squarish globe with pronounced high shoulders on the body whorl and a moderately low, but deeply sutured, spire. It is pale terra-cotta with oblique brown stripes across very fine spiral cords. The aperture is widely ovate with a thin, extended outer lip within which the pattern shows as a row of brown dots around the rim.

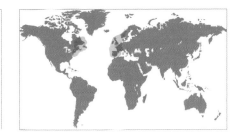

FAMILY	Littorinidae
SHELL SIZE RANGE	⅝ to 2⅛ in (16 to 53 mm)
DISTRIBUTION	Western Europe and northeastern America
ABUNDANCE	Abundant
DEPTH	Intertidal
HABITAT	Rocks
FEEDING HABIT	Herbivore
OPERCULUM	Corneous

SHELL SIZE RANGE
⅝ to 2⅛ in
(16 to 53 mm)

PHOTOGRAPHED SHELL
1¾ in
(44 mm)

LITTORINA LITTOREA

COMMON PERIWINKLE

(LINNAEUS, 1758)

277

Members of the family Littorinidae are generally small and not very colorful or sculptural, and they are not much appreciated by shell collectors. Theirs is a subtler beauty, mostly of tiny variations in spire and aperture. *Littorina littorea* is common on the intertidal rocks of all North Atlantic coasts, where it has been a popular edible variety for hundreds if not thousands of years.

RELATED SPECIES

Littorina littoralis Linnaeus, 1758, extends southward to New England and the Mediterranean. It is less than half the size, with a more rounded, smooth, swollen body and a flat spire. There is much variation in color of both body and banding.

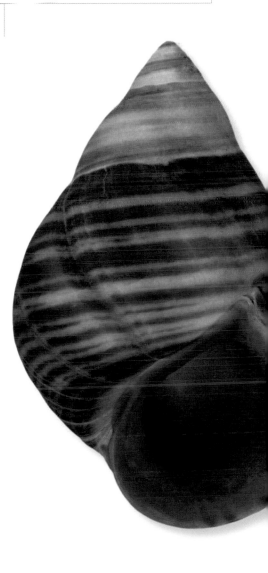

Actual size

The shell of the Common Periwinkle is variable in color but generally dark chestnut to dark gray-brown. Its spire is moderately low with a fine suture, and growth lines on the body may be quite deep axial grooves. It is usually decorated with fine spiral bands, and is paler toward the apex. It has a dark interior and normally white aperture, with a thin sharp outer lip and a short, angular posterior notch.

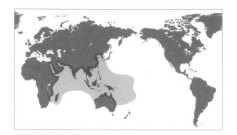

FAMILY	Littorinidae
SHELL SIZE RANGE	1¼ to 2⅝ in (30 to 65 mm)
DISTRIBUTION	Indo-West Pacific
ABUNDANCE	Common
DEPTH	Intertidal, supertidal
HABITAT	Rocks
FEEDING HABIT	Herbivore
OPERCULUM	Corneous

SHELL SIZE RANGE
1¼ to 2⅝ in
(30 to 65 mm)

PHOTOGRAPHED SHELL
2⅛ in
(55 mm)

278

TECTARIUS PAGODUS

PAGODA PRICKLY WINKLE

(LINNAEUS, 1758)

Tectarius pagodus is the largest species in the genus *Tectarius*, which includes many of the largest species within the family Littorinidae. Like all littorinids, *T. pagodus* is herbivorous, and inhabits the uppermost reaches of the intertidal zone on rocky shores. Thus, the animals are submerged only infrequently, and are resistant to dessication. Animals are active at night during periods of rainfall or high humidity. Species of *Tectarius* have various reproductive strategies. Most produce either feeding or non-feeding planktonic larvae, but two species retain the eggs in the female's body until they hatch.

RELATED SPECIES

Tectarius tectumpersicum (Linnaeus, 1758) is slightly smaller and shares the same distribution. It has a marginally narrower body whorl, and its sculpture is rougher, with fewer and less elevated, shoulder nodes. The spiral beading on the base is larger.

The shell of the Pagoda Prickly Winkle is white to cream, largely obscured above the keel by many bands of tan to dark brown. It is covered in fine uneven beaded spiral cords, undulating over pronounced axial ribs from the acute body keel upward. On the shoulders of its tall spire, the ribs terminate in large blunt upturned nodules; below the keel the cords are less fine and white. The aperture is white; the interior is pale tan and lined with wide grooves.

Actual size

FAMILY	Pickworthiidae
SHELL SIZE RANGE	less than ⅛ in (1 to 3 mm)
DISTRIBUTION	Florida to Puerto Rico and Gulf of Mexico
ABUNDANCE	Uncommon
DEPTH	16 to 2,340 ft (5 to 710 m)
HABITAT	Soft sediments
FEEDING HABIT	Unknown
OPERCULUM	Unknown

SANSONIA TUBERCULATA

TUBERCULATE SANSONIA

(WATSON, 1886)

SHELL SIZE RANGE
less than ⅛ in
(1 to 3 mm)

PHOTOGRAPHED SHELL
less than ⅛
(1 mm)

Sansonia tuberculata is a micromollusk that belongs to the family Pickworthiidae. Most species of pickworthiids have a maximum shell size of less than ⅕ in (5 mm). While empty shells are not uncommon in subtidal sediment, live specimens have rarely been collected, and therefore its biology remains poorly known. Shells in the family range from tall conical to nearly flattened disk-shaped shells, usually with strongly axial and spiral sculpture.

RELATED SPECIES

Sansonia alisonae Le Renard and Bouchet, 2003, from the Red Sea and Hawaii, has a shell that resembles *S. tuberculata*, but it has a larger shell with sharper sculpture and a pointed spire. *Sherbornia mirabilis* Iredale, 1917, from Christmas Island to Polynesia, Central Pacific, has an extremely well-developed winglike expansion of the lip that is larger than the entire shell.

The shell of the Tuberculate Sansonia is extremely small and top-shaped, with a tall spire. Its spire whorls have 2 rounded beaded spiral rows, while the body whorl has 3 rows. The suture is incised. The larval shell is rounded and has a different sculpture, consisting of fine spiral lines. The aperture is circular with a thickened lip, oriented at an angle of 45 degrees from the shell axis. The shell color is white, and the rounded beads are glossy.

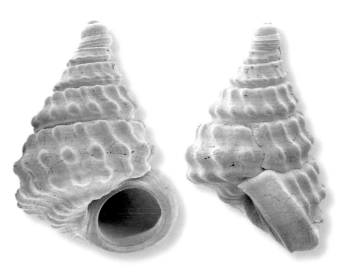

◁ Actual size

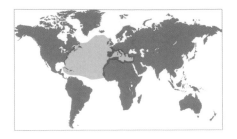

FAMILY	Skeneopsidae
SHELL SIZE RANGE	less than ⅛ in (1 to 3 mm)
DISTRIBUTION	Northern Atlantic; Mediterranean
ABUNDANCE	Common
DEPTH	Intertidal to 230 ft (70 m)
HABITAT	On algae in tide pools
FEEDING HABIT	Microalgae feeder
OPERCULUM	Corneous, circular

SHELL SIZE RANGE
less than ⅛ in
(1 to 3 mm)

PHOTOGRAPHED SHELL
less than ⅛ in
(1 mm)

280

SKENEOPSIS PLANORBIS
FLAT SKENEOPSIS
(FABRICIUS, 1780)

Skeneopsis planorbis is a minute gastropod that can be found on algae in tide pools and offshore, sometimes in large numbers. It occurs on both sides of the North Atlantic, ranging from Greenland to Florida, and from Iceland to the Azores, as well as into the Mediterranean. The shell is glossy and translucent, and disk-shaped with a short spire. Breeding can occur throughout the year, but is more common in the spring. The female lays tiny egg capsules attached to algal filaments. The embryos undergo direct development, and juveniles crawl out of the egg capsules. There are only a few living species known in the family Skeneopsidae.

RELATED SPECIES

Starkeyna starkeyae (Hedley, 1899), from New South Wales, Australia, has a minute shell similar to that of *S. planorbis*, but it has a closed umbilicus. The relationships of the Skeneopsidae are uncertain, but they appear to be related to the Littorinidae, despite the very different shell shapes. Molecular studies of the group may help elucidate the family's systematic position.

The shell of the Flat Skeneopsis is minute, thin, translucent, glossy, and disk-shaped. Its spire is short, with a well-impressed suture, a nearly flat apex, and rounded whorls. The surface is smooth to the naked eye, but under the electron microscope (as illustrated here) it has fine growth lines. The aperture is rounded, the outer lip thin, and the columella smooth. The umbilicus is wide and deep. The shell color is brownish when fresh, whitish in beached shells.

Actual size

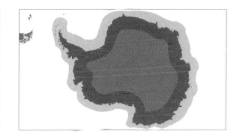

FAMILY	Eatoniellidae
SHELL SIZE RANGE	less than ⅛ to ⅙ in (2 to 4 mm)
DISTRIBUTION	Antarctica
ABUNDANCE	Common
DEPTH	33 to 850 ft (10 to 260 m)
HABITAT	Fine sand and mud bottoms
FEEDING HABIT	Herbivore
OPERCULUM	Small, with peg on inner surface

EATONIELLA KERGUELENSIS

KERGUELEN ISLAND EATONIELLA

(SMITH, 1915)

SHELL SIZE RANGE
less than ⅛ to ⅙ in
(2 to 4 mm)

PHOTOGRAPHED SHELL
⅛ in
(3 mm)

281

The family Eatoniellidae is characterized by a small, simple, conical shell with a high spire and rounded whorls. The operculum is small and has a distinctive, thickened, peglike projection from its inner surface. *Eatoniella kerguelensis* occurs in large colonies of hundreds to thousands of individuals at subtidal depths in Antarctic and subantarctic waters. Several subspecies have been described from different island groups. The animals graze on diatomaceous film, detritus, and algae.

RELATED SPECIES

As the name implies, *Eatoniella depressa* Ponder and Yoo, 1978, from the southern coast of Australia, is more flattened, with a lower spire and a larger, more circular aperture. *Eatoniella exigua* Ponder and Yoo, 1978, also from southern Australia, has a smaller, more inflated shell with fewer whorls. *Barleeia subtenuis* Carpenter, 1864, from Alaska to Baja California, has a slightly broader shell, with fewer whorls and a more elongated aperture.

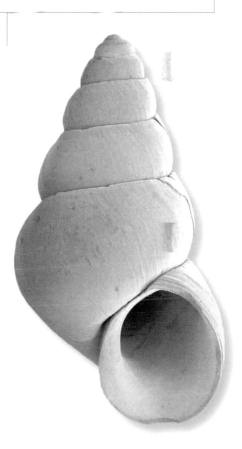

Actual size

The shell of the Kerguelen Island Eatoniella is conical, and high-spired, with a rounded anterior. The protoconch is smooth; the adult whorls are evenly rounded and smooth. The aperture is simple and slightly ovate, with a narrow rim. The columella has a thickened margin that may be flared anteriorly. The surface sculpture is limited to fine growth striae. The shell color is grayish, and the aperture is white.

FAMILY	Tornidae
SHELL SIZE RANGE	less than ⅛ in (2 to 3 mm)
DISTRIBUTION	North Carolina, U.S.A. to Yucatán, Mexico
ABUNDANCE	Common
DEPTH	70 to 3,800 ft (20 to 1,170 m)
HABITAT	Soft sediments
FEEDING HABIT	Detritivore
OPERCULUM	Corneous, rounded, multispiral

SHELL SIZE RANGE
less than ⅛ in
(2 to 3 mm)

PHOTOGRAPHED SHELL
less than ⅛ in
(2 mm)

TEINOSTOMA RECLUSUM

RECLUSE VITRINELLA

(DALL, 1889)

The shell of the Recluse Vitrinella is very small, smooth, and has a rounded top-shape. Its spire is slightly raised, and the spire and body whorls are rounded. The shell is smooth, and the aperture subcircular. In adult shells there is a callus at the base of the shell, but in juveniles there may be a depression at the base, the umbilicus. The shell color is white or cream.

Teinostoma reclusum is a tiny gastropod that belongs to a large group whose taxonomy is still in flux; even the name of the family has recently changed. Most members of the family Tornidae are very small, and usually have a flattened spire, and a disk-shaped and smooth shell, although some have spiral sculpture or a short spire. Some species are known to live with burrowing invertebrates. Because of their small size, their biology and anatomy are poorly known. There are probably hundreds of species of Tornidae worldwide, ranging from temperate to tropical waters. At least 45 species are recognized in the Gulf of Mexico alone.

RELATED SPECIES

Circulus texanus (Moore, 1965), from Texas, has a tiny translucent shell with a flat spire and nearly circular profile. It is about half of the size of *Teinostoma reclusum*. *Cyclostremiscus beauii* (Fischer, 1857), which ranges from North Carolina to Brazil and the Gulf of Mexico, is one of the larger American species, growing to about ½ in (13 mm), and is one of the few species whose biology has been studied.

Actual size

FAMILY	Barleeidae
SHELL SIZE RANGE	less than ⅛ in (2 to 3 mm)
DISTRIBUTION	Alaska to Baja California
ABUNDANCE	Common
DEPTH	Shallow subtidal
HABITAT	Sand and rock bottoms
FEEDING HABIT	Detritivore
OPERCULUM	Calcareous, with peg on inner surface

BARLEEIA SUBTENUIS

FRAGILE BARLEYSNAIL

(CARPENTER, 1864)

SHELL SIZE RANGE
less than ⅛ in
(2 to 3 mm)

PHOTOGRAPHED SHELL
⅛ in
(3 mm)

283

Like many of the families included in the large and diverse superfamily Rissooidea, members of the Barleeidae have minute, simple, relatively featureless shells. They are distinguished from the other families primarily by anatomical characters. Most species scrape decaying vegetation or a bacterial film from the substrates on which they live, or ingest mud that is rich in organic sediments. Maximum lifespan is about two years.

RELATED SPECIES

Barleeia haliotiphila (Carpenter, 1864), which shares much of the range of *B. subtenuis*, has a larger, broader, light brown shell with a somewhat angular periphery. It lives in slightly deeper water, on kelp holdfasts, rocks, and abalone shells. While similar in size and shape, *Irivadia trochlearis* (Gould, 1861) is easily distinguished by its prominent spiral cords.

The shell of the Fragile Barleysnail is very small, thin, conical, and high-spired, with an ovate anterior. The protoconch is small and minutely pitted. The adult whorls are flatly convex, with very weak spiral threads and axial growth lines. The aperture is oval, with an anteriorly flared rim and a thickened columellar margin. The umbilicus is shallow, narrow, and crescent shaped. The shell color ranges from dark brown to olive brown. The aperture is white.

Actual size

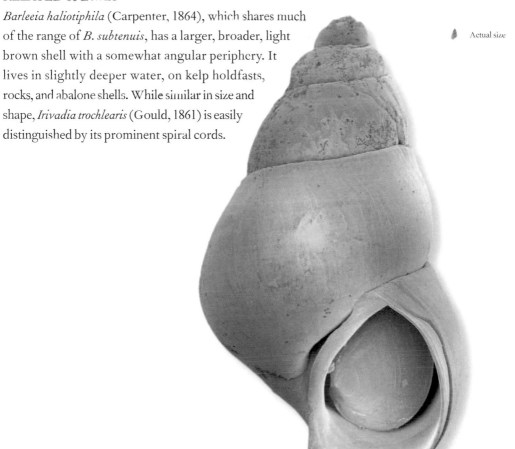

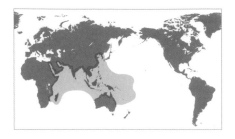

FAMILY	Iravadiidae
SHELL SIZE RANGE	around ⅛ in (3 to 4 mm)
DISTRIBUTION	Indo-West Pacific
ABUNDANCE	Common
DEPTH	Intertidal to subtidal
HABITAT	Fine sand and mud bottoms
FEEDING HABIT	Detritivore
OPERCULUM	Corneous

SHELL SIZE RANGE
around ⅛ in
(3 to 4 mm)

PHOTOGRAPHED SHELL
⅛ in
(4 mm)

284

IRAVADIA TROCHLEARIS
PULLEY IRAVADIA
(GOULD, 1861)

The shell of the Pulley Iravadia is a thick, conical, high-spired shell with a rounded anterior. The pronounced spiral bands of this shell superficially resemble a tiny version of the Grammatus Whelk (see page 413), which is not a close relative. The Pulley Iravadia has a large, smooth protoconch. After metamorphosis, the adult shell develops strong spiral cords. Upon reaching adulthood, a weakly flaring varix is produced, and the aperture is thickened. The shell is ash-colored, with a brown periostracum.

Members of the worldwide family Iravadiidae are characterized by small, solid, conical, high-spired shells. The protoconch is smooth and depressed, while the adult shell usually has strong spiral or reticulate sculpture, and a terminal varix marking adulthood. The animals are typically black in color. They live in bays and estuaries, buried in soft mud or sediment, which they consume in order to digest the organic components. Most species have a planktonic larval stage.

RELATED SPECIES
Iravadia quadrasi (Böttger, 1902), a western Pacific species, differs in having a broader shell with a thicker varix and a reticulate sculpture with beads at the intersections of spiral cords and axial ribs. *Iravadia yendoi* (Yokoyama, 1927) is endemic to Japan. It is easily recognized by its exceptionally tall and narrow shell, weak spiral sculpture, and reduced varix. *Rissopsis typica* Garrett, 1873, from the tropical western Pacific, is also extremely elongated and nearly cylindrical, with an aperture that is roughly triangular.

Actual size

FAMILY	Caecidae
SHELL SIZE RANGE	less than ⅛ in (2 to 4 mm)
DISTRIBUTION	Massachusetts to Brazil
ABUNDANCE	Common
DEPTH	0 to 330 ft (100 m)
HABITAT	Sandy bottoms
FEEDING HABIT	Microphagous feeder
OPERCULUM	Calcareous, circular, multispiral

SHELL SIZE RANGE
less than ⅛ in
(2 to 4 mm)

PHOTOGRAPHED SHELL
less than ⅛ in
(2 mm)

CAECUM PULCHELLUM

BEAUTIFUL CAECUM

STIMPSON, 1851

285

Caecum pulchellum is a tiny gastropod that, like most caecids, has a slightly curved, tubular shell as an adult. The larval shell is coiled, but after metamorphosis, the juvenile forms an uncoiled shell. At a certain point, the animal forms an apical plug or septum in the shell and the protoconch falls off. The shell continues to grow, and another decollation occurs. The adult shell has a mucro (spike), at the posterior end of the shell. Caecids usually live in algal mats on rocks in shallow waters. They are microphagous grazers, feeding on small particles. There are hundreds of species of caecids, distributed worldwide.

The shell of the Beautiful Caecum is minute, slightly curved, and tubular. The protoconch is coiled but lost in the adult. The adult shell is a slightly curved tube, with a circular, somewhat constricted, aperture. There is a low triangular projection (the mucro) at the posterior end. The surface sculpture consists of about 30 transverse rings separated by interspaces of about equal breadth and depth. The shell color ranges from white to light brown.

Actual size

RELATED SPECIES

Caecum clava Folin, 1867, from the Gulf of Mexico and the Caribbean, has a shell that is barely arched and is ornamented with longitudinal ribs; *C. imbricatum* Carpenter, 1858, from the Caribbean, has a tapering and slightly curved shell ornamented with both spiral and longitudinal ridges, giving it a cancellate texture; and *Meioceras nitidum* (Stimpson, 1851), from Florida to Uruguay, has a curved and smooth shell that is widest in the middle, and has a constricted aperture.

FAMILY	Strombidae
SHELL SIZE RANGE	¾ to 1¼ in (20 to 30 mm)
DISTRIBUTION	Taiwan to Indonesia
ABUNDANCE	Uncommon
DEPTH	50 to 400 ft (15 to 120 m)
HABITAT	Sandy bottoms
FEEDING HABIT	Grazer, feeds on algae
OPERCULUM	Corneous, elongated, claw-shaped

SHELL SIZE RANGE
¾ to 1¼ in
(20 to 30 mm)

PHOTOGRAPHED SHELL
¾ in
(21 mm)

286

VARICOSPIRA CRISPATA
NETTED TIBIA
(SOWERBY I, 1842)

Varicospira crispata is a small strombid related to the genus *Tibia* that lives on soft bottoms in depths up to 400 ft (120 m). It is most common in the Philippines, but is also found in Taiwan, Melanesia, and Indonesia. *Varicospira crispata* has been listed as rare by the government of the Philippines, and cannot now be exported from that country. There are four species recognized in the genus, all from the Indo-West Pacific. There are more than 70 species of strombids worldwide.

RELATED SPECIES

Varicospira cancellata (Lamarck, 1816), from the Indo-West Pacific, is the most common species in the genus *Varicospira*. Its shell is similar to *V. crispata* but it is more elongated, the cancellated sculpture less regular, and the posterior canal long and curved. *Tibia fusus* (Linnaeus, 1758), from the western Pacific, is a related species with a remarkably long siphonal canal.

Actual size

The shell of the Netted Tibia is small, thick, and fusiform. Its spire is tall and sharp, and the suture well marked. The sculpture consists of a cancellated pattern of evenly spaced, strong spiral cords and fine axial ribs. The aperture is narrow and lanceolate, and the posterior canal short and curled. The columella is smooth; the outer lip is thickened and bears many denticles. The shell color ranges from white to light brown; the aperture is tinted in brown.

FAMILY	Strombidae
SHELL SIZE RANGE	¾ to 2½ in (19 to 65 mm)
DISTRIBUTION	East Africa to central Pacific and Hawaii
ABUNDANCE	Uncommon
DEPTH	Intertidal to 260 ft (80 m)
HABITAT	Sandy bottoms near coral reefs
FEEDING HABIT	Grazer, feeds on algae
OPERCULUM	Corneous, elongated, claw-shaped

SHELL SIZE RANGE
¾ to 2½ in
(19 to 65 mm)

PHOTOGRAPHED SHELL
1¼ in
(45 mm)

STROMBUS DENTATUS

SAMAR CONCH

LINNAEUS, 1758

287

Strombus dentatus is an uncommon conch that has a wide distribution, ranging from East Africa to the central Pacific. It is usually found in shallow waters on sandy bottoms, especially near coral reefs. Its scientific name comes from the projections along the lower margin of the outer lip. Like many strombids, its shell is variable in size, shape, and coloration. Despite its small size, it is collected for food in some areas. The animal is mottled green with a dark green proboscis with cream spots.

RELATED SPECIES

Strombus urceus Linnaeus, 1758, from the western Pacific, is a common and variable species, whose shell resembles *S. dentatus* but has more angular whorls, a long aperture, and the outer lip is not dentate. *Strombus gibberulus* Linnaeus, 1758, from the tropical Indo-Pacific, is another conch that varies in size, shape, and coloration. It has an inflated shell with a long aperture, and asymmetrically coiled whorls.

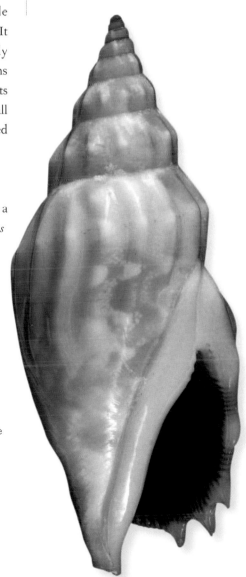

The shell of the Samar Conch is medium in size, glossy, solid, and elongated, with a tall spire. Its shell is variable in size, shape, and coloration. The whorl sculpture can be nearly smooth or have rounded axial ribs. The aperture is relatively small, and the outer lip is thick, with 3 or 4 pointed teeth, and many black spiral striae. The shell color is cream with brown maculations, and the outer lip and columella are white.

Actual size

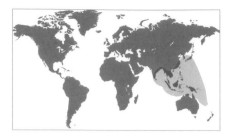

FAMILY	Strombidae
SHELL SIZE RANGE	1¼ to 4½ in (30 to 115 mm)
DISTRIBUTION	India to West Pacific
ABUNDANCE	Abundant
DEPTH	Intertidal to 180 ft (55 m)
HABITAT	Muddy sand bottoms
FEEDING HABIT	Herbivore, feeds on algae
OPERCULUM	Corneous, elongated, claw-shaped

SHELL SIZE RANGE
1¼ to 4½ in
(30 to 115 mm)

PHOTOGRAPHED SHELL
1¾ in
(46 mm)

288

STROMBUS CANARIUM
DOG CONCH
LINNAEUS, 1758

The shell of the Dog Conch is medium-sized, thick, heavy, and obesely fusiform. Its spire may be short and smooth, or high with angulated whorls and weak sculpture, but the large body whorl is usually smooth and rounded at the shoulder. The aperture is long. The outer lip is thickened and wing-shaped. The columella is smooth, shiny, and has a callus that is thicker anteriorly. The shell color ranges from white to light brown, with irregular, wavy, axial brown lines, and a white aperture.

Strombus canarium has a heavy and obese fusiform shell. It grows to about 4½ in (115 mm) in length, although most specimens collected are about half of that size. It is abundant on muddy sand and algae bottoms, ranging from the intertidal zone to offshore. It has a corneous, claw-shaped, and serrated operculum. There is a commercial fishery for this species in Southeast Asia, and in some locations its heavy shells are employed by fishermen as sinkers for nets. Many species of *Strombus* are widespread and abundant locally.

RELATED SPECIES

Strombus epidromis Linnaeus, 1758, from southern Japan to Australia and New Caledonia, has a more elongated shell, and the outer lip has a rounded profile. *Strombus plicatus* (Röding, 1798), from the Red Sea to the western Pacific, has a shell with a taller spire, and axial and spiral ridges.

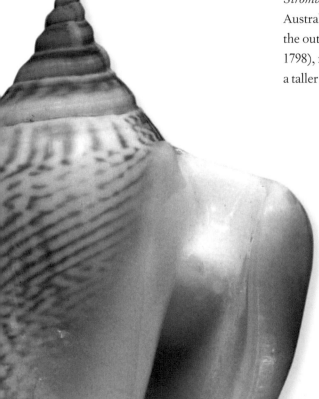

Actual size

FAMILY	Strombidae
SHELL SIZE RANGE	1¼ to 3 in (30 to 77 mm)
DISTRIBUTION	Red Sea to western Pacific
ABUNDANCE	Uncommon
DEPTH	Shallow subtidal to 300 ft (90 m)
HABITAT	Sandy bottoms
FEEDING HABIT	Herbivore, feeds on algae and detritus
OPERCULUM	Corneous, elongated, claw-shaped

SHELL SIZE RANGE
1¼ to 3 in
(30 to 77 mm)

PHOTOGRAPHED SHELL
2¼ in
(58 mm)

STROMBUS PLICATUS

PLICATE CONCH

(RÖDING, 1798)

289

Strombus plicatus has a variable shell, ranging from short and smooth to long and sculptured with spiral threads, as in the shell illustrated here. The name *plicatus* refers to the wrinkled appearance of the outer lip and columella. Four subspecies are recognized on the basis of shell shape differences; for example, *S. plicatus sibbaldi* Sowerby I, 1842, from the Gulf of Aden to Sri Lanka, is typically stunted and smooth, while *S. plicatus columba* Lamarck, 1822, from the western Indian Ocean, has a more elongate and wrinkled shell.

RELATED SPECIES

Strombus variabilis Swainson, 1821, from the Indo-West Pacific, as the name suggests, is also very variable in shape and coloration. *Strombus vittatus* Linnaeus, 1758, from the Indo-West Pacific, resembles an elongated *S. plicatus*.

The shell of the Plicate Conch is medium-sized, thick, and fusiform. Its spire is tall, stepped, and marked with axial ribs. The body whorl is large and swollen, bears blunt knobs at the shoulder, and has spiral ridges. The outer lip is thickened and flared, and in some shells has a wrinkled appearance, while in others it may be smooth. The aperture and the columella have lirations stained in brown. The shell color is white or cream, with light brown maculations.

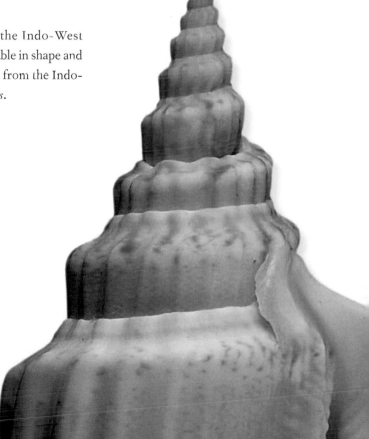

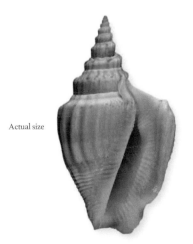

Actual size

FAMILY	Strombidae
SHELL SIZE RANGE	1¾ to 3 in (30 to 75 mm)
DISTRIBUTION	Tropical Indo-Pacific
ABUNDANCE	Abundant
DEPTH	Intertidal to 65 ft (20 m)
HABITAT	Sandy bottoms and seagrass beds
FEEDING HABIT	Grazer, feeds on algae
OPERCULUM	Corneous, elongated, claw-shaped

SHELL SIZE RANGE
1¾ to 3 in
(30 to 75 mm)

PHOTOGRAPHED SHELL
2¼ in
(59 mm)

290

STROMBUS GIBBERULUS
HUMPBACK CONCH
LINNAEUS, 1758

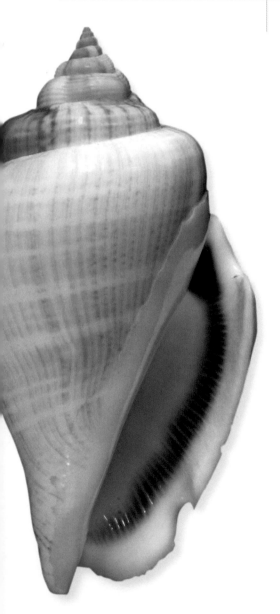

Strombus gibberulus is an abundant and very variable conch. It lives on sandy bottoms and seagrass beds in intertidal and shallow subtidal waters. Shells from deeper waters are usually smaller and have brighter colors. It is collected for its colorful shell and for food, especially in the Philippines and Fiji. Because it is an herbivore, *S. gibberulus* is sold in the aquarium industry as a good species to keep the substrate and rocks clean of algae.

RELATED SPECIES

Strombus luhuanus Linnaeus, 1758, from the western Pacific, has a conical shell with a low spire, a dark brown band on the smooth columella, and a rich orange or red aperture. *Strombus mutabilis* Swainson, 1821, from the tropical Indo-Pacific, also has a shell with an angular body whorl and a short spire.

The shell of the Humpback Conch is medium-sized, solid, and fusiform-inflated. Its spire is moderately high, and the whorls coil asymmetrically. The penultimate whorl bulges over the suture on the dorsum. The sculpture is mostly smooth, except for slightly raised spiral striae near the anterior end and outer lip margin. The aperture is long and the outer lip thickened and lirate inside. The shell color is variable, usually white with tan to brown spiral bands of various widths; the aperture is white, tinted with brown, orange, or purple.

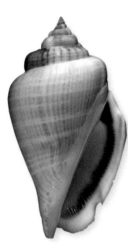

Actual size

FAMILY	Strombidae
SHELL SIZE RANGE	1¼ to 3¼ in (30 to 80 mm)
DISTRIBUTION	Western Pacific
ABUNDANCE	Abundant
DEPTH	Intertidal to 65 ft (20 m)
HABITAT	Sandy bottoms and seagrass beds
FEEDING HABIT	Herbivore, feeds on algae
OPERCULUM	Corneous, elongated, claw-shaped

SHELL SIZE RANGE
1¼ to 3¼ in
(30 to 80 mm)

PHOTOGRAPHED SHELL
2⅜ in
(59 mm)

STROMBUS LUHUANUS

STRAWBERRY CONCH

LINNAEUS, 1758

291

Despite being a very variable shell, *Strombus luhuanus* is easily recognizable by a dark brown or black band along its smooth columella. It is an abundant conch found on sandy bottoms near coral reefs, as well as on coral rubble and seagrass beds. Its shell is conical and superficially resembles shells in the family Conidae, but the presence of the stromboid notch, a deep U-shaped notch near the anterior margin of the lip, confirms that it belongs in the Strombidae.

RELATED SPECIES

Strombus decorus Röding, 1798, from the Indian Ocean, has a shell that resembles *S. luhuanus*, but lacks the dark brown band on the columella. *Strombus vittatus* Linnaeus, 1758, which ranges from the South China Sea to Fiji, has a tall spire that accounts for nearly half of the shell length.

The shell of the Strawberry Conch is medium sized, thick and conical, with a low spire. Its spire whorls coil asymmetrically, bulging over the suture in some specimens. The aperture is long and narrow, and the outer lip thickened. The shell surface is nearly smooth, except for shallow spiral grooves near the anterior end, and fine axial growth marks. Its shell color is usually white with varying tan to brown mottlings, the aperture rich orange, and a dark brown band lines the columella.

Actual size

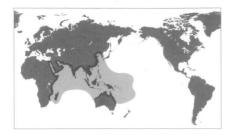

FAMILY	Strombidae
SHELL SIZE RANGE	2 to 4⅛ in (50 to 105 mm)
DISTRIBUTION	Tropical Indo-West Pacific
ABUNDANCE	Common
DEPTH	Intertidal to 13 ft (4 m)
HABITAT	Coral sand bottoms
FEEDING HABIT	Herbivore, feeds on algae
OPERCULUM	Corneous, elongated, claw-shaped

SHELL SIZE RANGE
2 to 4⅛ in
(50 to 105 mm)

PHOTOGRAPHED SHELL
3⅜ in
(87 mm)

292

STROMBUS LENTIGINOSUS
SILVER CONCH
LINNAEUS, 1758

The shell of the Silver Conch is medium-sized, thick and heavy, with a short or moderately high, pointed spire. Its spire whorls have a spiral row of nodulose ribs that become strong knobs on the body whorl. The aperture is long and narrow, and the outer lip is flared and thickened, with a wavy anterior edge. The anterior stromboid notch and the posterior anal canal are deep. The body whorl is broad, and the columella smooth and callused. The shell color is white, mottled with brown-gray, and the aperture is pinkish orange.

Strombus lentiginosus is a widely distributed conch in the tropical Indo-West Pacific. In Polynesia, it occurs only in some areas, but it is common wherever it becomes established. It is found from the intertidal zone to about 13 ft (4 m) deep, on barrier, fringing, and lagoon reefs, and on coral sand or seagrass bottoms, especially in clear waters. The animal is mottled with green, and the mantle has a yellow margin; the eyes are yellow, with a red border. It occurs in moderate to large colonies. Like other strombs, it is locally collected for food, and sold in markets in the Philippines. Its shell is commonly used in shellcraft.

RELATED SPECIES

Strombus pipus Röding, 1798, also from the tropical Indo-West Pacific, is the closest relative to *S. lentiginosus*, but it has a smaller shell with a less developed outer lip. *Strombus sinuatus* Humphrey, 1786, from the southwest Pacific, vaguely resembles *S. lentiginosus*, but it has a broad flared and thickened outer lip with four digitations at the posterior end.

Actual size

FAMILY	Strombidae
SHELL SIZE RANGE	1⅜ to 4 in (35 to 100 mm)
DISTRIBUTION	Okinawa to tropical west Pacific
ABUNDANCE	Common
DEPTH	Shallow subtidal to 165 ft (50 m)
HABITAT	Sandy mud bottoms
FEEDING HABIT	Herbivore, feeds on algae
OPERCULUM	Corneous, elongated, claw-shaped

STROMBUS VITTATUS

STRIPED CONCH

LINNAEUS, 1758

SHELL SIZE RANGE
1⅜ to 4 in
(35 to 100 mm)

PHOTOGRAPHED SHELL
3⅜ in
(93 mm)

293

Strombus vittatus is a common and variable conch, ranging from Okinawa to the tropical western Pacific. Three subspecies have been recognized on the basis of shell morphology and coloration. *Strombus vittatus* lives offshore on sandy mud bottoms. Strombid shells develop a flared and thickened outer lip when the animal matures. Juveniles lack the flared lip and look quite different from the adults. Some can be easily mistaken for shells in the family Conidae. The sexes are separate; and sexual dimorphism has been recorded in many species; females usually have a larger shell than males.

RELATED SPECIES

Strombus campbelli Griffith and Pidgeon, 1834, is endemic to Australia, resembles *S. vittatus*, but has a shorter spire, with the last three whorls smoother, and a broader body whorl. *Strombus listeri* Gray, 1852, from the Bay of Bengal and northwestern Indian Ocean, resembles an elongated *S. vittatus* but with a mostly smooth shell.

The shell of the Striped Conch is medium-sized, fusiform, elongated, and has a very high and pointed spire. This species is very variable in shell shape and coloration. Its spire is very high, reaching almost half of the shell length in some shells. Its spire whorls have strong axial ribs that attenuate on the body whorl; near the anterior end there are spiral threads. The aperture is narrow and the outer lip is thickened and flared. Its shell color is pale yellow-brown, and the aperture white.

Actual size

FAMILY	Strombidae
SHELL SIZE RANGE	2¾ to 5¾ in (70 to 145 mm)
DISTRIBUTION	Mauritius Islands (western Indian Ocean)
ABUNDANCE	Rare
DEPTH	13 to 83 ft (4 to 25 m)
HABITAT	Sandy bottoms
FEEDING HABIT	Grazer, feeds on filamentous algae
OPERCULUM	Corneous, elongated, claw-shaped

SHELL SIZE RANGE
2¾ to 5¾ in
(70 to 145 mm)

PHOTOGRAPHED SHELL
3¾ in
(95 mm)

294

LAMBIS VIOLACEA

VIOLET SPIDER CONCH

(SWAINSON, 1821)

Lambis violacea is one of the rarest species in the genus, and has a very restricted distribution, found only in Mauritius, in the western Indian Ocean. It is easily recognized by the deep, purple-tinted aperture that gives the shell its name. The color of the aperture seems to be quite stable, since museum specimens more than 100 years old have faded only slightly. There are only about ten species in the genus *Lambis*, all of which are restricted to the tropical waters of the Indian and Pacific oceans. They usually live in shallow waters on soft bottoms.

RELATED SPECIES

Lambis millipeda (Linnaeus, 1758), a common species from the southwest Pacific, and *L. digitata* (Perry, 1811), another rare spider conch from the Indo-West Pacific, have shells that resemble *L. violacea*, but they both have brown lirations in the aperture.

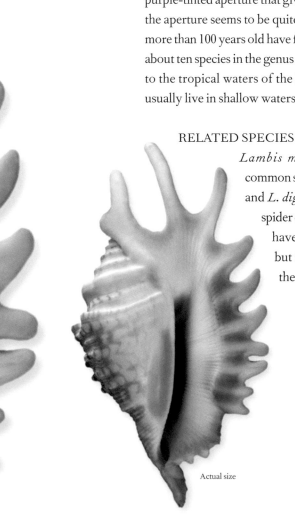

The shell of the Violet Spider Conch is medium-sized, thick, and has a broad flared outer lip bearing about 15 to 17 digitations. Such digitations vary in number and size, with those near the posterior end longer than those at or along the anterior end. The spire is long and pointed, and the siphonal canal long and recurved. The dorsal sculpture consists of many strong and nodulose spiral ribs. The shell color is white mottled with brown, the aperture purple, and the broad lip white.

Actual size

FAMILY	Strombidae
SHELL SIZE RANGE	3¼ to 5¾ in (80 to 145 mm)
DISTRIBUTION	Southwest Pacific
ABUNDANCE	Locally common
DEPTH	Intertidal to 70 ft (20 m)
HABITAT	Coral sand and algal bottoms
FEEDING HABIT	Herbivore, feeds on algae
OPERCULUM	Corneous, elongated, claw-shaped

SHELL SIZE RANGE
3¼ to 5¾ in
(80 to 145 mm)

PHOTOGRAPHED SHELL
4 in
(103 mm)

295

STROMBUS SINUATUS

LACINIATE CONCH

HUMPHREY, 1786

Strombus sinuatus is easily recognized by its purplish brown aperture and broad outer lip with four wavy lobes along the posterior part. It lives on coral sand and algal bottoms, from the intertidal zone to offshore. It varies in abundance from uncommon to locally abundant, as in the Bohol-Cebu area of the Philippines. Like other strombs, the shell grows until it reaches sexual maturity; then the outer lip thickens, and the shell ceases to grow in size, although it may become thicker.

RELATED SPECIES

Strombus thersites Swainson, 1823, from the western Pacific, is considered one of the rarest of the genus *Strombus*. It has a massive and broad shell with a tall and pointed spire. *Strombus taurus* Reeve, 1857, endemic to the Marshall Islands and Micronesia, is an uncommon conch with a solid shell with two spines on the posterior outer lip.

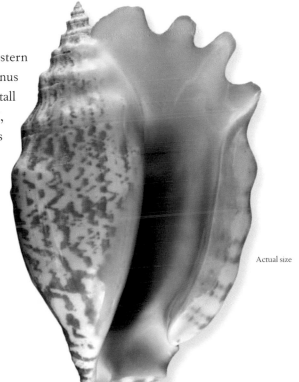

Actual size

The shell of the Laciniate Conch is of medium size, solid, moderately heavy, and fusiform with a broad flared outer lip and a tall spire. Its spire is stepped and knobbed, and the body whorl is broad with one spiral row of strong knobs at the shoulder, and fine spiral lines. There are 3 or 4 projections in the posterior region of the flared outer lip. The shell color is white or cream with yellow-brown spiral bands on the dorsum and zigzag vertical lines on the base; the aperture is purplish brown.

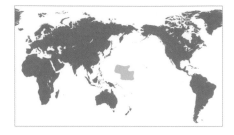

FAMILY	Strombidae
SHELL SIZE RANGE	3¼ to 5 in (80 to 130 mm)
DISTRIBUTION	Endemic to the Marshall and Mariana Islands
ABUNDANCE	Uncommon
DEPTH	50 to 80 ft (15 to 25 m)
HABITAT	Sand and coral rubble bottoms
FEEDING HABIT	Herbivore, feeds on algae
OPERCULUM	Corneous, elongated, claw-shaped

SHELL SIZE RANGE
3¼ to 5 in
(80 to 130 mm)

PHOTOGRAPHED SHELL
4⅛ in
(104 mm)

296

STROMBUS TAURUS
BULL CONCH
REEVE, 1857

Strombus taurus was once considered one of the rarest strombs. Originally, its locality was erroneously attributed to the Indian Ocean. However, with the advent of scuba diving, its true habitat was discovered in the late 1950s; it is endemic to the Marshall and Mariana Islands, in the central Pacific. It lives offshore, and because its shell is often encrusted with coralline algae, it blends well with coral rubble bottom. *Strombus taurus* is easily distinguished from other strombs by a thick, flared outer lip with only two to three projections, one of which is rather long.

RELATED SPECIES

Strombus oldi Emerson, 1965, has a restricted distribution, from Somalia and Oman, and is one of the rarest strombs. The body whorl has several raised spiral ridges that continue onto the outer lip, which is crenulated. *Strombus gallus* Linnaeus, 1758, from Florida to eastern Brazil and the West Indies, has a large shell with a broadly flared and undulating outer lip.

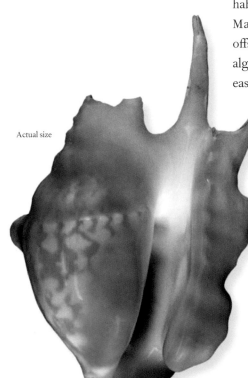

Actual size

The shell of the Bull Conch is medium-large, thick, heavy, glossy, and conical. Its spire is tall, the suture not well marked, but with a row of spiral knobs near the shoulder. The sculpture consists of a few low spiral ridges, a row of large knobs on the shoulder, and an oblique heavy knob on the dorsum. The aperture is narrow, the outer lip flared and thickened, with 2 or 3 posteriorly directed projections. One of the projections is longer than the spire. The outer color is white mottled with orange-brown, and the aperture is white with a purple-brown stain.

FAMILY	Strombidae
SHELL SIZE RANGE	3½ to 6¼ in (90 to 160 mm)
DISTRIBUTION	Bay of Bengal and northwestern Indian Ocean
ABUNDANCE	Common
DEPTH	165 to 400 ft (50 to 120 m)
HABITAT	Sandy bottoms
FEEDING HABIT	Herbivore, feeds on algae
OPERCULUM	Corneous, elongated, claw-shaped

SHELL SIZE RANGE
3½ to 6¼ in
(90 to 160 mm)

PHOTOGRAPHED SHELL
4¾ in
(122 mm)

STROMBUS LISTERI

LISTER'S CONCH

GRAY, 1852

297

Strombus listeri was once considered one of the rarest shells in collections, and only a few specimens were known until the 1960s, when its habitat was discovered in the northwestern Indian Ocean. It is one of the deepest dwelling strombs, ranging from about 165 to 400 ft (50 to 120 m). *Strombus listeri* has a beautifully elegant shell, and its broad flared outer lip has a wide, sinuous stromboid notch near the anterior end.

RELATED SPECIES

Strombus vittatus Linnaeus, 1758, from the tropical western Pacific, has a shell similar to *S. listeri* but with a narrower outer lip and strong axial ribs on the spire. *Tibia fusus* (Linnaeus, 1758), from the southwestern Pacific, is a related species that has an elongated fusiform shell with a very tall spire and very long siphonal canal.

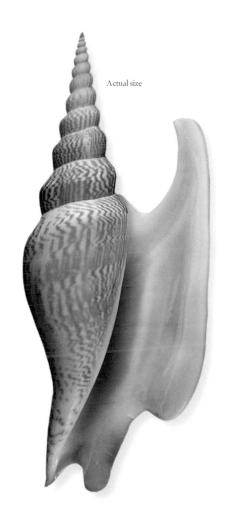

Actual size

The shell of the Lister's Conch is of medium size, fusiform, elongated, lightweight but strong, with a high spire and a broadly flared outer lip. The stepped high spire has axial ribs that become obsolete on the last 3 whorls. The body whorl has fine spiral lines and is mostly smooth. The aperture is long and narrow, and the outer lip is flared, with a long, flattened, narrow, posteriorly directed lobe. The shell color is white, covered with yellowish brown zigzag lines. The aperture and outer lip are white.

FAMILY	Strombidae
SHELL SIZE RANGE	3 to 7½ in (75 to 192 mm)
DISTRIBUTION	Florida to eastern Brazil; West Indies
ABUNDANCE	Uncommon
DEPTH	1 to 160 ft (0.3 to 48 m)
HABITAT	Sandy bottoms
FEEDING HABIT	Herbivore, feeds on algae
OPERCULUM	Corneous, elongated, claw-shaped

SHELL SIZE RANGE
3 to 7½ in
(75 to 192 mm)

PHOTOGRAPHED SHELL
6 in
(152 mm)

298

STROMBUS GALLUS
ROOSTER-TAIL CONCH
LINNAEUS, 1758

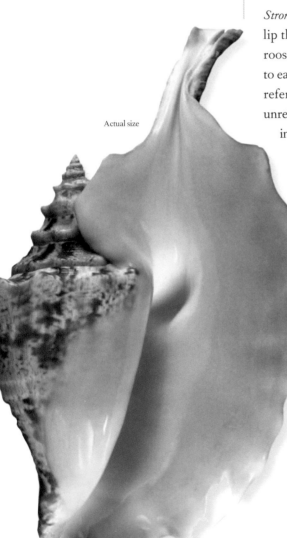

Actual size

Strombus gallus is a distinctive stromb with a large, flared outer lip that has one long projection posteriorly, which recalls a rooster's tail. It is an uncommon species ranging from Florida to eastern Brazil and the West Indies. The term conch usually refers to strombid gastropods, but it is also applied to many unrelated gastropods, usually edible and large, such as species in the families Buccinidae, Melongenidae, and Fasciolariidae. To differentiate them from other groups, the strombids are also known as "true conchs" or strombs. Many strombs are large and edible, and commercial fisheries exist for some species, such as *Strombus gigas* (see page 301).

RELATED SPECIES

Strombus peruvianus Swainson, 1823, from Peru to Mexico, is similar to *S. gallus*, but it is larger, the spire is shorter, the projection of the winged, flared lip is triangular in profile, and the outer lip is not wavy. *Strombus tricornis* Humphrey, 1786, from the Red Sea and Gulf of Aden, also resembles *S. gallus* but is smaller and broader, and the winged, flared, outer lip is shorter.

The shell of the Rooster-tail Conch is medium-sized, relatively lightweight, conical, and with a broad, flared outer lip. Its spire is high, with a spiral row of tubercles that become strong knobs on the shoulder of the body whorl; the body whorl has strong spiral threads. The outer lip in adult shells is flared and thickened, with an undulating margin and a long projection posteriorly that extends beyond the spire. The shell color is cream, mottled with orange-tan, and the aperture is pale to golden brown.

FAMILY	Strombidae
SHELL SIZE RANGE	6 to 12½ in (150 to 310 mm)
DISTRIBUTION	Japan to Indonesia
ABUNDANCE	Common
DEPTH	17 to 500 ft (5 to 150 m)
HABITAT	Muddy bottoms
FEEDING HABIT	Herbivore, feeds on algae
OPERCULUM	Corneous, lanceolate

SHELL SIZE RANGE
6 to 12½ in
(150 to 310 mm)

PHOTOGRAPHED SHELL
8⅛ in
(206 mm)

TIBIA FUSUS

SHINBONE TIBIA

(LINNAEUS, 1758)

Tibia fusus is an unmistakable species, with a fusiform shell that has one of the longest siphonal canal among all gastropods. The siphonal canal can be almost as long as the rest of the shell (it varies from about 30 to 45 percent of the shell length). It lives on muddy bottoms, generally in deep water, and is collected by trawling. It is amazing that shells with such a delicate canal are collected from deep water and brought to the surface in perfect condition. *Tibia fusus* has a narrow distribution in the southwest Pacific, and is most common around the Philippines.

RELATED SPECIES

Tibia martinii (Marrat, 1877), from off Taiwan to Indonesia, has a similar but broader shell, and with a shorter siphonal canal. The aperture is longer than in *T. fusus*. Formerly a rare species, it is now frequently trawled, suggesting it is common in its deepwater habitat. *Varicospira crispata* (Sowerby I, 1842), from Taiwan to Indonesia, has a small, cancellated shell. The aperture is somewhat similar to that in *T. fusus*, but the siphonal canal is short.

The shell of the Shinbone Tibia is long, slender, relatively lightweight, smooth, glossy, and fusiform. Its spire is very tall, with as many as 19 whorls, and an incised suture. The surface of the spire whorls has axial sculpture, but it fades toward the body whorl, which is mostly smooth, with fine spiral lines along its anterior portion. The aperture is lanceolate, the outer lip has 5 long digitations, and an extremely long, straight, or slightly, curved siphonal canal. The shell color is tan to brown, and the aperture white.

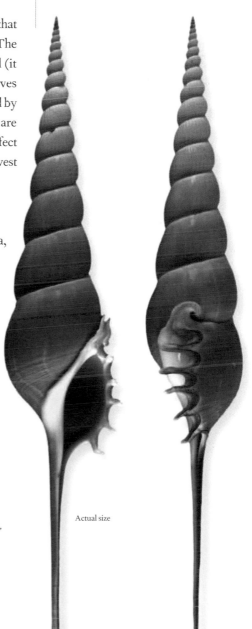

Actual size

FAMILY	Strombidae
SHELL SIZE RANGE	3⅜ to 13 in (85 to 330 mm)
DISTRIBUTION	Tropical Indo-Pacific
ABUNDANCE	Common
DEPTH	Intertidal to 83 ft (25 m)
HABITAT	Coarse sand and coral rubble
FEEDING HABIT	Grazer, feeds on filamentous algae
OPERCULUM	Corneous, elongated, claw-shaped

SHELL SIZE RANGE
3⅜ to 13 in
(85 to 330 mm)

PHOTOGRAPHED SHELL
10½ in
(259 mm)

300

LAMBIS CHIRAGRA

CHIRAGRA SPIDER CONCH

(LINNAEUS, 1758)

Lambis chiragra has a very distinct shell with long digitations including two recurved at right angles to the left side. It is a large and common species that is used for food. The shell of females is usually much larger than that of males. The operculum is corneous and claw-shaped, and used for locomotion. First, the animal stabs the pointed edge of the operculum into the substrate, extends the anterior part of the long foot, lifts the shell, then thrusts the shell forward, in a leaping movement.

RELATED SPECIES

Lambis truncata (Humphrey, 1786) from the Indo-Pacific is the largest species of the genus *Lambis*, and has a massive shell with a broad outer lip. *Lambis crocata* (Link, 1807), from the Indo-West Pacific, has a much smaller and more delicate shell.

Actual size

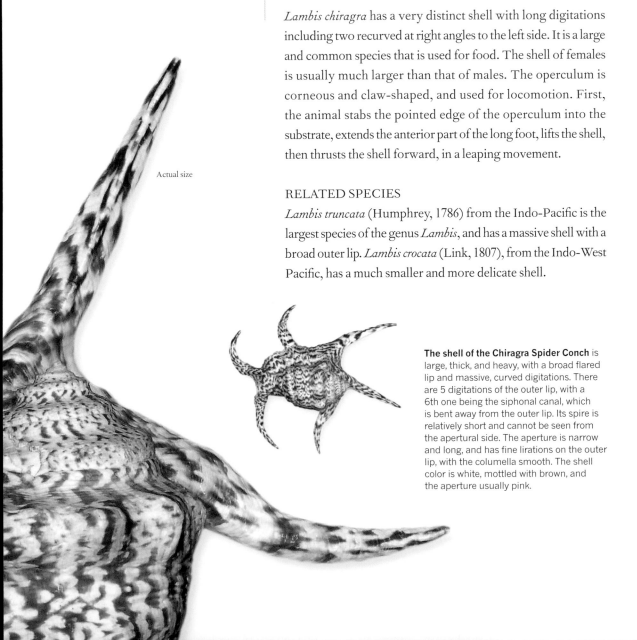

The shell of the Chiragra Spider Conch is large, thick, and heavy, with a broad flared lip and massive, curved digitations. There are 5 digitations of the outer lip, with a 6th one being the siphonal canal, which is bent away from the outer lip. Its spire is relatively short and cannot be seen from the apertural side. The aperture is narrow and long, and has fine lirations on the outer lip, with the columella smooth. The shell color is white, mottled with brown, and the aperture usually pink.

FAMILY	Strombidae
SHELL SIZE RANGE	6 to 14 in (150 to 350 mm)
DISTRIBUTION	Florida to Venezuela
ABUNDANCE	Common locally
DEPTH	1 to 65 ft (0.3 to 20 m)
HABITAT	Seagrass beds and sandy bottoms
FEEDING HABIT	Herbivore, feeds on algae
OPERCULUM	Corneous, elongated, claw-shaped

SHELL SIZE RANGE
6 to 14 in
(150 to 350 mm)

PHOTOGRAPHED SHELL
10¾ in
(270 mm)

STROMBUS GIGAS

QUEEN CONCH

LINNAEUS, 1758

301

Strombus gigas is one of the largest and most commercially important gastropods in the Caribbean. It lives on seagrass beds and sandy bottoms in shallow waters. Its populations are now declining or have already collapsed in many areas due to overharvesting. Collecting this species is now banned in the United States and in Yucatán, Mexico. Adults migrate to shallow waters to reproduce. Females lay up to almost half a million eggs in long gelatinous strings. They can live as long as 30 years.

RELATED SPECIES

Strombus goliath Schröter, 1805, which is endemic to northeastern Brazil, is the largest species in the genus *Strombus*. It has a thick and heavy shell with a huge flaring outer lip with a sweeping curve. *Strombus costatus* Gmelin, 1791, from North Carolina to eastern Brazil, has a shell that resembles *S. gigas* but it is smaller.

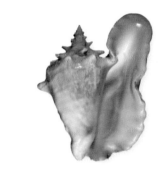

The shell of the Queen Conch is large, solid, thick, and has a wide flared lip. Its spire is relatively high, and has large knobs or blunt spines. Juvenile shells are biconic and lack the flaring lip. The body whorl is broad and sculpted with strong spines at the shoulder, and spiral ridges. The aperture is long and wide, and the outer lip is large and wavy. The shell color is cream, and the aperture can be bright or paler pink.

Actual size

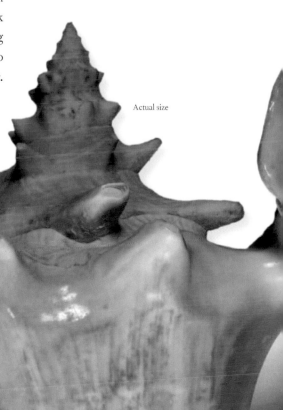

FAMILY	Strombidae
SHELL SIZE RANGE	8½ to 14⅛ in (220 to 360 mm)
DISTRIBUTION	Western Pacific
ABUNDANCE	Common
DEPTH	Shallow subtidal to 100 ft (30 m)
HABITAT	Sand near coral reefs
FEEDING HABIT	Herbivore
OPERCULUM	Corneous, unguiculate

SHELL SIZE RANGE
8½ to 14⅛ in
(220 to 360 mm)

PHOTOGRAPHED SHELL
13⅜ in
(360 mm)

LAMBIS TRUNCATA SEBAE
SEBA'S SPIDER CONCH
(KIENER, 1843)

The shell of the Seba's Spider Conch is very large and heavy. Its spire is moderately high. Its aperture is narrow and long with the siphonal canal bent toward the aperture. The labrum is greatly expanded and has 6 to 8 digitations or fingerlike projections. The dorsum is dull, covered with a very faint orangish periostracum. The aperture is glossy with a parietal shield, and the shell color is orange, purple, pink, or yellow.

Lambis truncata sebae has one of the largest shells in the genus *Lambis*. It is a common gastropod in shallow tropical waters of the Indo-Pacific and is used as food by indigenous peoples. The genus has a characteristic, expanded labrum with digitations ("fingers") that vary in number. Some species express sexual dimorphism in the shell, with female shells usually bearing longer digitations than those of males. As in other strombids, the long, clawlike operculum is used to anchor the shell during locomotion. A pair of colorful, stalked eyes peer up through the siphonal canal and stromboid notch. There are few species in the genus *Lambis*, all restricted to the tropical Indo-Pacific.

RELATED SPECIES
The subspecies *Lambis truncata truncata* Humphrey, 1786, ranges from East Africa to the West Pacific. It has a flattened spire, and a shell that can grow even larger than *L. truncata sebae*. *Lambis chiragra* (Linnaeus, 1758), from the Indo-West Pacific, has only a few digitations, but they are thick and long, two of which curved.

Actual size

FAMILY	Strombidae
SHELL SIZE RANGE	11 to 15 in (275 to 380 mm)
DISTRIBUTION	Endemic to northeastern Brazil
ABUNDANCE	Common
DEPTH	Intertidal to 165 ft (50 m)
HABITAT	Sand and seagrass
FEEDING HABIT	Herbivore, feeds on macroalgae
OPERCULUM	Corneous, elongate

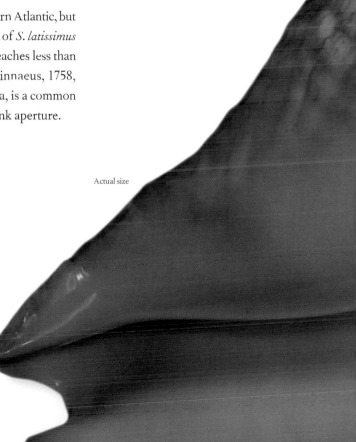

SHELL SIZE RANGE
11 to 15 in
(275 to 380 mm)

PHOTOGRAPHED SHELL
14½ in
(369 mm)

STROMBUS GOLIATH
GOLIATH CONCH
SCHRÖTER, 1805

303

Strombus goliath is appropriately named since it has the heaviest and most massive shell in the family. It is endemic to northeastern Brazil, living in shallow waters. It is often used as food, and although still common, there are concerns about its conservation. Juvenile shells look more like a *Conus*, with the spire comprising almost half of the shell length. However, as the animal grows the body whorl becomes more inflated, the shell thickens, and the labrum flares out and grows wide and thick.

RELATED SPECIES

There are several species of *Strombus* in the western Atlantic, but the shell of *S. goliath* most closely resembles that of *S. latissimus* Linnaeus, 1758, from the Indo-Pacific, which reaches less than half of the size of *S. goliath*. *Strombus gigas* Linnaeus, 1758, which ranges from South Carolina to Venezuela, is a common conch with a stepped spire, broad shell, and a pink aperture.

The shell of the Goliath Conch is massive, very heavy, and has a greatly expanded, smooth, rounded, and flared labrum. Its spire is pointed and relatively short. The body whorl is inflated and has strong knobs along the periphery. The dorsum is sculpted by radial ridges, and is covered by a light brown periostracum. The aperture is long and wide, and the siphonal canal is short and bent. A thick callus covers the columella; like the aperture, it is glossy orangish to light pink, fading to cream in older shells.

Actual size

FAMILY	Aporrhaidae
SHELL SIZE RANGE	1 to 2½ in (26 to 65 mm)
DISTRIBUTION	Norway and Iceland to Morocco and the Mediterranean
ABUNDANCE	Locally abundant
DEPTH	33 to 600 ft (10 to 180 m)
HABITAT	Muddy sand or mud bottoms
FEEDING HABIT	Browser and detritus feeder
OPERCULUM	Corneous, elongated

SHELL SIZE RANGE
1 to 2½ in
(26 to 65 mm)

PHOTOGRAPHED SHELL
1⅝ in
(43 mm)

304

APORRHAIS PESPELECANI

COMMON PELICAN'S FOOT

(LINNAEUS, 1758)

Aporrhais pespelecani has a distinctive shell that resembles a pelican's foot, hence the name. It has been known since at least the time of Aristotle, who first described its shell. *Aporrhais pespelecani* is abundant in the Adriatic Sea, where it is consumed as food. The family Aporrhaidae has a single genus and five living species, but there are many fossil species dating back to the Jurassic Period.

RELATED SPECIES

Aporrhais serresianus (Michaud, 1828) resembles *A. pespelecani* but has a more delicate shell and thinner digitations. *Aporrhais occidentalis* Beck, 1836, is the only aporrhaid in the northwestern Atlantic. The outer lip is expanded but there are no digitations.

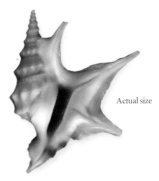

Actual size

The shell of the Common Pelican's Foot is small and has a flared outer lip with extensions that resemble a webbed foot. The spire is high, with a well-marked suture and a spiral row of beads. The body whorl is large and has 3 spiral beaded rows. In mature specimens the outer lip is thick and has 2 long digitations. Other smaller digitations may develop both near the spire and the siphonal canal. The shell color ranges from cream to brown, and the aperture is white.

FAMILY	Aporrhaidae
SHELL SIZE RANGE	1½ to 3 in (40 to 75 mm)
DISTRIBUTION	Greenland to North Carolina
ABUNDANCE	Common
DEPTH	33 to 6,500 ft (10 to 2,000 m)
HABITAT	Muddy gravel
FEEDING HABIT	Browser and detritus feeder
OPERCULUM	Corneous, elongated

SHELL SIZE RANGE
1½ to 3 in
(40 to 75 mm)

PHOTOGRAPHED SHELL
2⅛ in
(53 mm)

APORRHAIS OCCIDENTALIS

AMERICAN PELICAN'S FOOT

BECK, 1836

305

Aporrhais occidentalis has the thickest and heaviest shell in the family, although it lacks the characteristic digitations. It is considered the most primitive species in the family. Field studies revealed that it is active and moves on the surface of muddy gravel bottoms, feeding on diatoms and decaying brown algae, but buries in the sediment during the cold months.

RELATED SPECIES

Aporrhais serresianus (Michaud, 1828), from Norway and Iceland to the Mediterranean, lives on very fine mud bottoms. It has a smaller, lighter shell. *Aporrhais pesgallinae* Barnard, 1963, from southwest Africa to Angola, has a shell similar to *A. serresianus*.

Actual size

The shell of the American Pelican's Foot is thick and heavy, with a broad and expanded outer lip. It lacks the typical long digitations characteristic of the family. Its spire is high, with well-marked sutures. The sculpture of the shell consists of strong vertical ribs and fine spiral lines. The juvenile shell resembles a cerith, but mature specimens have a flared outer lip. The aperture is long and moderately wide. The shell color is cream or white.

FAMILY	Aporrhaidae
SHELL SIZE RANGE	1⅜ to 2½ in (35 to 60 mm)
DISTRIBUTION	Norway and Iceland to the Mediterranean
ABUNDANCE	Common
DEPTH	Offshore to 3,300 ft (1,000 m)
HABITAT	Fine mud bottoms
FEEDING HABIT	Browser and detritus feeder
OPERCULUM	Corneous, elongated

SHELL SIZE RANGE
1⅜ to 2½ in
(35 to 60 mm)

PHOTOGRAPHED SHELL
2¼ in
(57 mm)

306

APORRHAIS SERRESIANUS

MEDITERRANEAN PELICAN'S FOOT

(MICHAUD, 1828)

Aporrhais serresianus has the longest digitations among the living species in the family Aporrhaidae. Like other aporrhaids, the shells vary in the number and length of their digitations. The pattern of digitations is important in the identification of the species. These digitations help stabilize the shell on the soft sediment into which aporrhaids burrow. The longer the digitations, the finer the substrate the animal lives on.

RELATED SPECIES

Aporrhais pespelecani (Linnaeus, 1758), from Norway and Iceland to Morocco and the Mediterranean, is the most common species in the family. Its shell has long digitations. *Aporrhais senegalensis* Gray, 1838, from West Africa, has the smallest shell among the living aporrhaids.

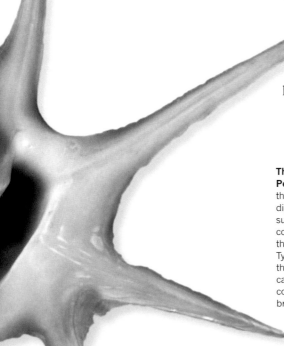

The shell of the Mediterranean Pelican's Foot is thin and light, and the flared outer lip has long digitations. The spire is high and the suture well marked. The sculpture consists of 1 spiral beaded row on the spire, and 3 on the body whorl. Typically, there are 4 long digitations that are webbed, and the siphonal canal is also very long. The shell color ranges from white to light brown, and the aperture is white.

Actual size

FAMILY	Seraphsidae
SHELL SIZE RANGE	1⅛ to 3 in (29 to 75 mm)
DISTRIBUTION	Indo-West Pacific
ABUNDANCE	Common
DEPTH	Shallow subtidal to 100 ft (30 m)
HABITAT	Sandy bottoms
FEEDING HABIT	Herbivore
OPERCULUM	Corneous, elongate, serrated edge

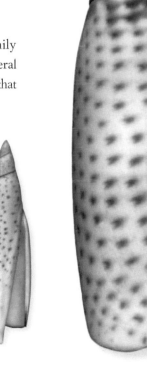

SHELL SIZE RANGE
1⅛ to 3 in
(29 to 75 mm)

PHOTOGRAPHED SHELL
2⅜ in
(61 mm)

TEREBELLUM TEREBELLUM

TEREBELLUM CONCH

(LINNAEUS, 1758)

307

Terebellum terebellum is the sole survivor of a fossil group related to strombids. The group first appeared during the Paleocene, and became most diverse during the Eocene Period. The species *T. terebellum* appeared in the fossil record during the Miocene. It has a torpedo-shaped shell that is well adapted for its habit of rapid burrowing in sand. Although it shell looks quite different from those of strombids, the long stalked eyes with bright irises, and an elongate operculum with a serrated edge reveal some stromboid features.

RELATED SPECIES

Terebellum terebellum is the only living species in the family Seraphsidae. Because its color and pattern vary widely, several species and subspecies have been proposed, but studies show that they are all populations of a single variable species.

The shell of the Terebellum Conch is medium-sized, glossy, narrow, elongate, streamlined, and torpedo-shaped. Its spire is moderately short, with a channeled suture, slightly convex whorls, and a pointed apex. The outer surface is smooth and glossy. The aperture is narrow, widest anteriorly, the columella smooth, and the outer lip smooth, thin, and truncated anteriorly. The outer shell color and pattern are very variable, usually with a light background and zigzag or spiral lines, dots, or blotches; often the suture is lined in brown. The aperture is white.

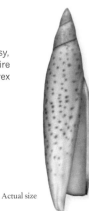

Actual size

FAMILY	Struthiolariidae
SHELL SIZE RANGE	2 to 3½ in (50 to 90 mm)
DISTRIBUTION	Endemic to New Zealand
ABUNDANCE	Common
DEPTH	Intertidal to 240 ft (75 m)
HABITAT	Fine sand or muddy bottoms
FEEDING HABIT	Ciliary mucous feeder
OPERCULUM	Chitinous, small, with a terminal spike

SHELL SIZE RANGE
2 to 3½ in
(50 to 90 mm)

PHOTOGRAPHED SHELL
3 in
(75 mm)

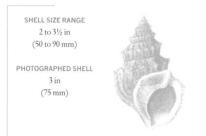

308

STRUTHIOLARIA PAPULOSA

LARGE OSTRICH-FOOT

(MARTYN, 1784)

The shell of the Large Ostrich-foot is small, solid, relatively heavy, angulated, and tall-spired. Its spire and body whorls are angular, and the shoulder has a spiral row of pointed nodules and fine spiral cords. The aperture is wide and the lip and columella are thick, usually with a strong callus. The shell color ranges from white to cream with brown irregular vertical stripes; the lip and callus are white, and the operculum small, pointed, and chitinous.

Struthiolaria papulosa is the largest species in a small group known as ostrich-foot shells. To escape from predatory starfish, it stretches its foot, uses its spiky operculum to dig into the sediment, and quickly contracts the foot, causing a violent jerky movement that sends the shell into a series of somersaults. There are only four known living species in the family Struthiolariidae.

RELATED SPECIES

Struthiolaria vermis (Martyn, 1784), endemic to the North Island of New Zealand, has a similar but smaller and smoother shell. *Perissodonta mirabilis* (Smith, 1875), from South Georgia and Kerguelen Islands, occurs in deeper waters. It has a smaller shell that lacks the thickened aperture of *S. papulosa*.

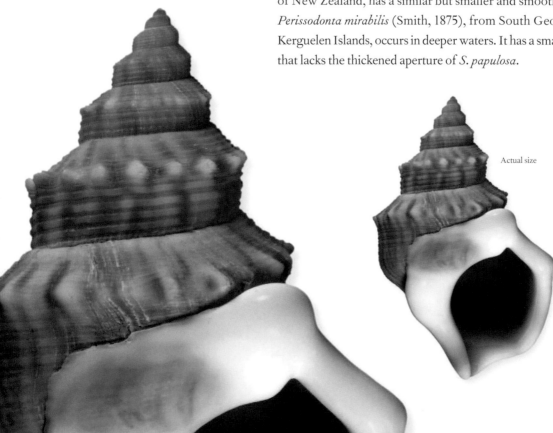

Actual size

FAMILY	Hipponicidae
SHELL SIZE RANGE	⅝ to 1⅝ in (15 to 40 mm)
DISTRIBUTION	Western Atlantic, western Africa; Indo-Pacific
ABUNDANCE	Common
DEPTH	Intertidal to 200 ft (60 m)
HABITAT	On rocks or other shells
FEEDING HABIT	Detritivore
OPERCULUM	Absent

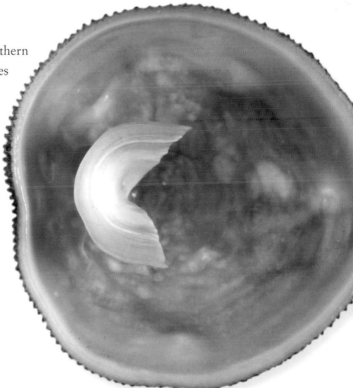

CHEILEA EQUESTRIS

FALSE CUP-AND-SAUCER

(LINNAEUS, 1758)

SHELL SIZE RANGE
⅝ to 1⅝ in
(15 to 40 mm)

PHOTOGRAPHED SHELL
1⅛ in
(30 mm)

Cheilea equestris has a very variable, limpetlike shell. It is a sedentary gastropod, and its aperture takes the shape of the surface onto which it lives attached. It has a large, half funnel-shaped internal projection of the shell; the superficially similar genus *Crucibulum* has a full funnel-shaped projection, which gives the group the popular name cup-and-saucer shells. Some settle on other shells, sometimes near the exhalant siphon, using extendable snout to collect fecal pellets from its host. At least some hipponicids start out as males and change sex to females.

The shell of the False Cup-and-Saucer is medium-sized, lightweight, and conical with an irregular, somewhat circular outline. Its spire is low, with a pointed and recurved apex, located near the center. The sculpture consists of concentric growth lines in the juvenile shell, and primarily radial ridges in the adult. The aperture is wide, with the outer margin crenulated. Inside the shell there is a large, half funnel-shaped projection. The shell color, inside and out, is white or grayish, with a brown periostracum.

RELATED SPECIES

Cheilea flindersi Cotton, 1935, endemic to southern Australia, is probably one of the largest species in the family, reaching over 2 in (53 mm) in diameter. *Sabia conica* (Schumacher, 1817), from the Red Sea and Indian Ocean to the western Pacific, has a strong, conical shell that lacks the internal expansion of the shell.

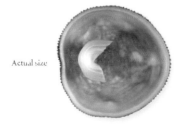

Actual size

FAMILY	Vanikoridae
SHELL SIZE RANGE	¼ to 1 in (7 to 25 mm)
DISTRIBUTION	Indo-Pacific, including Hawaii
ABUNDANCE	Common
DEPTH	Subtidal to 80 ft (25 m)
HABITAT	Hard bottoms
FEEDING HABIT	Detritivore
OPERCULUM	Corneous, thin, oval

SHELL SIZE RANGE
¼ to 1 in
(7 to 25 mm)

PHOTOGRAPHED SHELL
¾ in
(18 mm)

310

VANIKORO CANCELLATA
CANCELLATE VANIKORO
(LAMARCK, 1822)

Vanikoro cancellata has a wide distribution in the Indo-Pacific region, reaching Hawaii, where it occurs offshore. It lives on hard bottoms and among coral rubble. The animal uses its suckerlike foot to attach to hard substrates. If detached, some *Vanikoro* species apparently cannot re-attach themselves. Worldwide there may be as many as 70 living species in the family Vanikoridae, with almost 30 species in the western Atlantic alone. However, the taxonomy of this family is poorly known.

RELATED SPECIES

Vanikoro expansa (Sowerby I, 1842), from the Indo-West Pacific, has a similar but smaller shell that has a smoother sculpture when adult. *Macromphalina palmalitoris* Pilsbry and McGinty, 1950, from North Carolina to Texas and to Colombia, has a very small, translucent, compressed shell, with spiral sculpture.

Actual size

The shell of the Cancellate Vanikoro is small, thin, and globose. Its spire is very short, with a small apex and impressed suture. The body whorl is very large, and the sculpture in the adult shell is dominated by lamellate axial ribs, crossed by spiral threads; the axial ribs tend to become obsolete in large specimens. The aperture is large, accounting for more than half of the shell diameter; the outer lip is smooth and sharp, the columella smooth, and the umbilicus narrow and deep. The shell color, both inside and out, is white or off-white.

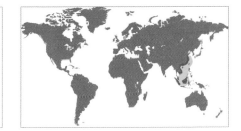

FAMILY	Calyptraeidae
SHELL SIZE RANGE	¾ to 1⅝ in (17 to 40 mm)
DISTRIBUTION	Malaysia, South and East China Sea
ABUNDANCE	Common
DEPTH	Subtidal to 100 ft (30 m)
HABITAT	On rocks or other shells
FEEDING HABIT	Filter feeder
OPERCULUM	Absent

SHELL SIZE RANGE
¾ to 1⅝ in
(17 to 40 mm)

PHOTOGRAPHED SHELL
1 in
(26 mm)

CALYPTRAEA EXTINCTORIUM

ASIAN CUP-AND-SAUCER

LAMARCK, 1822

311

Cup-and-saucer shells derive their common name from their
unusual interior. They have modified the spiral form, retaining
a thin residual blade to act as protection for their soft organs. In
many species, the plate curls around on itself enough to give the
impression of a small cup inside the very high-sided saucer
of the shell. *Calyptraeidae* are consecutive hermaphrodites:
they begin as males and change in later life to females.

RELATED SPECIES

Calyptraea chinensis (Linnaeus, 1758), is another
small member of the family, less than 1 in (25 mm)
in diameter, found (despite its name) in temperate
European waters, usually on subtidal rocks. It is dirty
cream, with a rounded apex. The internal blade is pleated
and curved, leaning toward the rim of the shell with which
it may connect.

Actual size

The shell of the Asian Cup-and-Saucer is conical, usually concavely so, with
a finely incised suture. Its exact shape varies according to what it is attached
to. There is a hard curtainlike structure inside, descending from the apex,
attached to the side, and curling back at the center. The exterior is cream to
pale tan, with very fine mid-tan to purple-brown axial stripes. Internally it is
walnut brown, paler at the rim.

FAMILY	Calyptraeidae
SHELL SIZE RANGE	1 to 2¾ in (25 to 70 mm)
DISTRIBUTION	Ecuador to Chile
ABUNDANCE	Moderately common
DEPTH	Offshore
HABITAT	Rocks
FEEDING HABIT	Filter feeder
OPERCULUM	Absent

SHELL SIZE RANGE
1 to 2¾ in
(25 to 70 mm)

PHOTOGRAPHED SHELL
1⅛ in
(28 mm)

312

TROCHITA TROCHIFORMIS
PERUVIAN HAT
(BORN, 1778)

The shell of the Peruvian Hat is a rough, low, pink-brown cone with a distinct suture. The flat whorls have a spiral sculpture of bold axial ribs that do not align with those on adjacent whorls. The interior is smooth, and a spiral shelf hangs from the apex, sloping downward and attached diagonally to the side, almost reaching the rim. It covers half the area of the shell.

Calyptraeidae are a fairly immobile family, and rather than scavenge for their diet of vegetable detritus they are content to let the food come to them, filter feeding more in the manner of bivalves than of most gastropods. *Trochita trochiformis* has a low, conical shell with a broad rim that extends well beyond and below the base of the shell. The exterior of the shell has oblique, axial ribs.

RELATED SPECIES

Crucibulum scutellatum (Gray, 1828) is found from Panama to the Gulf of California, beyond the northern end of the Peruvian Hat's range. It is the same size and has a similar rough ribbed exterior and shiny interior. However, the shelf is detached from the rim and recurved on itself to form a distinctively open, finely banded white cup.

Actual size

FAMILY	Calyptraeidae
SHELL SIZE RANGE	¾ to 2⅞ in (18 to 71 mm)
DISTRIBUTION	Southern California to Chile; Hawaii; Philippines
ABUNDANCE	Common
DEPTH	Subtidal to 200 ft (60 m)
HABITAT	On rocks and other shells
FEEDING HABIT	Filter feeder
OPERCULUM	Absent

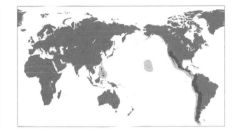

SHELL SIZE RANGE
¾ to 2⅞ in
(18 to 71 mm)

PHOTOGRAPHED SHELL
1¼ in
(29 mm)

CRUCIBULUM SPINOSUM

SPINY CUP-AND-SAUCER

(SOWERBY I, 1824)

313

It is no surprise that beached specimens of this extraordinarily sculptured shell are usually damaged. The living animal, however, occupies depths that are safe from the pounding effects of the waves above, clinging to submerged rocks and sometimes the less secure perch of dead and living shells. Like others of its family, its form varies considerably; it doesn't move much, and, therefore, has been able to adapt its shape in order to maximize its hold on its location. This is an eastern Pacific species that now has a much broader distribution, having been unintentionally introduced to Hawaii and the Philippines.

The shell of the Spiny Cup-and-Saucer is conical, with a rounded, recurving apex. It is cream to orange-yellow, sometimes with concentric purple bands, and may have fine to moderate radial ridges. It is covered (except for the apex) in fine perpendicular spines, sometimes tubular. The inside is shiny and white, often largely obscured by chestnut staining. There is a well-formed, semicircular cup, largely detached from the side and with the lip extended at one end.

RELATED SPECIES

Crucibulum serratum (Broderip, 1834) is a rare species confined between southern Mexico and Ecuador. The internal cup is distinctively flattened against the side. Externally it lacks spines but retains the radial ridges. A beautiful all-white variant at its northern extent, *C. s. concameratum* (Reeve, 1859), has ridges that radiate spirally and are crossed by concentric cords.

Actual size

FAMILY	Calyptraeidae
SHELL SIZE RANGE	⅝ to 1⅜ in (15 to 33 mm)
DISTRIBUTION	New Zealand
ABUNDANCE	Common
DEPTH	Offshore, 6 to 60 ft (2 to 20 m)
HABITAT	On rocks and other shells
FEEDING HABIT	Filter feeder
OPERCULUM	Absent

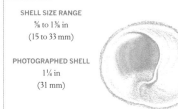

SHELL SIZE RANGE
⅝ to 1⅜ in
(15 to 33 mm)

PHOTOGRAPHED SHELL
1¼ in
(31 mm)

314

SIGAPATELLA NOVAEZELANDIAE
CIRCULAR SLIPPER
(LESSON, 1830)

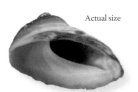

Actual size

Sigapatella is a genus of few species, most of which are scattered in the temperate seas around New Zealand and Australia. S*igapatella novaezelandiae* has a low, broad, coiled shell with a circular, limpetlike rim that conforms to the surface to which the animal is attached. The thin, translucent shelf spans three-quarters of the width of the shell, and produces an unusual, spirally coiled umbilicus.

RELATED SPECIES

Sigapatella calyptraeformis (Lamarck, 1822) is endemic to the southern half of Australia. Like *S. novaezelandiae*, the shell of the living animal has a coarse brown periostracum. It is variable in color, usually pale gray-brown with a white apex. The interior differs in being completely white and in showing less defined growth pleats on the shelf.

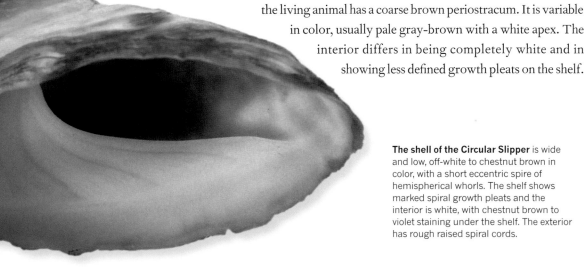

The shell of the Circular Slipper is wide and low, off-white to chestnut brown in color, with a short eccentric spire of hemispherical whorls. The shelf shows marked spiral growth pleats and the interior is white, with chestnut brown to violet staining under the shelf. The exterior has rough raised spiral cords.

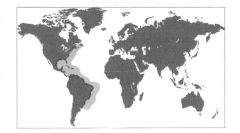

FAMILY	Calyptraeidae
SHELL SIZE RANGE	½ to 1¾ in (12 to 43 mm)
DISTRIBUTION	Nova Scotia to Brazil, Caribbean
ABUNDANCE	Common
DEPTH	Subtidal to 50 ft (15 m)
HABITAT	On rocks and other shells
FEEDING HABIT	Herbivore
OPERCULUM	Absent

SHELL SIZE RANGE
½ to 1¾ in
(12 to 43 mm)

PHOTOGRAPHED SHELL
1⅝ in
(40 mm)

CREPIDULA PLANA

EASTERN WHITE SLIPPER

SAY, 1822

Crepidula plana takes the classic slipper form: elongated, flat, and with a shallow shelf that protects its delicate tissues. It is frequently found stuck not to rocks but to other shells that may be dead or alive, preferring to attach to concave surfaces. It adheres to the shells of large, shallow-water muricids and buccinids, and even to the undersides of horseshoe crabs. Like all calyptraeids, this species is hermaphroditic. Small specimens are males, but change to females as they grow larger.

RELATED SPECIES

Crepidula fornicata (Linnaeus, 1758), originally from eastern North America, was introduced to Britain in the late nineteenth century and became a pest in oyster beds. It has a deeper internal shelf than *C. plana*, covering half the body. These mollusks have the unusual habit of living stacked up one on top of another, oldest and largest at the bottom.

The shell of the Eastern White Slipper is ovate, elongate, and very shallow, with an eccentric recurving apex situated above the rim at one end. It is white to pink, and unsculpted apart from fine, concentric growth lines. The white interior has a calyptraeid shelf below the apex, extending to less than half the length of the shell. Its rims are thin and sharp.

Actual size

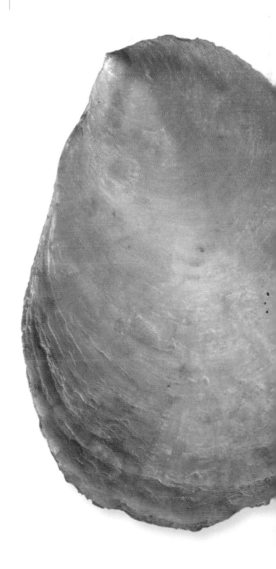

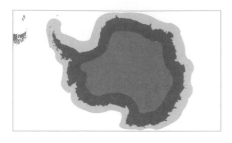

FAMILY	Capulidae
SHELL SIZE RANGE	1 to 2 in (25 to 50 mm)
DISTRIBUTION	Antarctica
ABUNDANCE	Uncommon
DEPTH	230 to 7,700 ft (70 to 2,350 m)
HABITAT	Soft bottoms
FEEDING HABIT	Filter feeder
OPERCULUM	Corneous, with terminal nucleus

SHELL SIZE RANGE
1 to 2 in
(25 to 50 mm)

PHOTOGRAPHED SHELL
1¼ in
(33 mm)

316

TORELLIA MIRABILIS

MIRACULOUS TORELLIA

(SMITH, 1907)

The shell of the Miraculous Torellia is medium in size, thin, lightweight, flexible, and globose. The spire is short, the smooth calcareous apex is pointed, and the suture is deep and channeled. Except for the protoconch, the entire shell is covered by a thick, chestnut periostracum, which has growth lines with dense axial rows of fine hair. The aperture is large and almost circular, with the columella slightly reflected. There is a wide and deep umbilicus in the base of the shell.

Capulidae, which may have shells that are either coiled or limpetlike, have a characteristic "pseudoproboscis"—an elongated extension of the mouth that has a slit along its dorsal surface. *Torellia mirabilis* is a deepwater Antarctic gastropod whose shell is almost completely composed of periostracum, which is thick and hairy. As a result of its mostly proteinaceous shell, it is flexible and tough. Its larval shell is calcareous, but as the shell grows, the proportion of calcium carbonate decreases and periostracum increases.

RELATED SPECIES

Torellia smithi Warén, 1986, from Antarctica, has a small shell with a taller spire that resembles a nerite. It has a thick and spongy periostracum that covers the entire shell. The related capulid, *Capulus ungaricus* (Linnaeus, 1758), ranging from Iceland to the Mediterranean, and from Greenland to Texas, lives attached to mollusks and rocks.

Actual size

FAMILY	Capulidae
SHELL SIZE RANGE	⅝ to 2½ in (15 to 60 mm)
DISTRIBUTION	Iceland to Mediterranean; Greenland to Texas
ABUNDANCE	Common
DEPTH	90 to 2,800 ft (25 to 850 m)
HABITAT	Offshore on rocks and molluscan shells
FEEDING HABIT	Filter feeds; can be parasitic on other mollusks
OPERCULUM	Absent

SHELL SIZE RANGE
⅝ to 2½ in
(15 to 60 mm)

PHOTOGRAPHED SHELL
2 in
(50 mm)

CAPULUS UNGARICUS
FOOL'S CAP
(LINNAEUS, 1758)

317

Capulus ungaricus has a beret-shaped shell that resembles a Hungarian cap, hence its name. It is a common species in temperate regions, living from shallow subtidal to deep waters. It lives attached to rocks or molluscan shells, especially bivalves such as scallops. *Capulus ungaricus* feeds by filtering water with its gills, but occasionally it drills a hole on a host's shell to feed on the food collected by the filter-feeding host. Small specimens are males, and turn into females when they grow larger. Smaller shells are usually more solid than larger ones, which become more fragile with increased size.

RELATED SPECIES

Capulus incurvatus (Gmelin, 1791), which ranges from North Carolina to Brazil, has a much smaller shell that sometimes uncoils. *Trichamantina nobilis* (Adams, 1867), from northern Japan Sea to Sakhalin Island, has a shell similar in size and shape, but with a thick periostracum that has two spiral ridges.

The shell of the Fool's Cap is small, thin, and cap-shaped; it can be either elevated or flattened. Its apex is spiral, inclined to one side, and curled posteriorly. The aperture is wide and rounded, and conformed to the surface to which the shell was attached. The body whorl comprises most of the shell. The shell color is yellowish white or pink, and inside it is glossy white or pink. The periostracum is brown and thick.

Actual size

FAMILY	Xenophoridae
SHELL SIZE RANGE	1½ to 4 in (38 to 100 mm)
DISTRIBUTION	Indo-West Pacific
ABUNDANCE	Common
DEPTH	65 to 1,150 ft (20 to 350 m)
HABITAT	Sandy mud bottoms
FEEDING HABIT	Feeds exclusively on foraminiferans
OPERCULUM	Corneous, oval

SHELL SIZE RANGE
1½ to 4 in
(38 to 100 mm)

PHOTOGRAPHED SHELL
2⅞ in
(74 mm)

318

ONUSTUS EXUTUS
BARREN CARRIER SHELL
(REEVE, 1842)

Onustus exutus is a large carrier shell that does not carry any attachments, except for some grains of sand near the apex, hence the name Barren Carrier Shell. Most xenophorids attach foreign objects to their shells but some like *O. exutus* do not. These attachments are believed to provide camouflage, add strength to the shell, or help stabilize the shell in fine mud substrates by increasing its diameter. The animal is active and moves rapidly in vaulting movements, by lodging its foot in the bottom and then propelling its shell forward.

RELATED SPECIES

Onustus longleyi Bartsch, 1931, ranging from North Carolina to Brazil, has a thick shell unlike most species in the family Xenophoridae. It may have none or few attachments, which are usually bivalve or gastropod shells or pieces. *Xenophora conchyliophora* (Born, 1780), from North Carolina to Brazil and the Caribbean, has most of its shell covered by attachments.

Actual size

The shell of the Barren Carrier Shell is large, thick, glossy, and broadly conical, with a low spire. There are no shells or rocks attached to its shell, but there may be some sand grains attached near the apex. The whorls have an irregular wavy margin or radial extensions. The dorsal sculpture consists of oblique axial ridges and wavy ribs that run at 90 degrees to these. The shell color is cream or light brown.

FAMILY	Xenophoridae
SHELL SIZE RANGE	1 to 3 in (25 to 77 mm)
DISTRIBUTION	North Carolina to Brazil; Caribbean
ABUNDANCE	Uncommon
DEPTH	Intertidal to 330 ft (100 m)
HABITAT	Sand and rubble bottoms
FEEDING HABIT	Herbivore, feeds on filamentous algae
OPERCULUM	Corneous, of variable shape

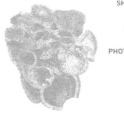

SHELL SIZE RANGE
1 to 3 in
(25 to 77 mm)

PHOTOGRAPHED SHELL
3 in
(77 mm)

XENOPHORA CONCHYLIOPHORA
ATLANTIC CARRIER SHELL
(BORN, 1780)

319

Xenophora conchyliophora is an interesting gastropod that can be considered the original shell collector: it cements shells, rocks, sand grains, and other foreign objects to its own shell, hence the name Carrier Shell. *Xenophora conchyliophora* picks up and maneuvers foreign objects with its snout and tentacle bases. Then the animal cleans its shell and cements the object to its shell with secretions from the mantle. Bivalve shells and pieces are positioned with the inner side facing upward, while gastropods are usually attached with the aperture facing up. This process can take up to an hour and a half to complete. After that, the animal remains motionless for up to ten hours to ensure its new attachment is secure.

The shell of the Atlantic Carrier Shell is medium in size, thin, and broadly conical, with a low spire. Most of the spire and dorsal surface of the shell are covered by shells, heavy pebbles, and other debris. Some attachments are added at the margin of the shell and extend well beyond the margin. There are no attachments to the base of the shell, which has obliquely radial sculpture and bands of light and darker brown.

RELATED SPECIES
Onustus caribaeus (Petit, 1857), from Florida to southern Brazil and the Caribbean, has a large, top-shaped shell that, in contrast to *X. conchyliophora*, has few attachments. *Stellaria solaris* (Linnaeus, 1767), from the Indo-Pacific, has a large shell, with about 17 to 19 long hollow spines per whorl at the periphery of the whorls. Foreign objects, if any, are only attached to the early whorls.

Actual size

FAMILY	Xenophoridae
SHELL SIZE RANGE	1¾ to 4 in (44 to 100 mm)
DISTRIBUTION	Florida to southern Brazil; Caribbean
ABUNDANCE	Common
DEPTH	65 to 2,100 ft (20 to 640 m)
HABITAT	Sandy and muddy bottoms
FEEDING HABIT	Detritivore
OPERCULUM	Corneous, thin, oval

SHELL SIZE RANGE
1¾ to 4 in
(44 to 100 mm)

PHOTOGRAPHED SHELL
4 in
(100 mm)

320

ONUSTUS CARIBAEUS

CARIBBEAN CARRIER SHELL

(PETIT, 1857)

Actual size

Onustus caribaeus has a wide latitudinal distribution, occurring from Florida to the Caribbean, and to southern Brazil. It lives in moderate to deep waters, on sandy or muddy bottoms. The animal attaches few objects, usually shells, to the margin of the broad "skirt" on the periphery of the shell. This skirt nearly doubles the diameter of the shell, which helps stabilize the shell on soft sediments, and provides protection to the animal against predators. In general, xenophorids that have few attachments on their shells live on soft bottoms, where large attachments would make their shell more conspicuous.

RELATED SPECIES

Onustus indicus (Gmelin, 1791), from the Indo-West Pacific, has a shell similar in size and shape to *O. caribaeus*, but with even fewer attachments, usually restricted to only the early whorls. It also possesses a wide skirt and an umbilicus. *Xenophora conchyliophora* (Born, 1820), from North Carolina to northeastern Brazil and the Caribbean, has a smaller shell with many attachments, usually whole shells or rocks. The shell lacks an umbilicus and a peripheral skirt.

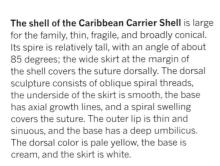

The shell of the Caribbean Carrier Shell is large for the family, thin, fragile, and broadly conical. Its spire is relatively tall, with an angle of about 85 degrees; the wide skirt at the margin of the shell covers the suture dorsally. The dorsal sculpture consists of oblique spiral threads, the underside of the skirt is smooth, the base has axial growth lines, and a spiral swelling covers the suture. The outer lip is thin and sinuous, and the base has a deep umbilicus. The dorsal color is pale yellow, the base is cream, and the skirt is white.

FAMILY	Xenophoridae
SHELL SIZE RANGE	2½ to 5¼ in (59 to 135 mm)
DISTRIBUTION	Tropical Indo-West Pacific
ABUNDANCE	Common
DEPTH	Shallow subtidal to 800 ft (250 m)
HABITAT	Sandy or muddy bottoms
FEEDING HABIT	Detritivore
OPERCULUM	Corneous, oval

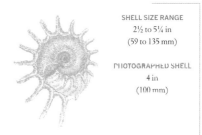

SHELL SIZE RANGE
2½ to 5¼ in
(59 to 135 mm)

PHOTOGRAPHED SHELL
4 in
(100 mm)

STELLARIA SOLARIS
SUNBURST CARRIER SHELL
(LINNAEUS, 1764)

321

Stellaria solaris is one of the most distinctive species in the family Xenophoridae because of its long radiating spines projecting from the margin of the whorls. The spines are angled downward and raise the shell from the substrate. The shell has small attachments cemented to only the early whorls. Like other xenophorids, it moves by lifting the shell with its muscular foot and then thrusting the shell forward by about half of the shell diameter. *Stellaria solaris* is often collected by shrimp trawlers, and its shell is used in shellcraft.

The shell of the Sunburst Carrier Shell is large, thin, lightweight, and broadly conical, with a low spire. Its distinguishing feature is the presence of long, hollow spines at the periphery of the whorls that protrude like spokes on a wheel. The ends of spines from early whorls are not attached to the subsequent whorls. The dorsal surface has fine oblique ribs, and the ventral surface has strong oblique radial ribs. The shell color is light brown.

RELATED SPECIES
Stellaria gigantea (Schepman, 1909), from the tropical Indo-Pacific, is the largest species in the family Xenophoridae. It has a moderately high spire and usually few attachments only at the margin of the whorls. *Onustus exutus* (Reeve, 1842), from the Indo-West Pacific, has no attachments to its shells, except for some grains of sand on the early whorls.

Actual size

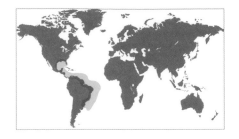

FAMILY	Vermetidae
SHELL SIZE RANGE	2 to 8 in (50 to 200 mm)
DISTRIBUTION	Florida to southern Brazil
ABUNDANCE	Common
DEPTH	Intertidal to 33 ft (10 m)
HABITAT	Cemented to hard substrates
FEEDING HABIT	Filter feeder
OPERCULUM	Corneous

SHELL SIZE RANGE
2 to 8 in
(50 to 200 mm)

PHOTOGRAPHED COLONY
12 in
(315 mm)
Individual shells in colony
up to 7 in (183 mm)

322

PETALOCONCHUS VARIANS

VARIABLE WORM SNAIL

(D'ORBIGNY, 1839)

Petaloconchus varians lives in dense colonies cemented to hard substrates. The larval shell is coiled for the first few whorls, but the adult shell is tubular and uncoiled. The juvenile crawls around until it finds a suitable substrate and settles, becoming permanently cemented. Shells of many vermetids resemble those of serpulid polychaete worms; there has been a lot of confusion between the two groups. Vermetid shells have three layers, while the shells of polychaetes have only two.

RELATED SPECIES

Petaloconchus innumerabilis Pilsbry and Olsson, 1935, from western Mexico to Peru, forms very dense colonies, with individual shells loosely coiled. *Serpulorbis oryzata* (Mörch, 1862), from western Mexico, does not form colonies like *P. varians*. The outer surface is wrinkled and granulose.

Actual size

The shell of the Variable Worm Snail is small and tubular. The snail grows in large colonies with hundreds or thousands of individuals cemented together, forming a dense, rounded mass. It is difficult to see the details of single shells, but they are tubular and irregular, with a round aperture. The shell color ranges from whitish to pale brown.

FAMILY	Vermetidae
SHELL SIZE RANGE	2 to 18½ in (50 to 470 mm)
DISTRIBUTION	Endemic to western Mexico
ABUNDANCE	Common
DEPTH	Intertidal to 133 ft (40 m)
HABITAT	Cemented to hard substrates
FEEDING HABIT	Filter feeder
OPERCULUM	Absent

SHELL SIZE RANGE
2 to 18½ in
(50 to 470 mm)

PHOTOGRAPHED SHELL
10¼ in
(259 mm)

SERPULORBIS ORYZATA
RICE WORM SNAIL
(MÖRCH, 1862)

323

Serpulorbis oryzata is one of the largest species in the family
Vermetidae. Most specimens grow to about 10 in (250 mm)
long, but it can reach over 18½ in (470 mm) in length. The first
whorls are loosely coiled and cemented to a hard substrate, but
adult shells become detached from the substrate, and may form a
broad curve or become almost straight. The males release sperm
packets, and the female broods the young in the mantle cavity.
The animal has a large, brightly colored, and extensible foot.
Some species, like *Petaloconchus varians*, are colonial, but
others, like *S. oryzata* are solitary.

RELATED SPECIES
Petaloconchus erectus (Dall, 1888), from Florida
to Brazil, has a small, solitary shell that has the first
whorls coiled, cemented to a hard substrate, and the last
part of the shell uncoiled and standing upright. *Petaloconchus
varians* (d'Orbigny, 1839) has a narrow, tubular shell and is
found in dense colonies, usually with hundreds or thousands
of individuals.

The shell of the Rice Worm Snail is large,
wrinkled, and tubular. Its first whorls are loosely
coiled, but most of the shell is not coiled, and
can be curved or nearly straight. The nuclear
whorls are usually eroded or broken. The surface
has axial folds and a nodulose texture. The
aperture is circular, up to about ⅝ in (15 mm)
in diameter, and the outer lip is thin. The shell
color is cream or tan.

Actual size

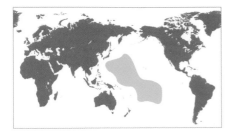

FAMILY	Cypraeidae
SHELL SIZE RANGE	¼ to ¾ in (7 to 20 mm)
DISTRIBUTION	Marianas to the Society Islands
ABUNDANCE	Uncommon
DEPTH	Shallow subtidal to 40 ft (12 m)
HABITAT	Coral rubble
FEEDING HABIT	Grazer, feeds on algae
OPERCULUM	Absent

SHELL SIZE RANGE
¼ to ¾ in
(7 to 20 mm)

PHOTOGRAPHED SHELL
½ in
(13 mm)

CYPRAEA GOODALLI

GOODALL'S COWRIE

SOWERBY I, 1832

Cypraea goodalli is one of the smallest species of cowrie. Its shell differs from other species in its complex by the presence of a dark orange-brown blotch on the dorsum. It is usually found in shallow water in coral rubble. The animal is white with a yellowish tint. The mantle is thin, with many minute, branched papillae. There are about 250 species of cowries around the world, with the highest diversity in the tropical Indo-Pacific.

RELATED SPECIES

Cypraea owenii Sowerby II, 1837, which is endemic to the Mauritius Islands, has fine marginal spotting; *C. stolida* Linnaeus, 1758, from the Indo-West Pacific, has a variable shell, usually with a rectangular brown blotch on the dorsum linked to four other blotches; and *C. ursellus* Gmelin, 1791, from the western Pacific, has a shell similar to *C. goodalli*, but with two black blotches on each extremity.

Actual size

The shell of the Goodall's Cowrie is small, lightweight, and ovately elongate. The spire is not exposed in the adult shell, and the area around the hidden spire is flattened. Both the anterior and posterior canals are thin. The aperture is narrow and long. The apertural teeth are longer on the labrum. The labrum is thick and forms a shallow labral groove on the side of the shell. The shell is white with a large, dark orange-brown blotch on the dorsum and minute brown marginal spots.

FAMILY	Cypraeidae
SHELL SIZE RANGE	⅜ to 1 in (8 to 24 mm)
DISTRIBUTION	Red Sea to Indo-Pacific
ABUNDANCE	Common
DEPTH	13 to 100 ft (4 to 30 m)
HABITAT	Crevices in coral reefs
FEEDING HABIT	Grazer, feeds on algae
OPERCULUM	Absent

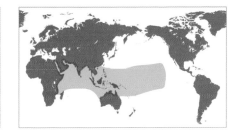

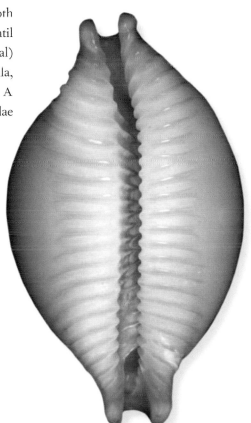

SHELL SIZE RANGE
⅜ to 1 in
(8 to 24 mm)

PHOTOGRAPHED SHELL
⅜ in
(10 mm)

CYPRAEA CICERCULA

CHICKPEA COWRIE

LINNAEUS, 1758

325

Cypraea cicercula is a small, common, and widespread cowrie. It is usually found in shallow waters. Most cowries have smooth shells but the shell of *C. cicercula* is normally pustulated. Until recently, cowrie taxonomy (and that of mollusks in general) was based primarily on shell characters and those of the radula, but molecular studies are increasingly more common. A recent comprehensive molecular phylogeny of the Cypraeidae confirms that *C. cicercula* and *C. margarita* are sister species.

RELATED SPECIES

Cypraea cicercula takahashii (Moretzsohn, 2007), from Hawaii and the Marshall Islands, has a smooth shell. *C. bistrinotata* Schilder and Schilder, 1937, from the western and central Pacific, has dorsal and ventral blotches, and its shell can be either pustulated or smooth.

Actual size

The shell of the Chickpea Cowrie is small, lightweight, and globose. The extremities are rostrated, fine, and have pointed tips. The spire is not exposed in adult shells, but is marked by a brown blotch. The apertural teeth are long and may reach the sides. Throughout most of its geographic range, the shell dorsum is typically pustulated and there is a dorsal groove; however, in Hawaii, most shells are smooth (like the photographed shell here) and pustulated ones are rare. The shell color is cream or light brown.

FAMILY	Cypraeidae
SHELL SIZE RANGE	½ to 1¼ in (11 to 31 mm)
DISTRIBUTION	Indo-Pacific
ABUNDANCE	Common
DEPTH	Intertidal to 80 ft (25 m)
HABITAT	Under coral and rubble
FEEDING HABIT	Grazer, feeds on algae
OPERCULUM	Absent

SHELL SIZE RANGE
½ to 1¼ in
(11 to 31 mm)

PHOTOGRAPHED SHELL
⅞ in
(22 mm)

CYPRAEA NUCLEUS
NUCLEUS COWRIE
LINNAEUS, 1758

The shell of the Nucleus Cowrie is medium in size for the family, solid and globose, with an ovate outline. The anterior and posterior extremities are extended and fine. The dorsum has many pustules that are sometimes connected by transverse ridges, and the base is convex and has transverse ridges that cross the base width. The shell is glossy, especially the spaces between pustules, and the shell color is tan, with white extremities.

Cypraea nucleus is one of just a few species of cowrie with a pustulose shell; most cowries have smooth and glossy shells. The dorsum of the shell has several rounded pustules, and the base has ridges that cross the entire base perpendicular to the aperture length. While in other cowries, the mantle deposits colored spots on the dorsum, in *C. nucleus* it deposits shell material until it becomes a three-dimensional spot. The animal of *C. nucleus* has some of the longest dorsal papillae in the family, which when fully extended, make the animal resemble a sea urchin.

RELATED SPECIES

Cypraea granulata Pease, 1862, a closely related species endemic to the Hawaiian Islands, has a broad oval shell with blunt extremities. It also has nodules on the dorsum, but it is the only cowrie with a rough and dull texture.

Actual size

FAMILY	Cypraeidae
SHELL SIZE RANGE	⅜ to 1¾ in (10 to 44 mm)
DISTRIBUTION	East Africa to central Pacific
ABUNDANCE	Abundant
DEPTH	Shallow subtidal
HABITAT	Tidepools and rocks near coral reefs
FEEDING HABIT	Grazer, feeds on algae
OPERCULUM	Absent

SHELL SIZE RANGE
⅜ to 1¾ in
(10 to 44 mm)

PHOTOGRAPHED SHELL
¾ in
(20 mm)

CYPRAEA MONETA

MONEY COWRIE

LINNAEUS, 1758

327

Cypraea moneta is the most abundant and widely distributed of all cowries. It occurs from East Africa and the Red Sea, throughout the tropical Indian and Pacific oceans to Cocos Island off Panama, and in all major islands in between. It was used as currency by indigenous peoples along the eastern coast of Africa and in many Pacific islands, hence its name. Today it is often used for decoration and jewelry. The animal of *C. moneta* is dazzling, with the mantle striped in black and yellow. Unlike most gastropods, once the cowrie reaches sexual maturity, the lips thicken and the shell no longer increases in size (although it may increase in shell thickness and become more callused).

RELATED SPECIES

Cypraea annulus Linnaeus, 1758, from the Indo-West Pacific, has a greenish yellow dorsum with a bright yellow or orange ring. *Cypraea obvelata* Lamarck, 1810, from French Polynesia, has a small shell that resembles *C. annulus* but has strong marginal calluses and a depression in the dorsum.

The shell of the Money Cowrie is small, strong, and morphologically variable. It is usually flattened with an oval outline, but it can be callused and angular; a rare variation has rostrate extremities. The base is flattened, and the aperture is narrow and long, surrounded by thick lips bearing few but strong teeth. Most shells have a faint yellow dorsum with three grayish bands; deep yellow shells are rare. The margins, base, and teeth are white or yellowish, and the interior of the shell is purplish.

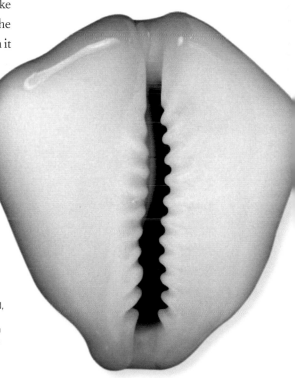

Actual size

FAMILY	Cypraeidae
SHELL SIZE RANGE	⅜ to 1⅝ in (10 to 43 mm)
DISTRIBUTION	Red Sea and Indian Ocean to central Pacific
ABUNDANCE	Common
DEPTH	Shallow subtidal to 100 ft (30 m)
HABITAT	On or near orange or red sponges
FEEDING HABIT	Grazer, feeds on algae, and sometimes on sponges
OPERCULUM	Absent

SHELL SIZE RANGE
⅜ to 1⅝ in
(10 to 43 mm)

PHOTOGRAPHED SHELL
1⅝ in
(36 mm)

328

CYPRAEA CRIBRARIA
SIEVE COWRIE
LINNAEUS, 1758

Cypraea cribraria is variable in size and coloration, and is the most common and widely ranging species in its species complex. All species in this group have red or orange mantles that blend in color with the sponges they live on or near. In adults, the mantle deposits an orange to brown pigmented layer on the dorsum, except under each dorsal papillae, leaving round holes in the pigmented layer that reveal the white juvenile shell underneath. Linnaeus aptly named this species *cribraria*, which means "sieve" in Latin.

RELATED SPECIES

Cypraea cumingii Sowerby I, 1832, from Eastern Polynesia, has a small, elongated shell, with a fine brown ring around dorsal spots; *C. gaskoinii* Reeve, 1846, endemic to Hawaii, has a globose shell with an orange-brown dorsum and black marginal spots; and *C. gravida* (Moretzsohn, 2002), from southeastern Australia, has elliptical dorsal spots.

The shell of the Sieve Cowrie is glossy, elliptical, and variable in size, coloration, and shape. The apex is depressed and often covered by a callus. The extremities are slightly rostrated and the anterior canal is bent upward. The labral teeth are thicker than the columellar teeth, and the aperture is long, narrow, and curved. The sides, extremities, and base are white, while the dorsum is red-brown, with round white spots. Marginal spots are common in shells from Sri Lanka, and rare elsewhere.

Actual size

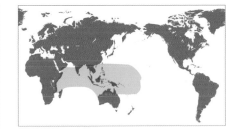

FAMILY	Cypraeidae
SHELL SIZE RANGE	⅝ to 1¾ in (15 to 46 mm)
DISTRIBUTION	Tropical Indo-West Pacific
ABUNDANCE	Uncommon
DEPTH	10 to 100 ft (3 to 30 m)
HABITAT	Coral reefs, under corals and stones
FEEDING HABIT	Grazer, feeds on algae
OPERCULUM	Absent

SHELL SIZE RANGE
⅝ to 1¾ in
(15 to 46 mm)

PHOTOGRAPHED SHELL
1⅝ in
(41 mm)

329

CYPRAEA STOLIDA

STOLID COWRIE

LINNAEUS, 1758

Cypraea stolida is one of 38 wide-ranging species of cowries that in New Caledonian waters are affected by a condition called melanism, in which the mantle secretes high amounts of melanin, causing the shell to become much darker than usual. Some cowries develop much longer than usual extremities to the point that the shell becomes rostrated (beaklike), as is the specimen shown here. Such specimens are most common in one bay in New Caledonia, rich in heavy metals, especially nickel, where they live next to typical specimens. Few shells are both melanistic and rostrated, and such extreme aberrations are highly prized.

RELATED SPECIES

Cypraea goodalli Sowerby I, 1832, from the Marianas to the Society Islands, has a small, ovately elongated shell with a light brown dorsal blotch. *Cypraea ursellus* Linnaeus, 1758, from the Indian Ocean to the central Pacific, has four dark brown blotches near the extremities, and lighter blotches on the dorsum.

The shell of a typical Stolid Cowrie is medium in size, inflated, and has an ovate outline. Its extremities are slightly thickened, and the dorsum has a large, squarish, light brown blotch. The teeth around the aperture are long and reach more than half of the base. Melanistic shells have a slightly darker to very dark brown or black dorsum; rostrated specimens have the extremities expanded, thickened, and in extreme cases, the anterior and posterior canals are bent upward as in the photograph shown here.

Actual size

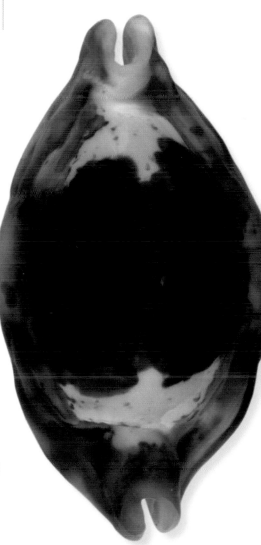

FAMILY	Cypraeidae
SHELL SIZE RANGE	2 to 3¼ in (50 to 80 mm)
DISTRIBUTION	Southeastern Africa
ABUNDANCE	Formerly rare
DEPTH	200 to 820 ft (60 to 250 m)
HABITAT	Unknown
FEEDING HABIT	Spongivore
OPERCULUM	Absent

SHELL SIZE RANGE
2 to 3¼ in
(50 to 80 mm)

PHOTOGRAPHED SHELL
2⅝ in
(65 mm)

330

CYPRAEA FULTONI
FULTON'S COWRIE
SOWERBY III, 1903

Cypraea fultoni is a deepwater cowrie that, until recently, was found only when collected from the stomachs of large fish such as the musselcracker. Within the past few decades, commercial trawlers started to bring back live specimens. All molluscan shells are secreted by the mantle, but in cowries the mantle covers the entire shell. In adults, the mantle continues to deposit thin pigmented and unpigmented layers onto the dorsum. In several cowries, such as *C. fultoni*, some layers are translucent, resulting in complex, three-dimensional color patterns on the dorsum.

RELATED SPECIES

Cypraea teulerei Cazanavette, 1846, from the Red Sea and Gulf of Oman, has a broad shell with a wide posterior extremity and nearly toothless lips. It is the only other living representative in this group, which contains several fossil species. The next closest living relative is *C. mus* Linnaeus, 1758, from the Caribbean coast of Colombia and Venezuela. It somewhat resembles *C. teulerei* but its shell has apertural teeth.

Actual size

The shell of the Fulton's Cowrie is medium in size, heavy, and ovate-pyriform. It varies in shape and coloration, and no two shells are alike. The dorsum is inflated and the base is flattened and slightly convex. The aperture is long and curved, the apertural teeth are thick. The dorsal color is light brown with irregular, three-dimensional patterns in dark brown; the margins have brown spots, and the base is beige.

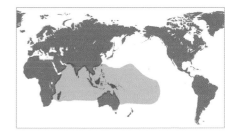

FAMILY	Cypraeidae
SHELL SIZE RANGE	1¾ to 4¼ in (46 to 110 mm)
DISTRIBUTION	East Africa to the central Pacific
ABUNDANCE	Common
DEPTH	3 to 33 ft (1 to 10 m)
HABITAT	Crevices near coral reefs
FEEDING HABIT	Grazer, feeds on algae
OPERCULUM	Absent

SHELL SIZE RANGE
1¾ to 4¼ in
(46 to 110 mm)

PHOTOGRAPHED SHELL
2¾ in
(70 mm)

CYPRAEA ARGUS

EYED COWRIE

LINNAEUS, 1758

331

Cypraea argus has one of the most distinctive cowrie shells, with dorsal rings or "eyes," and cannot be confused with any other species. It varies considerably in size and in the number of dorsal rings. The species name derives from the character Argus Panoptes from Greek mythology: Argus, said to have 100 eyes, was the "all-seeing" guardian of the heifer-nymph Io. The animal of *C. argus* is dark brown, with a thin mantle that does not obscure the dorsal shell pattern. There are many long, gray-brown branched papillae.

RELATED SPECIES

Cypraea leucodon Broderip, 1828, from the East Indian Ocean to the west Pacific, has thick apertural teeth and large dorsal spots; *C. aurantium* Gmelin, 1791 from the southwest and central Pacific, has a large and deep orange shell; and *C. porteri* Cate, 1966, from the Philippines to northwest Australia, has a medium-sized shell with large dorsal spots and an orange base.

Actual size

The shell of the Eyed Cowrie is large, heavy, cylindrical, and elongated. The sides are nearly parallel to one another. The aperture is long and narrow, and the apertural teeth long and thin. The apex area is flattened and partially covered by a thick callus. The shell background is beige, and the dorsum has 3 or 4 broad darker bands. The dorsum has brown rings of varying thickness. The base has 4 dark brown blotches.

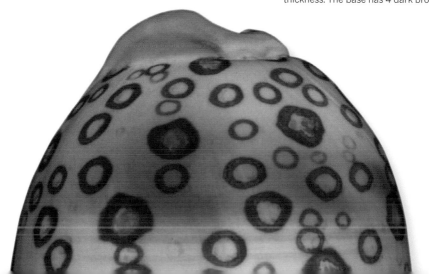

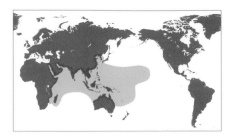

FAMILY	Cypraeidae
SHELL SIZE RANGE	2 to 4 in (50 to 100 mm)
DISTRIBUTION	Indo-West Pacific
ABUNDANCE	Common
DEPTH	15 to 120 ft (5 to 35 m)
HABITAT	Coral reefs, under corals and stones
FEEDING HABIT	Grazer, feeds on algae
OPERCULUM	Absent

SHELL SIZE RANGE
2 to 4 in
(50 to 100 mm)

PHOTOGRAPHED SHELL
2⅞ in
(72 mm)

332

MAP COWRIE

LINNAEUS, 1758

Cypraea mappa is appropriately named because its dorsum is reminiscent of a worn and burned map. Cowries are glossy because the shell is covered by the mantle; the area on the dorsum where the two lobes of the mantle meet is usually stained in a different color, often as a thin, straight or slightly curved line. In *C. mappa*, this dorsal line is thick and meandering, and sometimes resembles the course of a winding river.

RELATED SPECIES

Cypraea tigris Linnaeus, 1758, which has a wide Indo-Pacific distribution, is an abundant and beautiful shell. It varies widely in size, shape, and coloration. *Cypraea mauritiana* Linnaeus, 1758, another Indo-Pacific cowrie, has a large, heavy, and humpbacked shell, a chocolate brown dorsum with lighter brown spots.

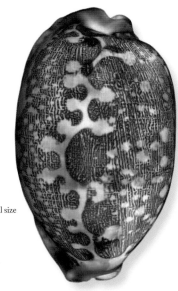

Actual size

The shell of the Map Cowrie is large for the family, inflated and humped with an ovate outline. Its extremities are thick and the margins callused. The aperture is narrow and long; the teeth do not extend much onto the base, and are sometimes stained in orange. The dorsum is light brown with orange-brown lines and reticulations, with some lighter color spots; the base and extremities are cream. The tan dorsal line is thick, meandering, and outlined in brown.

FAMILY	Cypraeidae
SHELL SIZE RANGE	1½ to 3 in (40 to 78 mm)
DISTRIBUTION	East Indian Ocean and west Pacific
ABUNDANCE	Uncommon
DEPTH	165 to 1,200 ft (50 to 360 m)
HABITAT	Sand and rubble
FEEDING HABIT	Grazer, feeds on algae
OPERCULUM	Absent

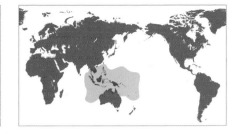

SHELL SIZE RANGE
1½ to 3 in
(40 to 78 mm)

PHOTOGRAPHED SHELL
3 in
(78 mm)

333

CYPRAEA GUTTATA

GREAT SPOTTED COWRIE

GMELIN, 1791

Cypraea guttata is easily recognized because of its distinctive dorsal markings and ribbed base, although it is variable in shell shape and coloration. When the animal is undisturbed, the opaque mantle covers the shell, obscuring the shell pattern. The mantle has two types of papillae: long, branched ones, and shorter, wartlike ones. This cowrie lives in deep water in tropical areas of the Indo-West Pacific, on sand and rubble bottoms.

RELATED SPECIES

Cypraea lamarckii Gray, 1825, from the Indian Ocean, has small dorsal spots, and its base is not ribbed; *C. helvola* Linnaeus, 1758, from the Indo-Pacific, has small, dense dorsal spots and a deep orange or red base; and *C. acicularis* Gmelin, 1791, from North Carolina to Brazil, has a marginal pitting with a white base.

The shell of the Great Spotted Cowrie is medium-sized, callused, and pyriform. The extremities are pointed and can be rostrated (beaklike). The base is unique in its strong ribbing; the ribs are a continuation of the apertural teeth and extend to the margins and anterior canal (sometimes also to the posterior canal). The base and the sides can have calluses that obscure the ribbing. The dorsal color is orange-brown, with large and small white spots. The base is white with brown ribs and calluses.

Actual size

FAMILY	Cypraeidae
SHELL SIZE RANGE	1⅝ to 4¼ in (42 to 107 mm)
DISTRIBUTION	Western and southwestern Australia
ABUNDANCE	Uncommon
DEPTH	16 to 330 ft (5 to 100 m)
HABITAT	Found on large sponges
FEEDING HABIT	Spongivore
OPERCULUM	Absent

SHELL SIZE RANGE
1⅝ to 4¼ in
(42 to 107 mm)

PHOTOGRAPHED SHELL
3¼ in
(81 mm)

334

CYPRAEA FRIENDII
FRIEND'S COWRIE
GRAY, 1831

The shell of the Friend's Cowrie is large, smooth, and very variable in shape, dentition, and color. It is ovate and elongate, rostrate, with a slight humped dorsum and flared anterior and posterior canals. The base is flattened, slightly convex, with a long and narrow aperture that spans the entire length of the shell. The labral teeth are more numerous than the columellar teeth, and some shells have almost no columellar teeth. The dorsal coloration varies from a light background with large brown blotches to dark brown. The base can be white to brown, and the margins are usually darker than the dorsum.

Cypraea friendii belongs to a group of cowries endemic to Australia (subgenus *Zoila*), that are very popular among shell collectors. Unlike most cowries that have a pelagic larva, these species have direct development, resulting in localized populations that are quite variable morphologically. They live and feed on large sponges, occurring from shallow subtidal to deep water. Because they do not hide in crevices, their shells often have scars or chips from encounters with predatory fishes.

Actual size

RELATED SPECIES
Cypraea rosselli Cotton, 1848, from southwestern Australia, has a triangular shell and humped dorsum that can be rich brown to black; *C. marginata* Gaskoin, 1849, from Western and southern Australia, has a shell similar to *C. friendii* but with expanded margins; and *C. thersites* Gaskoin, 1849, from southern Australia, has a humped dorsum and ovate outline.

FAMILY	Cypraeidae
SHELL SIZE RANGE	2¾ to 3¾ in (70 to 94 mm)
DISTRIBUTION	East Indian Ocean to the west Pacific
ABUNDANCE	Uncommon
DEPTH	100 to 1,000 ft (30 to 300 m)
HABITAT	Crevices on reefs
FEEDING HABIT	Grazer, feeds on algae
OPERCULUM	Absent

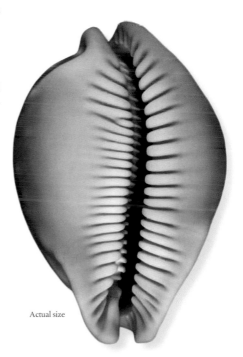

SHELL SIZE RANGE
2¾ to 3¾ in
(70 to 94 mm)

PHOTOGRAPHED SHELL
3¼ in
(86 mm)

CYPRAEA LEUCODON
WHITE-TOOTHED COWRIE
BRODERIP, 1828

335

Cypraea leucodon is so striking in appearance that the only two specimens available at the time it was discovered were sufficient to establish it as a new species; *leucodon* refers to the distinctive white teeth. It lives in crevices, under ledges, or caves on outer reefs in deep waters, and is found from the eastern Indian Ocean to the Philippines and Solomon Islands. It is supposedly common in its habitat, but because it lives so deep it is collected infrequently, especially in the Philippines. Several human lives have been lost in the pursuit of this species. Its shell varies in size and coloration, and a few subspecies have been described.

RELATED SPECIES
Cypraea aurantium Gmelin, 1791, from the southwest and central Pacific, has a solid, deep orange dorsum; *C. broderipii* Sowerby I, 1832, from Somalia to South Africa, has a dorsum with a brown, netlike pattern on a pinkish background; and *C. vitellus* Linnaeus, 1758, from southeastern Africa to Hawaii, somewhat resembles *C. leucodon*, but has a more slender shell with small and large dorsal spots and smaller teeth.

The shell of the White-toothed Cowrie is large, heavy, and ovately inflated. The anterior and posterior extremities are wide and thick. The dorsum is rounded and the base flattened and slightly convex. The aperture is long, narrow, and curved, and the numerous apertural teeth are long and thick. The dorsal color is light to chocolate brown with whitish dorsal spots of varying size, usually large, and a thick, curved dorsal line. The base color is beige, and the teeth are whitish.

Actual size

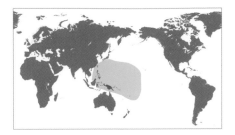

FAMILY	Cypraeidae
SHELL SIZE RANGE	2¼ to 4½ in (58 to 117 mm)
DISTRIBUTION	Southwest and central Pacific
ABUNDANCE	Uncommon
DEPTH	30 to 130 ft (10 to 40 m)
HABITAT	Caves and crevices in coral reefs
FEEDING HABIT	Grazer, feeds on algae
OPERCULUM	Absent

SHELL SIZE RANGE
2¼ to 4½ in
(58 to 117 mm)

PHOTOGRAPHED SHELL
3 ⅝ in
(93 mm)

336

CYPRAEA AURANTIUM
GOLDEN COWRIE
GMELIN, 1791

Cypraea aurantium has a striking, large, and orange shell that cannot be mistaken for any other cowrie, and is one of the more famous species in the family. When fresh, the dorsal color is a deep magenta, but it fades to a deep orange when exposed to strong sunlight. Like most cowries, *C. aurantium* is a nocturnal species, hiding in crevices during the day. The mantle in cowries is large and has two lobes that cover the shell. The mantle has projections, called papillae, that may help in respiration as well as camouflage. In *C. aurantium*, the mantle is orange-brown, with large, branched papillae and smaller, unbranched ones.

RELATED SPECIES

Some of the closest relatives include *Cypraea leucodon* Broderip, 1827, from the East Indian Ocean to the western Pacific, which has thick apertural teeth and large dorsal spots; *C. broderipii* Sowerby I, 1832, from Somalia to South Africa, which has a wide shell with a reticulated dorsum; and the common *C. lynx* Linnaeus, 1758, from the Red Sea, Indian Ocean to Hawaii, which has a variable shell with dark dorsal spots.

Actual size

The shell of the Golden Cowrie is large, heavy, and inflated. The deep orange dorsum is smooth and glossy, and small growth lines are common (shells with a flawless dorsum are rare). The aperture is narrow and long, with thickened lips bearing numerous teeth, more numerous on the columellar lip, and thicker, longer and more spaced out in the outer lip. The base, margins, and extremities are white to gray, and the lips are stained in orange near the aperture. Unlike in most cowries, the shell lacks a dorsal line, and instead has a solid, rich orange dorsum.

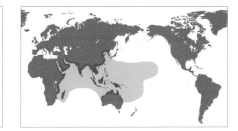

FAMILY	Cypraeidae
SHELL SIZE RANGE	1⅝ to 6 in (42 to 152 mm)
DISTRIBUTION	Indo-Pacific, including Hawaii
ABUNDANCE	Abundant
DEPTH	Shallow subtidal to 60 ft (18 m)
HABITAT	Tidepools and rocks near coral reefs
FEEDING HABIT	Grazer, feeds on algae
OPERCULUM	Absent

SHELL SIZE RANGE
1⅝ to 6 in
(42 to 152 mm)

PHOTOGRAPHED SHELL
4¾ in
(125 mm)

CYPRAEA TIGRIS

TIGER COWRIE

LINNAEUS, 1758

337

Cypraea tigris is one of the best known and arguably one of the most beautiful cowries. It is found in gift shops worldwide and often sold as a "local" shell, although most specimens probably come from the Philippines. *Cypraea tigris* is the type species of the genus *Cypraea* and the family Cypraeidae. It is one of the most variable cowries; no two shells are identical. It also varies widely in size with giant specimens from Hawaii being more than three times larger than the smallest, although both extremes are rare. Unlike most cowries, *C. tigris* usually does not hide in crevices but is commonly found in the open near reefs, and is active during the day.

RELATED SPECIES

Cypraea pantherina Lightfoot, 1786, from the Red Sea to Gulf of Aden, is the sister species, and has a similar but more elongate shell; other, not so closely related species from the Indo-West Pacific include *C. mauritiana* Linnaeus, 1758, with a humped dorsum; and *C. mappa* Linnaeus, 1758, which has a distinct meandering and thick dorsal line that resembles a map.

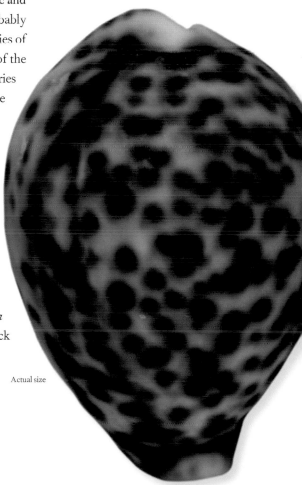

Actual size

The shell of the Tiger Cowrie is large, inflated, heavy, smooth, and very glossy. The aperture is narrow, long, and slightly curved, surrounded by thick lips with many strong teeth; columellar teeth extend onto the columella. Shell color and pattern are quite variable, usually with a white or bluish background and many large, dark irregular spots or blotches often edged with orange-yellow tints. The background color varies from almost white ("albino") to black (melanistic). A yellow or orange dorsal line, usually curved, crosses the shell near mid dorsum. The base, aperture, and teeth are white.

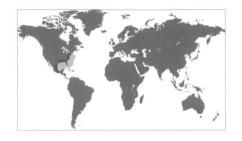

FAMILY	Cypraeidae
SHELL SIZE RANGE	1⅝ to 7½ in (42 to 190 mm)
DISTRIBUTION	North Carolina to the Gulf of Mexico
ABUNDANCE	Common
DEPTH	Shallow subtidal to 115 ft (35 m)
HABITAT	Under coral or rock slabs
FEEDING HABIT	Grazer, feeds on algae
OPERCULUM	Absent

SHELL SIZE RANGE
1⅝ to 7½ in
(42 to 190 mm)

PHOTOGRAPHED SHELL
7¼ in
(185 mm)

338

CYPRAEA CERVUS
DEER COWRIE
(LINNAEUS, 1771)

Cypraea cervus holds the distinction of being the largest of the living cowries (some extinct species grew to more than 12 in/300 mm long). There is an almost 5-fold difference in length between the smallest and largest known adult specimens of *C. cervus*. This species was once common in the shallow waters of the southeastern U.S.A., but is increasingly less common, and specimens collected are becoming smaller. Giant shells are usually from old collections or from deeper waters. In Texas, it occurs in offshore coral reefs and banks in deeper waters.

The shell of the Deer Cowrie is elongate and inflated, and very large for the family. It has an elongate-oval outline, with a smooth dorsum and slightly convex base. The aperture is as long as the shell, and widest near the anterior extremity. The extremities are elongated, and the siphonal canal is wide. Juvenile *C. cervus* and related species have the dorsum with 4 thick brown stripes over a fawn background. As the animal matures, new shell layers cover the stripes with a brown background and with hundreds of irregular fawn spots that span the dorsum. The apertural teeth are dark brown.

RELATED SPECIES
The closest species are *Cypraea cervinetta* Kiener, 1843, which ranges from west Mexico to Peru, and *C. zebra* Linnaeus, 1758, from North Carolina to southern Brazil. They both have similar shells to *C. cervus*, which can be distinguished by its larger size and more inflated shells, and by slightly coarser teeth along a wider aperture. *Cypraea cervinetta* has a more cylindrical outline, while *C. zebra* has ocellated marginal spots.

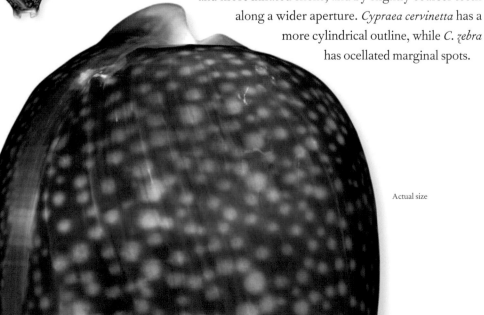

Actual size

FAMILY	Ovulidae
SHELL SIZE RANGE	½ to 1¼ in (11 to 33 mm)
DISTRIBUTION	Southern California to Ecuador; Galápagos Islands
ABUNDANCE	Common
DEPTH	Intertidal to 50 ft (15 m)
HABITAT	Under rocks, and on sponges and corals
FEEDING HABIT	Parasite on corals
OPERCULUM	Absent

SHELL SIZE RANGE
½ to 1¼ in
(11 to 33 mm)

PHOTOGRAPHED SHELL
⅝ in
(17 mm)

JENNERIA PUSTULATA

JENNER'S FALSE COWRIE

(LIGHTFOOT, 1786)

339

Jenneria pustulata has one of the most distinctive of all shells. It is the only known living ovulid to have tubercles on its shell. *Jenneria pustulata* parasitizes hard corals. The animal's mantle lobes bear long branched projections, and as in cowries, cover its shell. There are about 250 living species in the family Ovulidae worldwide, mostly in tropical and subtropical waters. The earliest known ovulid fossil dates from the Middle Cretaceous Period.

RELATED SPECIES

Pseudocypraea adamsonii (Sowerby II, 1832), from the Indo-Pacific, has a small cowrie-like shell, but unlike true cowries (family Cypraeidae), its shell has a reticulate dorsal sculpture. *Calpurnus verrucosus* (Linnaeus, 1758), from the Red Sea to the Indo-Pacific, is a common species that occurs on leather corals. Its shell also resembles a cowrie; it is white, with a humped dorsum and a rounded knob at both extremities.

The shell of the Jenner's False Cowrie is small, strong, glossy, and closely resembles a true cowrie in shape. Its spire and early whorls are internalized, and only the body whorl is seen. The aperture is narrow and surrounded by long, white, toothlike ridges on its base. The dorsum has many rounded tubercles, and a dorsal groove dividing the dorsum in two halves. The shell color is a gray or brown background, with orange-red tubercles, often outlined in black; the columella and aperture margins are white. The sculpture and color are very variable.

Actual size

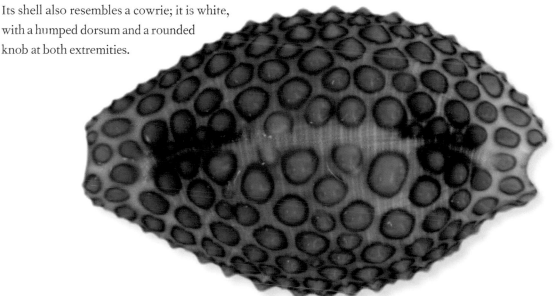

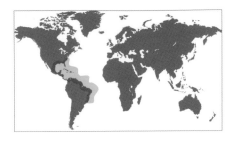

FAMILY	Ovulidae
SHELL SIZE RANGE	¾ to 1½ in (20 to 39 mm)
DISTRIBUTION	North Carolina to Brazil
ABUNDANCE	Abundant
DEPTH	Intertidal to 100 ft (30 m)
HABITAT	On gorgonians
FEEDING HABIT	Parasitic on gorgonians
OPERCULUM	Absent

SHELL SIZE RANGE
¾ to 1½ in
(20 to 39 mm)

PHOTOGRAPHED SHELL
⅞ in
(24 mm)

340

CYPHOMA GIBBOSUM
FLAMINGO TONGUE
(LINNAEUS, 1758)

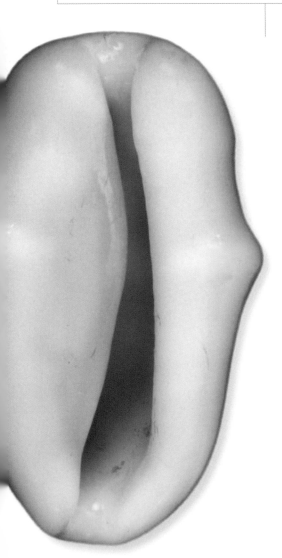

Cyphoma gibbosum is an abundant ovulid in shallow waters of the Caribbean. *Cyphoma gibbosum* lives on several host gorgonians, such as *Muricea* or *Plexaurella*. The shell is smooth and glossy, and varies in shape and color, ranging from white to orange. The animal has a striking and very colorful mantle and foot, with a white background and deep yellow blotches outlined in black. It is a favorite subject of underwater photographers, who can easily spot it against its purple hosts. However, it is becoming rare due to overcollecting.

RELATED SPECIES

The shells of species in the genus *Cyphoma* are similar, but the mantle coloration can easily distinguish species. *Cyphoma signata* Pilsbry and McGinty, 1939, from Florida to Brazil, has vertical yellow and red-violet bands, while *C. mcgintyi* Pilsbry, 1939, from the Gulf of Mexico and Puerto Rico, has brown blotches.

Actual size

The shell of the Flamingo Tongue is medium in size, smooth, thick, heavy, glossy, and ovate to rhomboidal in outline. Its spire and spire whorls are internalized and only the body whorl is seen in the adult shell. The surface is smooth, with an angular and broad ridge that transverses the shell. The extremities are broad and rounded, and the aperture is widest anteriorly. The shell color ranges from white to orange, sometimes with orange stains; the aperture is white.

FAMILY	Ovulidae
SHELL SIZE RANGE	⅜ to 1½ in (10 to 40 mm)
DISTRIBUTION	Red Sea to Indo-Pacific
ABUNDANCE	Common
DEPTH	Intertidal to 65 ft (20 m)
HABITAT	On soft corals
FEEDING HABIT	Parasitic on soft corals
OPERCULUM	Absent

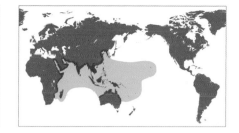

SHELL SIZE RANGE
⅜ to 1½ in
(10 to 40 mm)

PHOTOGRAPHED SHELL
1¼ in
(31 mm)

CALPURNUS VERRUCOSUS

UMBILICAL OVULA

LINNAEUS, 1758

341

Calpurnus verrucosus is one of the better known ovulids. It has a very distinctive shell, with a round knob near each of the two extremities. The shell is white, with pink stains at both extremities. There is very little variation in shell shape and color. It lives and feeds on soft corals of the genus *Sarcophyton* and *Lobophytum* in shallow tropical waters. The animal has two flaps of tissue, the mantle, that enclose the shell when fully extended. The animal is white, and peppered with brown or black spots, and has a short siphon and lacks cephalic tentacles.

RELATED SPECIES

Rotaovula hirohitoi Cate and Azuma, 1973, from off Japan to the Philippines, has a small but unmistakable shell, which is yellow and purple. *Volva volva* (Linnaeus, 1758), from the Indo-West Pacific, also has a distinctive shell. It is large, spindle-shaped, with very long extremities.

Actual size

The shell of the Umbilical Ovula is medium in size, thick, inflated, glossy, and oval, resembling a cowrie. Like other ovulids, the spire and spire whorls are internalized and not seen in the adult. Its distinguishing feature is the presence of rounded knobs near the anterior and posterior extremities. The dorsum is humped, with an angled ridge near the middle of the shell. The aperture is narrow and the outer lip thick, with many teeth, while the columella is smooth, with a notch anteriorly. The shell color is white, stained in pink near extremities.

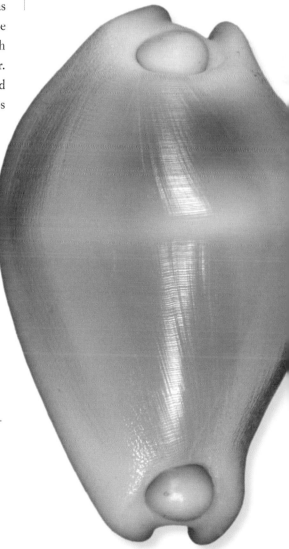

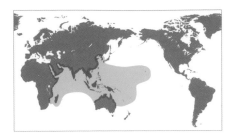

FAMILY	Ovulidae
SHELL SIZE RANGE	1¼ to 4½ in (32 to 120 mm)
DISTRIBUTION	Red Sea to Indo-Pacific
ABUNDANCE	Common
DEPTH	Intertidal to 65 ft (20 m)
HABITAT	On soft corals (several species)
FEEDING HABIT	Parasitic on soft corals
OPERCULUM	Absent

SHELL SIZE RANGE
1¼ to 4½ in
(32 to 120 mm)

PHOTOGRAPHED SHELL
3 in
(78 mm)

OVULA OVUM
COMMON EGG COWRIE
(LINNAEUS, 1758)

Actual size

Ovula ovum is the largest ovulid in terms of animal size, although other species, such as *Volva volva* (Linnaeus, 1758) can have longer shells. Both words in the scientific name refer to "egg," which is a good description of its shiny, white egg-shaped shell. The shell is used for decoration as well as for tribal symbols in Melanesia and Polynesia. The animal is jet black, and the velvety mantle has small, white raised spots. It feeds on several species of soft corals. Like many ovulids, *Ovula ovum* resembles true cowries (family Cypraeidae); many species were originally classified as cypraeids.

RELATED SPECIES

Sphaerocypraea incomparabilis (Briano, 1993), from Somalia to Mozambique, is currently the rarest of the ovulids, and an important recent discovery; only a few specimens are known. Its shell resembles *Ovula ovum* in shape, although it has a dark red-brown outer color, a wide aperture, and the outer lip has strong, equally spaced, white teeth. *Jenneria pustulata* (Lightfoot, 1786), from the eastern Pacific, has a shell with rounded tubercles.

The shell of the Common Egg Cowrie is large (for the family), thick, heavy, glossy, inflated, and egg-shaped. Its extremities are elongated, with the anterior one broader than the posterior. The surface is smooth and glossy. The aperture is narrow and long, and widest anteriorly; the outer lip is folded toward the aperture, and irregularly crenulated. The columella is smooth and curved. The shell color is porcelain white, and the interior is red-brown.

FAMILY	Ovulidae
SHELL SIZE RANGE	1¾ to 7⅜ in (45 to 186 mm)
DISTRIBUTION	Indo-Pacific
ABUNDANCE	Common
DEPTH	33 to 330 ft (10 to 100 m)
HABITAT	On sandy muddy bottoms
FEEDING HABIT	Carnivore, feeds on sea pens
OPERCULUM	Absent

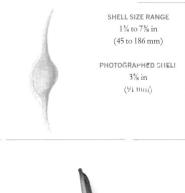

SHELL SIZE RANGE
1¾ to 7⅜ in
(45 to 186 mm)

PHOTOGRAPHED SHELL
3⅝ in
(91 mm)

VOLVA VOLVA

SHUTTLECOCK VOLVA

(LINNAEUS, 17580)

343

Volva volva is one of the most distinctive ovulids, with a globose-ovate body whorl, and very long, narrow canals. The body whorl is only about one-third or less of the shell length. *Volva volva* differs from most ovulids by the fact that it crawls across sandy mud bottoms, instead of living on its prey. It feeds on sea pens (pennatulaceans) of the genus *Actinoptilum*, as well as other sand-dwelling cnidarians. Like other ovulids, it accumulates some of the host's toxic chemicals, and uses them for its defense. The mantle has dark brown blotches and short projections.

RELATED SPECIES

Contrasimnia xanthochila (Kuroda, 1928), which ranges from Japan to New Caledonia, has a thin, oval-elongate, and translucent shell; the outer lip, extremities, and columella are tinted with yellow. The animal is gray-white with white spots and black bars and spots. *Ovula ovum* (Linnaeus, 1758), from the Red Sea and Indo-Pacific, has a large, glossy, white egg-shaped shell with a red-brown interior.

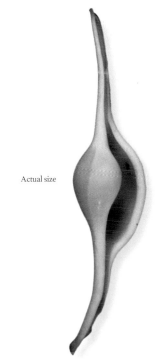

Actual size

The shell of the Shuttlecock Volva is medium large, glossy, inflated, and oval-elongate, with extremely long and narrow anterior and posterior canals that can be either straight or curved. Its body whorl is oval and inflated, and accounts for about one-third of the shell length. The surface is smooth, with incised spiral lines. The aperture is narrow and long, the outer lip thickened, and the columella smooth. The shell color ranges from pure white to beige or pink, with terminals stained in orange. The interior is white.

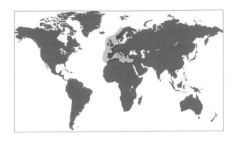

FAMILY	Triviidae
SHELL SIZE RANGE	⅛ to ½ in (3 to 12 mm)
DISTRIBUTION	Norway to the Canary Islands; Mediterranean
ABUNDANCE	Common
DEPTH	50 to 500 ft (15 to 150 m)
HABITAT	Hard bottoms, rocks and rubble
FEEDING HABIT	Carnivore, feeds on ascidians
OPERCULUM	Absent

SHELL SIZE RANGE
⅛ to ½ in
(3 to 12 mm)

PHOTOGRAPHED SHELL
½ in
(12 mm)

344

ERATO VOLUTA
VOLUTE ERATO
(MONTAGU, 1803)

Erato voluta is a small, common triviid that lives offshore. It is a carnivore and preys on ascidians (sea squirts). Species in the genus of *Erato* and related genera had been classified as a separate family, but are now considered a subfamily of the Triviidae. Eratos usually have smooth shells, while most triviines have ribbed shells. Eratos are often confused with marginellids, but the dentition of the inner lip distinguishes them. There are about 180 living species in the family Triviidae, including 30 species of eratos. The oldest fossils date from the Eocene Epoch.

RELATED SPECIES
Erato grata Cossignani and Cossignani, 1997, from the Philippines, has a small shell that resembles *E. voluta* in shape, but has a granulose outer surface. *Trivia pediculus* (Linnaeus, 1758), from North Carolina to southeastern Brazil, has an oval shell with a hidden spire, strong ribs, and a dorsal groove.

Actual size

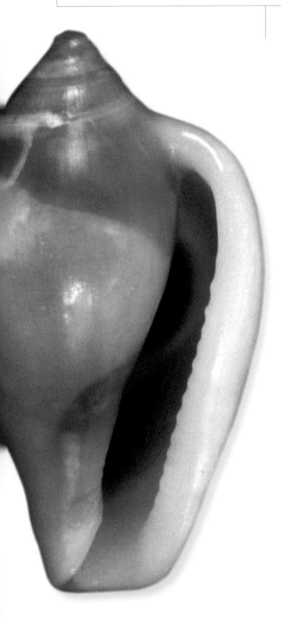

The shell of the Volute Erato is small, solid, globose, and pear-shaped. Its spire is short, the apex pointed, and the suture poorly marked. The surface is smooth and glossy, with fine growth lines. The shell is broader posteriorly, and tapers anteriorly. Its aperture is narrow and long, the outer lip thickened and bearing 12 to 18 denticles, and the columella has small folds. The shell color is grayish to cream, the outer lip cream, and the interior grayish.

FAMILY	Triviidae
SHELL SIZE RANGE	¼ to ⅞ in (7 to 22 mm)
DISTRIBUTION	North Carolina to southeastern Brazil
ABUNDANCE	Common
DEPTH	Intertidal to 420 ft (130 m)
HABITAT	Under rocks
FEEDING HABIT	Carnivore, feeds on ascidians
OPERCULUM	Absent

SHELL SIZE RANGE
¼ to ⅞ in
(7 to 22 mm)

PHOTOGRAPHED SHELL
½ in
(13 mm)

TRIVIA PEDICULUS

COFFEE BEAN TRIVIA

(LINNAEUS, 1758)

345

Trivia pediculus has a small, oval shell that resembles a coffee bean in shape and color, hence the popular name. It is a common species, and is often found under rocks or near coral reefs in shallow subtidal waters, but can also occur offshore. Although it can reach about ⅞ in (22 mm) in length, most shells are about half of the length. Triviids cover their shells with the two lobes of the mantle. The mantle of *T. pediculus* can be clear or opaque, with varying colors, bluish or greenish gray, and has fingerlike papillae.

RELATED SPECIES

Trivia monacha (Costa, 1778), from western Europe and the Mediterranean, has a small shell that resembles *T. pediculus*, but lacks a dorsal groove. *Trivia solandri* (Sowerby II, 1832), from southern California to Peru and the Galápagos Islands, has fewer, stronger, and unbeaded ribs, and larger, rounded nodules along the dorsal groove.

Actual size

The shell of the Coffee Bean Trivia is small, ribbed, and cowrie-shaped. Its spire is hidden in adult shells by the large body whorl. The sculpture consists of 15 to 18 ribs that extend from the dorsum to the aperture; some ribs do not reach the aperture. The ribs are continuous on the base, but beaded on the dorsum. There is a groove that divides the dorsum in two. The aperture is narrow and long, and the outer and inner lip denticulate. The shell color is usually brown, but some can be tan or pinkish, with 6 brown dorsal spots, and a whitish aperture.

FAMILY	Triviidae
SHELL SIZE RANGE	⅜ to ⅞ in (10 to 21 mm)
DISTRIBUTION	California to Peru, and Galápagos Islands
ABUNDANCE	Common
DEPTH	Intertidal to 115 ft (35 m)
HABITAT	Rocky bottoms
FEEDING HABIT	Carnivore, feeds on ascidians
OPERCULUM	Absent

SHELL SIZE RANGE
⅜ to ⅞ in
(10 to 21 mm)

PHOTOGRAPHED SHELL
¾ in
(18 mm)

346

TRIVIA SOLANDRI
SOLANDER'S TRIVIA
(SOWERBY II, 1832)

The shell of the Solander's Trivia is large for the family, solid, ribbed, and oval in outline. Its spire is hidden in adult shells by the large body whorl. The sculpture consists of about 11 to 14 strong, continuous ribs. The ribs have rounded nodules along the dorsal groove. The aperture is narrow, widest anteriorly, and both outer and inner lips are denticulate. The shell color is brown to reddish-brown, with 2 dark bands along the dorsum; the ribs and aperture are whitish.

Actual size

Trivia solandri is one of the largest triviids from the eastern Pacific. It is common under rocks in shallow waters in the southern part of its distribution, and less common in the north. This species has a brownish mantle, peppered with small black and white spots, and short fingerlike, orange-brown papillae. The mantle coloration and texture may provide camouflage when the gastropod is on its colonial ascidian prey. The animal also has a large siphon and short cephalic tentacles, with an eye on a swollen tubercle at the base of each tentacle.

RELATED SPECIES

Triviella calveriola (Kilburn, 1980), is one of many triviids endemic to South Africa, and one of the largest species in the family. Like several species from the region, it has a globose and smooth shell, with a denticulate aperture. *Trivia pediculus* (Linnaeus, 1758), from North Carolina to southeastern Brazil, has a similar shell, but with smaller nodules along the dorsal groove, and beaded ribs on the dorsum.

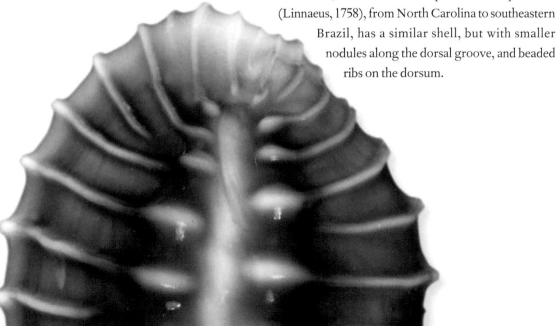

FAMILY	Velutinidae
SHELL SIZE RANGE	1 to 1½ in (25 to 37 mm)
DISTRIBUTION	Tropical Indo-West Pacific
ABUNDANCE	Uncommon
DEPTH	Intertidal
HABITAT	Under rocks and coral slabs
FEEDING HABIT	Carnivore, feeds on ascidians
OPERCULUM	Absent

347

SHELL SIZE RANGE
1 to 1½ in
(25 to 37 mm)

PHOTOGRAPHED SHELL
1⅛ in
(30 mm)

CORIOCELLA NIGRA

BLACK CORIOCELLA

(BLAINVILLE, 1824)

The shell of *Coriocella nigra* is entirely enveloped in the fleshy mantle of the large (to 4 in / 100 mm), sluglike animal that may have multiple lobes or folds on its dorsal surface. Its tissue color is usually black, but may be brown, reddish, yellowish, or even blue, sometimes with a spotted or netlike pattern. It feeds on colonial ascidians. While some velutinids mimic the color or their prey, *Coriocella* species may be brightly colored, signaling to predators that they contain distasteful chemical defenses.

The shell of the Black Coriocella is small compared to the size of the animal, very thin, low, and ear-shaped. It consists of a few, rapidly expanding whorls, and has a small, low spire and a large, elliptical aperture. The surface is glossy, but not smooth; its sculpture limited to uneven growth striations. The shell is uniformly white, with a thin, orange-brown periostracum that produces an orange band at the suture.

RELATED SPECIES

Lamellaria perspicua (Linnaeus, 1758), from the Mediterranean and northeastern Atlantic, has a similar shell that is much smaller (less than ½ in / 12 mm) and proportionally taller and narrower. *Marseniopsis mollis* (Smith, 1902), a common Antarctic species, has a small, vestigial, thin, transparent shell. The animal has a nearly spherical body.

Actual size

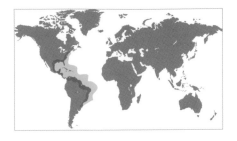

FAMILY	Naticidae
SHELL SIZE RANGE	⅛ to ⅜ in (2 to 8 mm)
DISTRIBUTION	Maine to Brazil, Caribbean
ABUNDANCE	Common
DEPTH	Intertidal to 150 ft (50 m)
HABITAT	Sand
FEEDING HABIT	Carnivore
OPERCULUM	Calcareous

SHELL SIZE RANGE
⅛ to ⅜ in
(2 to 8 mm)

PHOTOGRAPHED SHELL
⅛ in
(2 mm)

348

TECTONATICA PUSILLA

MINIATURE MOON SNAIL

(SAY, 1822)

This tiny member of the Naticidae family is hard to see with the naked eye, but nevertheless exhibits all the family's characteristics, both in form and in habit. All species live on sandy substrates, and search just below the surface for other mollusks, which they envelope in a disproportionately large foot. Naticids have an accessory organ from which they secrete a softening chemical onto the shell of their prey before drilling a very neat, beveled hole through it with the radula. They then extract the contents with their proboscis and digest it.

RELATED SPECIES

Tectonatica micra (Hass, 1953) is endemic to the waters of central Brazil. The largest specimens found to date have not exceeded ¼ in (5 mm). It may have a slightly less flattened spire, and fewer patches of brown on the white surface of the shell.

● Actual size

The shell of the Miniature Moon Snail is globular with a short spire of rounded whorls and a blunt apex. The suture is moderately impressed, but otherwise there is no sculpture except for growth lines. The umbilicus is more or less closed by the columellar callus, and there is a thin, sharp, outer lip. The shell is white overall, with broad and much broken spiral bands of mid brown.

FAMILY	Naticidae
SHELL SIZE RANGE	¾ to 1⅜ in (20 to 29 mm)
DISTRIBUTION	South Georgia, South Sandwich, South Orkney and South Shetland Islands, northern Antarctic peninsula
ABUNDANCE	Uncommon
DEPTH	75 to 1,500 ft (25 to 400 m)
HABITAT	Sand
FEEDING HABIT	Carnivore
OPERCULUM	Corneous

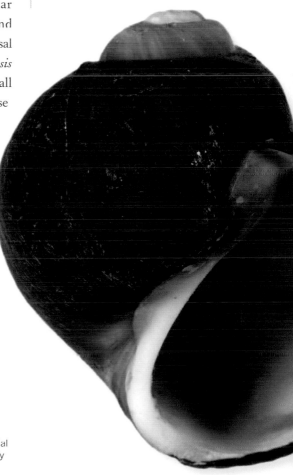

SHELL SIZE RANGE
¾ to 1⅜ in
(20 to 29 mm)

PHOTOGRAPHED SHELL
¾ in
(20 mm)

AMAUROPSIS AUREOLUTEA

GOLDEN AMAUROPSIS

(STREBEL, 1908)

349

Amauropsis is a cold-water naticid genus with a bipolar distribution. Greatest diversity occurs in Antarctic and subantarctic regions, but the genus is represented in the abyssal fauna of the North Atlantic, as well as in the Arctic. *Amauropsis aureolutea* lives on sandy and muddy bottoms where, like all Naticidae, it lays eggs in broad, collar-shaped capsules. These capsules are made of sand grains cemented with mucus, with the inner edge conforming to the curve of the aperture.

RELATED SPECIES

Amauropsis islandica (Gmelin, 1791) is the lone Arctic *Amauropsis*, occurring commonly around the Arctic Circle from Alaska to northern Europe and as far south as Virginia. It is hazelnut brown, with a white aperture and pale tan to lilac interior.

Actual size

The shell of Golden Amauropsis is globular. Its very short spire has hemispherical whorls and a rounded white apex. It is mid tan to chestnut brown in color, paler on the fasciole and spiral whorls, with faint darker axial and spiral striations. The interior is white. The aperture is ovate with a fairly straight columella extending as a callus to close the umbilicus.

FAMILY	Naticidae
SHELL SIZE RANGE	½ to 1⅜ in (13 to 35 mm)
DISTRIBUTION	Southeastern Africa to central Pacific
ABUNDANCE	Common
DEPTH	Shallow subtidal
HABITAT	Sand
FEEDING HABIT	Carnivore
OPERCULUM	Calcareous

SHELL SIZE RANGE
½ to 1⅜ in
(13 to 35 mm)

PHOTOGRAPHED SHELL
¾ in
(20 mm)

350

GLYPHEPITHEMA ALAPAPILIONIS

BUTTERFLY MOON

(RÖDING, 1798)

Glyphepithema alapapilionis is an uncommon species that has a very broad range throughout the tropical Indo-Pacific, ranging from Natal, South Africa to Fiji. Like other naticids, it lives in sandy bottoms, where it hunts for bivalves and even other naticids by plowing below the surface using its extremely large foot. Despite appearances, the large foot can be completely retracted into the shell, where it is protected by a very close-fitting operculum.

RELATED SPECIES

The distinctive spiral bands of *Glyphepithema alapapilionis* also appear, with variations, on several other related species from western Africa to western America. *Natica turtoni* (Smith, 1890) and *N. caneloensis* (Herlein & Strong, 1955), from those respective extremes, are remarkably similar and even have the same opercular spiral grooves. The latter has fewer bands on the body, and a smaller umbilicus.

The shell of the Butterfly Moon is globular; the body is very swollen, and the spire small. There are faint radial grooves below the suture, and fine axial striations of white to terra-cotta are overlaid with narrow spiral bands of white and chestnut brown dashes, 4 on the body and 1 on the spiral whorl. There is a modest umbilicus. The aperture is white and the interior is dusky pink.

Actual size

FAMILY	Naticidae
SHELL SIZE RANGE	½ to 1 in (12 to 25 mm)
DISTRIBUTION	Philippines to Queensland
ABUNDANCE	Uncommon
DEPTH	Subtidal to 65 ft (20 m)
HABITAT	Sand
FEEDING HABIT	Carnivore
OPERCULUM	Calcareous

SHELL SIZE RANGE
½ to 1 in
(12 to 25 mm)

PHOTOGRAPHED SHELL
⅞ in
(21 mm)

TECTONATICA VIOLACEA

VIOLET MOON

(SOWERBY I, 1825)

351

The family Naticidae is an extremely diverse group that spans all marine habitats, from pole to pole, and from the intertidal to abyssal depths. Nearly 300 species are living today, although the group traces its origins to the Triassic Period. Species of the genus *Tectonatica* tend to be small and inhabit fairly shallow subtidal depths. They do not burrow deeply into the sand, and often leave trails that can be followed to reveal the specimen buried at one end. *Tectonatica violacea* has a distinctive violet columella and parietal shield.

Actual size

RELATED SPECIES

Natica arachnoidea (Gmelin, 1791) is found in shallower water throughout the Indo-Pacific. Its aperture is not as downward-sloping as that of *T. violacea*. The coloration of *N. arachnoidea* is extremely variable, usually with a white to tan base color, and dark brown, tentlike markings that may be arranged in spiral bands of varying width and density.

The shell of the Violet Moon is shiny and globular with a short spire and a fine impressed suture. It is off-white, with an uneven pattern of light chestnut, spiral blotches that are often narrower and more regular at the keel. The interior is white, with a fine rounded outer lip defining a D-shaped, downward-sloping aperture. The columella is stained violet, and its callus may completely cover a very narrow umbilicus.

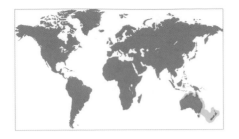

FAMILY	Naticidae
SHELL SIZE RANGE	⅝ to 2⅛ in (15 to 53 mm)
DISTRIBUTION	Australia, Tasmania, New Zealand
ABUNDANCE	Moderately common
DEPTH	Intertidal to subtidal
HABITAT	Sand
FEEDING HABIT	Carnivore
OPERCULUM	Corneous

SHELL SIZE RANGE
⅝ to 2⅛ in
(15 to 53 mm)

PHOTOGRAPHED SHELL
⅞ in
(21 mm)

352

CONUBER CONICUS
CONICAL MOON
(LAMARCK, 1822)

Conuber conicus is restricted to the waters off Australia, including Tasmania, and New Zealand, where it is less common. This species inhabits sandy bottoms within estuaries, where its broad trails can be seen in the sand and mud at low tide.

RELATED SPECIES

The other three eastern Australian members of the *Conuber* genus are (largest first) *C. sordidus* (Swainson, 1821), *C. melastomus* (Swainson, 1821), and *C. putealis* (Garrard, 1961). The latter, from deep water, is rare and all off-white; the other two are estuarine and pale gray-brown to tan, with interiors of red-brown and orange-brown respectively.

Actual size

The shell of the Conical Moon is ovate with a small spire of rounded whorls and a pointed apex. It is slightly concave immediately below the suture and of variable color, off-white to tan with axial striations of chestnut to purple-brown. The aperture is deep and there is orange staining around the posterior notch and the umbilicus.

FAMILY	Naticidae
SHELL SIZE RANGE	¾ to 1⅝ in (20 to 40 mm)
DISTRIBUTION	East Africa to southeastern Australia
ABUNDANCE	Common
DEPTH	Offshore, 33 to 100 ft (10 to 30 m)
HABITAT	Sand
FEEDING HABIT	Carnivore
OPERCULUM	Calcareous

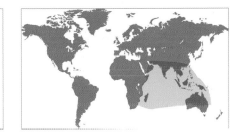

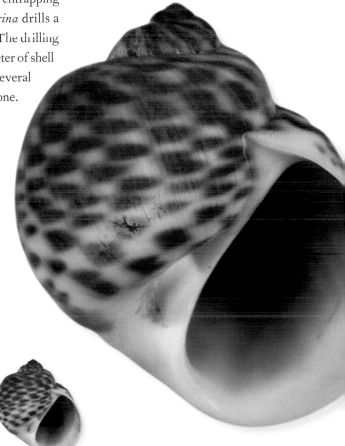

SHELL SIZE RANGE
¾ to 1⅝ in
(20 to 40 mm)

PHOTOGRAPHED SHELL
1 in
(26 mm)

353

NOTOCOCHLIS TIGRINA

TIGER MOON

(RÖDING, 1798)

Named because of its surface pattern, *Notocochlis tigrina* has a shell that is more reminiscent of a cheetah's coat than a tiger's. Like the big cats, it is a predatory carnivore. While entrapping its prey with its broad, enveloping foot, *N. trigrina* drills a hole in the shell of its victim to feed on its tissues. The drilling process is slow, penetrating as little as half a millimeter of shell a day. Often shells of prey species are found with several incomplete drill holes in addition to the successful one.

RELATED SPECIES

Natica variolaria (Récluz, 1844), from the western coast of Africa, is similar in color pattern although generally a little smaller. Its spots are smaller than those of *Notocochlis tigrina*, more dots than dashes; and it has a much larger umbilical opening.

The shell of the Tiger Moon is globular. It has a low narrow spire of rounded whorls and an angular suture. It is white to cream with a blurred spiral pattern of mid to dark brown, triangular dashes. There is a narrow umbilicus. The outer lip is thin and the off-white operculum has 3 ridges at its outer edge. The interior is white.

Actual size

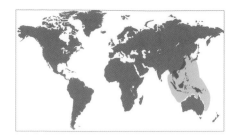

FAMILY	Naticidae
SHELL SIZE RANGE	¾ to 2¼ in (20 to 58 mm)
DISTRIBUTION	Philippines to Australia
ABUNDANCE	Common
DEPTH	Intertidal to 60 ft (20 m)
HABITAT	Sand
FEEDING HABIT	Carnivore
OPERCULUM	Corneous

SHELL SIZE RANGE
¾ to 2¼ in
(20 to 58 mm)

PHOTOGRAPHED SHELL
1⅛ in
(29 mm)

354

NATICA AURANTIA
GOLDEN MOON
(RÖDING, 1798)

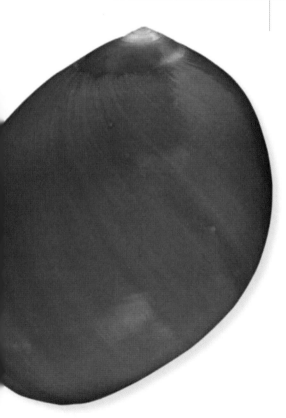

When spawning, female naticids deposit their eggs in a sand collar—a distinctive, horseshoe-shaped structure with a raised internal edge and a wavy external margin. Sand collars are composed of gelatinous matrix and sand, with individual eggs embedded between the sand grains. Other snails sometimes attach their egg cases to sand collars. *Natica aurantia* has a distinctive orange shell, with a white protoconch and early whorls.

RELATED SPECIES
Natica stellata Chenu, 1845, lives in the same depths and distribution as *N. aurantia*, but also extends into the Indian Ocean. It, too, is shiny and pale orange, but it is smaller with slightly squarer shoulders. The aperture is more open and the umbilicus never covered by the columella. The shell has spiral bands and axial flames of deeper orange.

The shell of the Golden Moon is brightly colored, pale to deep orange with a pure white interior. It is shiny and globular, with a pointed spire that is white toward the apex. The outer lip is thin and sharp, and the white columellar callus often completely covers the umbilicus.

Actual size

FAMILY	Naticidae
SHELL SIZE RANGE	¾ to 2 in (18 to 51 mm)
DISTRIBUTION	Eastern Africa to western Pacific
ABUNDANCE	Common
DEPTH	Intertidal to shallow subtidal
HABITAT	Sand
FEEDING HABIT	Carnivore
OPERCULUM	Corneous

SHELL SIZE RANGE
¼ to 2 in
(18 to 51 mm)

PHOTOGRAPHED SHELL
1⅛ in
(29 mm)

MAMILLA MELANOSTOMA
BLACK MOUTH MOON
(GMELIN, 1791)

355

Mamilla melanostoma is a common predator in shallow, sandy bottoms throughout the Indian and western Pacific Oceans. Like all naticids, it envelops its prey with its foot, and drills a broad rimmed hole through its shell in order to feed. It is easily recognized by its broad, D-shaped aperture and dark, chocolate brown columella and umbilicus.

RELATED SPECIES

Globularia fluctuata (Sowerby I, 1825) is an uncommon naticid in deeper waters of the western Pacific. It is superficially similar, but the spiral bands are broken by axial zigzags of the shell color. The aperture is much more open, and the dark brown stripe of the columellar callus across the umbilicus is preceded by an equal stripe of white columella at the inner lip. The interior is a darker gray-brown.

Actual size

The shell of the Black Mouth Moon is globular with an aperture that extends downward and a very small spire. It is pearl white to pale gray-brown, and has spiral bands of varying widths in pale to mid chestnut-brown, which are crossed by axial striations. The pattern may be seen internally. The operculum is dark red-brown, and there is a very dark brown columellar callus that tends to cover the umbilicus.

FAMILY	Naticidae
SHELL SIZE RANGE	1 to 2¾ in (25 to 70 mm)
DISTRIBUTION	Panama to Chile, Galápagos Islands
ABUNDANCE	Moderately common
DEPTH	Shallow water, subtidal to 30 ft (10 m)
HABITAT	Sand
FEEDING HABIT	Carnivore
OPERCULUM	Absent

SHELL SIZE RANGE
1 to 2¾ in
(25 to 70 mm)

PHOTOGRAPHED SHELL
1⅛ in
(29 mm)

SINUM CYMBA
BOAT EAR MOON
(MENKE, 1828)

356

The shell of the Boat Ear Moon is large, fairly thick, and somewhat elongated. The aperture is large and tilted relative to the coiling axis of the shell. The columella is short, forming a narrow callus that does not cover the umbilical region. Sculpture is limited to coarse spiral cords. The color is chestnut brown nearer the apex, becoming whitish anteriorly and toward the outer lip. The aperture is brownish.

Sinum cymba has one of the most inflated shells in the genus *Sinum*. Most species of the genus *Sinum* have flattened shells with large, sharply angled apertures. The animals are too large to fit within the shell, and lack an operculum. The animal burrows through sand with its large, broad muscular foot, which also envelops the edges of the shell. *Sinum cymba* ranges from Panama to Chile, and the Galápagos Islands. It lives on sandy bottoms in shallow waters.

RELATED SPECIES

Sinum javanicum (Griffith and Pidgeon, 1834) lives in deep water on the other side of the Pacific, from Japan to Indonesia. It is slightly smaller than *S. cymba* and pale yellow, with a faint purple, spiral band that becomes stronger as it reaches and covers the protoconch. The interior is very glossy and white.

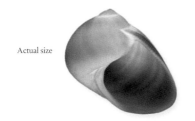

Actual size

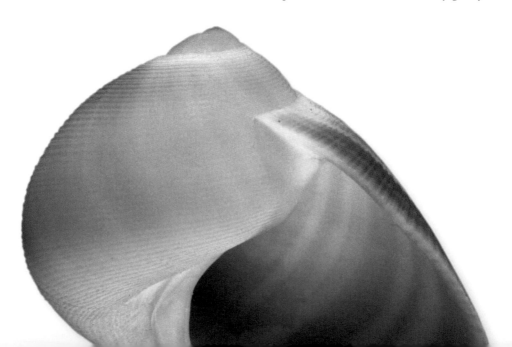

FAMILY	Naticidae
SHELL SIZE RANGE	¾ to 2 in (20 to 51 mm)
DISTRIBUTION	Maryland to Brazil, Caribbean
ABUNDANCE	Common
DEPTH	Intertidal to 33 ft (10 m)
HABITAT	Sand flats
FEEDING HABIT	Carnivore
OPERCULUM	Absent

SHELL SIZE RANGE
¾ to 2 in
(20 to 51 mm)

PHOTOGRAPHED SHELL
1⅜ in
(35 mm)

SINUM PERSPECTIVUM

BABY'S EAR MOON
(SAY, 1831)

357

Sinum perspectivum has a small, flat, greatly reduced shell that is completely enveloped by the cream-colored animal. The front of the foot has a broad, plowlike shield of tissue, called the propodium, which can be expanded by the uptake of seawater into a special system of sinuses within its tissues. The animal crawls along sandy bottoms, leaving a broad track that can be followed at low tide to reveal where the animal has burrowed.

The shell of the Baby's Ear Moon is small, thin, and greatly flattened, resembling an ear. The aperture is proportionally very large and nearly perpendicular to the coiling axis of the shell, which is visible from within the aperture. The columella is extremely short. The shell has spiral sculpture of numerous fine threads. The axial sculpture is limited to growth lines of varying intensity. The shell color is uniform white. The aperture is white and glazed within.

RELATED SPECIES

Sinum maculatum (Say, 1831) is less common and has a narrower range from North Carolina to the Caribbean. It is similar to *S. perspectivum*, but has a taller, more inflated shell with weaker spiral sculpture. The shell color is brown or yellowish brown, the animal is whitish, with purple spots.

Actual size

FAMILY	Naticidae
SHELL SIZE RANGE	1⅛ to 1⅜ in (28 to 35 mm)
DISTRIBUTION	Western Pacific, China to eastern Australia
ABUNDANCE	Uncommon
DEPTH	Subtidal to offshore
HABITAT	Sand
FEEDING HABIT	Carnivore
OPERCULUM	Absent

SHELL SIZE RANGE
1⅛ to 1⅜ in
(28 to 35 mm)

PHOTOGRAPHED SHELL
1¾ in
(44 mm)

358

SINUM INCISUM
INCISED MOON
(REEVE, 1864)

Sinum incisum, like *S. perspectivum*, has a greatly flattened shell, in which the coiling axis is within the aperture. The shell was substantially reduced in size in the course of evolution, and can no longer contain the entire animal. Rather, it encases and protects the delicate gonad and liver, which are situated at the tip of the coiled body.

RELATED SPECIES

Sinum concavum (Lamarck, 1822) is one of a few species of *Sinum* to be found off eastern Atlantic coasts, where it is confined to western Africa. It is similar in size and sculpture to *S. incisum*, but is cream to pale hazel in color, with a white bottom and apex.

Actual size

The shell of the Incised Moon is ovate and flattened, the whorls enlarging rapidly from the apex to produce a very wide aperture. It is polished white inside, matt white externally with a sculpture of many fine spiral grooves, which become wavy toward the sharp, thin outer lip. The apex is sometimes purple.

FAMILY	Naticidae
SHELL SIZE RANGE	⅞ to 2½ in (22 to 65 mm)
DISTRIBUTION	North Carolina to Brazil, Caribbean
ABUNDANCE	Common
DEPTH	Offshore to 200 ft (60 m)
HABITAT	Sand
FEEDING HABIT	Carnivore
OPERCULUM	Calcareous, paucispiral

SHELL SIZE RANGE
⅞ to 2½ in
(22 to 65 mm)

PHOTOGRAPHED SHELL
2 in
(51 mm)

NATICARIUS CANRENA
COLORFUL ATLANTIC MOON
(LINNAEUS, 1758)

359

Naticarius canrena is a large, colorful naticid that is popular with shell collectors. The animal is nearly four times the length of the shell, and has a conspicuous mottled pattern on the sides and rear of the foot as well as a propodium with multiple parallel lines. Its operculum is thick, white, and calcified, with a complex pattern of about ten parallel grooves.

RELATED SPECIES

Naticarius hebraeus (Martyn, 1786) is a common Mediterranean naticid that is unusual in having adapted to a coarser gravel habitat. It is white to cream in color, with alternating broad spiral bands of tiny chestnut spots and larger chestnut splotches, the larger ones running together near the lip. It has a fairly broad umbilicus, with a pronounced narrow callus.

Actual size

The shell of the Colorful Atlantic Moon is smooth, glossy, and roughly spherical, with a rounded spire and a large, D-shaped aperture. The umbilicus is broad, with a broad callus along the midline of the columella. The shell has a complex pattern of narrow white bands on a chestnut brown background, crossed by wavy, dark brown axial lines.

FAMILY	Naticidae
SHELL SIZE RANGE	2¼ to 6½ in (57 to 166 mm)
DISTRIBUTION	Vancouver Island to Mexico
ABUNDANCE	Common
DEPTH	Intertidal to 165 ft (50 m)
HABITAT	Sand
FEEDING HABIT	Carnivore
OPERCULUM	Corneous

SHELL SIZE RANGE
2¼ to 6½ in
(57 to 166 mm)

PHOTOGRAPHED SHELL
5 in
(128 mm)

360

EUSPIRA LEWISII
LEWIS' MOON
(GOULD, 1847)

Euspira lewisii is among the largest of the living Naticidae. The female is generally larger than the male and longer lived, with a lifespan of as much as 14 years. Lewis' Moon shell is a dominant predator on sand flats in its habitat, where it feeds on clams and oysters, as well as other gastropods, including other naticids. Shells of this species were common in middens left by Native Americans. The tissues can become toxic when the animals feed on bivalves affected by the red tide.

RELATED SPECIES

Euspira heros (Say, 1822) is almost as large and very similar in appearance but distributed on the Atlantic coast of North America. It is pale tan with the same frequent growth lines but its umbilical stains are paler not darker, and it lacks the concave channel around the body that is found in *E. lewisii*.

The shell of Lewis' Moon is thick and nearly spherical, and covered in frequent, rough, growth striations. It has a low spire of rounded whorls and a darkly stained umbilicus partly covered by a white columellar callus. There is a wide, shallow, spiral depression below the suture ending in the top half of the outer lip. The shell is pale beige to chestnut in color, paler still inside.

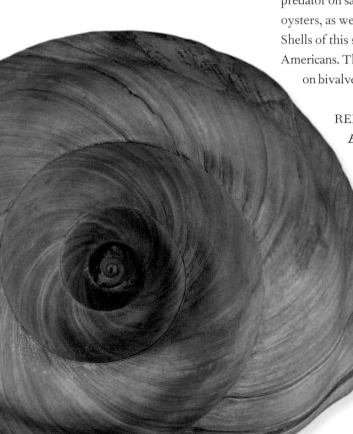

Actual size

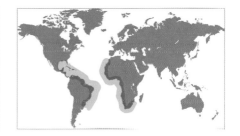

FAMILY	Bursidae
SHELL SIZE RANGE	1⅜ to 3 in (35 to 75 mm)
DISTRIBUTION	Florida to Brazil, Canary Islands to South Africa
ABUNDANCE	Uncommon
DEPTH	100 to 900 ft (30 to 275 m)
HABITAT	Rocky and gravel bottoms
FEEDING HABIT	Carnivore, feeds on polychaetes
OPERCULUM	Corneous, ovate, with central nucleus

SHELL SIZE RANGE
1⅜ to 3 in
(35 to 75 mm)

PHOTOGRAPHED SHELL
2⅜ in
(60 mm)

BURSA RANELLOIDES TENUISCULPTA

FINE-SCULPTURED FROG SHELL

(DAUTZENBERG AND FISCHER, 1906)

361

Bursa ranelloides tenuisculpta has a more elongate shell with lighter sculpture than *B. ranelloides ranelloides*, from the Indo-Pacific; the name *tenuisculpta* refers to its light sculpture. The animal is cream, mottled with orange and white, and the cephalic tentacles are yellow with black bands. There are about 60 living species in the family Bursidae worldwide, including species in the tropics and subtropics. The oldest bursid fossil dates from the mid-Cretaceous Period.

RELATED SPECIES

Tutufa bardeyi (Jousseaume, 1894), from the Gulf of Aden to Kenya, is the largest living bursid; its shell can reach more than 16 in (400 mm) in length. *Bufonaria bufo* (Bruguière, 1792), from North Carolina to Brazil, has a laterally compressed, oval shell with spiral beaded rows.

The shell of the Fine-sculptured Frog Shell is medium-sized, relatively lightweight, and ovate-conical. Its spire is tall, with a well-impressed suture, and a smooth apex. The sculpture consists of 5 to 7 spiral rows of small tubercles per whorl, with 1 row of heavier nodules at the shoulder. The aperture is oval, the outer lip thickened and dentate, and the columella has several folds. A varix appears at about every two-thirds of a whorl. The shell color is cream or reddish-tan, with a white aperture.

Actual size

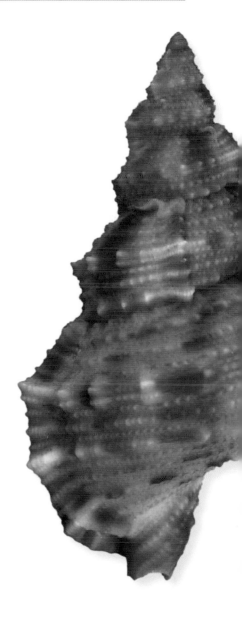

FAMILY	Bursidae
SHELL SIZE RANGE	¾ to 2¾ in (20 to 70 mm)
DISTRIBUTION	North Carolina to northeastern Brazil
ABUNDANCE	Common
DEPTH	Intertidal to 330 ft (100 m)
HABITAT	Rocky, sandy, and muddy bottoms
FEEDING HABIT	Carnivore, feeds on polychaetes
OPERCULUM	Corneous, oval, with central nucleus

SHELL SIZE RANGE
¾ to 2¾ in
(20 to 70 mm)

PHOTOGRAPHED SHELL
2½ in
(64 mm)

362

BUFONARIA BUFO
CHESTNUT FROG SHELL
(BRUGUIÈRE, 1792)

Bufonaria bufo has a flattened shell with varices at every half whorl that are nearly aligned with the varix from the previous whorl. Like other bursids, it has an anterior siphonal canal as well as a constriction at the posterior margin of the aperture, the anal canal, just beneath the suture. The former anal canals can be seen at each varix, but only the last one, on the aperture, is open. *Bufonaria bufo* is rare in Florida but common to abundant offshore in the Caribbean. Bursids are known to feed on polychaete and sipunculan worms, which are captured and ingested quickly by the snail.

RELATED SPECIES

Bufonaria echinata (Link, 1807), from the Indo-West Pacific, has a similar shell but with spines that grow outwardly from the varices. These projections can be as long as the spire height, but in some specimens they are short. *Bufonaria foliata* (Broderip, 1825), from Somalia to South Africa, has a larger shell than *B. bufo*, and is recognized by its wide, reflected, dentate outer lip and columellar shield stained in red-orange.

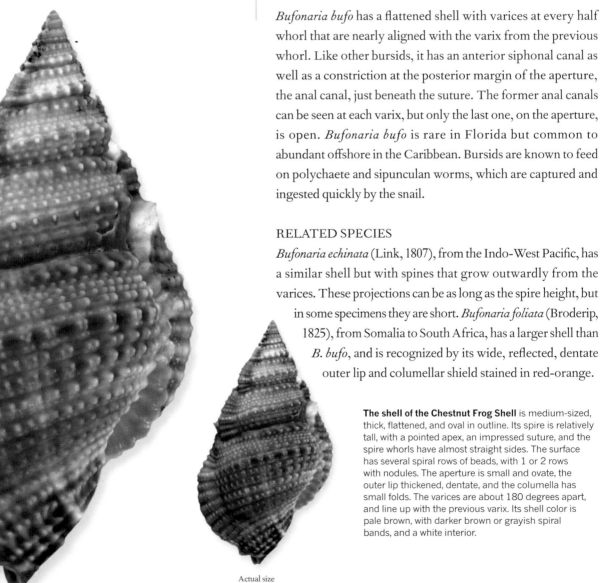

The shell of the Chestnut Frog Shell is medium-sized, thick, flattened, and oval in outline. Its spire is relatively tall, with a pointed apex, an impressed suture, and the spire whorls have almost straight sides. The surface has several spiral rows of beads, with 1 or 2 rows with nodules. The aperture is small and ovate, the outer lip thickened, dentate, and the columella has small folds. The varices are about 180 degrees apart, and line up with the previous varix. Its shell color is pale brown, with darker brown or grayish spiral bands, and a white interior.

Actual size

FAMILY	Bursidae
SHELL SIZE RANGE	1¼ to 4½ in (30 to 115 mm)
DISTRIBUTION	Somalia to South Africa, and western Indian Ocean
ABUNDANCE	Uncommon
DEPTH	85 to 100 ft (25 to 30 m)
HABITAT	Rocky and muddy bottoms
FEEDING HABIT	Carnivore, feeds on polychaetes
OPERCULUM	Corneous, oval

SHELL SIZE RANGE
1¼ to 4½ in
(30 to 115 mm)

PHOTOGRAPHED SHELL
2¾ in
(71 mm)

BUFONARIA FOLIATA

FRILLED FROG SHELL

(BRODERIP, 1825)

363

Bufonaria foliata is one of the most striking bursids because of its wide, denticulate, reflected outer lip and columella stained in bright orange-red, which contrast with the normally pale shell. Its varices are separated by about 180 degrees and each lines up with a varix from the previous whorl. It is a rare species in South Africa, where it lives offshore, but may be more common elsewhere. Some species of bursids are known to display sexual dimorphism of apertural features, with egg-laying females having a more flared aperture than non-breeding females or males.

RELATED SPECIES

Bufonaria borisbeckeri Parth, 1996, from the Philippines, has a similar but smaller shell, with long posterior canals and a white aperture. The species was named in honor of the German tennis player Boris Becker. *Bufonaria bufo* (Bruguière, 1792), from North Carolina to Brazil, has a laterally flattened shell with a surface covered by spiral rows of beads.

The shell of the Frilled Frog Shell is medium-sized, relatively thin, compressed, and ovale. Its spire is moderately short, with a pointed apex. Several spiral cords bear tubercles, with stronger and pointed spines along the shoulder. The aperture is lanceolate, with a denticulate outer lip, and a wide, lirate columellar shield. The posterior part of the aperture has a long anal canal. The shell color is usually cream or tan, sometimes pinkish, with the periphery of the aperture brightly colored in orange-red.

Actual size

FAMILY	Bursidae
SHELL SIZE RANGE	2 to 9½ in (50 to 240 mm)
DISTRIBUTION	Red Sea to Indo-Pacific
ABUNDANCE	Common
DEPTH	Intertidal to 660 ft (200 m)
HABITAT	Rocky bottoms
FEEDING HABIT	Carnivore, feeds on polychaetes
OPERCULUM	Corneous, ovate, with terminal nucleus

SHELL SIZE RANGE
2 to 9½ in
(50 to 240 mm)

PHOTOGRAPHED SHELL
8 in
(197 mm)

364

TUTUFA BUFO
RED-RINGED FROG SHELL
(RÖDING, 1798)

Actual size

Tutufa bufo is a large representative of the family Bursidae, also known as frog shells, because the knobbed sculpture of its shell resembles frog skin. It is widely distributed, ranging from the Red Sea to the central Pacific, including subtropical regions, such as northern New Zealand, and as far as Hawaii. It also has a wide depth range, occurring from the intertidal zone to about 660 ft (200m). *Tutufa bufo* is typically found in shallow waters in coral reefs. It is an active predator with a specialized diet of polychaete worms.

RELATED SPECIES

Tutufa bardeyi (Jousseaume, 1894), from the Gulf of Aden to Kenya, is the largest bursid; some shells can be almost twice as long as a large *T. bufo*, reaching about 17 in (430 mm) in length. *Bursa corrugata* (Perry, 1811), from Florida to Brazil, has a much smaller shell characterized by an orange-brown, broad, flared, and corrugated lip bearing light-colored teeth.

The shell of the Red-ringed Frog Shell is large, solid, heavy, and broadly fusiform. Its spire is high, with 2 to 3 spiral ribs bearing tubercles, while others bear weak nodules. There are 4 or 5 spiral ribs that swell to form thick ridges. Its aperture is large and ovate, the outer lip flared and scalloped, and the columella smooth; there is a thin columellar shield that can be broad. The anterior and posterior canals are well developed. The shell color is white or fawn, and the outer lip white or pink, rust colored near the aperture.

FAMILY	Cassidae
SHELL SIZE RANGE	1¼ to 4 in (30 to 100 mm)
DISTRIBUTION	Red Sea and Indo-Pacific to Hawaii
ABUNDANCE	Common
DEPTH	Low intertidal to 330 ft (100 m)
HABITAT	Sandy bottoms
FEEDING HABIT	Carnivore, feeds on echinoderms
OPERCULUM	Corneous, small

SHELL SIZE RANGE
1¼ to 4 in
(30 to 100 mm)

PHOTOGRAPHED SHELL
1½ in
(38 mm)

CASMARIA PONDEROSA
HEAVY BONNET
(GMELIN, 1791)

365

Casmaria ponderosa is aptly named: its shell can be quite massive and heavy, such as the illustrated here. However, it has a variable shell, and several subspecies and forms have been named, including some with thin shells. Like other cassids, it is a carnivore with a specialized diet of echinoderms, feeding primarily at night. Cassids have a long extensible proboscis that can reach the test (shell) of a sea urchin, even in long-spined urchins such as *Diadema*. There are 70 living species in the family Cassidae, occurring in tropical and temperate waters.

RELATED SPECIES

Casmaria vibex (Linnaeus, 1758), from the Indo-West Pacific, has a smaller, thinner, and smoother shell. The outer lip is thickened but not denticulate within. *Phalium glaucum* (Linnaeus, 1758), from the Indian Ocean and western Pacific, has a larger, thinner, and globose shell. The color is a solid gray, and the outer lip is flared, with three or four spines anteriorly.

The shell of the Heavy Bonnet is medium in size, thick, massive, glossy, and oval-elongate in outline. Its spire is moderately tall, with a pointed apex, well-marked suture, and convex whorls. The surface is smooth and glossy, but the body whorl may have knobs at the shoulder (as illustrated here). The aperture is narrow, both the outer and inner lip thickened, callused and denticulate within, and the outer lip has a row of pointed denticles at the margin. The shell color is whitish or cream, with a row of brown blotches near the suture; the aperture is white.

Actual size

FAMILY	Cassidae
SHELL SIZE RANGE	2 to 4½ in (50 to 110 mm)
DISTRIBUTION	Mediterranean; western Africa
ABUNDANCE	Common
DEPTH	100 to 500 ft (30 to 150 m)
HABITAT	Muddy, sandy bottoms
FEEDING HABIT	Carnivore, feeds on seas urchins
OPERCULUM	Corneous, thin, semicircular

SHELL SIZE RANGE
2 to 4½ in
(50 to 110 mm)

PHOTOGRAPHED SHELL
2⅜ in
(60 mm)

366

GALEODEA ECHINOPHORA
SPINY BONNET
(LINNAEUS, 1758)

Although common, particularly in the Adriatic Sea, the Spiny Bonnet is generally only found in small numbers. It feeds on sea urchins by first clearing a small area of spines, before using its proboscis to drill out a narrow hole, through which all the urchin's edible flesh is removed. The Spiny Bonnet is eaten in certain European countries, such as Spain and Italy, and is also favored by shell collectors for its shape and surface sculpturing.

RELATED SPECIES

Phalium saburon (Bruguière, 1792), also from the Mediterranean to western Africa is more inflated with a shorter, broader siphonal canal, and lacks the distinctive nodes of *Galeodea echinophora*. *Galeodea rugosa* (Linnaeus, 1758) also shares a similar range, but can occur farther north; it has a less angulate shoulder and is also less nodulose.

Actual size

The shell of the Spiny Bonnet has a medium spire and a short but upturned siphonal canal. The distinctive body whorl features between 4 and 6 spiral ridges, with posterior ridges more nodulose than anterior ones. The nodes are often quite pointed, giving this species of helmet shell its common name. The creamy white, slightly thickened, and flared lip features remnants of teeth. The shell color ranges from light to dark brown.

FAMILY	Cassidae
SHELL SIZE RANGE	2½ to 5¾ in (60 to 147 mm)
DISTRIBUTION	East Africa to western Pacific
ABUNDANCE	Common
DEPTH	Intertidal to 200 ft (60 m)
HABITAT	Sandy bottoms
FEEDING HABIT	Carnivore, feeds on echinoderms
OPERCULUM	Corneous, fan-shaped, elongate

SHELL SIZE RANGE
2½ to 5¾ in
(60 to 147 mm)

PHOTOGRAPHED SHELL
3 in
(77 mm)

PHALIUM GLAUCUM
GRAY BONNET
(LINNAEUS, 1758)

367

Phalium glaucum is a large and inflated species of *Phalium*, with a solid gray dorsum and three or four pointed spines on the anterior margin of the outer lip, making it easily recognizable. It is a carnivore that feeds on sand dollars and sea urchins, on sandy bottoms and sand flats, from the intertidal zone to offshore. During the spawning season, several females spawn together, laying irregular egg masses. It is collected for food and the shell trade. The operculum of some cassids is fan-shaped and elongate, with radial ridges and grooves.

RELATED SPECIES

Phalium flammiferum (Röding, 1798), from Japan to Vietnam, has a very distinct and handsome shell with axial flames. The shell is glossy, with incised spiral lines and a varix every two-thirds of a whorl. *Galeodea echinophora* (Linnaeus, 1758), from the Mediterranean and western Africa, has a globose shell with an elongate siphonal canal. The surface has several spiral rows of tubercles.

Actual size

The shell of the Gray Bonnet is moderately large, smooth, globular, and oval in outline. Its spire is short, with a pointed apex, well-marked suture, and angular whorls. The surface is mostly smooth, with faint spiral ribs crossed by axial growth lines. There is a varix every two-thirds of a whorl, and the shoulder of the last 2 whorls is angulated and has small tubercles. The aperture is semicircular and elongate, the outer lip thickened and denticulate within, and the anterior margin has 3 to 4 projecting spines. The shell color is a solid gray, the aperture is brownish, and the outer lip pale orange.

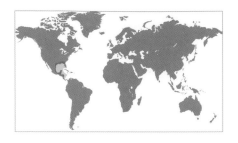

FAMILY	Cassidae
SHELL SIZE RANGE	3½ to 5¼ in (90 to 135 mm)
DISTRIBUTION	Florida to Gulf of Mexico
ABUNDANCE	Uncommon
DEPTH	420 to 3,000 ft (130 to 900 m)
HABITAT	Sandy or muddy bottoms
FEEDING HABIT	Carnivore
OPERCULUM	Corneous, oval

SHELL SIZE RANGE
3½ to 5¼ in
(90 to 135 mm)

PHOTOGRAPHED SHELL
4 in
(100 mm)

OOCORYS BARTSCHI
BARTSCH'S FALSE TUN
REHDER, 1943

The shell of the Bartsch's False Tun is moderately large, thin-walled but strong, inflated, and broadly spindle-shaped. Its spire is short, with well-marked suture, convex whorls, and a pointed apex. The sculpture consists of about 40 evenly spaced, flat spiral ribs on the body whorl. Smaller shells appear to lack a varix, but larger ones may have a few. The aperture is large and oval, the outer lip reflected and crenulated, and the columella curved and smooth. The shell color is peach or pale orange, and the interior cream.

Oocorys bartschi is a deepwater cassid, living on sandy muddy bottoms off the continental shelf. It is not collected often, therefore not much about its biology is known. It has a corneous, oval operculum that is too small to plug the aperture. It is easily recognized by its globose shell, sculptured with about 40 spiral flat ribs. There are about 15 living species in the subfamily Oocorythinae ("false tuns") worldwide; in contrast with most cassids, oocorythines tend to occur in deep waters. They were once considered a separate family, but anatomical features such as the radula demonstrate they belong in the Cassidae.

RELATED SPECIES
Oocorys sulcata Fischer, 1883, from North Carolina to the West Indies, and also from the Indian Ocean, is a common species in deep waters below 3,300 ft (1,000 m). It has a similar but smaller shell, with a narrower aperture, and fewer spiral ribs. *Cypraecassis rufa* (Linnaeus, 1758), from East Africa to Polynesia, has a large, solid shell with a thick, oval apertural shield stained in orange. It is one of the shells used by artists to produce cameo.

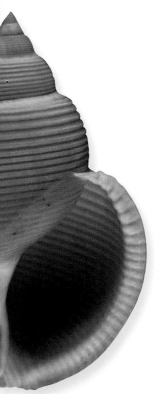
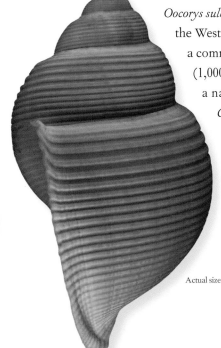

Actual size

FAMILY	Cassidae
SHELL SIZE RANGE	2½ to 8 in (65 to 200 mm)
DISTRIBUTION	Indo-West Pacific
ABUNDANCE	Common
DEPTH	Intertidal to 40 ft (12 m)
HABITAT	Near coral reefs
FEEDING HABIT	Carnivore, feeds on sea urchins
OPERCULUM	Corneous, thin

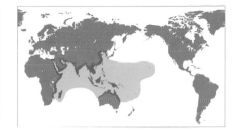

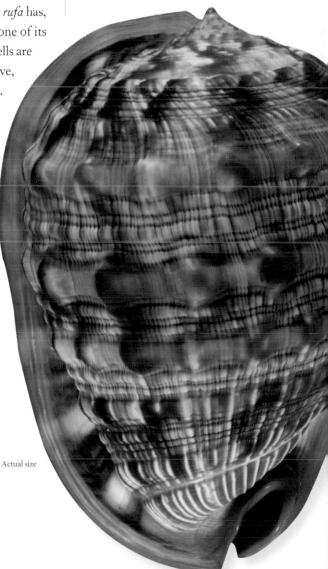

SHELL SIZE RANGE
2½ to 8 in
(65 to 200 mm)

PHOTOGRAPHED SHELL
6½ in
(166 mm)

CYPRAECASSIS RUFA

BULLMOUTH HELMET

(LINNAEUS, 1758)

369

Like other large species of helmet shells, *Cypraecassis rufa* has, for centuries, been used for making cameos, hence one of its alternative common names, the Cameo Shell. The shells are transported primarily from East Africa, where it is native, to Italy where the carving of the cameos takes place. Like all Cassidae, the Bullmouth, or Red Helmet, is carnivorous, using its radula and acidic secretions to make a hole in the shells of sea urchins to gain access to the edible flesh within.

RELATED SPECIES

Cassis flammea (Linnaeus, 1758), from Bermuda and Florida to the Lesser Antilles, is smaller, has a paler, less red parietal shield and outer lip, with only up to ten teeth, compared with *Cypraecassis rufa's* 22 to 24. *Cassis cornuta* (Linnaeus, 1758), from the Indo-West Pacific, features much larger shoulder nodes (which in males can be hornlike), is generally gray to white, and also has fewer teeth on the outer lip.

The shell of the Bullmouth Helmet is thick and heavy, with a low spire. The angulate shoulder leads to a body whorl with 3 or 4 thick nodulose spiral ridges, interspersed with smaller spiral bands with small knobs. Anterior pale, axial riblets lead to a sharply upturned, red-brown siphonal canal. The red columella features white teeth with dark-brown interstices. Around 22 to 24 pale, distinctive teeth run along the outer lip.

Actual size

FAMILY	Cassidae
SHELL SIZE RANGE	5 to 15½ in (50 to 390 mm)
DISTRIBUTION	Indo-West Pacific
ABUNDANCE	Common
DEPTH	Shallow water, 6 to 100 ft (2 to 30 m)
HABITAT	Near coral reefs
FEEDING HABIT	Carnivore, feeds on sea urchins
OPERCULUM	Corneous, thin

SHELL SIZE RANGE
5 to 15½ in
(50 to 390 mm)

PHOTOGRAPHED SHELL
8 in
(200 mm)

370

CASSIS CORNUTA
HORNED HELMET
(LINNAEUS, 1758)

Despite its large, thick shell, *Cassis cornuta* is less commonly used in the manufacture of cameos than other large helmet shells. It is the largest helmet shell in its range, and is used as a water carrier by the indigenous peoples of the region. *Cassis cornuta* live in colonies near the coral reefs of the Indo-West Pacific; they are widespread on the sandy bottom, seeking out sea urchins among the fragments of coral. As with other species of the Cassidae family, males tend to be smaller than females.

RELATED SPECIES

Cassis madagascariensis (Lamarck, 1822), from the Caribbean and Florida to the Lesser Antilles, is smaller, and has pronounced ribs and folds with black interstices running the length of the columella. *Cassis flammea* (Linnaeus, 1758), from the Caribbean, is much smaller and features dark, axial zigzag markings on the body whorl.

Actual size

The shell of the Horned Helmet
has a short spire comprising around 7 whorls. The angulate shoulder has between 5 and 7 distinctive knobs, which in males can be protruding and hornlike. The body whorl features 3 spiral bands on which blunt knobs grow, with the anterior knobs being smaller. The wide parietal shield reveals color and sculpture at the posterior, while the anterior two-thirds has strongly indented columella. The outer lip is indistinctly banded, and has a row of up to 12 blunt teeth, with the most prominent at the center.

FAMILY	Cassidae
SHELL SIZE RANGE	8 to 16¼ in (200 to 410 mm)
DISTRIBUTION	Southeastern U.S.A. to Barbados and Gulf of Mexico
ABUNDANCE	Common
DEPTH	10 to 90 ft (3 to 27 m)
HABITAT	Sand and seagrass
FEEDING HABIT	Carnivore, feeds on echinoderms
OPERCULUM	Corneous, very elongate

SHELL SIZE RANGE
8 to 16¼ in
(200 to 410 mm)

PHOTOGRAPHED SHELL
13½ in
(337 mm)

CASSIS MADAGASCARIENSIS SPINELLA

CLENCH'S HELMET

CLENCH, 1944

Cassis madagascariensis spinella is the largest of the helmet shells. It is a fairly common species in shallow seagrass meadows of the southeastern U.S.A., the Gulf of Mexico, and the Antilles. Its shell is a favorite material to carve for relief-shell cameos and brooches. The artist uses the differently colored shell layers to produce intricate designs. Helmet shells feed on echinoderms such as sea urchins and sand dollars.

RELATED SPECIES

Cassis tuberosa (Linnaeus, 1758), from the temperate and tropical western Atlantic, has a large, thick, nearly triangular parietal shield; *C. cornuta* (Linnaeus, 1758), from the Indo-Pacific, is among the largest helmets, and has an inflated body whorl; and *Cypraecassis rufa* (Linnaeus, 1758), also from the tropical Indo-Pacific, has a solid, reddish shell with thick outer lip, and is another species traditionally used for carving cameos.

The shell of the Clench's Helmet is very large and solid. It appears inflated and has a thick, expanded parietal shield that is ovately triangular in outline. Its spire is short, with thick axial varices spaced three-quarters of a whorl apart. The shell surface has 3 rows of low, rounded knobs, the most prominent at the shoulder, and numerous weaker spiral cords crossed by axial lines. The aperture is long, narrow, and constricted by the thickened outer lip with large teeth. The shell is white or cream in color, with the aperture and parietal shield a glossy, orange tan, with dark brown streaks between the teeth.

Actual size

FAMILY	Ficidae
SHELL SIZE RANGE	2½ to 6½ in (60 to 165 mm)
DISTRIBUTION	North Carolina to Venezuela; Gulf of Mexico
ABUNDANCE	Common
DEPTH	Intertidal to 580 ft (175 m)
HABITAT	Sandy bottoms
FEEDING HABIT	Carnivore, feeds on invertebrates
OPERCULUM	Absent

SHELL SIZE RANGE
2½ to 6½ in
(60 to 165 mm)

PHOTOGRAPHED SHELL
2⅞ in
(73 mm)

372

FICUS COMMUNIS
COMMON FIG SHELL
RÖDING, 1798

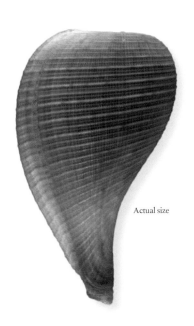

Actual size

Ficus communis is a common species in shallow warm waters from North Carolina to Venezuela and the Gulf of Mexico. Sometimes it washes ashore in great numbers. It is a sand dweller, and spends much of the time buried in sand, hunting for polychaetes and other worms. The animal of *F. communis* is large and the mantle has two lobes that cover most of the shell. *Ficus communis* is sometimes confused with *Busycotypus spiratus*, which has a vaguely similar shell, although the former is thinner, lighter and has flat spiral threads.

RELATED SPECIES

Ficus gracilis Sowerby I, 1825, from the Indo-West Pacific, has a shell similar to *F. communis* but it has a taller spire, is more elongate, and can grow larger. *Ficus subintermedia* d'Orbigny, 1852, from the Indo-Pacific, has one of the most colorful shells in the family, pinkish brown, with a cancellate sculpture.

The shell of the Common Fig Shell is elongate, thin, very fragile, and fig-shaped. Its spire is very low, and the suture is well impressed. The body whorl is large, the aperture wide and long, and there is no operculum. The siphonal canal is drawn out, tapering toward the anterior end. The sculpture consists of stronger, flat, spiral threads intercalated with weaker ones, and thinner axial lines. The shell color is pinkish-white or beige, with a white aperture and light brown interior.

FAMILY	Ficidae
SHELL SIZE RANGE	3¼ to 8 in (80 to 200 mm)
DISTRIBUTION	Indo-West Pacific
ABUNDANCE	Common
DEPTH	Shallow subtidal to 665 ft (200 m)
HABITAT	Sand or mud bottoms
FEEDING HABIT	Carnivore, feeds on invertebrates
OPERCULUM	Absent

SHELL SIZE RANGE
3¼ to 8 in
(80 to 200 mm)

PHOTOGRAPHED SHELL
5½ in
(143 mm)

FICUS GRACILIS

GRACEFUL FIG SHELL

(SOWERBY I, 1825)

Ficus gracilis has the largest shell in the family Ficidae. It lives in sandy or muddy bottoms in tropical and warm waters at continental shelf depths. The animal has a large arrowhead-shaped foot, and a very long proboscis used to hunt for tube-dwelling worms. Two flaps of tissue cover parts of the shell when the animal is expanded. Like other fig shells, shell shape, sculpture, and coloration do not vary much, and most species have similar shells. The Ficidae are a small family, with only about a dozen recognized species worldwide.

RELATED SPECIES

Ficus communis Röding, 1798, is a common species ranging from North Carolina to Venezuela, and the Gulf of Mexico. It resembles *F. gracilis*, but it has a nearly flat spire and a slightly wider shell. *Thalassocyon bonus* Barnard, 1960, the single species in its genus, is an abyssal species found from South Africa to New Zealand, and has a small and strongly keeled shell. Unlike other ficids, it has an operculum.

The shell of the Graceful Fig Shell is elongate, thin, fragile, and fig-shaped. Its spire is low, and the suture is deeply impressed. The body whorl is large and inflated, with a wide, long aperture and a long, tapering, slender siphonal canal. The sculpture is fine, with strong low spiral ribs crossed by fine axial lines. The outer lip is thickened at the top. The shell coloration ranges from orange to light brown, with faint, vertical zigzag markings. The aperture can be rich brown to orangish, fading to off-white at the margin of the aperture.

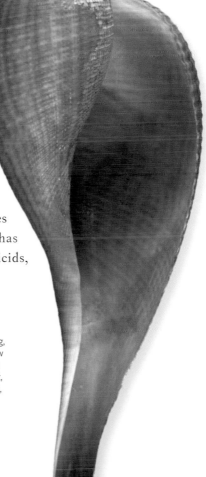

Actual size

FAMILY	Personidae
SHELL SIZE RANGE	1⅜ to 4 in (33 to 100 mm)
DISTRIBUTION	Indo-Pacific
ABUNDANCE	Uncommon
DEPTH	Intertidal to 100 ft (30 m)
HABITAT	Under coral
FEEDING HABIT	Carnivore
OPERCULUM	Corneous, thin, small

SHELL SIZE RANGE
1⅜ to 4 in
(33 to 100 mm)

PHOTOGRAPHED SHELL
2¾ in
(68 mm)

374

DISTORSIO ANUS

COMMON DISTORSIO

(LINNAEUS, 1758)

Actual size

The shell of the Common Distorsio is large for the family, inflated, distorted, and fusiform. Its spire is moderately tall, with a pointed apex and a wavering suture. The surface has spiral and axial ribs forming a rough knobby sculpture. The aperture is narrow and constricted, the outer lip thickened, with about 7 teeth, and the columella has strong teeth. There is a broad apertural shield with sharp, meandering edges. The shell color is cream with brown bands.

Distorsio anus has one of the largest, most distorted, and most colorful shells in the family Personidae. The animal is very colorful, red or orange, with irregular white blotches, and a pair of small black eyes at the base of long tentacles. It uses its extremely long proboscis to reach into crevices in search of its polychaete prey. The genus name is a good descriptor of the shells of personids, most of which are moderately to severely distorted. There are about 20 living species in the family Personidae worldwide.

RELATED SPECIES

Distorsio kurzi Petuch and Harasewych, 1980, from the Indo-West Pacific, is perhaps the most distorted of the personids. The axis of the columella seems to change direction with each new whorl added, and the body whorl has a large bulge opposite the aperture. *Distorsio clathrata* (Lamarck, 1816), which ranges from North Carolina to northeastern Brazil, has a relatively less distorted shell. Its surface has a reticulated pattern.

FAMILY	Personidae
SHELL SIZE RANGE	¾ to 4 in (19 to 100 mm)
DISTRIBUTION	North Carolina to Texas, and to Brazil
ABUNDANCE	Common
DEPTH	Shallow subtidal to 1,000 ft (300 m)
HABITAT	Sandy bottoms and under coral
FEEDING HABIT	Carnivore
OPERCULUM	Corneous, thin, small

SHELL SIZE RANGE
¾ to 4 in
(19 to 100 mm)

PHOTOGRAPHED SHELL
3¼ in
(83 mm)

DISTORSIO CLATHRATA

ATLANTIC DISTORSIO

(LAMARCK, 1816)

375

Distorsio clathrata is a common personid that ranges from the western Atlantic and Gulf of Mexico to Brazil. It is usually found in shallow subtidal depths but can also be located in deep water. Its shell is less distorted than most species in the family Personidae, and has an evenly rounded dorsum. It has a varix every 270 degrees (three-quarter whorl). When live, the shell is covered with a hairy periostracum. It has a fossil record that ranges from the Miocene Epoch of Colombia and Mexico. It is more broadly represented in younger deposits.

The shell of the Atlantic Distorsio is slightly large for the family, distorted, and fusiform. Its spire is moderately tall, with a pointed apex and a wavering suture. The surface has a reticulate pattern formed by crossed spiral and axial ribs, with the intersections elevated. The aperture is narrow and constricted. The outer lip is thickened, with teeth on both outer and inner lips, and there is a deep notch on the columella. The parietal shield is large and glossy. The shell color is white to yellowish, the parietal shield is stained in orange, and the aperture is white.

RELATED SPECIES

Distorsio anus (Linnaeus, 1758), from the Indo-Pacific, is quite distorted and has a larger parietal shield and more colorful shell than *D. clathrata*. It also has a curved siphonal canal. *Distorsio burgessi* Lewis, 1972, is a rare personid endemic to Hawaii. Its shell is broad and the surface has a reticulated texture somewhat similar to *D. clathrata*, but the checkerboard parietal shield has incised brown lines, and the shell is very distorted.

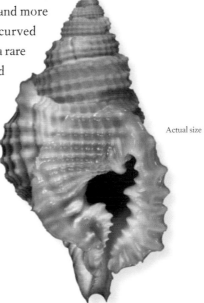

Actual size

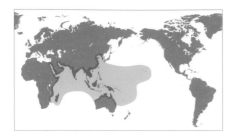

FAMILY	Ranellidae
SHELL SIZE RANGE	1¼ to 4 in (31 to 100 mm)
DISTRIBUTION	Indo-Pacific
ABUNDANCE	Common
DEPTH	Deep, 165 to 4,000 ft (50 to 1,200 m)
HABITAT	Soft sediments
FEEDING HABIT	Carnivore
OPERCULUM	Corneous, concentric

SHELL SIZE RANGE
1¼ t to 4 in
(31 to 100 mm)

PHOTOGRAPHED SHELL
2⅜ in
(59 mm)

376

BIPLEX PERCA
MAPLE LEAF TRITON
(PERRY, 1811)

Actual size

Biplex perca and a few related species of tritons have a shell with a distinct outline resembling a maple leaf, making them easy to recognize. The winged varices align with those of previous whorls; they are offset from the prior varix by nearly 180 degrees, making them spiral slightly around the shell. Large specimens of *B. perca* used to be commonly fished off Taiwan. Since the trawlers moved elsewhere, large specimens are rarely collected. Currently, much smaller specimens are taken in the Philippines. *Biplex perca* is the type species of the genus, which has a few species with similar shells.

RELATED SPECIES
Biplex pulchra (Gray, 1836), from Japan to Australia, has a shell similar to *B. perca* , but it is smaller, with larger granules on the sculpture. *Cymatium parthenopeum* (Von Salis, 1793), which has a cosmopolitan distribution in warm waters, has a thick and elongate shell with strong varices and a lirate columella. The periostracum is thick and hairy.

The shell of the Maple Leaf Triton is flattened dorsoventrally, and has long winged varices that align with the wings from previous whorls. It has a tall spire and very impressed sutures. The sculpture consists of spiral and axial threads, forming rounded white beads at intersections. Spiral threads extend onto the flattened, winged varices on opposite sides of the each whorl. The aperture is nearly circular, with 2 teeth posteriorly and a recurved siphonal canal. The shell color ranges from pale brown to gray; the aperture is white.

FAMILY	Ranellidae
SHELL SIZE RANGE	1¼ to 3¼ in (30 to 80 mm)
DISTRIBUTION	Indo-Pacific; northeastern Brazil
ABUNDANCE	Uncommon
DEPTH	Subtidal to 165 ft (50 m)
HABITAT	Rocky bottoms
FEEDING HABIT	Carnivore, feeds on bivalves
OPERCULUM	Corneous, thin, small

CYMATIUM SUCCINCTUM
LESSER GIRDLED TRITON
(LINNAEUS, 1771)

SHELL SIZE RANGE
1¼ to 3¼ in
(30 to 80 mm)

PHOTOGRAPHED SHELL
2⅜ in
(59 mm)

377

Cymatium succinctum has a handsome shell with glossy brown spiral ribs. Unlike other ranellids, most shells do not develop axial varices, but some specimens may have an occasional varix. The shell is covered by a dense, membranous (but not hairy) periostracum that in life appears to have veins like an insect wing. A membrane of the periostracum extends from each spiral rib. When the periostracum dries, it eventually peels off the shell. In Hawaii, *C. succinctum* had been common on the formerly abundant *Pinna* beds. The female lays a spherical egg mass.

RELATED SPECIES

Cymatium cingulatum Lamarck, 1822, has a wide distribution in the Indo-West Pacific, western Atlantic, and northwestern Africa. It has a broad, globose shell with a wide aperture and crenulated outer lip. *Ranella olearium* (Linnaeus, 1758) also has a wide distribution; the shell is large and thick with a tall spire.

Actual size

The shell of the Lesser Girdled Triton is medium in size, thin, globose, and fusiform in outline. Its spire is moderately tall, the whorls rounded, and the suture well impressed. The sculpture consists of about 12 to 13 flat, glossy, and evenly spaced ribs. Most shells do not have axial varices, but some do. The aperture is semicircular and the outer lip thickened, with denticles corresponding to the spiral ribs. The columella has a strong tooth posteriorly. The shell color is yellow-brown, with brown axial ribs, and a white aperture.

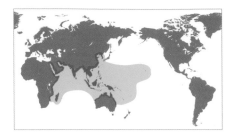

FAMILY	Ranellidae
SHELL SIZE RANGE	2 to 5 in (50 to 130 mm)
DISTRIBUTION	Red Sea, Indo-Pacific
ABUNDANCE	Uncommon
DEPTH	Intertidal to 90 ft (28 m)
HABITAT	On coral and sandy bottoms
FEEDING HABIT	Carnivore, feeds on invertebrates
OPERCULUM	Thick, corneous, with a nucleus near margin

SHELL SIZE RANGE
2 to 5 in
(50 to 130 mm)

PHOTOGRAPHED SHELL
3⅝ in
(93 mm)

RANULARIA PYRUM
PEAR TRITON
(LINNAEUS, 1758)

Ranularia pyrum has a solid and nodulose shell, with a long and irregularly curved siphonal canal. It has a strong varix every two-thirds of a whorl. Like other ranellids, it is a predatory gastropod, and feeds primarily on other gastropods, but also on other invertebrates, such as tube worms and sea cucumbers. It is found from the intertidal zone to offshore, on sand and near coral reefs; in Hawaii it is found mostly in deeper waters. It is collected for food and the shell trade. The family had been known as Cymatiidae, but Ranellidae is the oldest name.

RELATED SPECIES

Ranularia oblitum Lewis and Beu, 1976, from the Philippines to Australia, has a smaller, club-shaped shell with a very long siphonal canal that can account for more than half of the shell length. The length of the siphonal canal varies, and it can be straight or bent. *Cymatium succinctum* (Linnaeus, 1771), from the Indo-West Pacific and Atlantic, has a handsome shell with fine, raised, glossy brown spiral ribs on a fawn background.

Actual size

The shell of the Pear Triton is medium in size, thick, solid, nodulose, and pyriform in outline. Its spire is short, with 2 spiral rows of tubercles, and an impressed suture. The sculpture consists of spiral and axial ribs, forming tubercles at intersections of the main ribs. The body whorl is large, with angulated whorls, and 2 strong varices. The aperture is oval and the outer lip is thickened, with 7 denticles. The curved columella has small folds. The siphonal canal is long and curved. The shell color is orange-brown or red-brown, and the aperture and denticles white.

FAMILY	Ranellidae
SHELL SIZE RANGE	2½ to 9½ in (60 to 240 mm)
DISTRIBUTION	Florida to southeastern Brazil
ABUNDANCE	Common
DEPTH	2 to 500 ft (0.6 to 150 m)
HABITAT	Sandy bottoms near sea grasses
FEEDING HABIT	Carnivore, feeds on invertebrates
OPERCULUM	Corneous, thick, elongate

SHELL SIZE RANGE
2½ to 9½ in
(60 to 240 mm)

PHOTOGRAPHED SHELL
5 in
(127 mm)

CYMATIUM FEMORALE

ANGULAR TRITON

(LINNAEUS, 1758)

379

Cymatium femorale is a large ranellid or triton with a distinctive angular, nearly triangular profile. It has two strong, winged varices per whorl; when viewed from the apex, the shell has a triangular profile. *Cymatium femorale* lives on rubble bottoms and near eelgrass beds, ranging from shallow subtidal waters to offshore depths. Young shells are usually more brightly colored than adults. Like other ranellids, it is a carnivore, and feeds on other mollusks, sea-cucumbers (holothurians), and tube worms.

RELATED SPECIES

Cymatium ranzinii (Bianconi, 1851), from the Red Sea to Mozambique, has a shell that resembles *C. femorale*, but it is smaller, the aperture larger, the varices are not as upturned, and in general it has a paler color. *Charonia tritonis* (Linnaeus, 1758), from the Indo-West Pacific, and Galápagos Islands, is the largest species in the family Ranellidae. It has an inflated body whorl and a long spire with uneven varices and large aperture.

Actual size

The shell of the Angular Triton is large, thick, solid, angular, and nearly triangular in outline. Its spire is moderately tall, with angular whorls, a well-marked suture, and a long and narrow apex that is often missing in adults. The sculpture consists of several strong nodulose spiral ribs, with smaller ones between them. The aperture is long and wide, the outer lip is thickened and toothed in adults; the varices are thick and upturned. The siphonal canal is long and recurved. The shell color is reddish brown, with white nodules on the varices, and the interior is white.

FAMILY	Ranellidae
SHELL SIZE RANGE	3½ to 8½ in (90 to 220 mm)
DISTRIBUTION	Mediterranean; south and central Atlantic, Indian Ocean, southwest Pacific
ABUNDANCE	Uncommon
DEPTH	135 to 1,350 ft (40 to 410 m)
HABITAT	Sandy, muddy or shell hash bottoms
FEEDING HABIT	Carnivore, feeds on invertebrates
OPERCULUM	Corneous, thick, oval

SHELL SIZE RANGE
3½ to 8½ in
(90 to 220 mm)

PHOTOGRAPHED SHELL
6⅞ in
(175 mm)

380

RANELLA OLEARIUM

WANDERING TRITON

(LINNAEUS, 1758)

Ranella olearium has a wide distribution covering most of the world's oceans, hence the popular name. Like other ranellids, it has a larva with a long planktonic life, during which it is transported over long distances by ocean currents. Thus many ranellids have exceptionally broad distributions. *Ranella olearium* varies in size, shell thickness, and color, but is rather constant in shape, despite its wide distribution. In Europe, it occurs in deeper waters north of France, and tends to have a smaller shell than those from shallower waters in the south.

Actual size

RELATED SPECIES

Fusitriton magellanicus (Röding, 1798), from the southwestern Atlantic and southeastern Pacific, has a smaller shell with more globose whorls and a cancellate sculpture, covered by a thick periostracum with short hairs. *Ranularia pyrum* (Linnaeus, 1758), from the Red Sea and Indo-West Pacific, has an orange-brown solid shell with an irregularly twisted, long siphonal canal.

The shell of the Wandering Triton is large, thick, and fusiform in outline. Its spire is tall, with convex whorls, and a well-impressed suture. The sculpture consists of many spiral ribs, some of which are nodulose, crossed by fine growth lines. The aperture is large and oval, the outer lip is thickened and denticulate, and the columella curved and smooth, with one fold anteriorly. The siphonal canal is long, and the anal canal short. The shell color varies from white to pale brown, and the aperture is white.

FAMILY	Ranellidae
SHELL SIZE RANGE	4 to 20 in (100 to 490 mm)
DISTRIBUTION	Indo-Pacific; Galápagos Islands
ABUNDANCE	Common locally
DEPTH	Intertidal to 100 ft (30 m)
HABITAT	Coral reefs
FEEDING HABIT	Carnivore, feeds on echinoderms
OPERCULUM	Corneous, concentric, oval

SHELL SIZE RANGE
4 to 20 in
(100 to 490 mm)

PHOTOGRAPHED SHELL
17¼ in
(437 mm)

CHARONIA TRITONIS

TRUMPET TRITON

(LINNAEUS, 1758)

381

Charonia tritonis is the largest species in the family Ranellidae. It has been collected for centuries for food, as well as for its beautiful shell. In many areas, it has been used (among other shells) as a trumpet, by having a hole drilled into the early whorls. *Charonia tritonis* lives near coral reefs in shallow, tropical waters nearly worldwide. It is a voracious predator, feeding on echinoderms. It is well known as one of the few predators of the large, coral-eating Crown-of-Thorns starfish, *Acanthaster planci*, which can grow to 3 ft (1 m) in diameter.

RELATED SPECIES

Charonia variegata (Lamarck, 1816), the Atlantic Trumpet Triton, ranging from North Carolina to Brazil, has a similar but smaller and more squat shell. The lip in *C. variegata* usually has paired teeth with black interspaces. *Cymatium femorale* (Linnaeus, 1758), which ranges from South Florida to Brazil, has a thick shell with distinctive, angular, winglike varices.

The shell of the Trumpet Triton is very large, with a high and pointed spire, and an inflated body whorl. The ovate aperture is large, almost half the shell length, with a flared outer lip bearing well-defined teeth. The thickened lip forms an axial ridge that repeats every two-thirds of a whorl, so the varices of every other whorl align. The whorls are rounded and have coarse spiral cords with a single narrow cord between adjacent broad cords. The columella is thick and strongly lirate. The shell color is cream with brown crescents and blotches, the aperture is orange, and the inner lip is white, with brown bands.

Actual size

FAMILY	Tonnidae
SHELL SIZE RANGE	2 to 6 in (50 to 150 mm)
DISTRIBUTION	Red Sea to Indo-West Pacific
ABUNDANCE	Uncommon
DEPTH	33 to 230 ft (10 to 70 m)
HABITAT	Fine sand bottoms
FEEDING HABIT	Carnivore, feeds on echinoderms
OPERCULUM	Absent

SHELL SIZE RANGE
2 to 6 in
(50 to 150 mm)

PHOTOGRAPHED SHELL
2⅝ in
(68 mm)

382

TONNA SULCOSA
BANDED TUN
(BORN, 1778)

Actual size

The shell of the Banded Tun is medium-large, relatively thick for the family, and globose-ovate in outline. Its spire is low, with a pointed purplish apex, slightly convex whorls, and a channeled suture. The sculpture consists of about 20 flattened spiral ribs on the large body whorl. The aperture is wide, the outer lip thickened and denticulate, and the columella twisted. The shell color is white, with 3 to 4 evenly spaced, light brown spiral bands, covered with a dark brown periostracum, and a white aperture.

Tonna sulcosa has a very distinctive, globose-ovate, creamy white shell with three or four brown spiral bands and many spiral ribs. Like other tonnids, this species lacks an operculum. It lives from shallow subtidal waters to offshore depths, on fine sand and mud bottoms. *Tonna sulcosa* has a thicker shell than most tonnids, which are usually thin-shelled. Tonnids have a long-lived larva that can remain in the plankton as long as six months. There are about 30 living species in the family Tonnidae worldwide. Tonnids have a poor fossil record, probably because of their thin shells; the oldest known fossil dates from the Cretaceous Period.

RELATED SPECIES

Tonna allium (Dillwyn, 1817), from the Indo-West Pacific, has a smaller, more globose shell with about 13 strong, rounded spiral ribs. Some shells have interrupted tan markings on the spiral ribs. The aperture is thickened in mature specimens. *Tonna perdix* (Linnaeus, 1758), from the Indo-West Pacific and the Galápagos Islands, has a large, relatively more elongated tan or brownish shell, decorated with white crescent markings on the spiral ribs. The outer lip is thin and sharp.

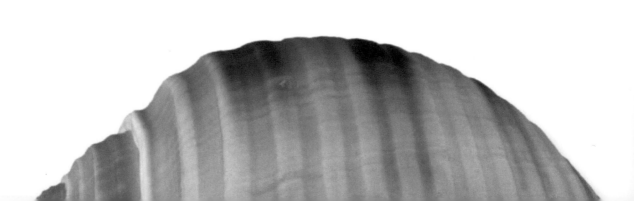

FAMILY	Tonnidae
SHELL SIZE RANGE	2¾ to 9 in (70 to 227 mm)
DISTRIBUTION	Red Sea to Indo-Pacific reaching Hawaii
ABUNDANCE	Common
DEPTH	Intertidal to 65 ft (20 m)
HABITAT	Sandy bottoms
FEEDING HABIT	Carnivore, feeds on holothurians
OPERCULUM	Absent

SHELL SIZE RANGE
2¾ to 9 in
(70 to 227 mm)

PHOTOGRAPHED SHELL
5¼ in
(132 mm)

TONNA PERDIX

PACIFIC PARTRIDGE TUN

(LINNAEUS, 1758)

383

Tonna perdix probably has the most colorful shell among the tun shells. Its coloration pattern is evocative of the plumage of the European partridge, hence the name. Like some tonnids, it has a wide distribution, ranging from the Red Sea, throughout the Indian Ocean, to the central Pacific, including Hawaii. It is most common in shallow waters, on sandy bottoms, where it burrows when inactive. It is a voracious predator of holothurians (sea cucumbers). The animal has a wide but thin foot, and a very wide proboscis. It is caught in trawl nets and fish traps, and is sometimes sold in local markets in the Philippines.

RELATED SPECIES

Tonna sulcosa (Born, 1778), from the Red Sea to the Indo-West Pacific, has an easily recognized shell with handsome banding. *Malea ringens* (Swainson, 1822), from western Mexico to Peru, and the Galápagos Islands, has the thickest and heaviest shell in the family Tonnidae. It has a unique columella with a deep notch at the center, and strong dentations.

The shell of the Pacific Partridge Tun is large, thin, lightweight, fragile, and globose-elongate. Its spire is relatively tall, with a pointed apex and an incised suture. The sculpture is smooth; the body whorl has about 20 shallow, rounded spiral ribs separated by grooves. The aperture is very large, the outer lip thin and sharp, and the columella smooth. The shell color is brown, with white crescent markings on the spiral ribs, and a yellow-brown interior.

Actual size

FAMILY	Tonnidae
SHELL SIZE RANGE	2½ to 9½ in (60 to 240 mm)
DISTRIBUTION	West Mexico to Peru; Galápagos Islands
ABUNDANCE	Common
DEPTH	Intertidal to 175 ft (55 m)
HABITAT	Sand bars or under rocky ledges
FEEDING HABIT	Carnivore, feeds on echinoderms
OPERCULUM	Absent

SHELL SIZE RANGE
2½ to 9½ in
(60 to 240 mm)

PHOTOGRAPHED SHELL
6¾ in
(174 mm)

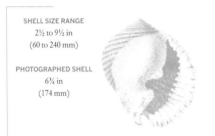

384

MALEA RINGENS
GRINNING TUN
(SWAINSON, 1822)

The shell of the Grinning Tun is large, solid, heavy, and globose. The spire is low and pointed, and the suture is shallow. The body whorl is large, with regularly spaced, broad, flat spiral ribs. The outer lip is reflected and thickened, with a crenulate outer edge and large teeth on the inner margin. The columella has a deep notch at the midpoint, with a few folds above and below it. The siphonal canal is short and curved. The shell color ranges from dirty beige to brown. The aperture is orange.

Malea ringens has the thickest and heaviest shell, and the thickest outer lip in the family Tonnidae. Most tuns have thin shells, but that does not deter them from being voracious predators of echinoderms. They use sulfuric acid, produced by the salivary glands, to dissolve a hole in the test (shell) of sea urchins and eat them. Other species feed on sea cucumbers.

RELATED SPECIES

Malea pomum (Linnaeus, 1758), from the Indo-West Pacific, is the smallest species in the family; *Tonna galea* (Linnaeus, 1758), widely distributed in the Atlantic Ocean, is one of the largest in the family; and *T. perdix* (Linnaeus, 1758), Indo-Pacific, has a large shell with a high spire and spiral rows of brown maculations.

Actual size

FAMILY	Atlantidae
SHELL SIZE RANGE	less than ⅛ to ½ (2 to 11 mm)
DISTRIBUTION	Worldwide in warm waters
ABUNDANCE	Common
DEPTH	0 to 165 ft (50 m)
HABITAT	Planktonic
FEEDING HABIT	Carnivore, feeds on other pelagic gastropods
OPERCULUM	Corneous, thin, trapezoid

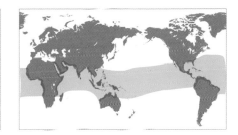

SHELL SIZE RANGE
less than ⅛ to ½
(2 to 11 mm)

PHOTOGRAPHED SHELL
⅜ in
(9 mm)

ATLANTA PERONI
PERON'S SEA BUTTERFLY
LESUEUR, 1817

Atlanta peroni is a holoplanktonic gastropod, meaning it spends its entire life floating and swimming near the surface of the ocean. It is carried by ocean currents, and has a worldwide distribution in warm waters. The family Atlantidae belongs to a group of planktonic gastropods called Heteropoda. Members of the genus *Atlanta* have a flattened, discoidal shell with a keel that helps stabilize swimming. The shell and the animal are transparent, making them less visible to predators. There are 16 living species in the family Atlantidae worldwide.

RELATED SPECIES

Species in the family Atlantidae are very similar and identification often depends on microscopic details of the larval shell. *Atlanta turriculata* d'Orbigny, 1836, from the Indo-Pacific, is one of the easiest to recognize because of its small size and relatively tall spire.

Actual size

The shell of the Peron's Sea Butterfly is small, thin, fragile, transparent, flattened, and dextrally coiled. Its spire is very short, with an impressed suture, and a pointed apex. The surface is smooth, with faint growth lines. There is a large keel at the periphery of the body whorl. The aperture is elongate oval, the outer lip thin, and the inner lip reflected. Its shell is transparent when fresh, and whitish when dried; there is a brown band at the base of the keel.

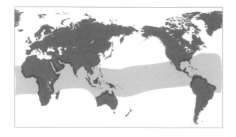

FAMILY	Carinariidae
SHELL SIZE RANGE	1¼ to 2½ (30 to 60 mm)
DISTRIBUTION	Worldwide in warm seas
ABUNDANCE	Uncommon
DEPTH	80 to 2,200 ft (25 to 670 m)
HABITAT	Pelagic
FEEDING HABIT	Carnivore, feeds on small planktonic animals
OPERCULUM	Absent

SHELL SIZE RANGE
1¼ to 2½
(30 to 60 mm)

PHOTOGRAPHED SHELL
1½ in
(38 mm)

386

CARINARIA LAMARCKI
LAMARCK'S GLASSY CARINARIA
(PÉRON AND LESUEUR, 1810)

Carinaria lamarcki is a strange-looking, pelagic gastropod. Its cap-shaped shell is thin and transparent, and covers only about 20 percent of the body, protecting the viscera. Like atlantids, carinariids spend their entire lives swimming or floating in the water column, although usually not close to the surface. Carinariids are also voracious predators, and feed on zooplankton, including other planktonic gastropods, small fish, and crustaceans. Lamarck's Glassy Carinaria is uncommonly collected, and because its shell is so fragile, it is rarely seen in good condition in collections. There are nine living species in the family Carinariidae worldwide. The oldest carinariid fossil dates from the Jurassic Period.

RELATED SPECIES

Carinaria cristata (Linnaeus, 1767), from the Indo-West Pacific, is the largest species in the family Carinariidae. Its shell is similar but taller than *C. lamarcki*, and can reach 2¾ in (70 mm), and the body up to 20 in (500 mm). *Carinaria japonica* Okutani, 1955, from Japan to California, has a tall triangular shell, and a body size that can reach 6 in (150 mm) in length.

The shell of the Lamarck's Glassy Carinaria is medium in size, thin, extremely fragile, laterally compressed, and cap-shaped, with a triangular profile. Its larval shell is globular, spirally coiled, and small; the body whorl is very large. The surface of the shell has axial folds, and the periphery of the body whorl has a well-developed keel that increases in height toward the aperture. The aperture is narrow and elongate. The shell is transparent.

Actual size

FAMILY	Triphoridae
SHELL SIZE RANGE	⅛ to ¼ in (3 to 6 mm)
DISTRIBUTION	Florida to Colombia
ABUNDANCE	Common
DEPTH	3 to 65 ft (1 to 20 m)
HABITAT	On and near sponges and sandy bottoms
FEEDING HABIT	Carnivore, feeds on sponges
OPERCULUM	Corneous, thin, circular

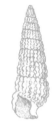

SHELL SIZE RANGE
⅛ to ¼ in
(3 to 6 mm)

PHOTOGRAPHED SHELL
¼ in
(5 mm)

MARSHALLORA MODESTA

MODEST TRIPHORA

(C. B. ADAMS, 1850)

387

Marshallora modesta is among the most common triphorids in Florida. It has a very small, brown shell with intricate sculpture best appreciated under a microscope (illustrated here under the scanning electron microscope). Most triphorids are sinistral (left-handed) micromollusks; relatively fewer are right-handed, and some may grow larger. Triphorids are specialized predators of sponges, and are usually found near, on, or within a particular sponge species. They are very diverse in the Indo-Pacific. As many as 80 species have been collected in a single sample. There are probably more than 1,000 living species in the family Triphoridae worldwide.

RELATED SPECIES

Tetraphora princeps (Sowerby III, 1904), from the Philippines, is the largest species in the family; the world record size is 2⅝ in (66 mm), which is very large for the family. It has a very tall spire, brown color, and four beaded spiral rows per whorl. *Inella asperrima* (Hinds, 1843), from the Indo-West Pacific, has a large shell that is needle-shaped, extremely narrow, and tall. It has a white shell with two granulated spiral cords per whorl.

The shell of the Modest Triphora is very small, glossy, beaded, sinistral, and conical-cylindrical. Its spire is tall, with a pointed, microscopically sculptured apex, well-marked suture, and straight-sided whorls. The sculpture consists of beaded spiral cords; there are 2 in early whorls, but after the 6th or 7th whorl, there are 3 cords per whorl. The aperture is squarish, and the outer lip thin, with a projection folding over the short siphonal canal. The shell color is chocolate brown, with the beads lighter brown.

Actual size

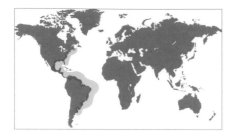

FAMILY	Cerithiopsidae
SHELL SIZE RANGE	⅛ to ½ in (4 to 13 mm)
DISTRIBUTION	Massachusetts to Uruguay
ABUNDANCE	Common
DEPTH	Intertidal to 260 ft (80 m)
HABITAT	On and near sponges and sandy bottoms
FEEDING HABIT	Carnivore, feeds on sponges
OPERCULUM	Corneous, thin, circular

SHELL SIZE RANGE
⅛ to ½ in
(4 to 13 mm)

PHOTOGRAPHED SHELL
⅛ in
(4 mm)

SEILA ADAMSII

ADAMS' MINIATURE CERITH

(LEA, 1845)

Seila adamsii is a widespread and common species, and is easily recognizable by its size and sculpture. Members of the Cerithiopsidae have shells that resemble those of Cerithiidae, hence the popular name. Most shells in the family are small, less than ⅜ in (10 mm) in length and thus considered by some authors as micromollusks, but *S. adamsii* is larger than most. Like the related Triphoridae, cerithiopsids are carnivores specialized for feeding on sponges. The protoconch is critical for identification in both families, but it is often missing in adult shells.

RELATED SPECIES

Seila marmorata (Tate, 1893), from southern Australia, has a shell similar to *S. adamsii* but with five spiral cords per whorl and more slender profile. The related triphorid, *Viriola incisa* (Pease, 1861), from the Indo-West Pacific, which is the most common triphorid in Hawaii, has a shell that resembles *S. adamsii* but it is sinistral.

 Actual size

The shell of the Adams' Miniature Cerith is very small, turriculate, and with a high spire. The protoconch is abrupt and bulbous, and the spire and body whorls have 3 strong, squarish spiral threads, with fine axial lines between each cord. The suture is not impressed, making it difficult to distinguish whorls. The aperture is subquadrate, the columella smooth and sinuous, and the siphonal canal short. The color of the larval shell is white, while the adult shell ranges from orange to dark brown in color.

FAMILY	Janthinidae
SHELL SIZE RANGE	½ to 1½ in (10 to 40 mm)
DISTRIBUTION	Worldwide in warm waters
ABUNDANCE	Common
DEPTH	Floats at the ocean surface
HABITAT	Pelagic; free swimming
FEEDING HABIT	Feeds on other floating sea creatures, plankton
OPERCULUM	Absent

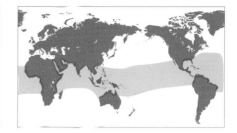

SHELL SIZE RANGE
½ to 1½ in
(10 to 40 mm)

PHOTOGRAPHED SHELL
¼ in
(20 mm)

JANTHINA GLOBOSA

ELONGATE JANTHINA

SWAINSON, 1822

389

Janthina globosa is not the most globose of the five species in the genus *Janthina*; it actually has a taller spire and more elongate shell than other species. Like other janthinids, it builds a float of mucus and air bubbles and drifts at the ocean surface. Some species, such as *J. globosa*, can be very abundant, and form shoals or groups; some shoals have been reported as being more than 200 nautical miles (370 km) across. On occasion, after storms, the shells are washed ashore in large numbers. There are about eight living species in the family Janthinidae worldwide, found floating at the surface in tropical and warm waters.

RELATED SPECIES

Most janthinids have cosmopolitan distributions. *Janthina janthina* (Linnaeus, 1758) has a slightly larger and wider shell with two tones: pale and bright purple. *Recluzia rollandiana* Petit, 1853, has a smaller shell that is similar in shape to *J. globosa*, but is brown. It looks like a miniature apple snail, a freshwater gastropod.

Actual size

The shell of the Elongate Janthina is small-medium, thin, fragile, and globose. Its spire is moderately tall, with a deep suture. Its spire whorls and body whorl are rounded, and bear fine diagonal lines forming a chevron pattern on the body whorl. The aperture is wide and slightly elongate, with a thin, smooth outer lip. The columella is straight, and continues down to form a point at the bottom, where it meets the outer lip. The shell color is pale to bright violet, and white just below the suture.

FAMILY	Janthinidae
SHELL SIZE RANGE	1 to 1½ in (25 to 40 mm)
DISTRIBUTION	Worldwide in warm waters
ABUNDANCE	Common
DEPTH	Floats at the ocean surface
HABITAT	Pelagic
FEEDING HABIT	Feeds on pelagic jellyfish
OPERCULUM	Absent

SHELL SIZE RANGE
1 to 1½ in
(25 to 40 mm)

PHOTOGRAPHED SHELL
1⅜ in
(34 mm)

390

JANTHINA JANTHINA
COMMON JANTHINA
(LINNAEUS, 1758)

Janthina janthina is one of a few pelagic predatory gastropods. It makes a raft by trapping air bubbles in mucus, which hardens like plastic. It then floats attached to the underside of this raft. It feeds on pelagic cnidarians such as *Vellela* and the Portuguese Man-of-War (*Physalia*). *Janthina janthina* has no eyes, and cannot control where its raft goes; instead, it drifts in the ocean, and feeds only when it happens upon its prey. The pelagic nudibranch snail *Glaucus* feeds on *Janthina* as well as on cnidarians.

RELATED SPECIES

Most species of janthinids have a circumtropical distribution, including *Janthina globosa* (Linnaeus, 1758), which despite its name, has a more elongate shell than *J. janthina*. It has a pale to bright purple or bluish shell. *Janthina pallida* (Thompson, 1840), as the name suggests, has a pale lavender shell.

Actual size

The shell of the Common Janthina is small-medium, thin, fragile, and globose in outline. Its spire is depressed and the suture well marked. The whorls are rounded and the sculpture consists of fine spiral striae and fine growth marks. The aperture is wide and rounded, the outer lip thin and sharp, and the columella is long and twisted. There is no umbilicus nor operculum. The upper part of the shell is pale violet, and the base is deep purple.

FAMILY	Epitoniidae
SHELL SIZE RANGE	¾ to 2⅜ in (20 to 61 mm)
DISTRIBUTION	Circumboreal
ABUNDANCE	Common
DEPTH	53 to 1,000 ft (16 to 300 m)
HABITAT	Sandy mud bottoms
FEEDING HABIT	Parasite on sea anemones
OPERCULUM	Corneous, oval, multispiral

SHELL SIZE RANGE
¾ to 2⅜ in
(20 to 61 mm)

PHOTOGRAPHED SHELL
⅞ in
(22 mm)

EPITONIUM GREENLANDICUM

GREENLAND WENTLETRAP

(PERRY, 1811)

391

Epitonium greenlandicum is a relatively large and common epitoniid with a circumboreal distribution. In the western Atlantic, its southernmost records are from New York. Epitoniids usually have a white shell with a tall spire and many axial ribs that represent former positions of the outer lip. It lives from subtidal water to the deep ocean. It buries itself in soft mud during periods of inactivity. There are at least 250 living species in the family Epitoniidae worldwide.

RELATED SPECIES

Epitonium ulu Pilsbry, 1921, from the Hawaiian Islands, is an example of an epitoniid associated with hard corals. It has a small, fragile shell with fine axial costae. *Epitonium clathrum* (Linnaeus, 1758), from western Europe and the Mediterranean, is common in shallow waters. It has a colorful chestnut shell, with thick axial lamellae.

Actual size

The shell of the Greenland Wentletrap is large for the family, thick, elongate, and conical. Its spire is tall, with well-marked sutures, and the whorls are convex. The sculpture consists of about 12 to 14 thick axial varices, and 7 broad, flat spiral ridges in the space between the axial lamellae. The aperture is oval, the outer lip is thickened, and there is no umbilicus. The shell color is chalky white to beige, the aperture white, and the operculum black.

FAMILY	Epitoniidae
SHELL SIZE RANGE	½ to 1½ in (13 to 40 mm)
DISTRIBUTION	Western Europe; Mediterranean
ABUNDANCE	Common
DEPTH	3 to 230 ft (1 to 70 m)
HABITAT	Sandy and muddy bottoms
FEEDING HABIT	Parasitic on sea anemones
OPERCULUM	Corneous, oval, multispiral

SHELL SIZE RANGE
½ to 1½ in
(13 to 40 mm)

PHOTOGRAPHED SHELL
1¼ in
(31 mm)

392

EPITONIUM CLATHRUM
COMMON WENTLETRAP
(LINNAEUS, 1758)

Epitonium clathrum has an attractive shell with many thick axial varices and cinnamon spiral bands or speckles. It usually lives in shallow subtidal depths on sandy or muddy bottoms, but in the spring it moves close to the shore to lay eggs. It is a consecutive hermaphrodite, and alternates between male and female each season. Like most epitoniids, it is parasitic on large anemones; its host is *Anemonia sulcata*. It is one of the most common epitoniids in Europe.

RELATED SPECIES

Epitonium krebsii (Mörch, 1875), which ranges from North Carolina to São Paulo, Brazil, has a short and stout shell, and no spiral sculpture. *Epitonium turtonis* (Turton, 1919), from northern Europe to the Canary Islands, has a more slender shell.

The shell of the Common Wentletrap is medium in size, elongated, and tower-shaped. Its spire is tall, with about 15 whorls, a pointed apex (although often missing), and convex whorls. The sculpture consists of about 9 thick axial varices per whorl, and smooth interspaces. The aperture is oval, the outer lip is thickened and reflected backward, and there is no umbilicum. The shell color varies from white (often on the Atlantic side) to cinnamon spiral bands or speckled with brown (in deeper waters or in the Mediterranean). The aperture has a similar color.

Actual size

FAMILY	Epitoniidae
SHELL SIZE RANGE	⅞ to 1¾ in (21 to 44 mm)
DISTRIBUTION	Northern Europe to the Canary Islands; Mediterranean
ABUNDANCE	Common
DEPTH	17 to 200 ft (5 to 60 m)
HABITAT	Sandy and muddy bottoms
FEEDING HABIT	Parasitic on sea anemones
OPERCULUM	Corneous, oval, multispiral

SHELL SIZE RANGE
⅞ to 1¾ in
(21 to 44 mm)

PHOTOGRAPHED SHELL
1⅜ in
(34 mm)

EPITONIUM TURTONIS

TURTON'S WENTLETRAP

(TURTON, 1819)

393

Epitonium turtonis has an elegant, slender shell. Some specimens, such as the one illustrated here, show considerable contrast between its violet-brown background and the cream-colored axial varices. It is a parasite of sea anemones, and lives in shallow subtidal waters. It is most common in the Mediterranean, where it often lives in deeper waters than in the Atlantic. Shells from the Mediterranean are often larger than their Atlantic counterparts.

RELATED SPECIES

Epitonium clathrum (Linnaeus, 1758), from western Europe and the Mediterranean, is similar in size, coloration, and sculpture, but is less slender and has fewer axial varices per whorl; the varices are erect. *Epitonium imperialis* (Sowerby II, 1844), from the southwest Pacific, has a broad, bulbous, and thin shell with about 30 closely spaced axial varices on the body whorl.

The shell of the Turton's Wentletrap is medium in size, moderately thick, slender, elongated, and tower-shaped. Its spire is tall, with about 12 to 15 whorls, a deep suture, and a pointed apex. The sculpture consists of 12 axial varices on the body whorl, and smooth interspaces. The axial varices are bent backward and lie close to the whorl surface. Some varices are rather broad. The aperture is oval, and the outer lip thickened. The shell color is pale brown to violet-brown, with 2 spiral reddish bands, the varices lighter or cream, and a brown aperture.

Actual size

FAMILY	Epitoniidae
SHELL SIZE RANGE	1¼ to 2¾ in (30 to 70 mm)
DISTRIBUTION	Indo-West Pacific
ABUNDANCE	Uncommon
DEPTH	Intertidal to 525 ft (160 m)
HABITAT	Sandy bottoms
FEEDING HABIT	Parasitic on sea anemones
OPERCULUM	Corneous, oval, multispiral

SHELL SIZE RANGE
1¼ to 2¾ in
(30 to 70 mm)

PHOTOGRAPHED SHELL
1⅜ in
(35 mm)

394

CIRSOTREMA VARICOSUM
VARICOSE WENTLETRAP
(LAMARCK, 1822)

Cirsotrema varicosum is a well-known epitoniid from the Indo-West Pacific. In most places, its shell grows to about 1⅜ in (35 mm), but in the Japonic region it can reach twice that size. It ranges from the intertidal zone to deep water, but typically lives subtidally, on sandy bottoms. Its shell is readily recognized by its trellis texture, formed by crenulated axial ribs set close together. The seas around Japan are very rich in epitoniids, with more than 120 species recorded.

RELATED SPECIES

Cirsotrema rugosa Kuroda and Ito, 1961, from the western Pacific, is the largest species in the genus *Cirsotrema*. It has a tall spire, sharp and erect axial lamellae that have an angular shoulder, and a well-marked suture. *Amaea magnifica* (Sowerby II, 1844), which ranges from Japan to Australia, is the largest living epitoniid. It has a relatively thin shell with spiral and axial ribs.

The shell of the Varicose Wentletrap is medium-sized, thick, solid, rough, slender, and tower-shaped. Its spire is tall, with rounded whorls, pointed apex, and a well-marked suture. The sculpture consists of about 25 axial, oblique, crenulated ribs that give the appearance of a trellis texture; 2 axial ribs per whorl are strong and irregularly spaced. The aperture is rounded, the outer lip thick, and there is no umbilicus. The shell color is dirty white or pale gray, the aperture white, and the operculum red-brown.

Actual size

FAMILY	Epitoniidae
SHELL SIZE RANGE	1 to 1¾ in (25 to 46 mm)
DISTRIBUTION	North Carolina to Barbados
ABUNDANCE	Rare
DEPTH	300 to 4,850 ft (90 to 1,480 m)
HABITAT	Sandy and rocky bottoms
FEEDING HABIT	Parasitic on sea anemones
OPERCULUM	Corneous, circular, multispiral

SHELL SIZE RANGE
1 to 1¾ in
(25 to 46 mm)

PHOTOGRAPHED SHELL
1¾ in
(46 mm)

STHENORYTIS PERNOBILIS

NOBLE WENTLETRAP

(FISCHER AND BERNARDI, 1857)

395

Sthenorytis pernobilis is a rare, deepwater epitoniid, and therefore its life history is not well known. In general, epitoniids with a thin shell and short axial ribs are usually parasitic on large sea anemones; they bury under the host and use it for its protection. Epitoniids with thicker shells and large axial ribs, such as *S. pernobilis*, often forage and prey upon small sea anemones on rocky bottoms. Large epitoniids may even swallow small anemones whole.

RELATED SPECIES

Epitonium krebsii (Mörch, 1875), which ranges from North Carolina to São Paulo, Brazil, resembles *S. pernobilis*, but it is smaller, has fewer, less angulated axial ribs, and is slightly more slender. *Sthenorytis turbinum* Dall, 1908, a species endemic to the Galápagos Islands, is very similar, differing in being slightly larger, broader, and in having longer, more pointed axial ribs.

The shell of the Noble Wentletrap is medium in size, sturdy, inflated, broad, and conical. Its spire is moderately tall, with about 6 to 7 whorls, a well-marked suture, and a pointed apex. The angle of the spire is about 50 degrees. The sculpture is dominated by about 12 to 15 erect, bladelike, axial ribs on each whorl. The aperture is nearly circular and oblique, and the outer lip thickened, with an angular lamella; there is no umbilicus. The shell color is white to grayish, and the operculum black.

Actual size

FAMILY	Epitoniidae
SHELL SIZE RANGE	1 to 2⅞ in (25 to 72 mm)
DISTRIBUTION	Indo-West Pacific
ABUNDANCE	Common
DEPTH	65 to 400 ft (20 to 120 m)
HABITAT	Sandy mud bottoms
FEEDING HABIT	Parasitic on sea anemones
OPERCULUM	Corneous, oval, multispiral

SHELL SIZE RANGE
1 to 2⅞ in
(25 to 72 mm)

PHOTOGRAPHED SHELL
2¼ in
(57 mm)

396

EPITONIUM SCALARE
PRECIOUS WENTLETRAP
(LINNAEUS, 1758)

Epitonium scalare is the best known epitoniid, and prized by collectors. Its exquisite shell is unusual in that its whorls are loosely coiled and do not touch each other, but rather only the bladelike axial varices touch the adjacent whorls. The axial ribs are aligned, which means that the number of ribs per whorl is the same in each whorl. Studies suggest that species with few ribs per whorl, such as *E. scalare*, may have a short life.

RELATED SPECIES

Epitonium albidum (d'Orbigny, 1824), which ranges from North Carolina to Uruguay, has a slender shell with about 12 to 14 axial ribs per whorl. Its whorls do not touch each other, as in *E. scalare*. *Epitonium clathrum* (Linnaeus, 1758), from western Europe and the Mediterranean, has thick axial ribs that may or may not align with ribs from previous whorls.

The shell of the Precious Wentletrap is medium-sized, thin, lightweight, broad, and conical. Its spire is tall, the whorls rounded, and the suture very deep, because the whorls do not touch, but are joined by the axial ribs. The sculpture consists of about 10 to 11 evenly spaced axial ribs per whorl, and smooth interspaces between ribs. The aperture is oval, the outer lip is thickened, and the umbilicus is deep and wide. The shell color is white or beige, with white axial ribs, a white aperture, and a black operculum.

Actual size

FAMILY	Epitoniidae
SHELL SIZE RANGE	2½ to 5 in (60 to 130 mm)
DISTRIBUTION	Japan to Australia
ABUNDANCE	Uncommon
DEPTH	100 to 650 ft (30 to 200 m)
HABITAT	Sandy or muddy bottoms
FEEDING HABIT	Parasitic on sea anemones
OPERCULUM	Corneous, oval, multispiral

AMAEA MAGNIFICA

MAGNIFICENT WENTLETRAP

(SOWERBY II, 1844)

SHELL SIZE RANGE
2½ to 5 in
(60 to 130 mm)

PHOTOGRAPHED SHELL
3⅜ in
(85 mm)

397

Amaea magnifica is the largest epitoniid or wentletrap. It usually
lives in deep water, and although it appears not to be rare in its
habitat, perfect specimens (such as the one illustrated here) are
seldom seen in collections—most specimens have a broken apex
or healed scars. *Amaea magnifica* used to be trawled in numbers
off Taiwan, but since trawlers moved to other locations, it is now
less available. Its relatively thin shell suggests that it is a parasite
of sea anemones, and buries in the sand under its host.

RELATED SPECIES

Amaea mitchelli (Dall, 1896), from Texas to Surinam,
is one of the largest epitoniids in the Atlantic Ocean.
It looks like a more colorful miniature *A. magnifica*.
Cirsotrema varicosum (Lamarck, 1822), from the
Indo-West Pacific, has erect, crenulated axial ribs
that form a trellis or honeycombed texture, and
about two strong ribs per whorl.

The shell of the Magnificent Wentletrap is large
for the family, relatively thin, delicate, slender, and
conical. Its spire is tall, with about 10 to 12 convex
whorls, and a deep suture. Its sculpture consists
of evenly spaced spiral ribs and thin axial ribs,
with some odd, strong axial ribs, producing a
cancellated texture. The early whorls are
smoothish, and have a pale brown spiral band.
The aperture is oval, and the outer lip and
columella are smooth; there is no umbilicus. The
shell color is chalky white, the aperture is white
stained in light brown, and the operculum is beige.

Actual size

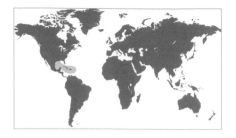

FAMILY	Eulimidae
SHELL SIZE RANGE	⅜ to ⅞ in (8 to 22 mm)
DISTRIBUTION	Texas to the Caribbean
ABUNDANCE	Uncommon
DEPTH	Intertidal to 300 ft (90 m)
HABITAT	Near coral reefs
FEEDING HABIT	Parasitic on echinoderms
OPERCULUM	Corneous, thin, oval

SHELL SIZE RANGE
⅜ to ⅞ in
(8 to 22 mm)

PHOTOGRAPHED SHELL
⅜ in
(10 mm)

398

SCALENOSTOMA SUBULATA
DISTORTED EULIMA
(BRODERIP, 1832)

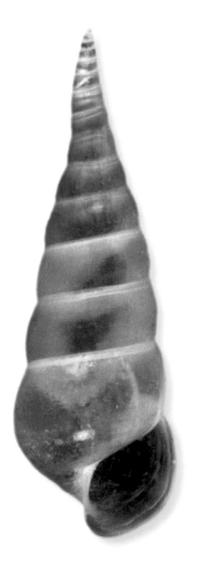

Scalenostoma subulata has a distorted shell, with the spire often bent, and the spire whorls of unequal height. As in many eulimids, there is a sexual dimorphism in shell shape, with females having a broader and larger shell than males. Eulimids are specialized parasites of echinoderms; some are free-living and can move from one host to another, but others are endoparasites, and live deeply embedded in the host. In some extreme cases, the shell is lost. There are probably several thousand living species in the family Eulimidae worldwide, including many undescribed ones.

RELATED SPECIES

Scalenostoma carinata Deshayes, 1863, from Réunion Island (Indian Ocean) to French Polynesia and Hawaii, has a larger, distorted shell that is slightly broader anteriorly. *Thyca crystallina* (Gould, 1846), from the western Pacific, has a very different shell to that of most eulimids, and looks more like a horn-shaped *Hipponix*. It is very small, white, broad, with a short spire, and spiral ribs. It parasitizes starfishes.

Actual size

The shell of the Distorted Eulima is small, thin, fragile, translucent, and conical. Its spire is tall and distorted, with a long and narrow larval shell, and convex whorls in the juvenile and adult shell. The surface is smooth and glossy, the body whorl inflated, and the suture is well marked. The aperture is oval, the outer lip thin, and the columella smooth. The shell color is translucent whitish, with the suture white.

FAMILY	Eulimidae
SHELL SIZE RANGE	⅛ to ⅝ in (4 to 14 mm)
DISTRIBUTION	Western Pacific
ABUNDANCE	Common
DEPTH	Intertidal to shallow subtidal
HABITAT	Coral reefs
FEEDING HABIT	Parasitic on echinoderms
OPERCULUM	Corneous, thin

SHELL SIZE RANGE
⅛ to ⅝ in
(4 to 14 mm)

PHOTOGRAPHED SHELL
⅝ in
(14 mm)

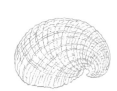

THYCA CRYSTALLINA

CRYSTALLINE THYCA

(GOULD, 1846)

399

Thyca crystallina has an unusually globose, cap-shaped shell among the eulimids, but like other eulimids, it is a parasite of echinoderms. In particular, it parasitizes the beautiful, bright blue starfish *Linckia laevigata*, which inhabits shallow, tropical waters of the Indo-Pacific, as well as other species in the genus *Linckia*. *Thyca crystallina* penetrates the skin of the starfish and becomes permanently embedded in the host's body. It acts like a shelled mosquito, sucking the haemolymph ("blood") of the starfish. It can occur in relatively large numbers, and several snails can parasitize a single starfish.

RELATED SPECIES

Thyca nardoafrianti (Yamamoto and Habe, 1976), from Japan, has a smaller shell with higher spire. It has nodulose spiral ribs. *Niso tricolor* Dall, 1889, from North Carolina, U.S.A., has a larger, smooth pyramidal shell. The shell is beige with a caramel spiral band on the suture, and a few weak varices of the same caramel color.

The shell of the Crystalline Thyca is small, strong, granulose, globose, and cap-shaped. Its spire is short, with a coiled apex. The sculpture consists of strong, nodulose spiral ribs crossed by growth marks. The shell is glossy and translucent when juvenile, but becomes duller and opaque as it grows larger. The aperture is large and rounded, the outer lip is slightly thickened, and the columella has a central notch. The shell color ranges from translucent whitish in juveniles to off-white in adults.

Actual size

FAMILY	Eulimidae
SHELL SIZE RANGE	¾ to 1 in (18 to 24 mm)
DISTRIBUTION	North Carolina, U.S.A.
ABUNDANCE	Uncommon
DEPTH	90 to 640 ft (27 to 196 m)
HABITAT	Offshore, on rocky bottoms
FEEDING HABIT	Parasitic on echinoderms
OPERCULUM	Corneous, thin

SHELL SIZE RANGE
¾ to 1 in
(18 to 24 mm)

PHOTOGRAPHED SHELL
1 in
(24 mm)

NISO TRICOLOR

TRICOLOR NISO

DALL, 1889

Niso tricolor has a relatively large shell for the family Eulimidae. It has been dredged from North Carolina, along with other large eulimids, from rocky bottoms at shallow to moderate depths. Many eulimids have conical, smooth, and white shells, although some, including *N. tricolor*, have colored shells. The species name *tricolor* refers to the three colors on its shell: the pale beige background, the brown spiral band, and the white umbilical area.

RELATED SPECIES

Niso splendidula (Sowerby, 1834), which ranges from the Gulf of California to Ecuador, is one of the largest and most colorful species in the family Eulimidae. Its shell resembles *N. tricolor* in shape, but the color pattern is more complex. *Scalenostoma subulata* (Broderip, 1832), from Texas to the Caribbean, has a small shell with a tall and distorted spire.

Actual size

The shell of the Tricolor Niso is relatively large for the family, thin, smooth, glossy, and conical. Its spire is tall, with about 12 to 14 whorls, an incised suture, and a pointed apex (eroded in the shell illustrated here). The surface is smooth, with axial growth marks, and the occasional weak varix. The aperture is leaf-shaped, the outer lip thin, and the columella smooth, with a reflected inner lip. The umbilicus is deep and smooth. The shell color is beige with a dark caramel spiral band along the suture; the umbilical area is white.

FAMILY	Buccinidae
SHELL SIZE RANGE	⅜ to ¾ in (10 to 20 mm)
DISTRIBUTION	Indo-West Pacific
ABUNDANCE	Common
DEPTH	Intertidal
HABITAT	Under rocks
FEEDING HABIT	Scavenging carnivore
OPERCULUM	Corneous, oval

ENGINA MENDICARIA
STRIPED ENGINA
(LINNAEUS, 1758)

SHELL SIZE RANGE
⅜ to ¾ in
(10 to 20 mm)

PHOTOGRAPHED SHELL
¾ in
(18 mm)

401

The Buccinidae is a very large family, with representatives in the polar seas as well as in tropical waters. Many have striking spiral sculpture, and those in warmer environments can be very colorful. They are all carnivores: some prey on bivalves, but many, including *Engina mendicaria*, are scavengers of dead fish. The family Buccinidae are not widely collected; the cold-water species can be drab, occurring in remote and inaccessible habitats.

RELATED SPECIES

Engina zonalis (Lamarck, 1822) enjoys the same tropical distribution and similar spiral banding, but always in black and white. It is a more slender shell with a steeper spire, and its outer lip is less prominent at the top. The rim of the aperture is stained ruby orange.

Actual size

The shell of the Striped Engina is small, fusiform, glossy, and black with a cream apex. It has very bold off-white to yellow spiral bands, usually 3 on the body whorl and one on the shoulders of the spire. Spiral sculpture of moderate nodules is also more prominent on the shoulders. The outer lip is deeply dentate, and thick; and the aperture is lined in yellow-orange.

FAMILY	Buccinidae
SHELL SIZE RANGE	¾ to 1¾ in (20 to 45 mm)
DISTRIBUTION	Western Pacific
ABUNDANCE	Abundant
DEPTH	Intertidal
HABITAT	Under rocks and coral
FEEDING HABIT	Scavenging carnivore
OPERCULUM	Corneous, oval

SHELL SIZE RANGE
¾ to 1¾ in
(20 to 45 mm)

PHOTOGRAPHED SHELL
1¼ in
(33 mm)

402

CANTHARUS UNDOSUS
WAVED GOBLET
(LINNAEUS, 1758)

Cantharus undosus feeds on a varied diet of bivalves, worms, and carrion, all of which are abundant in its choice of habitat—muddy rocks, and the rubble of broken, dead coral. In life it has a thick brown periostracum. Like other buccinids of warmer waters it has a bright colorful appearance; its polar relatives are much duller. The bright apertural margin and the strong spiral cords are particularly typical of goblet shells, which also frequently exhibit strong axial folds.

RELATED SPECIES

Cantharus wagneri (Anton, 1839) is a rather rare goblet from the tropical Pacific. It has similarly tall spiral whorls, but they are separated by a very deep suture. It is cream with broad spiral bands of mid brown, darker between the axial folds. The folds are much more pronounced, lending a wavy appearance to the typical goblet spiral cords.

The shell of the Waved Goblet is solid and fusiform, with a moderately rounded body. It is white to beige, with faint axial folds and well-defined, chestnut to dark brown spiral cords, becoming obsolete at the apex. The spire is tall, with a fine suture. There are dental folds on both outer lip and columella, and a delicate orange edge to the white aperture. The siphonal canal is short and broad.

Actual size

FAMILY	Buccinidae
SHELL SIZE RANGE	¾ to 2 in (19 to 51 mm)
DISTRIBUTION	Florida to Brazil, Ascension Island
ABUNDANCE	Moderately common
DEPTH	Intertidal
HABITAT	Corals and rocks
FEEDING HABIT	Scavenging carnivore
OPERCULUM	Corneous, oval

SHELL SIZE RANGE
¾ to 2 in
(19 to 51 mm)

PHOTOGRAPHED SHELL
1⅜ in
(37 mm)

PISANIA PUSIO

MINIATURE TRITON TRUMPET

(LINNAEUS, 1758)

403

Miniature Triton Trumpets are almost always found in pairs, rarely more than 1 ft (30 cm) apart from each other. They scavenge among rocks and coral for scraps of carrion, but rarely hunt actively. In turn, they are hunted by several predators; in particular, it comprises a significant proportion of the diet of a highly selective and efficient inshore predator, *Octopus insularis*, a newly described species of octopus found off Brazil.

RELATED SPECIES

Pisania ignea (Gmelin, 1791) is a very striking species of the genus from shallow water around Japan and the western Pacific. It has a similar form; the spiral whorls are all slightly rounded, however, and it is a smaller, slighter shell. The lip is thinner and the lirae less pronounced. The shell is shiny, and bright orange-yellow with axial flammules of mid orange.

Actual size

The shell of the Miniature Triton Trumpet
is solid and roundly fusiform. The spire sides
are nearly flat, apart from the penultimate
whorl, which is slightly convex. The aperture
is ovate, with a long siphonal canal and spiral
folds on both columella and lip, which both
become distinctly dental at the deep posterior
notch. The shell is cream to dusky pink with a
sculpture of fine spiral cords and spiral bands
of small, triangular chestnut blotches. A narrow
white band defines the shoulder.

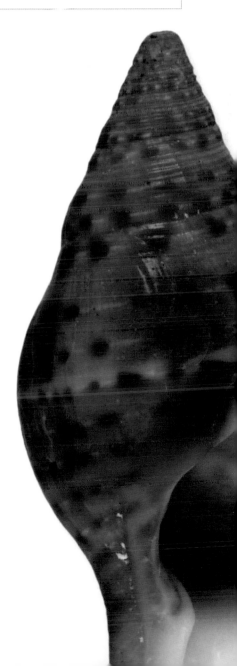

FAMILY	Buccinidae
SHELL SIZE RANGE	1⅜ to 2 in (35 to 50 mm)
DISTRIBUTION	Southern Japan
ABUNDANCE	Common
DEPTH	30 to 200 ft (10 to 60 m)
HABITAT	Rocks
FEEDING HABIT	Scavenging carnivore
OPERCULUM	Corneous, oval

SHELL SIZE RANGE
1⅜ to 2 in
(35 to 50 mm)

PHOTOGRAPHED SHELL
1¾ in
(45 mm)

404

SIPHONALIA PFEFFERI

PFEFFER'S WHELK

SOWERBY III, 1900

Actual size

Siphonalia pfefferi has an attractive, inflated fusiform shell, with raised spiral cords peppered with chestnut brown dots. Its shell varies in shape; some specimens, such as the one in the engraving above, have a long, curved siphonal canal. *Siphonalia pfefferi* lives offshore on sandy bottoms. Like some species in the genus *Siphonalia*, it is endemic to Japan, where it has a narrow distribution in the southern part of the country. Buccinids are particularly diverse in Japan and surrounding waters, with more than 200 species recorded.

RELATED SPECIES

Siphonalia callizona (Kuroda and Habe, 1961) is another *Siphonalia* endemic to Japan. It has a thin lip and a less swollen body whorl than *S. pfefferi*, a very tall spire, and a spiral sculpture of fine cords with deep blunt nodules on the shoulders. It is creamy yellow, with a spiral band of orange running across the nodules.

The shell of Pfeffer's Whelk is rounded and fusiform, with an enlarged ovate aperture and a relatively tall spire of convex whorls defined by an incised suture. There is a varix near the thickened lip, and a moderately long anterior canal. The shell is white, with flattened spiral cords of various widths with chestnut brown dots. The aperture is pink, with spiral lirae on the lip and upper half of the callused columella.

FAMILY	Buccinidae
SHELL SIZE RANGE	2 to 3¼ in (50 to 80 mm)
DISTRIBUTION	West Mexico to Ecuador
ABUNDANCE	Common
DEPTH	25 to 120 ft (7 to 35 m)
HABITAT	Muddy bottoms
FEEDING HABIT	Carnivore and scavenger
OPERCULUM	Corneous, claw-shaped

SHELL SIZE RANGE
2 to 3¼ in
(50 to 80 mm)

PHOTOGRAPHED SHELL
1⅞ in
(49 mm)

NORTHIA PRISTIS

NORTH'S LONG WHELK

(DESHAYES *IN* LAMARCK, 1844)

Northia pristis has a smooth body whorl with sharp spines on the outer lip. It is a common species in shallow offshore waters on soft bottoms. Although it has a superficial resemblance with shells in the family Nassariidae, its radula shows that it belongs in the Buccinidae. Buccinids have separate sexes and internal fertilization. Females lay several dozen eggs in each of dozens of leathery egg capsules. Each egg capsule has a pre-formed exit aperture with a plug that dissolves when the young are ready to hatch.

Actual size

RELATED SPECIES

Northia northia (Griffith and Pidgeon, 1834), from southern Mexico to Panama, has a very similar but slightly broader shell and a smoother outer lip. *Engina alveolata* (Kiener, 1836), from the Indo-Pacific, has a small shell, with a white background and rounded beads in black, orange, or yellow.

The shell of the North's Long Whelk is solid, smooth, shiny, thick, and fusiform. It has a high spire and a pointed apex. The suture is well marked; the spire whorls have both spiral and axial sculpture that fade toward the body whorl, which is smooth. The aperture is lanceolate and shorter than the spire, the outer lip is thickened and dentate, and the columella smooth. The shell color is grayish tan to brown, and the aperture is white.

FAMILY	Buccinidae
SHELL SIZE RANGE	1⅜ to 3¼ in (35 to 87 mm)
DISTRIBUTION	Gulf of California to southern Mexico
ABUNDANCE	Common
DEPTH	Intertidal
HABITAT	Rocks
FEEDING HABIT	Scavenging carnivore
OPERCULUM	Corneous, oval

SHELL SIZE RANGE
1⅜ to 3¼ in
(35 to 87 mm)

PHOTOGRAPHED SHELL
2½ in
(65 mm)

406

MACRON AETHIOPS
RIBBED MACRON
(REEVE, 1847)

The shell of the Ribbed Macron is heavy and fusiform, distinguished by its deeply impressed suture and wide, flat spiral cords, the one nearest the fasciole being greatly raised. The cords are separated by deep channels, and end at a roundly crenulated outer lip, inside which there are faint spiral folds. The color of the shell is porcelain white, but specimens often retain the olive brown periostracum.

The genus *Macron* is endemic to western America, mostly around the Gulf of California. The scientific name of the Ribbed Macron (also known as the Ethiopian Macron) refers to the color of its periostracum, which is notable not only for being brown but also for its exceptional thickness. It sticks to the shell (which is white) and is a defining characteristic of the genus, along with a wide, deep siphonal notch.

RELATED SPECIES

Many macrons originally thought to be separate species are now believed to be gradations of *Macron aethiops*. *Macron lividus* (Adams, 1855) is distinct—it is about half the size, its interior is orange, and it lacks exterior sculpture save for pleating on the fasciole. The apex is rounded and its whorls are slightly convex.

Actual size

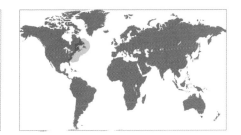

FAMILY	Buccinidae
SHELL SIZE RANGE	2 to 4½ in (50 to 115 mm)
DISTRIBUTION	New Foundland, Canada to North Carolina
ABUNDANCE	Common
DEPTH	13 to 2,160 ft (4 to 660 m)
HABITAT	Soft bottoms
FEEDING HABIT	Carnivore
OPERCULUM	Corneous, claw-shaped, with terminal nucleus

SHELL SIZE RANGE
2 to 4½ in
(50 to 115 mm)

PHOTOGRAPHED SHELL
3 in
(76 mm)

NEPTUNEA LYRATA DECEMCOSTATA

NEW ENGLAND NEPTUNE

(SAY, 1826)

407

Neptunea lyrata decemcostata has a handsome shell that is easily recognized by its strong, reddish brown, spiral cords on a grayish background. It is a common cold-water whelk that is usually found offshore, although it can also occur in quite deep water. Rarely, a few specimens wash ashore after a storm. In 1987, the New England Neptune became the official state shell of Massachusetts.

RELATED SPECIES

Neptunea elegantula Ito and Habe, 1965, from Japan and Korea, vaguely resembles *N. l. decemcostata*, with about 14 strong spiral cords, but the shell is more elongated and the cords are cream in color. *Buccinum undatum* Linnaeus, 1758, from northeastern U.S.A. and western Europe, is one the most abundant whelks in its region, and supports a commercial fishery.

The shell of the New England Neptune is medium in size, strong, thick, heavy, and fusiform. Its spire is tall and stepped, and the suture is incised. Its diagnostic feature, as the species name says, is the presence of 10 (actually between 7 and 10) strong spiral cords. Except for the spiral cords, the shell is mostly smooth. The aperture is large and lanceolate, the outer lip thin, and the siphonal canal relatively short and strong. The shell color is white or gray, with reddish brown spiral cords, and a white aperture.

Actual size

FAMILY	Buccinidae
SHELL SIZE RANGE	2 to 3¼ in (50 to 83 mm)
DISTRIBUTION	Southeast India
ABUNDANCE	Uncommon
DEPTH	Intertidal to 65 ft (20 m)
HABITAT	Sandy bottoms
FEEDING HABIT	Carnivore
OPERCULUM	Corneous, ovate, with terminal nucleus

SHELL SIZE RANGE
2 to 3¼ in
(50 to 83 mm)

PHOTOGRAPHED SHELL
3¼ in
(83 mm)

408

TUDICLA SPIRILLUS
SPIRAL TUDICLA
(LINNAEUS, 1767)

Tudicla spirillus has a very distinct shell with a wide and globose body whorl, a short spire, and a large, rounded apex. It is an uncommon species, and endemic to Sri Lanka and southeastern India, where it is protected by the Indian Wildlife Act of 1972. Its biology is not well studied. *Tudicla spirillus* is the single species in its genus. The group is well represented in the fossil record, with the earliest records from the Cretaceous Period.

RELATED SPECIES

The superficially similar *Tudivasum inermis* (Angas, 1878), from western and northern Australia, was once thought to be closely related, but belongs in the family Turbinellidae. *Afer cumingii* (Reeve, 1844), from Japan and Taiwan, has a robust shell with a tall spire and long siphonal canal.

The shell of the Spiral Tudicla is medium in size, thick, and turnip-shaped. It has a very short spire, a rounded apex, a globose body whorl, and a long, sometimes curved, siphonal canal. The body whorl has a sharp ridge at the shoulder and a spiral row of large, blunt tubercles at the lower part of the whorl. The aperture is large and ovate, the outer lip has lirations within, and the columella is smooth. The shell color is pale orange to gray, and the aperture white.

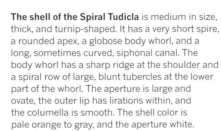

Actual size

FAMILY	Buccinidae
SHELL SIZE RANGE	2 to 3½ in (50 to 90 mm)
DISTRIBUTION	Japan and Taiwan
ABUNDANCE	Common
DEPTH	Intertidal to 165 ft (50 m)
HABITAT	Sandy and muddy bottoms
FEEDING HABIT	Carnivore
OPERCULUM	Corneous, ovate, with terminal nucleus

SHELL SIZE RANGE
2 to 3½ in
(50 to 90 mm)

PHOTOGRAPHED SHELL
3⅜ in
(87 mm)

AFER CUMINGII
CUMING'S AFER
(REEVE, 1844)

Afer cumingii is easily distinguished from other species from around Japan by its large rounded body whorl and long siphonal canal. It is a common buccinid in shallow waters on fine sandy or muddy bottoms. *Afer cumingii* is one of many shells named after Hugh Cuming, the foremost English shell collector of the nineteenth century, known as the "Prince of Collectors." He traveled the world to collect, buy, and exchange seashells (and other natural history objects, including orchids), and amassed the finest collection in the world, with nearly 83,000 specimens. Upon his death in 1865, his collection was sold to the British Museum.

RELATED SPECIES

Afer porphyrostoma (Adams and Reeve, 1847), from the Canary Islands to Mauritania, is smaller with a shorter siphonal canal and a purple aperture. *Tudicla spirillus* (Linnaeus, 1758), from southeastern India and Sri Lanka, has an inflated body whorl, a short spire with a rounded apex, and a long siphonal canal.

The shell of the Cuming's Afer is medium in size, thick, and globose spindle-shape. It has a tall, stepped spire with a well-impressed suture. Its sculpture consists of strong spiral cords and a spiral row of blunt nodules on the shoulder of each whorl. The aperture is ovate, the outer lip has internal lirations, and the columella is smooth. The siphonal canal is long and nearly closed. The shell color is yellowish brown, mottled with darker brown and white, and the aperture is white.

Actual size

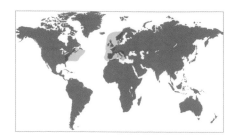

FAMILY	Buccinidae
SHELL SIZE RANGE	1¼ to 5 in (30 to 130 mm)
DISTRIBUTION	Northeastern U.S.A. and western Europe
ABUNDANCE	Abundant
DEPTH	Intertidal to 4,000 ft (1,200 m)
HABITAT	Sandy, muddy and rocky bottoms
FEEDING HABIT	Carnivore
OPERCULUM	Corneous

SHELL SIZE RANGE
1¼ to 5 in
(30 to 130 mm)

PHOTOGRAPHED SHELL
3½ in
(90 mm)

410

BUCCINUM UNDATUM
COMMON NORTHERN BUCCINUM
LINNAEUS, 1758

Actual size

Buccinum undatum is an abundant gastropod in cold water on both sides of the North Atlantic, and one of the predominant species in shallow coastal waters. It is commercially fished in Europe, and has been collected for food in northwestern Europe since prehistoric times. *Buccinum undatum* is a generalist carnivore and feeds on worms and bivalves. Females often spawn in large groups, and lay egg masses in clusters that are attached to hard substrates. In some places, such as in Belgium and the Netherlands, egg masses can cover miles of sandy beaches.

RELATED SPECIES

Neptunea antiqua (Linnaeus, 1758), which ranges from Scandinavia to France, has a shell similar to *B. undatum*, but it lacks vertical ridges. *Neptunea contraria* (Linnaeus, 1771), which ranges from northern Spain to Morocco and the Mediterranean, may vaguely resemble *B. undatum*, but it is sinistral (coiled counterclockwise).

The shell of the Common Northern Buccinum is medium-sized, thick, inflated, and pear-shaped. It has a tall spire with a pointed apex, and a well-marked suture. The aperture is oval, the outer lip and columella smooth, and the siphonal canal short. The sculpture consists of spiral ribs, which are crossed by growth lines, and sometimes oblique folds. The shell is variable in size, weight, sculpture, and color. Its coloration is usually dirty white, gray, or cream, with a white or cream interior.

FAMILY	Buccinidae
SHELL SIZE RANGE	3 to 4½ in (75 to 114 mm)
DISTRIBUTION	British Columbia to southern California
ABUNDANCE	Moderately common
DEPTH	130 to 1,300 ft (40 to 400 m)
HABITAT	Sand and mud
FEEDING HABIT	Scavenging carnivore
OPERCULUM	Corneous, oval

SHELL SIZE RANGE
3 to 4½ in
(75 to 114 mm)

PHOTOGRAPHED SHELL
3¾ in
(95 mm)

NEPTUNEA TABULATA
TABLED NEPTUNE
(BAIRD, 1863)

411

Neptunes are confined to the cold and temperate oceans of the northern hemisphere. They are characteristically heavy with rounded whorls and, as with other buccinids of those waters, they are usually fairly restrained in color. Many bear spiral cords, and all neptunes have a broad and open siphonal canal; that of *Neptunea tabulata* is untypically long. The genus, and this species in particular, have long been a favorite of collectors.

RELATED SPECIES

Neptunea contraria (Linnaeus, 1771) from northern Spain to Morocco and the Mediterranean, bears all the hallmarks of a typical neptune—drab white to cream color, rounded whorls, narrow spiral cords, short and open siphonal canal—but with a twist, as it were: it is sinistral, coiled counterclockwise, the only one of its genus to be so.

The shell of the Tabled Neptune is white to cream, and elongated with a tall spire and moderately long anterior canal. Its suture is sharply impressed to create a flat shelf on top of each slightly convex whorl, with a rough carinate ridge at the periphery and a small ramp at the inner edge. It bears regular, well-rounded spiral cords, some raised higher than others.

Actual size

FAMILY	Buccinidae
SHELL SIZE RANGE	2¾ to 7⅝ in (70 to 195 mm)
DISTRIBUTION	Bay of Campeche, east Mexico
ABUNDANCE	Uncommon
DEPTH	150 to 300 ft (45 to 90 m)
HABITAT	Offshore
FEEDING HABIT	Carnivore, feeds on bivalves
OPERCULUM	Corneous, large, with terminal nucleus

SHELL SIZE RANGE
2¾ to 7⅝ in
(70 to 195 mm)

PHOTOGRAPHED SHELL
4¼ in
(112 mm)

412

BUSYCON COARCTATUM
TURNIP WHELK
(SOWERBY I, 1825)

Busycon coarctatum is recognized by its turnip or club shape, and a row of small spines along the shoulder of the body whorl. It lives offshore and was considered very rare until 1950, when shrimp trawling in the Gulf of Mexico produced many specimens. *Busycon coarctatum* is a carnivore that feeds on bivalves but it also scavenges for carrion. It uses an olfactory organ and its long siphonal canal to detect the scent of its bivalve prey. Once the prey is found, the whelk's large, muscular foot holds the bivalve, and, using the edge of its own shell like a crowbar, it forces the clam open.

RELATED SPECIES

Busycon perversum (Linnaeus, 1758), from New Jersey to the Gulf of Mexico and Yucatán, has a very large, left-handed shell. It is the state shell of Texas. Juveniles have a pattern of lightning bolts on their shells, hence the popular name, Lightning Whelk. *Busycotypus spiratum* (Lamarck, 1816), from North Carolina to Yucatán, eastern Mexico, has a similar shell in size and shape, but with a larger aperture and a channeled suture.

Actual size

The shell of the Turnip Whelk is medium-large, thick, strong, and club-shaped. Its spire is short or flattened, with a pointed apex, angular whorls, and a suture that is not channeled. The sculpture consists of fine spiral lines, crossed by weak growth lines, and spiral lirations inside the aperture. The aperture is large, and the siphonal canal long and narrow. The shell color is cream or off-white with vertical brown lines, and the aperture is yellow.

FAMILY	Buccinidae
SHELL SIZE RANGE	3¾ to 4¾ in (95 to 120 mm)
DISTRIBUTION	Japan
ABUNDANCE	Rare
DEPTH	1,450 to 2,600 ft (450 to 800 m)
HABITAT	Mud
FEEDING HABIT	Carnivore
OPERCULUM	Corneous, oval

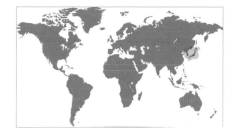

SHELL SIZE RANGE
3¾ to 4¾ in
(95 to 120 mm)

PHOTOGRAPHED SHELL
4¾ in
(120 mm)

ANCISTROLEPIS GRAMMATUS

GRAMMATUS WHELK

(DALL, 1907)

413

The Grammatus Whelk is rare, endemic to the deep waters off the northern Japanese island of Hokkaido. In life it is covered, like all species of *Ancistrolepis*, with a thick, brown, adherent periostracum that completely conceals the porcelain white shell beneath. The genus is typically very striking in sculpture, with a bold suture and a greatly flared aperture surrounding a thick operculum. All its species are rare and confined to waters of the northern Pacific.

RELATED SPECIES

Ancistrolepis unicum (Pilsbry, 1905), also uncommon and endemic to Japan shows a typically deep suture, which forms a flat table connected to the vertical whorl below by a sloping shelf. There is a raised spiral cord at the lower edge of the slope and a few lesser cords on the body. *Ancistrolepis vietnamensis* Sirenko and Goryachev, 1990, from the South China Sea is a smaller species with more rounded whorls and more numerous spiral cords.

Actual size

The shell of the Grammatus Whelk is large, moderately rounded, and fusiform, with a very tall spire. Its suture is deeply impressed especially at the base of the spire, and it is decorated with greatly raised spiral cords, flat or round, separated by wide flat channels. It has a deep flared aperture with a deep wide siphonal notch, and fine spiral folds inside. It is pure white, although before cleaning it retains an olive brown periostracum.

FAMILY	Buccinidae
SHELL SIZE RANGE	2½ to 16 in (60 to 400 mm)
DISTRIBUTION	New Jersey to Gulf of Mexico and Yucatán
ABUNDANCE	Common
DEPTH	Intertidal to 65 ft (20 m)
HABITAT	Estuaries, bays and oyster reefs
FEEDING HABIT	Carnivore, feeds on bivalves
OPERCULUM	Corneous, large and unguiculate, concentric

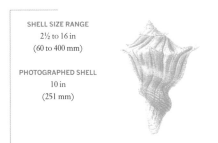

SHELL SIZE RANGE
2½ to 16 in
(60 to 400 mm)

PHOTOGRAPHED SHELL
10 in
(251 mm)

414

BUSYCON PERVERSUM
LIGHTNING WHELK
(LINNAEUS, 1758)

The shell of the Lightning Whelk is heavy, sinistral, and very large. It is pyriform, almost triangular in profile, with a short spire and broad shoulders that may bear large to small spines, knobs, or low tubercles. The aperture is long and to the left of the smooth columella. A large, corneous operculum protects the animal when it is withdrawn into the shell. The siphonal canal is long and tapering. The shell color varies from a light orange-tan with dark brown stripes in juveniles, to light tan or gray in adults.

Busycon perversum is the state shell of Texas. It is a large edible gastropod common along the eastern and southern coasts of the U.S.A. that has been used as food by Native Americans for thousands of years. Its popular name originates from the lightning-bolt pattern seen in juvenile shells, but the pattern fades in larger shells. *Busycon* species were traditionally classified in the family Melongenidae, but recent studies demonstrate they are buccinids. There are only about ten living species of *Busycon*, although the genus has a long and diverse fossil record.

RELATED SPECIES

Busycon carica Gmelin, 1791, ranges from Massachusetts to northeastern Florida, and most closely resembles *B. perversum*, but it is its mirror-image (that is, right-handed); *B. candelabrum* (Lamarck, 1816), and *B. coarctatum* (Sowerby I, 1825) live in deeper waters of the Gulf of Mexico. Both are right-handed.

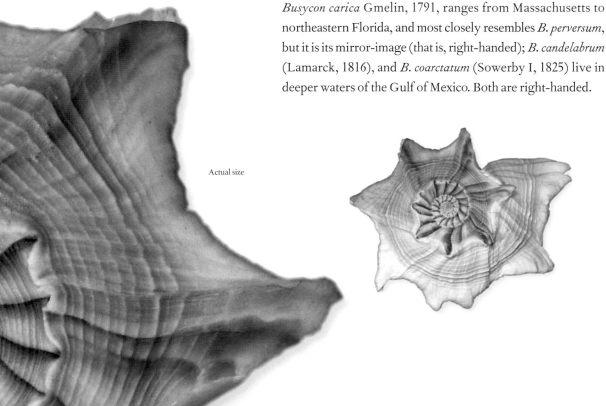

Actual size

FAMILY	Colubrariidae
SHELL SIZE RANGE	1⅜ to 3⅛ in (35 to 79 mm)
DISTRIBUTION	Japan to Australia
ABUNDANCE	Uncommon
DEPTH	50 ft (15 m) to deep water
HABITAT	Under rocks
FEEDING HABIT	Parasite of fishes
OPERCULUM	Corneous

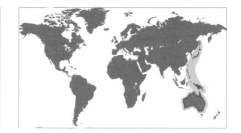

SHELL SIZE RANGE
1⅜ to 3⅛ in
(35 to 79 mm)

PHOTOGRAPHED SHELL
2¼ in
(55 mm)

COLUBRARIA CASTANEA
CHESTNUT DWARF TRITON
KURODA AND HABE, 1952

415

Colubrariids have thick shells distinguished by their intricate sculpture, particularly the deep varices on their whorls. They have a parietal callus extending from the inner lip, a short, recurved anterior canal, and a fairly narrow aperture. The spire is acute and very long, comprising around ten whorls; it gives the shell a dramatic appearance in life and in the collector's cabinet. None of the species of colubrariids is very common.

RELATED SPECIES
Colubraria obscura (Reeve, 1844) is uncommon, found on shallow reefs from Florida to Brazil. It is very similar in appearance but paler overall: white to gray-brown, with darker patches of color tending to occur in indistinct spiral bands. The aperture is narrow, the lip is a little thicker, and the columella if anything more callused.

The shell of the Chestnut Dwarf Triton is fusiform with a very tall spire, usually off-white to pale chestnut in color with lighter and darker patches and a paler, relatively narrow, ovate aperture. The sculpture is reticulate, with fine spiral beading distorted by irregular varices on the whorls and before the thick, denticulate outer lip. The columella is smooth except for one tooth at the posterior notch. A parietal shield extends across the body and fasciole.

Actual size

FAMILY	Colubrariidae
SHELL SIZE RANGE	1¾ to 4½ in (45 to 112 mm)
DISTRIBUTION	Indo-Pacific
ABUNDANCE	Uncommon
DEPTH	Intertidal
HABITAT	Rocks and reefs
FEEDING HABIT	Parasite, feeds on fish blood
OPERCULUM	Corneous

SHELL SIZE RANGE
1¾ to 4½ in
(45 to 112 mm)

PHOTOGRAPHED SHELL
3½ in
(88 mm)

416

COLUBRARIA MURICATA
MACULATED DWARF TRITON
(LIGHTFOOT, 1786)

Colubrariids have a ghoulish feeding habit: they drink the blood of sleeping fish. These animals extend their very long proboscis into the soft tissues of the fish. Salivary glands produce and anti-coagulant to prevent the blood from clotting while they ingest it. Because of this hematophagous feeding habit, colubrariids are also known as the vampire shells. The fish itself is not sedated in any way, and will escape when it wakens. *Colubraria muricata* finds its prey haphazardly while moving among reefs in the shallows of the Indo-Pacific region.

RELATED SPECIES

Colubraria tortuosa (Reeve 1844) inhabits the reefs of the western Pacific. It has a leaning spire, the result of a slightly twisted axis, and the appearance is exaggerated by the swollen varices on previous whorls. It has spiral bands of brown splashes on a cream to beige body, which is sculpted with fine, square spiral beads.

Actual size

The shell of the Maculated Dwarf Triton is fusiform. It is white to cream with broken patches of mid and dark chestnut, sometimes organized in spiral bands or short flames. Its tall spire has several irregularly spaced varices on each whorl. Another large varix sits on the thick outer lip, which has about 10 teeth. The aperture is narrow and white, with a moderately wide parietal shield and a deep, open, recurved anterior canal.

FAMILY	Columbellidae
SHELL SIZE RANGE	⅜ to ⅝ in (11 to 16 mm)
DISTRIBUTION	Southeastern Florida and the West Indies to Brazil
ABUNDANCE	Common
DEPTH	Intertidal to 6 ft (2 m)
HABITAT	Under rocks
FEEDING HABIT	Herbivore
OPERCULUM	Corneous, rounded, thin

SHELL SIZE RANGE
⅜ to ⅝ in
(9 to 16 mm)

PHOTOGRAPHED SHELL
½ in
(13 mm)

NITIDELLA NITIDA

GLOSSY DOVE SHELL

(LAMARCK 1822)

417

Nitidella nitida generally lives in tightly packed groups beneath rocks lying on sandy beds in warm shallow water. Most columbellids appear to be opportunistic, but judging from stomach contents, some like *N. nitida* are herbivorous. There are between 400 and 500 species of Columbellidae, occurring throughout the world, particularly in tropical seas. Most are less than ½ in (12 mm) long, and few species are more than 2 in (5 cm) in length.

RELATED SPECIES

Nitidella laevigata (Linnaeus, 1758) occurs in the same region as *N. nitida*, but is, if anything, more abundant. It has a wider aperture, a thinner outer lip, and a marked pattern on the whorl of axial orange zigzag stripes on a white base. *Nitidella laevigata* shares the columellar twin ridges of *N. nitida*, but they are nearer the siphonal canal.

The shell of the Glossy Dove Shell is, like most columbellids, highly glossy. The color is reddish to purplish brown speckled with white, the speckles accumulating in faint bands at the base of the spire and across the whorl. The spire is roughly equilateral, with a faint suture. The aperture is long and thin with a thickened outer lip, and there are 2 little ridges on the columella below its center.

Actual size

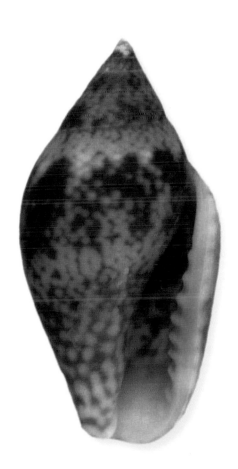

FAMILY	Columbellidae
SHELL SIZE RANGE	⅝ to 1 in (14 to 26 mm)
DISTRIBUTION	Red Sea, Indo-West Pacific
ABUNDANCE	Moderately common
DEPTH	Intertidal to 16 ft (5 m)
HABITAT	In sand under coral and rocks
FEEDING HABIT	Carnivore
OPERCULUM	Corneous, narrow

SHELL SIZE RANGE
⅝ to 1 in
(14 to 26 mm)

PHOTOGRAPHED SHELL
⅞ in
(20 mm)

418

PYRENE PUNCTATA
TELESCOPED DOVE SHELL
(BRUGUIÈRE, 1789)

The shell of the Telescoped Dove Shell is light to dark tan, most commonly a reddish brown, and glossy. The spire is convex with a deeply impressed suture, and the apex is often damaged or eroded on beached specimens. The color on the spire is interrupted subsuturally by large white spots, and the whorl bears distinctive triangular markings. There are narrow folds on the fasciole and the lower columella is also plicate.

Pyrene punctata derives its common name from the pronounced, almost steplike suture, which gives an impression of the series of sections of a telescope. It is fairly abundant in the tropical zones of its range, but can less commonly be found as far south as New South Wales, southeastern Australia. The *Pyrene* and *Columbella* genera of Columbellidae are very similar, and the two names have been retained principally to distinguish between the family's eastern Pacific and Indo-Pacific species, respectively.

RELATED SPECIES

Pyrene flava (Bruguière, 1789) has the same range but is more common. It is the same size, and has the same white blotches on the spire (which also occur on the whorl) but is generally of a paler tan color. The suture is less deeply impressed than on *P. punctata*, and the lip is thicker.

Actual size

FAMILY	Columbellidae
SHELL SIZE RANGE	½ to 1 in (14 to 25 mm)
DISTRIBUTION	Sea of Cortéz to Ecuador and Galápagos Islands
ABUNDANCE	Moderately common
DEPTH	Intertidal
HABITAT	Under rocks
FEEDING HABIT	Herbivore
OPERCULUM	Corneous, narrow

SHELL SIZE RANGE
½ to 1 in
(14 to 25 mm)

PHOTOGRAPHED SHELL
⅞ in
(23 mm)

COLUMBELLA HAEMASTOMA

BLOODSTAINED DOVE SHELL

SOWERBY 1, 1832

419

While most columbellids appear to be opportunistic, *Columbella haemastoma* and other species in the genus are herbivorous. It is particularly active at night, when it can be observed searching the floor of a muddy or sandy tidal pool for food. The sexes are separate, and females lay hemispherical eggs on hard substrates or on algae.

The shell of the Bloodstained Dove Shell is dark red-brown to chocolate, with large white blotches below the sutures becoming all-white at the protoconch. Its whorl is rounded, with a lip that is thickened and concave at its center. The aperture is narrow and the areas around it including columella, lip, and fasciole, are widely stained pale to mid orange.

RELATED SPECIES

Columbella strombiformis (Lamarck, 1822) has the same concavity of the outer lip, but is slightly larger and more common. The body whorl appears more swollen, and bears a distinctive, white, zigzag axial stripe pattern. The siphonal canal is longer than that of *C. haemastoma*, and although the aperture bears the same orange tinting it is less widespread across the lip and fasciole.

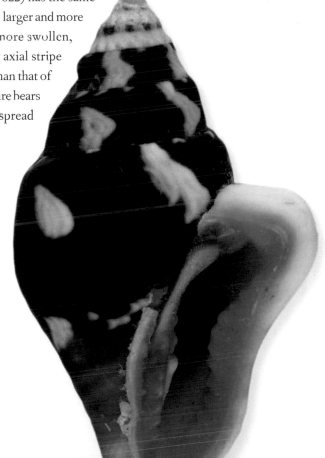

Actual size

FAMILY	Columbellidae
SHELL SIZE RANGE	⅜ to 1 in (10 to 24 mm)
DISTRIBUTION	Florida to Brazil and West Indies
ABUNDANCE	Abundant
DEPTH	Intertidal to 265 ft (80 m)
HABITAT	On and under rocks
FEEDING HABIT	Herbivore
OPERCULUM	Corneous, smaller than aperture

SHELL SIZE RANGE
⅜ to 1 in
(10 to 24 mm)

PHOTOGRAPHED SHELL
¼ in
(18 mm)

420

COLUMBELLA MERCATORIA

COMMON DOVE SHELL
(LINNAEUS, 1758)

One of the commonest of the Columbellidae, *Columbella mercatoria* is found in large numbers particularly in the Caribbean on the blades of sea grasses. The shell is often overgrown by algae and, when alive, it has a thin periostracum. The wide variations of color and pattern have given rise to a great many synonyms. Unlike most columbellids, which are opportunistic scavengers by night, *C. mercatoria* is a herbivore and feeds on algae.

RELATED SPECIES

Columbella rustica (Linnaeus, 1758), found in northwest Africa and the Mediterranean, is of a very similar appearance. It has the same variety of markings and color, the same narrow aperture, toothed lip, and plicate columella. Its spire is however slightly taller, and the spiral threads on the body whorl are much finer.

Actual size

The shell of the Common Dove Shell is distinguished by fine spiral cords and very fine axial ribs. Its spire is slightly short of medium. The aperture is narrow, the thickened lip and columella both strongly dentate. Although extremely variable both in color and marking, a typical shell is white with orange to dark brown axial stripes or zigzags.

FAMILY	Columbellidae
SHELL SIZE RANGE	⅞ to 1⅛ in (20 to 27 mm)
DISTRIBUTION	Gulf of California to Panama
ABUNDANCE	Uncommon
DEPTH	Subtidal to 330 ft (100 m)
HABITAT	Muddy bottoms
FEEDING HABIT	Carnivore, scavenger
OPERCULUM	Corneous, oval

SHELL SIZE RANGE
¼ to 1⅛ in
(20 to 27 mm)

PHOTOGRAPHED SHELL
1⅛ in
(27 mm)

421

STROMBINA MACULOSA
BLOTCHY STROMBINA
(SOWERBY I, 1832)

Members of the *Strombina* genus have evolved to cope with the deep muddy bottoms off the shores of tropical western America, to which all but one of their species are confined. There, they live at depths of up to 330 ft (100 m), feeding on the tiny organisms of that habitat. They are characterized by their tall, slender spires, and vary sculpturally—some have smooth whorls and relatively fine sutures, while others such as *Strombina maculosa* have nodulose shoulders.

RELATED SPECIES

Strombina angularis (Sowerby I, 1832) is a rarer, slightly larger shell from the same distribution and habitat as *S. maculosa*. It has a relatively shorter, blunter spire, on which the nodes combine to form rough axial ribs. Although the reticulate mid-brown patterns are similar, *S. angularis* displays them on an off-white to yellowish background.

The shell of the Blotchy Strombina is narrow with a long tapering spire. There is a row of well-defined nodes or tubercles on the shoulders on the spiral whorls. A moderately long, deep canal runs from the aperture, which is thicker behind the clearly dentate outer lip. The shell is white, marked all over with orange-brown maculations.

Actual size

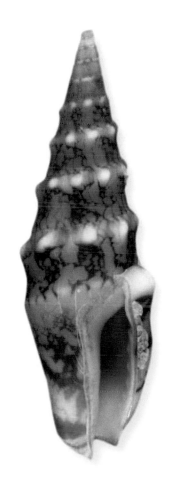

FAMILY	Columbellidae
SHELL SIZE RANGE	¾ to 1¼ in (20 to 30 mm)
DISTRIBUTION	Gulf of California to Peru
ABUNDANCE	Moderately common
DEPTH	Subtidal to 420 ft (128 m)
HABITAT	Mudflats
FEEDING HABIT	Carnivore
OPERCULUM	Corneous, oval

SHELL SIZE RANGE
¾ to 1¼ in
(20 to 30 mm)

PHOTOGRAPHED SHELL
1¼ in
(29 mm)

422

STROMBINA RECURVA
RECURVED STROMBINA
(SOWERBY I, 1832)

Strombina recurva is one of the more highly sculpted species of its genus, and one of the most widely distributed, particularly in its southern extent. It is also a relatively large member of the family Columbellidae, few of whose species extend to more than 1 in (25 mm). *Strombina recurva* lives at depths of up to 420 ft (128 m) on warm offshore mudflats that are rich in the small invertebrates on which this species feeds.

RELATED SPECIES
Strombina fusinoidea (Dall, 1916) resembles *S. recurva* in form and coloring, although its distribution is more restricted at both ends of the range. It is significantly larger, up to 2 in (50 mm) in length. Moreover, it is a much smoother shell, with more rounded shoulders and without the nodal sculpture of *S. recurva*.

Actual size

The shell of the Recurved Strombina is slender with an acute tapering tall spire. It is yellowish to orangey brown in color, and white inside the aperture, which swells behind the outer lip. The siphonal canal is long and recurved. Eight to 10 lirations run from the columella across the body whorl to the lip. The shoulders of the spiral whorls are marked by pronounced nodes that extend toward the apex, forming continuous axial ribs.

FAMILY	Nassariidae
SHELL SIZE RANGE	¼ to 1 in long (5 to 25 mm)
DISTRIBUTION	Southwest Europe, Mediterranean, Black Sea
ABUNDANCE	Common
DEPTH	Subtidal
HABITAT	Mud and sand substrates
FEEDING HABIT	Carnivore
OPERCULUM	Corneous, yellowish

SHELL SIZE RANGE
¼ to 1 in
(5 to 25 mm)

PHOTOGRAPHED SHELL
½ in
(13 mm)

CYCLOPE NERITEA

NERITE MUD SNAIL

(LINNAEUS, 1758)

423

Cyclope neritea is a member of the large family Nassariidae, also known as dog whelks, basket shells, or nassa mud snails. The family's hundreds of species are distributed throughout the world's temperate and tropical seas. Although a few live in deep water, most of the species prefer the warmer shallow waters of intertidal mudflats. *Cyclope neritea* has been expanding its range westward from the Mediterranean, and since the 1970s it has been found on the Atlantic coasts of Portugal and France.

RELATED SPECIES

Cyclope pelucida (Risso, 1826) is a tiny Mediterranean relative of *C. neritea*, only ¼ to ½ in (5 to 12 mm) in length. It is pale tan, with frequent irregular larger or smaller white patches. When larger, the patches may be defined by a darker brown outline, especially on the left-hand edge. The pattern shows through inside the aperture. The thickened lip and columella are pure white.

The shell of the Nerite Mud Snail is lenticular and glossy, white to yellowish with chestnut or darker brown speckling. The short spire is almost completely enveloped by the wide whorl of the body. The aperture is round, with a well-developed parietal shield extending across almost the whole base of the shell. The lip has very small internal folds.

Actual size

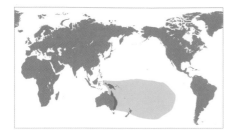

FAMILY	Nassariidae
SHELL SIZE RANGE	⅜ to 1⅛ in (9 to 28 mm)
DISTRIBUTION	Southwest Pacific
ABUNDANCE	Abundant
DEPTH	Intertidal
HABITAT	Mudflats
FEEDING HABIT	Carnivore, scavenger
OPERCULUM	Corneous, thin

SHELL SIZE RANGE
⅜ to 1⅛ in
(9 to 28 mm)

PHOTOGRAPHED SHELL
½ in
(13 mm)

424

NASSARIUS GLOBOSUS
GLOBOSE NASSA
(QUOY AND GAIMARD, 1833)

The shell of the Globose Nassa is chestnut brown in color. It has a moderately short spire with a fairly deep suture and the whorl is almost hemispherical. The whole dorsum is crisscrossed by closely packed axial ribs and spiral cords, giving an impression of netting. Its aperture is oval and surrounded by a well-developed, white parietal shield and outer lip that completely cover the base of the shell.

Nassas are efficient, carnivorous scavengers. They lie in wait below the surface of the mudflats which are their habitat, using their siphons as extensible sensors, the only visible sign above the mud of their presence. Through them they can detect food at distances of up to 100 ft (30 m). *Nassarius globosus* is abundant in large colonies throughout the islands of the southwest Pacific, where such warm, shallow, intertidal flats are plentiful.

RELATED SPECIES
Nassarius pullus (Linnaeus, 1758) is a comparable nassa with a wider distribution across the Indo-West Pacific region. Axial ribs are not broken up by spiral cords as they are on *N. globosus*, nor is the parietal shield quite as extended. The shield is creamier, and there are distinct cream and dark chestnut spiral bands on the body whorl.

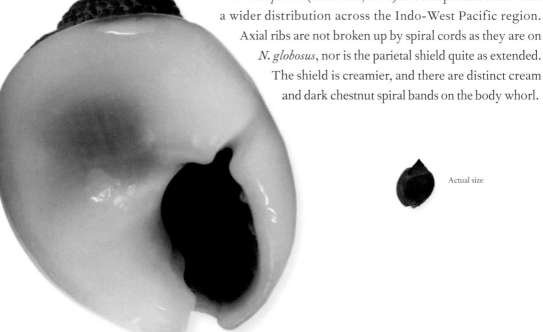

Actual size

FAMILY	Nassariidae
SHELL SIZE RANGE	⅝ to ¾ in (15 to 20 mm)
DISTRIBUTION	Eastern Mediterranean
ABUNDANCE	Common
DEPTH	Intertidal
HABITAT	Sand
FEEDING HABIT	Carnivore, scavenger
OPERCULUM	Corneous, thin, ovate, smaller than the aperture

NASSARIUS GIBBOSULUS
SWOLLEN NASSA
(LINNAEUS, 1758)

SHELL SIZE RANGE
⅝ to ¾ in
(15 to 20 mm)

PHOTOGRAPHED SHELL
⅝ in
(15 mm)

425

Nassarius gibbosulus illustrates the classic feature of the *Nassa* genus, the deep fossa around the base of the shell, which is formed by the recurving outer lip and large parietal shield. It also exhibits the deep siphonal canal typical of the genus. Nassas thrive in warm shallow seas. They are not only efficient opportunistic scavengers of carrion and other detritus, they actively attack bivalves and other snails.

RELATED SPECIES

Nassarius mutabilis (Linnaeus, 1758) shares the same Mediterranean range but is also found off western Africa and the Black Sea. It prefers deeper waters offshore, and the swollen appearance extends to the spiral whorls as well as the body. The fossa is present but less developed and with a smaller parietal shield.

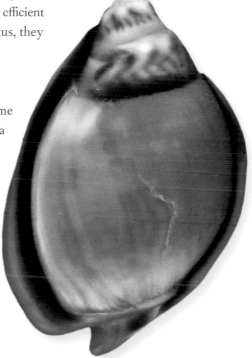

Actual size

The shell of the Swollen Nassa is usually light brown, with a white, much-extended parietal shield, columella, and outer lip that recurve to form a well-defined, dark brown fossa around the base of the shell. Its spire is quite small with a moderately impressed suture, and the body whorl is well rounded and smooth except for growth lines.

FAMILY	Nassariidae
SHELL SIZE RANGE	½ to 1⅛ in (14 to 29 mm)
DISTRIBUTION	Newfoundland to eastern Florida
ABUNDANCE	Common
DEPTH	Offshore to 165 ft (50 m)
HABITAT	Sand
FEEDING HABIT	Carnivore, scavenger
OPERCULUM	Corneous

SHELL SIZE RANGE
½ to 1⅛ in
(14 to 29 mm)

PHOTOGRAPHED SHELL
1⅛ in
(28 mm)

426

NASSARIUS TRIVITTATUS
NEW ENGLAND NASSA
(SAY, 1822)

Nassarius trivittatus is a common species along the Eastern Seaboard. It prefers clean sandy bottoms, where it lives in shallow waters. Its shell has variable coloration; some show the three spiral bands that give it its name. Although common as a beach shell, it is rarely collected alive in the New York City area. Most beached shells have a hole bored on the last whorl, indicating predation by other gastropods, such as moon snails or muricids. Other predators of *N. trivittatus* include crabs and ducks.

RELATED SPECIES

Nassarius clathratus (Born, 1778) is of a similar size and slightly paler color, with a similar sculpture. The axial ribs are more dominant by virtue of being fewer and more widely spaced, and the spiral cords are finer. Furthermore, *N. clathratus* is confined to the other side of the Atlantic, from northwestern Africa into the Mediterranean.

The shell of the New England Nassa is distinctly sculptural, with axial ribs and spiral cords of equal width creating a regular nodular pattern across both the whorl and the tall spire. It is white, tinged pale orange on some extremities. The aperture is nearly circular with a thin outer lip and columella, and a deep siphonal canal.

Actual size

FAMILY	Nassariidae
SHELL SIZE RANGE	¾ to 1⅝ in (18 to 40 mm)
DISTRIBUTION	Indo-West Pacific
ABUNDANCE	Very common
DEPTH	Intertidal to 6 ft (2 m)
HABITAT	Sandy bays and mudflats
FEEDING HABIT	Carnivore, scavenger
OPERCULUM	Corneous

NASSARIUS ARCULARIUS

CAKE NASSA
(LINNAEUS, 1758)

SHELL SIZE RANGE
¾ to 1⅝ in
(18 to 40 mm)

PHOTOGRAPHED SHELL
1⅝ in
(33 mm)

427

The distribution of *Nassarius arcularius* reflects clearly the preference of most nassa species for warm shallow water. The protected sandy bays of the islands of the western Pacific are ideal for these scavengers. They use their divided foot to plow through the sand floor in search of detritus and small organisms, and to conceal themselves below the surface, where they lie in wait for passing food. Shallow bays minimize the chances of disturbance by waves and exposure.

RELATED SPECIES

Nassarius hirtus (Kiener, 1834) to the east, in depths of up to 60 ft (20 m) in Polynesia and Hawaii, and *N. distortus* (Adams, 1852) to the west, in shallow water in the Indo-Pacific, share the coloring and the prominent axial ribs. Both shells have taller spires than *N. arcularius*, and neither have as developed a parietal shield as that species.

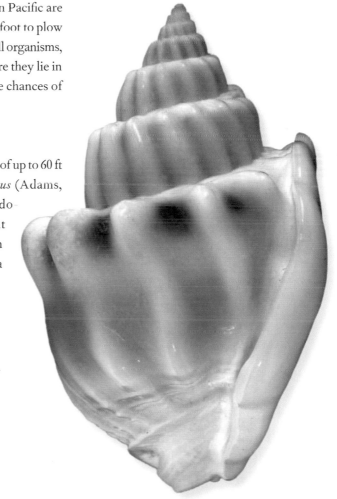

The shell of the Cake Nassa is distinguished by deep axial ribs across the whole body and spire. Well-defined sutures and the very glossy white surface give the impression of an iced, multi-tiered cake or jelly mold. The parietal shield covers the whole base of the shell. The inner lip is strongly lirate, and the interior is a streaky brown.

Actual size

FAMILY	Nassariidae
SHELL SIZE RANGE	⅝ to 2 in (15 to 50 mm)
DISTRIBUTION	Indo-Pacific, Réunion
ABUNDANCE	Common
DEPTH	Intertidal
HABITAT	Sand and mud flats
FEEDING HABIT	Carnivore, scavenger
OPERCULUM	Corneous

SHELL SIZE RANGE
⅝ to 2 in
(15 to 50 mm)

PHOTOGRAPHED SHELL
1½ in
(36 mm)

428

NASSARIUS GLANS
GLANS NASSA
(LINNAEUS, 1758)

Nassarius glans is one of the larger species of the genus and one of the most boldly patterned. Its shiny, tall spire may have given rise to its species name, but it is the distinctive bold red spiral line that is most readily identifiable. Like all nassas, *N. glans* is a predator as well as a scavenger, feeding on other snails, bivalves, and opistobranch eggs as well as algae and rotting flesh (for which it has a particularly keen sense of smell).

RELATED SPECIES

Several of the tulip genus *Fasciolaria*, found in deep water in the northern Gulf of Mexico, have similar spiral patterns to *N. glans*. The tulip form is, however, very different, with a leaf-shaped aperture and much extended anterior canal. In addition, tulips have no labial teeth at the ends of the spiral lines.

The shell of the Glans Nassa is a glossy cream in color with some orange staining. Its body whorl is bulbous, with a moderately tall spire that has a fairly deeply impressed suture. The higher spire whorls have narrow axial ribs and the apex is dark red. The whole shell is patterned with dark, reddish brown, spiral lines which begin from small lip teeth. There are about 6 fine cords on the fasciole.

Actual size

FAMILY	Nassariidae
SHELL SIZE RANGE	1 to 2 in (27 to 50 mm)
DISTRIBUTION	Southern Africa, Indian Ocean, western Australia
ABUNDANCE	Common
DEPTH	Intertidal
HABITAT	Mudflats
FEEDING HABIT	Carnivore
OPERCULUM	Corneous

SHELL SIZE RANGE
1 to 2 in
(27 to 50 mm)

PHOTOGRAPHED SHELL
1½ in
(38 mm)

BULLIA LIVIDA

RIBBON BULLIA

REEVE, 1846

Bullia livida is widely distributed from southern Africa to western Australia around the rim of the Indian Ocean. This broad distribution has given rise to more synonyms than most; the shell is also known as *B. plicata*, *B. vittata*, *Ancilla alba*, *Eburna monilis*, and *Terebra buccinoidea*, and by other common names. *Bullia livida* has a tall spire with a distinctive spiral band of short but strong axial ribs along the suture. This band is sometimes crossed by an incised spiral groove.

RELATED SPECIES

Many other bullias are locally common in parts of *B. livida*'s range. *Bullia tenuis* (Reeve, 1846), *B. annulata* (Lamarck, 1816), and *B. callosa* (Wood, 1828), for example, are all found in deeper water off the shores of South Africa. *Bullia kurrachensis* (Angas, 1877), the Karachi bullia, is found predominantly in shallow water near that city. Another group of bullias is distributed along the eastern shores of South America.

The shell of the Ribbon Bullia is variable in color from pale yellow to gray-lilac. The body whorl follows the line of the tall spire. There is a fairly deeply impressed suture, immediately below which the shoulders are decorated with a continuous row of short axial ribs like the tasseled edge of a ribbon. These ribs are terminated at their lower end by a fine spiral groove. The thickened lip forms a modest fossa, and the fasciole is pleated.

Actual size

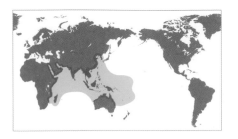

FAMILY	Nassariidae
SHELL SIZE RANGE	1¼ to 2 in (30 to 52 mm)
DISTRIBUTION	Indo-West Pacific
ABUNDANCE	Moderately common
DEPTH	Intertidal to 30 ft (10 m)
HABITAT	Mudflats
FEEDING HABIT	Carnivore, scavenger
OPERCULUM	Corneous

SHELL SIZE RANGE
1¼ to 2 in
(30 to 52 mm)

PHOTOGRAPHED SHELL
1¾ in
(44 mm)

430

NASSARIUS PAPILLOSUS
PIMPLED NASSA
(LINNAEUS, 1758)

Nassarius papillosus is a common nassariid, with a wide Indo-Pacific distribution. It ranges from the intertidal zone to shallow offshore depths. Some nassariids are used as food in southeast Asia. However, a recent case of shellfish poisoning in Taiwan revealed that *N. papillosus* can sometimes contain potentially dangerous levels of the potent neurotoxin Tetrodotoxin.

RELATED SPECIES

Nassarius coronatus (Bruguière, 1789) from the western half of *N. papillosus*' distribution, shares its pimpled sculpturing. It has a smaller, squatter shell with a fairly large parietal shield covered with columellar folds, echoed in the labial folds opposite it.

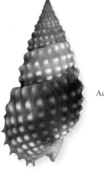

Actual size

The shell of the Pimpled Nassa is covered in deep axial ribs, which are cut through by equally deep spiral grooves that rise from labial teeth. This yields a regular sculptural grid of white pimples behind which the glossy shell can be stained light or chestnut brown. The spire is quite tall; the spiral whorls are defined by a suture slightly deeper than the spiral grooves. The aperture is almost circular.

FAMILY	Melongenidae
SHELL SIZE RANGE	1 to 2¾ in (25 to 70 mm)
DISTRIBUTION	Indian Ocean
ABUNDANCE	Abundant
DEPTH	Intertidal to 6 ft (2 m)
HABITAT	Sand and mud flats
FEEDING HABIT	Carnivore
OPERCULUM	Corneous, oval, large

VOLEMA PARADISIACA

PEAR MELONGENA

RÖDING, 1798

SHELL SIZE RANGE
1 to 2¾ in
(25 to 70 mm)

PHOTOGRAPHED SHELL
1⅞ in
(47 mm)

431

Volema paradisiaca belongs to the relatively small family Melongenidae, known as the melon conchs. There about 30 species worldwide, mostly in tropical waters, although some live also in temperate regions. *Volema paradisiaca* is more common along the coast of eastern Africa. It shares the genus' preference for muddy or brackish water, often in or near mangroves. Like all Melongenidae it is a voracious carnivore, although some species also feed on carrion.

RELATED SPECIES

Volema myristica (Röding, 1798) is a slightly larger but similar form. The spire is slightly taller, and the shoulder knobs have developed into small spines. The spiral grooves are deeper and the wavy edge of the lip is also more clearly defined. *Volema myristica* occurs farther east than *V. paradisiaca*, in the southwestern Pacific.

Actual size

The shell of the Pear Melongena is pear-shaped, with a moderately low spire. Its body whorl is covered in shallow spiral grooves, and the shoulder bears more or less poorly defined knobs. Its lip is thin with a wavy edge. The aperture is large, ovate, and orange with a well-defined narrow posterior canal and a smooth columella. The overall color varies from yellowish to reddish brown, sometimes with faint spiral banding.

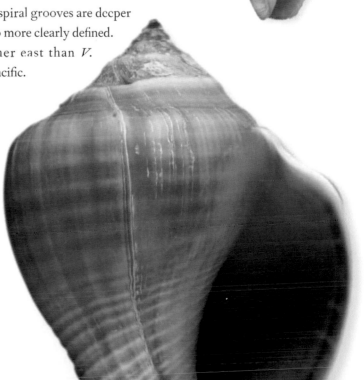

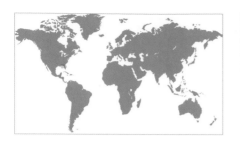

FAMILY	Melongenidae
SHELL SIZE RANGE	1 to 8 in (25 to 205 mm)
DISTRIBUTION	Alabama to northeastern Florida
ABUNDANCE	Locally very common
DEPTH	Intertidal
HABITAT	Mangroves
FEEDING HABIT	Carnivore
OPERCULUM	Corneous, oval, large

SHELL SIZE RANGE
1 to 8 in
(25 to 205 mm)

PHOTOGRAPHED SHELL
2¼ in
(56 mm)

432

MELONGENA CORONA

COMMON CROWN CONCH

(GMELIN 1791)

The shell of the Common Crown Conch is bulbous with a medium spire. The suture is well impressed and the whorls angulated, with 2 spiral rows of curved open spines. A further row of spines occurs on the lower half of the body. The shell has faint axial ridges and is usually decorated with spiral bands of variable width and color from light to purplish brown (although all-cream specimens do occur).

This spectacular shell has a relatively narrow distribution ranging from northeastern Florida, around the Florida peninsula to Mobile Bay, Alabama, but is locally very common in the brackish water of mangrove areas. It is an aggressive predator, attacking clams, oysters, and other bivalves. Although the family Melongenidae is small in number, its 30 or so species are farflung, occurring on both sides of the tropical and subtropical Indian, Pacific, and Atlantic oceans.

RELATED SPECIES

Melongena melongena (Linnaeus, 1758) is a large species from the West Indies. The spire is very small and lacks spines altogether, but the body whorl can have up to four spiral rows of them. The parietal wall is more extended than on *M. corona*. Also carnivore, its diet includes gastropods.

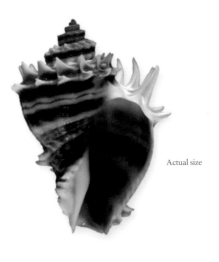

Actual size

FAMILY	Melongenidae
SHELL SIZE RANGE	3 to 10½ in (75 to 270 mm)
DISTRIBUTION	Trinidad to Brazil; Mauritania to Angola
ABUNDANCE	Common
DEPTH	Intertidal to shallow subtidal
HABITAT	Mangroves
FEEDING HABIT	Carnivore
OPERCULUM	Corneous, oval, large

SHELL SIZE RANGE
3 to 10½ in
(75 to 270 mm)

PHOTOGRAPHED SHELL
3½ in
(89 mm)

PUGILINA MORIO

GIANT HAIRY MELONGENA

(LINNAEUS, 1758)

433

Pugilinia morio has an unusual distribution, occurring on both sides of the Atlantic Ocean. It thrives in the shallow, muddy waters of mangrove zones, where it feeds on bivalves and carrion. Females usually have wider and more nodulose shells (like the shell illustrated here), while males have smoother shells. The popular name derives from the thick and hairy periostracum that covers the shell of live specimens.

RELATED SPECIES

Busycon carica (Gmelin, 1791), with a similar form, is another melongenid. It is larger, creamy or orangey white, and has a longer anterior canal, but the aperture and distribution of nodes are much the same. It is found from Florida to Massachusetts, where beads of its columella were used as money by the Native American population.

The shell of the Giant Hairy Melongena is dark brown to greenish black, and spindle-like. Its spire is moderately tall with steplike whorls, the shoulders emphasized by large nodules, more particularly on younger specimens. The whole shell is covered in rough, narrow spiral cords and a few pale to mid brown bands. There are labial spiral ridges inside the wide oval aperture. The anterior canal is moderately long.

Actual size

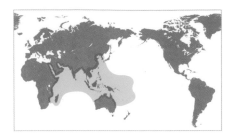

FAMILY	Melongenidae
SHELL SIZE RANGE	2⅜ to 6 in (60 to 150 mm)
DISTRIBUTION	Indo-West Pacific
ABUNDANCE	Common
DEPTH	Intertidal to 6 ft (2 m)
HABITAT	Muddy water
FEEDING HABIT	Carnivore
OPERCULUM	Corneous

SHELL SIZE RANGE
2⅜ to 6 in
(60 to 150 mm)

PHOTOGRAPHED SHELL
4¾ in
(119 mm)

434

PUGILINA COCHLIDIUM
SPIRAL MELONGENA
(LINNAEUS, 1758)

Like other members of Melongenidae, *Pugilina cochlidium* is covered by a thick periostracum when alive. The habitat of shallow muddy waters is rich in nutrition for the bivalves that are the prey of all species of the aggressively carnivorous melongenid. *Pugilina cochlidium* gets its common name from the sequence of nodes on its shoulders from the body whorl to the apex. While not particularly steplike, their even spacing suggests a spiral staircase, albeit a narrowing, tightening one.

RELATED SPECIES
Volema myristica (Röding, 1798) is a smaller but similar form. The shoulder nodules are less pronounced and there is no spine at the lip. Instead, the lip has a wavy edge which gives rise to spiral cords on the body whorl.

The shell of the Spiral Melongena is very thick and heavy, and orange to chestnut brown, sometimes with darker, axial growth lines. The shape is spindle-like, the body bulbous with faint spiral cords, tapering to a fairly long anterior canal. It has a moderate spire, with evenly spaced ribs or nodules on the shoulders of all whorls ending in a spine at the top of the lip. The aperture is smooth with faint labial folds within; the fasciole also folded.

Actual size

FAMILY	Melongenidae
SHELL SIZE RANGE	2¾ to 10 in (70 to 250 mm)
DISTRIBUTION	Indo-West Pacific
ABUNDANCE	Common
DEPTH	Shallow subtidal to 133 ft (40 m)
HABITAT	Sandy bottoms
FEEDING HABIT	Carnivore
OPERCULUM	Corneous, elongate, curved

SHELL SIZE RANGE
2¾ to 10 in
(70 to 250 mm)

PHOTOGRAPHED SHELL
7¼ in
(183 mm)

HEMIFUSUS CRASSICAUDUS
THICK-TAIL FALSE FUSUS
(PHILIPPI, 1848)

435

Hemifusus crassicaudus is one of the largest species in the family Melongenidae. While most specimens will reach up to 10 in (250 mm) in length, the world record shell size is more than 16 in (410 mm) long. *Hemifusus crassicaudus* is a common melongenid throughout the Indo-Pacific. This species is a predator of bivalves, and is often found on bivalve beds.

RELATED SPECIES

Hemifusus colosseus (Lamarck, 1822), which ranges from Japan to the East China Sea, has a large and slender shell with a tall spire and even more elongate aperture than *H. crassicaudus*. *Pugilina cochlideum* (Linnaeus, 1758), from the Indian Ocean, northern Australia, and the Philippines, has a solid and heavy shell that resembles *H. crassicaudus* but is smaller and broader, with a broader siphonal canal.

The shell of the Thick-tail False Fusus is large, thick, heavy, rough, and elongately fusiform. Its spire is tall, with well-marked suture, a pointed apex, and angular whorls with pointed nodules at the shoulder. Its sculpture consists of wide spiral ribs, crossed by growth marks. The aperture is large, elongate, and drawn into an open, long siphonal canal. The outer lip is thin and slightly crenulated, and the columella is smooth, with small folds in the siphonal canal. The shell is pinkish white, covered with a thick brown periostracum. The aperture is pinkish.

Actual size

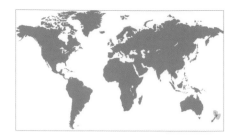

FAMILY	Fasciolariidae
SHELL SIZE RANGE	½ to ¾ in (12 to 19 mm)
DISTRIBUTION	Endemic to the northern island of New Zealand
ABUNDANCE	Very common
DEPTH	Intertidal
HABITAT	On and under rocks
FEEDING HABIT	Carnivore
OPERCULUM	Corneous, leaf-shaped

SHELL SIZE RANGE
½ to ¾ in
(12 to 19 mm)

PHOTOGRAPHED SHELL
½ in
(12 mm)

436

TARON DUBIUS
DUBIUS SPINDLE SHELL
(HUTTON, 1878)

The family Fasciolariidae comprises a diverse group of predatory snails that have fusiform shells with a siphonal canal aligned with the shell axis. Most inhabit tropical and temperate regions, at intertidal to bathyal depths. Smaller species tend to feed on polychaete worms, large species on bivalves and other gastropods. *Taron dubius* is among the smallest members of the Fasciolariidae. It is the type species of the genus, which is endemic to the northern island of New Zealand.

RELATED SPECIES

Taron mouatae Powell, 1968, has a larger, thinner, more elongate shell with a higher spire and finer, more numerous spiral cords and axial ribs. The shell color is a darker brown. *Taron albocastus* Ponder, 1968, is the smallest member of this genus. Its shell is narrower than that of *T. dubius*, and is yellowish white, with dark brown spiral cords.

Actual size

The shell of the Dubius Spindle Shell is small and broadly fusiform, with a conical spire that comprises half the shell length, a simple ovate aperture with a short, broad, siphonal canal. The protoconch is smooth. Early whorls of the adult shell have weak spiral cords that become stronger and crossed by progressively more prominent axial ribs. The outer lip of the aperture is thin and brown, the columella is smooth and white, and the siphonal canal orange. Like most fasciolariids, the animal is bright red.

FAMILY	Fasciolariidae
SHELL SIZE RANGE	1 to 1½ in (25 to 37 mm)
DISTRIBUTION	Tropical Indo-West Pacific
ABUNDANCE	Common
DEPTH	Intertidal to subtidal
HABITAT	On coral reef flats
FEEDING HABIT	Carnivore
OPERCULUM	Corneous, leaf-shaped

SHELL SIZE RANGE
1 to 1½ in
(25 to 37 mm)

PHOTOGRAPHED SHELL
1¼ in
(32 mm)

PERISTERNIA NASSATULA

NETTED PERISTERNIA

(LAMARCK, 1822)

Species of the *Peristernia* genus have heavy, colorful, strongly ribbed, nodulose shells that are often pseudo-umbilicate. The columella typically has two or three transverse folds in addition to the siphonal fold. Species are subtidal and epifaunal, inhabiting rocks and reefs, where they feed primarily on tube-dwelling polychaete worms and sipunculans. Most specimens are heavily encrusted with other reef organisms.

RELATED SPECIES

Peristernia australiensis (Reeve, 1847), from Queensland, is a slightly smaller shell with fewer, larger, and more rounded axial folds and a more quadrate aperture. The Indo-Pacific species *P. fastigium* (Reeve, 1847) is more elongated and has a longer siphonal canal. *Peristernia chlorostoma* (Sowerby I, 1825), from the Indo-Pacific, is much smaller with a shorter spire and a more rounded shell.

Actual size

The shell of the Netted Peristernia has a broadly biconical, sharply shouldered shell with a tall, even spire, an ovate aperture, and a short siphonal canal deflected to the left. Shell sculpture consists of numerous broad, closely spaced axial ribs that extend evenly from the suture to the siphonal canal. Sharp, narrow spiral cords run across the axial ribs. Finer spiral threads may occur between adjacent spiral cords. The aperture has an anteriorly reflected outer lip, a posterior notch, and 2 weak folds just posterior to the siphonal fold. The shell color is variable, ranging from white, to yellow, purplish, or brown, and may be banded.

FAMILY	Fasciolariidae
SHELL SIZE RANGE	1 to 2½ in (25 to 65 mm)
DISTRIBUTION	Japan to Philippines
ABUNDANCE	Uncommon
DEPTH	165 to 1,000 ft (50 to 300 m)
HABITAT	Sandy bottoms
FEEDING HABIT	Carnivore
OPERCULUM	Corneous, small, round

SHELL SIZE RANGE
1 to 2½ in
(25 to 65 mm)

PHOTOGRAPHED SHELL
1¾ in
(47 mm)

438

GRANULIFUSUS NIPONICUS
GRANULAR SPINDLE
(SMITH, 1879)

The shell of the Granular Spindle is thin, broadly fusiform, and moderate in size, with a tall spire, roundly ovate aperture, and a long, narrow, axial siphonal canal. The whorls are evenly rounded without a sharp shoulder. Axial ribs are broad, but the spaces between them are slightly broader. The spiral cords are thickest where they cross the axial ribs and at the outer lip, where they produce a crenulated edge. The shell color is tan, occasionally with brown spiral bands. The nodes where the axial and spiral sculpture intersect are white.

The 21 living species of the genus *Granulifusus* are distributed throughout the tropical Indo-West Pacific, usually at bathyal depths. They live on sand bottoms, where they prey on a wide range of invertebrates, including worms and mollusks. The group is distinguished from *Fusinus* by its granulated surface sculpture, and by having a small, round operculum. Unlike most fasciolariids, the animal of *Granulifusus* is white or yellowish rather than red. The round operculum is vestigial.

RELATED SPECIES

The shell of *Granulifusus rubrolineatus* (Sowerby II, 1870), an eastern African species, is similar in shape, but smaller, with narrower, finer spiral cords. The Japanese species *Granulifusus hayashii* Habe, 1961, has a much narrower shell, with a proportionally longer siphonal canal that is deflected to the left.

Actual size

FAMILY	Fasciolariidae
SHELL SIZE RANGE	1½ to 2 in (37 to 50 mm)
DISTRIBUTION	Tropical Indo-West Pacific
ABUNDANCE	Moderately common
DEPTH	Subtidal to 60 ft (20 m)
HABITAT	Rocky bottoms
FEEDING HABIT	Carnivore
OPERCULUM	Corneous, leaf-shaped

SHELL SIZE RANGE
1½ to 2 in
(37 to 50 mm)

PHOTOGRAPHED SHELL
1¾ in
(47 mm)

TURRILATIRUS TURRITUS

TOWER LATIRUS

(GMELIN, 1791)

439

The genus *Turrilatirus* comprises a small distinctive group of species that have a thick shell with a tall spire and a short siphonal canal. It ranges throughout the tropical Indo-Pacific, from eastern Africa to Polynesia. The earliest known records of this genus are from the Pliocene Period. These animals have a short, broad, bright red foot with a thick, corneous operculum, and live on rocky bottoms in fairly shallow water.

RELATED SPECIES

Turrilatirus nagasakiensis (Smith, 1880), from southern Japan, is similar in size, but has a more elongated aperture and a longer, axial siphonal canal. *Latirus belcheri* (Reeve, 1847) may reach twice the size, and is distinguished by its rectangular aperture, sharp shoulder, and basal angulations connected by rounded axial ribs, and a short, sharply demarcated, axial siphonal canal.

The shell of the Tower Latirus is stout, elongated, and of moderate size. The spire is tall and conical, comprising more than half the shell length. The aperture is elliptical, with a constriction at its juncture with the short, narrow siphonal canal, which is deflected to the left. The aperture has multiple, paired, spiral lirae within the outer lip and 2 weak folds above the siphonal fold. There is a short siphonal fasciole. The sculpture consists of low, broad axial ribs and broad, raised spiral cords. The base color is yellowish orange; the spiral cords are dark brown.

Actual size

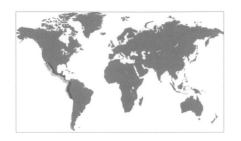

FAMILY	Fasciolariidae
SHELL SIZE RANGE	1 to 3 in (25 to 75 mm)
DISTRIBUTION	Baja California to Peru
ABUNDANCE	Common
DEPTH	Intertidal to shallow subtidal
HABITAT	Rocky shores
FEEDING HABIT	Carnivore, feeds on barnacles and bivalves
OPERCULUM	Thick, corneous

SHELL SIZE RANGE
1 to 3 in
(25 to 75 mm)

PHOTOGRAPHED SHELL
2⅛ in
(53 mm)

440

OPEATOSTOMA PSEUDODON
THORN LATIRUS
(BURROW, 1815)

Actual size

While most fasciolariids have elongate fusiform shells, *Opeatostoma pseudodon* has a broad and stout shell, with alternating brown and white spiral bands. It is unique among the fasciolariids in having a long, curved spine on the anterior margin of the outer lip that makes it easily recognized. It is a predator of clams and barnacles, and is found from the intertidal zone to offshore, on rocky shores. There are over 200 living species in the family Fasciolariidae worldwide, in tropical and subtropical regions. The fossil record of the family extends back to the Cretaceous Period.

RELATED SPECIES

Opeatostoma pseudodon is the only species in its genus. Other related species include *Leucozonia cerata* Wood, 1828, from the Galápagos Islands and western Central America, which has a nodulose fusiform shell. *Cyrtulus serotinus* Hinds, 1844, which is endemic to the Marquesas Island, has a uniquely shaped shell, with a narrow and tall spire, and a thick body whorl.

The shell of the Thorn Latirus is medium in size, solid, thick, heavy, inflated, and broadly fusiform. Its spire is relatively short, with suture with a callus. The whorls have an angular shoulder, and the sculpture consists of slightly elevated, dark brown, spiral cords on a white shell. The aperture is wide, the outer lip crenulate, and the columella has 2 to 3 folds. There is a single, short or long, curved spine at the bottom of the outer lip. Its periostracum is tan, and the aperture is white.

FAMILY	Fasciolariidae
SHELL SIZE RANGE	1¼ to 3½ in (30 to 90 mm)
DISTRIBUTION	Indo-West Pacific
ABUNDANCE	Common
DEPTH	Intertidal to 60 ft (18 m)
HABITAT	Rocks and corals
FEEDING HABIT	Carnivore, feeds on other invertebrates
OPERCULUM	Corneous, thick

SHELL SIZE RANGE
1¼ to 3½ in
(30 to 90 mm)

PHOTOGRAPHED SHELL
2¼ in
(56 mm)

LATIRUS BELCHERI

BELCHER'S LATIRUS

(REEVE, 1847)

441

Latirus belcheri is a common intertidal species in Australia, where it lives among rocks and corals. It is also found in the Indo-West Pacific, in depths of up to 60 ft (18 m). This species is one of several named after Sir Edward Belcher, a British admiral and an avid shell collector. Fasciolariids are carnivores and feed on other invertebrates; the smaller species feed on polychaetes, and the larger ones prey on bivalves and gastropods. Cannibalism is known in some species. Females lay stalked egg capsules with numerous eggs in each capsule.

RELATED SPECIES

Latirus infundibulum (Gmelin, 1791), which ranges from Florida to Brazil, has a tall-spired, thick shell. It has a golden background with raised axial nodules crossed by brown spiral cords. *Opeatostoma pseudodon* (Burrow, 1815), from Baja California to Peru, has a long tooth extending from the lower outer lip. Its white shell is lined with many elevated, dark brown spiral cords.

The shell of the Belcher's Latirus is medium in size, thick, solid, and broadly fusiform in outline. Its spire is tall, with a poorly marked suture. Its spire whorls have 1 row of large tubercles, and the body whorl has 2 rows of large tubercles and several spiral ribs. The aperture is large and squarish, with 2 sharp angles, the columella has 3 to 4 small folds. The siphonal canal is wide and slightly recurved. The shell color is white or cream with brown blotches, and the aperture is white with a dark margin.

Actual size

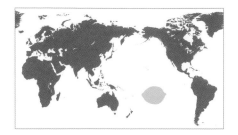

FAMILY	Fasciolariidae
SHELL SIZE RANGE	2 to 3¾ in (50 to 94 mm)
DISTRIBUTION	Polynesia
ABUNDANCE	Uncommon
DEPTH	60 to 100 ft (18 to 30 m)
HABITAT	Fine sand and silt bottoms
FEEDING HABIT	Carnivore, feeds on other mollusks
OPERCULUM	Corneous, concentric, elliptical

SHELL SIZE RANGE
2 to 3¾ in
(50 to 94 mm)

PHOTOGRAPHED SHELL
3 in
(75 mm)

CYRTULUS SEROTINUS
CYRTULUS SPINDLE
HINDS, 1844

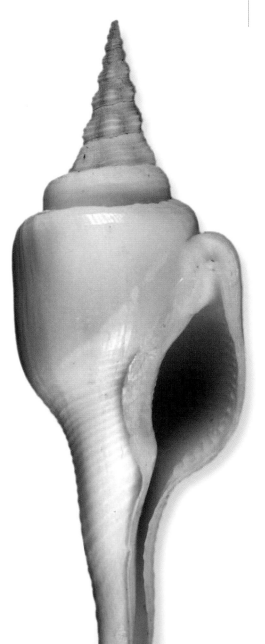

Cyrtulus serotinus has an uncommon and strange-looking shell. Juvenile specimens look like the fusiform *Fusinus colus* in shape and sculpture, but after about seven or eight whorls, the whorls thicken, become irregular, obliterating the features of the juvenile whorls. Adult specimens look more like a turbinellid, resembling a skinny Indian Chank with a *Fusinus*-like high spire. *Cyrtulus serotinus* lives on fine sand or silt bottoms offshore in Polynesia. It is the only species in its genus.

RELATED SPECIES

Other related species have a more typical fusiform shape, such as *Fusinus colus* (Linnaeus, 1758), from the Indo-West Pacific; *F. longissimus* (Gmelin, 1791), from Japan and West Pacific, one of the largest species in its genus, has a row of rounded nodules on the shoulder; and *F. syracusanus* (Linnaeus, 1758), from the Mediterranean to the Canary Islands, has a colorful shell, an exception in a family of mostly white shells.

The shell of the Cyrtulus Spindle is heavy, solid, and club-shaped but with a high spire, which is typical of *Fusinus* species. After 7 to 8 whorls, the whorls become thicker and lose all sculpture of the spire, wrapping around the shell irregularly. The adult shell has an angular body whorl that is cylindrical, with nearly parallel sides in the middle third of the shell, tapering toward a long and thick siphonal canal. Its aperture is elliptically elongate with a thick lip. The columella is smooth and thick, the inner lip reflected. The shell color is cream; the aperture is white.

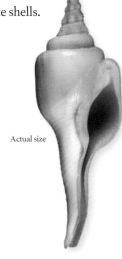

Actual size

FAMILY	Fasciolariidae
SHELL SIZE RANGE	3 to 7 in (75 to 180 mm)
DISTRIBUTION	Indo-West Pacific
ABUNDANCE	Common
DEPTH	Shallow subtidal, to 130 ft (40 m)
HABITAT	Sandy bottoms
FEEDING HABIT	Carnivore, feeds on other mollusks
OPERCULUM	Corneous

SHELL SIZE RANGE
3 to 7 in
(75 to 180 mm)

PHOTOGRAPHED SHELL
3¾ in
(96 mm)

FUSINUS NICOBARICUS
NICOBAR SPINDLE
(RÖDING, 1798)

443

With its brown axial markings, the Nicobar Spindle is one of the most colorful of the *Fusinus* genus. These markings combined with its long, elegant shape make *F. nicobaricus* a popular species with shell collectors. A carnivorous snail, *F. nicobaricus* feeds on smaller mollusks, which it finds on generally sandy bottoms amid rock and coral debris. It is commonly found in pairs.

RELATED SPECIES
Fusinus colus (Linnaeus, 1758), shares a similar range but is generally longer and thinner with fewer markings. The thinner siphonal canal is often brown, growing darker toward the tip. *Fusinus marmoratus* (Philippi, 1851), from eastern Brazil, the Mediterranean, and the Red Sea, is generally shorter and darker overall. The prominent axial ribs continue through to later body whorls.

The shell of the Nicobar Spindle has a long spire and siphonal canal, with the canal being slightly shorter than the spire. Posterior whorls have axial ridges, making them appear somewhat rounded; in anterior whorls the axial ridges are more nodulose, giving the shell an angular appearance. The aperture is white, with ridges showing through the thin columella callus. The lip is finely dentate.

Actual size

FAMILY	Fasciolariidae
SHELL SIZE RANGE	2½ to 10 in (60 to 250 mm)
DISTRIBUTION	North Carolina to Northern Brazil
ABUNDANCE	Common
DEPTH	Intertidal to 240 ft (75 m)
HABITAT	Sandy bottoms and seagrass beds
FEEDING HABIT	Carnivore, feeds on other mollusks
OPERCULUM	Corneous, thick

SHELL SIZE RANGE
2½ to 10 in
(60 to 250 mm)

PHOTOGRAPHED SHELL
5½ in
(141 mm)

444

FASCIOLARIA TULIPA
TRUE TULIP
(LINNAEUS, 1758)

Fasciolaria tulipa is a large fasciolariid that is commonly found on seagrass beds in bays and estuaries. It is a voracious carnivore and feeds on other mollusks, including other species of *Fasciolaria*, such as *F. lillium*, or even juveniles of *Strombus gigas*. The animal ranges in color from bright orange to red. Its flesh is edible, and collected and consumed locally. In the winter, females lay clustered, urn-shaped egg capsules. The operculum is corneous and thick, and fits snugly inside the aperture. Shells of the True Tulip are among the most common shells used by hermit crabs in bays.

RELATED SPECIES

Fasciolaria lillium Fischer, 1807, from Texas to Quintana Roo, Mexico, has a similar but smaller shell, which has fine, dark brown lines and pinkish or brownish blotches on a cream background. *Triplofusus giganteus* (Kiener, 1840), from North Carolina to Quintana Roo, Mexico, is the largest fasciolariid, and the largest gastropod in the Atlantic Ocean. It is the state shell of Florida.

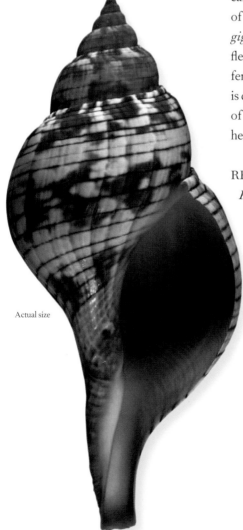

Actual size

The shell of the True Tulip is medium-large, slightly swollen, and fusiform in outline. Its spire is tall, with convex whorls and well-marked suture. The aperture is elliptical, the outer lip is dentate, and the columella has 2 spiral ridges. The siphonal canal is relatively short and broad. The surface is smooth, with very fine growth lines, and low spiral ridges at the base. The shell color is variable, with a cream background and light brown to reddish orange blotches, and fine, black spiral lines. The aperture is white or orange.

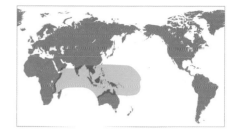

FAMILY	Fasciolariidae
SHELL SIZE RANGE	3¾ to 8⅜ in (95 to 220 mm)
DISTRIBUTION	Tropical Indo-West Pacific
ABUNDANCE	Uncommon
DEPTH	165 to 400 ft (50 to 120 m)
HABITAT	Sandy bottoms
FEEDING HABIT	Carnivore
OPERCULUM	Corneous, leaf-shaped

SHELL SIZE RANGE
3¾ to 8⅜ in
(95 to 220 mm)

PHOTOGRAPHED SHELL
6⅜ in
(161 mm)

FUSINUS CRASSIPLICATUS

RIBBED SPINDLE

KIRA, 1959

The genus *Fusinus* ranges throughout the temperate and tropical seas of the world. Many species are restricted to deeper waters along the outer continental shelf or continental slope, where they inhabit sandy bottoms. The animals have a red foot, often with yellow dots near the margin. Females deposit eggs in leathery, urn-shaped egg capsules that are attached to hard objects such as rocks or shell fragments.

RELATED SPECIES

Fusinus salisburyi Fulton, 1930, also from the western Pacific, is a larger species that has a pronounced shoulder with a prominent cord. *Fusinus perplexus* (A. Adams, 1864), from Japan and Taiwan, is slightly smaller, with a proportionally shorter siphonal canal and larger aperture. It lacks prominent axial sculpture on the final whorl and has a weak shoulder.

The shell of the Ribbed Spindle is very large, narrow, and fusiform with a tall spire, and a long, axial siphonal canal that is half the length of the shell. The aperture is small and ovate. The columella is smooth, and may have a raised fold in some specimens. Prominent axial folds that do not extend onto the siphonal canal are the dominant shell sculpture. Sharp spiral cords run over the folds on the body whorl, and are the only sculpture on the siphonal canal. The shell is white, but may have bands of tan between adjacent axial folds.

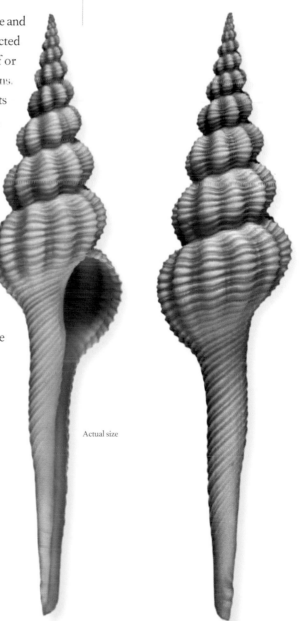

Actual size

FAMILY	Fasciolariidae
SHELL SIZE RANGE	4 to 10 in (100 to 250 mm)
DISTRIBUTION	Baja California to Ecuador
ABUNDANCE	Moderately common
DEPTH	Intertidal to 330 ft (100 m)
HABITAT	Sandy or clay bottoms
FEEDING HABIT	Carnivore, feeds on other mollusks
OPERCULUM	Corneous, thick, ovate

SHELL SIZE RANGE
4 to 10 in
(100 to 250 mm)

PHOTOGRAPHED SHELL
8⅞ in
(223 mm)

446

FUSINUS DUPETITTHOUARSI
DU PETIT'S SPINDLE
KIENER, 1840

Fusinus dupetitthouarsi has one of the largest shell in the genus *Fusinus*. In certain regions, notably Mexico, it is used as fishing bait, but rarely as a source of food. Fresh specimens are notable for a greenish brown, thin periostracum layer that covers the outside of the shell. A carnivore, the Du Petit's Spindle feeds on small mollusks that it finds in sandy or clay substrates.

RELATED SPECIES

Fusinus panamensis (Dall, 1908), from Mexico to Peru, is stouter with more regular brown markings on knobs of axial ridges. *Fusinus stegeri* Lyons, 1978, from the Gulf of Mexico and the Florida Keys, has a taller spire and more rounded whorls.

Actual size

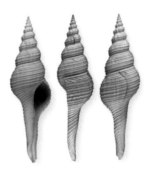

The shell of the Du Petit's Spindle has a long spire and siphonal canal. The siphonal canal and all the whorls have clearly defined spiral ridges, which on larger anterior whorls are most prominent at the periphery, where they may appear nodulose. Faint traces of spiral ribs are apparent inside the white aperture and on the otherwise smooth columella. The shell is white with occasional pale brown streaking, growing darker at the tip of the spire.

FAMILY	Fasciolariidae
SHELL SIZE RANGE	16 to 24 in (400 to 609 mm)
DISTRIBUTION	North Carolina to the Gulf of Mexico
ABUNDANCE	Common
DEPTH	Intertidal to 100 ft (30 m)
HABITAT	Seagrass beds, mud flats and sandy bottoms
FEEDING HABIT	Carnivore, feeds on other mollusks
OPERCULUM	Corneous, concentric, with a terminal nucleus

SHELL SIZE RANGE
16 to 24 in
(400 to 609 mm)

PHOTOGRAPHED SHELL
18¾ in
(474 mm)

TRIPLOFUSUS GIGANTEUS
FLORIDA HORSE CONCH
(KIENER, 1840)

447

Triplofusus giganteus has the largest shell in the Atlantic Ocean and is the second largest gastropod shell in the world. It is a voracious, predatory species that feeds on other mollusks, such as clams, snails, and even smaller members of its own species. It is common in shallow-water seagrass meadows, mud flats, and sandy bottoms. It is the state shell of Florida. The broad and muscular foot is red, and consumed as food in Mexico. The shell was also used as a trumpet by Native Americans.

RELATED SPECIES

Pleuroploca princeps (Sowerby I, 1825), which ranges from western Mexico to Ecuador, has a very large shell that is similar to *T. giganteus*. The related *Fasciolaria tulipa* (Linnaeus, 1758), from North Carolina to Brazil, also has a fusiform shell, but it is smooth and colorful.

Actual size

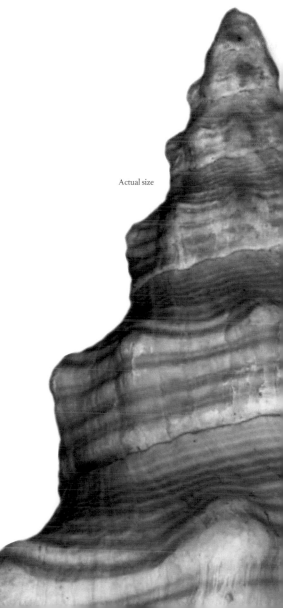

The shell of the Florida Horse Conch is large, heavy, thick, elongate, and fusiform. The spire is tall and the whorls are angular, with spiral cords and weak axial ribs forming large blunt knobs on the shoulder. Its body whorl is large, and the aperture is wide and lanceolate with a long, tapering siphonal canal. The columella has 3 oblique folds. The shells of juveniles are orange, but fade in color as they grow, becoming pale orange or grayish in adults. The color inside the aperture is orange. A brown deciduous periostracum covers the shell.

FAMILY	Muricidae
SHELL SIZE RANGE	1 to 4 in (25 to 100 mm)
DISTRIBUTION	Patagonia
ABUNDANCE	Common
DEPTH	Intertidal and subtidal
HABITAT	Rocky bottoms, mussel beds
FEEDING HABIT	Carnivore
OPERCULUM	Present

SHELL SIZE RANGE
1 to 4 in
(25 to 100 mm)

PHOTOGRAPHED SHELL
1½ in
(36 mm)

448

TROPHON GEVERSIANUS
GEVERS'S TROPHON
(PALLAS, 1774)

Trophon geversianus is a common species found on the large communities of mussels that inhabit the intertidal and subtidal rocky shores of Patagonia. Like most muricids, it is a predator that drills a hole through the shell of its prey by applying a secretion from a specialized organ in the sole of its foot. When the borehole is completed, it extends its proboscis through it to consume the mussel. The shells range from small and smooth to large and elaborately sculptured, often with broad, frilly lamellae. The larger, frillier specimens live subtidally.

RELATED SPECIES

Trophon geversianus strongly resembles several species of muricids from Antarctic waters in terms of size, color, and the presence of pronounced axial lamellae. However, it is actually more closely related to *Forreria belcheri* (Hinds, 1843), from California, which has a much larger, thicker shell with spines along its shoulder.

The shell of Gevers's Trophon is broad, with a high spire, large rounded aperture, strong shoulder, narrow umbilicus, and a short, narrow siphonal canal. It surface has multiple, closely spaced spiral cords, which may be coarse or fine, and numerous axial lamellae that range from short and inconspicuous to broad and bladelike. Specimens in which both spiral cords and axial lamellae are prominent have a distinctive cancellated appearance. The shell color may range from white to chestnut brown.

Actual size

FAMILY	Muricidae
SHELL SIZE RANGE	1 to 2 in (25 to 50 mm)
DISTRIBUTION	Florida to Honduras, Caribbean
ABUNDANCE	Uncommon
DEPTH	670 to 2000 ft (205 to 618 m)
HABITAT	Sandy and fine rubble bottoms
FEEDING HABIT	Carnivore
OPERCULUM	Corneous, ovate

SHELL SIZE RANGE
1 to 2 in
(25 to 50 mm)

PHOTOGRAPHED SHELL
1½ in
(39 mm)

449

PAZIELLA PAZI

PAZ'S MUREX

(CROSSE, 1869)

Paziella pazi has a very spinose shell, with long, nearly straight, hollow spines at the shoulder. Some spines on the lower body whorl and siphonal canal are curved. Its shell bears from six to eight spinose varices per whorl. It is an uncommon muricid, and lives in deep waters off the coasts of Florida to Honduras. *Paziella pazi* and a few related species are considered among the most primitive muricids. They resemble shells from 60 million years ago. The spines on its shell indicate that it does not burrow in sand. It is the type species of the genus *Paziella*.

RELATED SPECIES

Poirieria zelandica (Quoy and Gaimard, 1833), from off New Zealand, has a similar shell, but differs from *P. pazi* by having a larger shell with shorter spire, and slightly thicker spines. *Hexaplex trunculus* (Linnaeus, 1758), from the Mediterranean and northeast Africa, has a broad fusiform shell with several strong varices, and alternating brown and white spiral bands.

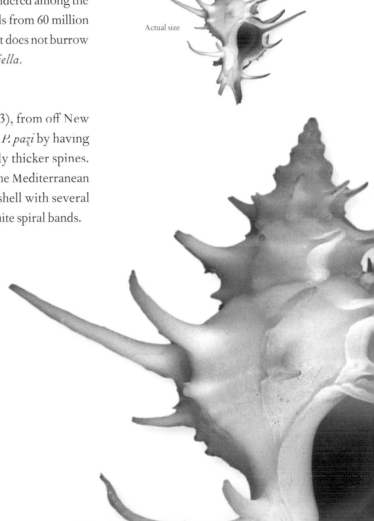

Actual size

The shell of the Paz's Murex is medium in size, relatively thin, spinose, and fusiform. Its spire is tall, with a well-marked suture, a pointed apex, and angulated spire whorls that become progressively more rounded toward the body whorl. A large and small, hollow spine occur along the shoulder of the whorls; the body whorl and the long, open siphonal canal bear more spines. The aperture is ovate and the outer lip is thin, with one large spine. The columella is smooth, with a reflected inner lip. The shell color is uniform white or grayish.

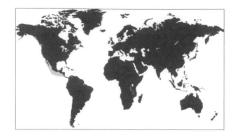

FAMILY	Muricidae
SHELL SIZE RANGE	1 to 1½ in (25 to 40 mm)
DISTRIBUTION	Gulf of California to Panama
ABUNDANCE	Uncommon
DEPTH	20 to 330 ft (6 to 100 m)
HABITAT	Muddy and rocky bottoms
FEEDING HABIT	Carnivore, feeds on invertebrates
OPERCULUM	Corneous, thin, oval

SHELL SIZE RANGE
1 to 1½ in
(25 to 40 mm)

PHOTOGRAPHED SHELL
1½ in
(39 mm)

450

GRAND TYPHIS

(ADAMS, 1855)

Typhisala grandis has one of the largest shells in the subfamily Typhinae. It is easily recognized by the flared outer lip that forms a broad apertural shield. There are several anal canals at the shoulder, but only the most recent one, which opens behind the apertural shield, is functional. *Typhisala grandis* lives offshore and is trawled infrequently from rocky and muddy bottoms.

RELATED SPECIES

Trubatsa pavlova (Iredale, 1936), from deep water off New Caledonia and Australia, has a distinctive shell with two very long canals (often broken): the anterior siphonal and the anal canal at the shoulder. *Vitularia salebrosa* (King and Broderip, 1832), from Baja California to Peru, and the Galápagos Islands, has a pyriform shell with keeled whorls and a long, open siphonal canal.

The shell of the Grand Typhis is large for the subfamily, massive, broad, and fusiform in outline. Its spire is moderately tall, with keeled whorls bearing open tubes; the suture is deeply impressed, but partly obscured. The sculpture consists of many spiral ribs, and 4 heavy varices per whorl. The aperture is oval, the outer lip is widely flared, webbed, and joined to the closed siphonal canal, forming a broad shield. The shell color ranges from white to tan, with the tips of the canals stained in purple-brown. The aperture is white.

Actual size

FAMILY	Muricidae
SHELL SIZE RANGE	⅝ to 2½ in (16 to 65 mm)
DISTRIBUTION	Western Mexico to Peru
ABUNDANCE	Common
DEPTH	Intertidal and shallow subtidal
HABITAT	Mangrove and oyster reefs
FEEDING HABIT	Carnivore, feeds on oysters
OPERCULUM	Corneous, with terminal nucleus

SHELL SIZE RANGE
⅝ to 2½ in
(16 to 65 mm)

PHOTOGRAPHED SHELL
1⅝ in
(40 mm)

STRAMONITA KIOSQUIFORMIS
KIOSK ROCK-SHELL
(DUCLOS, 1832)

451

Stramonita kiosquiformis is easily recognized by the presence of a frilly suture and by its angulate outline. It is a common inhabitant of mangrove and oyster reefs, where it is a very active predator of oysters. Like many other muricids, it uses its radula to drill a hole in the oyster shell and inject a secretion to narcotize its prey. The secretion has a milky appearance but turns purple when exposed to air. *Stramonita kiosquiformis* has a corneous operculum with a thickened ridge on the inner side.

The shell of the Kiosk Rock-shell is medium in size, thick, solid, and fusiform, with a tall spire. Its spire has several whorls, with a pointed apex, and a frilled suture. The shell has weak spiral sculpture, with the exception of a row of strong, knobby, or spinose tubercles at the periphery of the whorls. The aperture is semicircular, the outer lip denticulate, and the columella smooth. The shell color is chocolate brown with white spiral bands; the aperture is white and brown.

RELATED SPECIES
Thais armigera (Link, 1807), from the Indo-West Pacific, has a thick, heavy shell with strong, blunt spines arranged in a few spiral rows. *Concholepas concholepas* (Bruguière, 1792), from Peru to Chile and Argentina, has a thick, limpetlike shell. It is harvested for food, and its flesh is considered a delicacy.

Actual size

FAMILY	Muricidae
SHELL SIZE RANGE	2 to 5 in (50 to 130 mm)
DISTRIBUTION	Peru to Chile and to the Falkland Islands, Argentina
ABUNDANCE	Common
DEPTH	Intertidal to 130 ft (40 m)
HABITAT	Rocky shore
FEEDING HABIT	Carnivore, feeds on bivalves and barnacles
OPERCULUM	Corneous

SHELL SIZE RANGE
2 to 5 in
(50 to 130 mm)

PHOTOGRAPHED SHELL
2⅛ in
(53 mm)

452

CONCHOLEPAS CONCHOLEPAS
BARNACLE ROCK SHELL
(BRUGUIÈRE, 1792)

Concholepas concholepas is a muricid gastropod that converges on the limpet shell shape. Like limpets or abalones, it has a strong muscular foot that clamps the animal to rocks, where it feeds on mussels and barnacles. It is ecologically important as the top predator in its habitat. Although it has a limpet shape, this species still retains an operculum. Its flesh is used as food and considered a delicacy; it is sold worldwide under the misleading name "Chilean Abalone." It is one of the most important species fished in Chile, where it is known as "loco." Because of overharvesting, its fishery is now controlled. There is a single living species in the genus *Concholepas*.

RELATED SPECIES
Plicopurpura patula (Linnaeus, 1758), from the West Indies, has a shell that somewhat resembles that of *C. concholepas*, but has a taller spire, more inflated shell, and spiral rows of tubercles. *Plicopurpura patula* is still used by Native Central Americans to extract a purple dye for clothes. *Rapana venosa* (Valenciennes, 1846), is a large species originally from Japan and China, and now introduced to many places, including Chesapeake Bay in the United States.

Actual size

The shell of the Barnacle Rock Shell is massive and heavy, almost limpetlike in shape. It has a wide aperture and greatly expanded body whorl. Its spire is short, barely showing above the columellar shield. The columella is thick and smooth, and the siphonal canal is short. The sculpture consists of strong spiral ridges that cross concentric growth lines; some shells have frilly scales. The dorsal color ranges from brown to white, but larger shells are often encrusted with barnacles. Its aperture has 2 large teeth anteriorly; the interior is cream, with brown lirae on the margin.

FAMILY	Muricidae
SHELL SIZE RANGE	1 to 4¼ in (25 to 107 mm)
DISTRIBUTION	Baja California to Peru; Galápagos Islands
ABUNDANCE	Common
DEPTH	Intertidal to shallow subtidal
HABITAT	Rocky intertidal, beneath boulders
FEEDING HABIT	Parasitic on other mollusks
OPERCULUM	Corneous, elongate, multispiral

SHELL SIZE RANGE
1 to 4¼ in
(25 to 107 mm)

PHOTOGRAPHED SHELL
2¼ in
(56 mm)

VITULARIA SALEBROSA
RUGGED VITULARIA
(KING AND BRODERIP, 1832)

453

Vitularia salebrosa has one of the slowest feeding behaviors recorded in the family Muricidae. Like other muricids, it drills a hole through its molluscan prey and feeds on the tissues, but its feeding attack is extremely slow. A recent study estimates that such attacks last from 90 to 230 days. The snail targets renewable resources such as blood and digestive glands, to delay the death of the host. Therefore, *Vitularia salebrosa* should be considered an ectoparasite, not a predator. Adaptations to ectoparasitism include an elongated proboscis and the formation of a feeding tube within the host's shell.

RELATED SPECIES

Vitularia miliaris (Gmelin, 1791), from the Red Sea and Indo-West Pacific, is also an ectoparasite on other mollusks. It has a smaller shell with a shorter spire. *Trophon geversianus* (Pallas, 1774), from Argentina to the Straits of Magellan, has an inflated, fusiform shell that varies in size and sculpture.

The shell of the Rugged Vitularia is medium-large, thick, and pyriform in outline. Its spire is moderately tall, with sharply keeled whorls and deeply impressed suture. The sculpture varies from nearly smooth to finely wrinkled with many varices; whorls have keels that change to knobs on shoulders on later whorls. The aperture is subovate, the outer lip is thickened, with 12 to 16 denticles, and the columella is smooth. The siphonal canal is long and open. The shell color ranges from white to brown; the aperture is orange, and the interior white.

Actual size

FAMILY	Muricidae
SHELL SIZE RANGE	1¼ to 2½ in (30 to 65 mm)
DISTRIBUTION	Red Sea to Indo-West Pacific
ABUNDANCE	Common
DEPTH	Intertidal to 300 ft (90 m)
HABITAT	Rocky bottoms
FEEDING HABIT	Carnivore, feeds on mollusks and barnacles
OPERCULUM	Corneous, with subcentral nucleus

SHELL SIZE RANGE
1¼ to 2½ in
(30 to 65 mm)

PHOTOGRAPHED SHELL
2¼ in
(57 mm)

454

HOMALOCANTHA SCORPIO
SCORPION MUREX
(LINNAEUS, 1758)

Homalocantha scorpio is a carnivore that feeds on mollusks and barnacles, and is usually found in shallow waters. Its shell is often heavily encrusted with lime and marine growth. Specimens from old collections usually have four spines on the body whorl, in contrast with five in specimens collected recently. Most shells are brown or have brown spines; albino shells are uncommon. The genus includes species with palmately digitate projections, like *H. scorpio*, and non-palmate spines, as in *H. melanomathos*, from West Africa. The earliest representatives of the genus appeared in the early Cretaceous Period.

RELATED SPECIES

Homalocantha anatomica (Perry, 1811), has a similar distribution and shell to *H. scorpio*, with fewer but relatively larger foliose spines. The species *Murex pele* Pilsbry, 1920, considered by collectors as a Hawaiian endemic, is a synonym. *Paziella pazi* (Crosse, 1869), from Florida to Honduras, has a fusiform shell with long, solid spines.

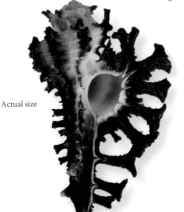

Actual size

The shell of the Scorpion Murex is moderately large for the genus, thick, and fusiform. Its spire can be blunt to moderately high, and the suture is wide and deeply excavated. The aperture is subcircular, and the siphonal canal long and straight. There are 6 to 7 varices with 4 or 5 large flattened foliose spines per whorl, and 2 to 3 large spines on the siphonal canal. The shell color can range from white to brown, with spines usually darker, and the interior is light gray or purplish.

FAMILY	Muricidae
SHELL SIZE RANGE	¾ to 2½ in (20 to 60 mm)
DISTRIBUTION	Indo-West Pacific
ABUNDANCE	Uncommon
DEPTH	Intertidal to 80 ft (25 m)
HABITAT	Coral reefs
FEEDING HABIT	Carnivore, feeds on polychaete worms and small crustaceans
OPERCULUM	Corneous, with lateral nucleus

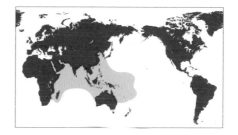

SHELL SIZE RANGE
¾ to 2½ in
(20 to 60 mm)

PHOTOGRAPHED SHELL
2¼ in
(58 mm)

DRUPA RUBUSIDAEUS
STRAWBERRY DRUPE
RÖDING, 1798

The Strawberry or Rose Drupe, also less well known as the Porcupine Castor Bean, is a coral-reef dweller. Throughout most of its range it feeds on polychaete worms, small crustacea, and even small fish, but around the Maldives, this species has been known to feed on sponges. The points of the spines of juveniles tend to be black, fading to white-gray with a creamy brown base as the individual matures. The longest spines are found nearest the aperture, which is usually bright pink in color.

RELATED SPECIES

Drupa clathrata (Lamarck, 1816) shares a stocky, spiny shell and is of similar size. It also shares a dentate lip, but the brown between teeth is more pronounced. *Drupa ricinus* (Linnaeus, 1758) is smaller, has purple-black spines, and a white aperture, often with a broken, pale yellow ring around the outside.

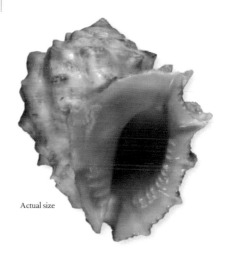

Actual size

The shell of the Strawberry Drupe is tan to off-white in color, rounded with a short spire, and almost flat in more mature individuals. Five rows of short, evenly spaced dorsal spines sit on low axial ribs and are interspersed with scaly ridges. The graduated pink-colored aperture opens to a dentate lip with between 10 and 12 white teeth and a columella with 3 plicae.

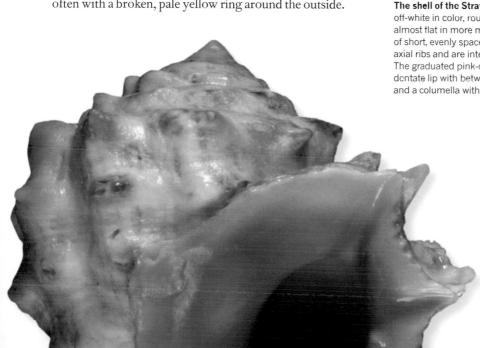

FAMILY	Muricidae
SHELL SIZE RANGE	1½ to 3 in (40 to 75 mm)
DISTRIBUTION	Northern Australia, Indonesia, and Papua New Guinea
ABUNDANCE	Common
DEPTH	Intertidal to 600 ft (180 m)
HABITAT	Rocky bottoms
FEEDING HABIT	Carnivore
OPERCULUM	Corneous, oval

SHELL SIZE RANGE
1½ to 3 in
(40 to 75 mm)

PHOTOGRAPHED SHELL
2¼ in
(58 mm)

456

CHICOREUS CERVICORNIS
DEER ANTLER MUREX
(LAMARCK, 1822)

The shell of the Deer Antler Murex is medium in size, relatively thin-walled, and fusiform. Its spire is relatively high, with an impressed suture, a pointed apex, and rounded whorls. The sculpture is dominated by the 2 long, recurved and branched spines per varix; there are 3 varices per whorl, and several spiral cords. The aperture is oval, the outer lip erect and slightly crenulate, and the columella is smooth and curved. The siphonal canal is very long and closed. The shell color ranges from white to pale orange. The aperture is white.

Chicoreus cervicornis is named for its long spines, which are curved and branched like deer antlers. The longest spines can be as long as the siphonal canal. It is restricted to the northern shore of Australia, Indonesia, and Papua New Guinea. The genus *Chicoreus* is among the most diverse in the family Muricidae, represented by at least 30 species in tropical waters worldwide; about half live in Australia. It is characterized by the presence of three foliaceous or spinose varices per whorl.

RELATED SPECIES
Chicoreus longicornis (Dunker, 1864), which ranges from the central Indian Ocean to the west Pacific, has a shell that is similar in size and shape, but its long spines are not branched as in *C. cervicornis*. *Chicoreus ramosus* (Linnaeus, 1758), ranging from the Red Sea to Indo-West Pacific, has the largest shell in the family Muricidae. It has several frondose spines at the varices, with the largest at the shoulder.

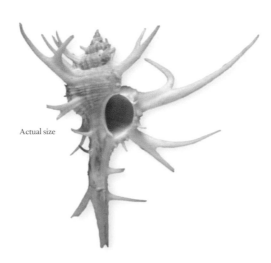

Actual size

FAMILY	Muricidae
SHELL SIZE RANGE	1½ to 4¼ in (35 to 108 mm)
DISTRIBUTION	Mediterranean and northeast Africa
ABUNDANCE	Common
DEPTH	Shallow subtidal
HABITAT	Rocky, sandy, and muddy bottoms
FEEDING HABIT	Carnivore, feeds on bivalves
OPERCULUM	Corneous, concentric

SHELL SIZE RANGE
1½ to 4¼ in
(35 to 108 mm)

PHOTOGRAPHED SHELL
2¾ in
(68 mm)

457

HEXAPLEX TRUNCULUS

TRUNCULUS MUREX

(LINNAEUS, 1758)

Hexaplex trunculus is a common gastropod from the Mediterranean. It was one of two muricid species used by the Phoenicians to extract a purple-blue dye, called Tyrian Purple, from its mucus. Because of the large number of shells needed to produce the dye, an estimated 12,000 shells for a single tunic, it was costly, and used only for the clothing of the aristocracy. The dye was lightfast and would not fade, and much superior to plant-based dyes. Currently *H. trunculus* is not used for its dye, but it remains a popular seafood item in many places.

The shell of the Trunculus Murex is of medium sized and thick, with a large body whorl and a tall, pointed spire. The sculpture consists of strong varices, with a heavy spine at the shoulder, coarse spiral ribs, and fine spiral threads. Its siphonal canal is short, broad, and recurved, and there is a deep umbilicus. The aperture is ovate, the lip thick and wavy, and the columella smooth, with a narrow columellar shield. The shell color is variable, ranging from yellowish to light brown, with alternating brown and white spiral bands.

RELATED SPECIES

Hexaplex radix (Gmelin, 1791), from Panama to Ecuador, has a large, heavy, and inflated shell bearing many short black spines. *Haustellum brandaris* (Linnaeus, 1758), from the Mediterranean, was also used by the Phoenicians to extract a purple-blue dye for clothing.

Actual size

FAMILY	Muricidae
SHELL SIZE RANGE	2 to 6¼ in (50 to 155 mm)
DISTRIBUTION	Panama to southern Ecuador
ABUNDANCE	Common
DEPTH	Intertidal
HABITAT	Rocky bottoms
FEEDING HABIT	Carnivore, feeds on mollusks
OPERCULUM	Corneous, thick

SHELL SIZE RANGE
2 to 6¼ in
(50 to 155 mm)

PHOTOGRAPHED SHELL
3½ in
(88 mm)

458

HEXAPLEX RADIX
RADISH MUREX
(GMELIN, 1791)

Hexaplex radix is one of the densest and spiniest species of the Muricidae family. The often upward-curving spines are foliated, helping to distinguish it from the similar but larger *Hexaplex nigritus* (see below). The inflated white body whorl of *H. radix* is covered with purple, almost black spines, which protect the animal from predators and give the shell a dramatic appearance, making it popular with collectors. *Hexaplex radix* is carnivorous, feeding on small mollusks, which it seeks under rocks.

RELATED SPECIES

Hexaplex nigritus (Phillipi, 1845) is very similar, and considered by some to be a northerly form of *H. radix*. *Hexaplex nigritus*, however, is generally larger and more elongated; in addition, the large black shoulder areas and wide black spiral bands give *H. nigritus* a darker overall appearance, hence its species name.

Actual size

The shell of the Radish Murex is extremely bulbous, with the off-white spire ranging from short and pointed to almost flat. The white body whorl has numerous varices, ranging between 6 and 10 in number, along which protrude purple-black spines, many of which are frilled. The white aperture is bounded on one side by a smooth, mostly white columella, featuring a posterior tooth and a jagged outer lip.

FAMILY	Muricidae
SHELL SIZE RANGE	2½ to 5 in (65 to 130 mm)
DISTRIBUTION	Sri Lanka to southwestern Pacific
ABUNDANCE	Uncommon
DEPTH	Intertidal to 300 ft (90 m)
HABITAT	Rocky bottoms
FEEDING HABIT	Carnivore, feeds on bivalves and gastropods
OPERCULUM	Corneous, with subcentral nucleus

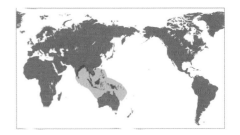

SHELL SIZE RANGE
2½ to 5 in
(65 to 130 mm)

PHOTOGRAPHED SHELL
4⅜ in
(113 mm)

CHICOREUS PALMAROSAE

ROSE-BRANCH MUREX
(LAMARCK, 1822)

459

Chicoreus palmarosae is an attractive shell with distinctive fronds. It is usually found on rocky bottoms in offshore waters. Like other muricids, it is a voracious carnivore and feeds on other mollusks, including giant clams. *Chicoreus palmarosae* has a variable shell throughout its range; shells from Sri Lanka often have purplish or pinkish tipped fronds, while those from the Philippines have dark brown and shorter fronds. Some shell dealers dip the fronds into a pink or violet dye to enhance their color and increase the shell's value. The shells are heavily encrusted with marine debris, and difficult to clean.

RELATED SPECIES
Chicoreus cervicornis (Lamarck, 1822), from western Papua New Guinea to northern Australia, has a long siphonal canal and long, distinctive, bifurcating spines that resemble deer antlers. *Chicoreus cnissodus* (Euthyme, 1889), from India to New Caledonia and Japan, has frondose spines of variable length, and an off-white shell.

The shell of the Rose-branch Murex is large and fusiform with a long and relatively broad siphonal canal. Its spire is tall, with about 9 whorls; the suture is well impressed; and there are 3 varices per whorl. The body whorl and spire have long, hollow, foliose spines, with the longest occurring at the shoulder. The aperture is ovate and small, with blunt teeth on the inside edge of the outer lip. The shell color is brownish red with brown ridges and a white aperture. Shells from Sri Lanka have pinkish fronds.

Actual size

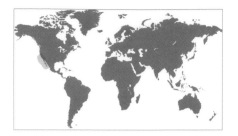

FAMILY	Muricidae
SHELL SIZE RANGE	2½ to 7¼ in (64 to 187 mm)
DISTRIBUTION	California to Baja California
ABUNDANCE	Common
DEPTH	Intertidal to 90 ft (27 m)
HABITAT	Oyster beds and sandy bottoms
FEEDING HABIT	Carnivore, feeds on oysters
OPERCULUM	Corneous, D-shaped

SHELL SIZE RANGE
2½ to 7¼ in
(64 to 187 mm)

PHOTOGRAPHED SHELL
4½ in
(114 mm)

460

FORRERIA BELCHERI
GIANT FORRERIA
(HINDS, 1843)

Forreria belcheri is the largest muricid in California. It is a predator of oysters, and is found near oyster beds or on sandy bottoms, ranging from the intertidal zone to offshore. Like most muricids, the animal has an organ on its foot that is used to drill through its prey's shell. This organ has been known since the late 1800s; originally it was thought to be a sucker used to hold the prey's shell. When it was studied, scientists discovered that it produces an acid secretion that dissolves the shell; the predator's radula is also used in the drilling process.

RELATED SPECIES

Austrotrophon catalinensis Oldroyd, 1927, which has a similar distribution as *Forreria belcheri*, has a much smaller shell, with an elongated siphonal canal and large lamellose spines pointed upward, along the shoulder. *Zacotrophon beebei* Hertlein and Strong, 1948, from deep waters off the southern Gulf of California, Mexico, has a smaller, thinner shell with loose coiled whorls, a tall spire, and a row of short spines along the shoulder.

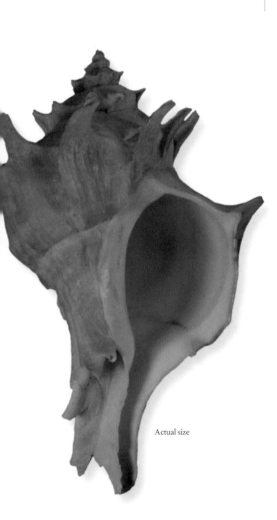

Actual size

The shell of the Giant Forreria is large, thick, heavy, and elongately fusiform. Its spire is short compared with the large body whorl; the suture is impressed, the apex pointed, and the whorls angulated. The sculpture consists of axial growth lines, and a row of large, open spines along the shoulder. The aperture is large and oval, the outer lip sharp, the columella smooth, and the siphonal canal long and open. The operculum is large and D-shaped. Shell color ranges from cream to light brown. The aperture is white or off-white.

FAMILY	Muricidae
SHELL SIZE RANGE	3 to 7½ in (75 to 190 mm)
DISTRIBUTION	Indo-Pacific
ABUNDANCE	Common
DEPTH	33 to 165 ft (10 to 50 m)
HABITAT	Sandy and muddy bottoms
FEEDING HABIT	Carnivore
OPERCULUM	Corneous, concentric

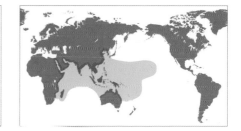

MUREX PECTEN

VENUS COMB MUREX

LIGHTFOOT, 1786

SHELL SIZE RANGE
3 to 7½ in
(75 to 190 mm)

PHOTOGRAPHED SHELL
4½ in
(119 mm)

461

Murex pecten is the most spectacular of the long-spined murexes, and a collector's favorite. It is a widely distributed species, and common on shallow soft bottoms. However, perfect specimens with intact spines are uncommon. It is the most spinose murex; its shell has more than 100 long and fragile spines, which provide protection to the animal while preventing it from sinking into soft sediments. As the animal grows, the mantle resorbs spines that would otherwise block the aperture, and secretes new ones to accommodate for shell growth.

RELATED SPECIES

Murex troschelli Lischke, 1868, from the Indo-West Pacific, has a similar shell with brown spiral lines but it has fewer spines and can grow larger than *M. pecten*. *Haustellum haustellum* (Linnaeus, 1758), from the Red Sea and Indo-Pacific, has a globose shell with a long, straight siphonal canal, low spire, but lacks spines.

The shell of the Venus Comb Murex is large and thin, with a long, straight siphonal canal, and many long, delicate, and evenly spaced, slightly curved spines. There are 3 varices per whorl, each with many closed spines. The spines on the siphonal canal form an angle of about 90 degrees to the canal axis. The siphonal canal is almost completely closed and very long. The shell color is white to light brown, and the aperture is white.

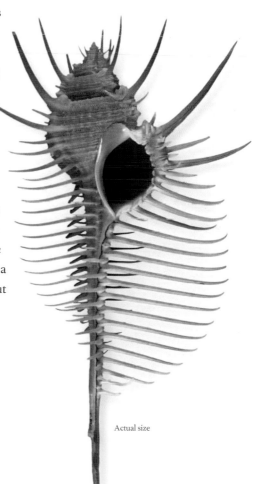

Actual size

FAMILY	Muricidae
SHELL SIZE RANGE	2½ to 7¼ in (65 to 185 mm)
DISTRIBUTION	Red Sea to Indo-Pacific
ABUNDANCE	Common
DEPTH	10 to 330 ft (3 to 100 m)
HABITAT	Sand and coral rubble
FEEDING HABIT	Carnivore, but also feeds on carrion
OPERCULUM	Corneous, with subcentral nucleus

SHELL SIZE RANGE
2½ to 7¼ in
(65 to 185 mm)

PHOTOGRAPHED SHELL
5 in
(126 mm)

462

HAUSTELLUM HAUSTELLUM

SNIPE'S BILL

(LINNAEUS, 1758)

Haustellum haustellum is a characteristically club-shaped shell with a very long siphonal canal that makes it resemble the bill of the Snipe bird, hence its popular name. It is a common species, usually found in soft sediment in shallow subtidal waters to 330 ft (100 m). *Haustellum haustellum* is an unselective carnivore, and sometimes feeds on carrion. It is collected locally for food and the shell trade. Its shell is variable, and a few subspecies have been named on the basis of shell characters such as the presence of more prominent sculpture or the length of the siphonal canal.

RELATED SPECIES

Haustellum hirasei (Shikama, 1973), from Japan to New Caledonia, also has a club-shaped shell, which has a taller spire and strong spiral ribs. *Murex pecten* Lightfoot, 1786, has a long siphonal canal that bears short and long, slightly curved spines.

Actual size

The shell of the Snipe's Bill is medium-large, solid, compact, and club-shaped, with a very long siphonal canal. Its spire is low and globose, the suture well impressed, and the aperture wide and ovate. There are 3 or 4 spiral rows of blunt spines on the axial ribs. The spiral bands are chestnut brown and evenly spaced on the body. The shell color is creamy or pink, with dark brown dashes or blotches and a peach pink aperture.

FAMILY	Muricidae
SHELL SIZE RANGE	1¾ to 6 in (45 to 153 mm)
DISTRIBUTION	Gulf of California to Peru
ABUNDANCE	Abundant
DEPTH	Intertidal to 1,000 ft (300 m)
HABITAT	Among rocks in shallow water
FEEDING HABIT	Carnivore, feeds on gastropods and bivalves
OPERCULUM	Corneous, ovate, multispiral

SHELL SIZE RANGE
1¾ to 6 in
(45 to 153 mm)

PHOTOGRAPHED SHELL
5⅜ in
(136 mm)

PHYLLONOTUS ERYTHROSTOMUS

PINK-MOUTH MUREX

(SWAINSON, 1831)

463

Phyllonotus erythrostomus is a large and globose muricid that once was the most abundant large gastropod in the Gulf of California. However, due to overharvesting, it is now far less common, and restricted to subtidal depths. It is often collected by shrimp trawlers. It has four or five varices per whorl, while species in the related genus *Hexaplex* have five to seven varices per whorl.

RELATED SPECIES

Phyllonotus pomum (Gmelin, 1791), from North Carolina to Brazil, has a smaller but thick, solid, and globose fusiform shell with three or four strong varices per whorl. *Hexaplex fulvescens* (Sowerby I, 1834), from North Carolina to Mexico, has a similar, larger shell with more numerous and longer spines.

The shell of the Pink-mouth Murex is moderately large for the genus, thick, solid, heavy, and globose-ovate. Its spire is low, the apex pointed, and the suture obscured by the succeeding whorl. The outer surface is rough, with axial varices with open and closed spines, and a smooth, glossy interior. The aperture is large and ovate, and the outer lip crenulate. The columella is smooth, and the siphonal canal large and closed. The shell color is white or creamy pink, and the interior is rich pink.

Actual size

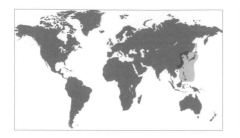

FAMILY	Muricidae
SHELL SIZE RANGE	4 to 8½ in (100 to 220 mm)
DISTRIBUTION	Japan to Philippines
ABUNDANCE	Uncommon
DEPTH	Deep water, from 165 to 800 ft (50 to 250 m)
HABITAT	Sandy gravel bottoms
FEEDING HABIT	Carnivore
OPERCULUM	Corneous, concentric

SHELL SIZE RANGE
4 to 8½ in
(100 to 220 mm)

PHOTOGRAPHED SHELL
5½ in
(141 mm)

464

SIRATUS ALABASTER
ALABASTER MUREX
(REEVE, 1845)

The Alabaster Murex is perhaps one of the prettiest murexes, and a favorite among collectors. It was once a great rarity. Until a few decades ago, specimens with delicate webbed varices were unknown (as shown in the nineteenth-century engraving above). Initially, very large specimens were obtained by Taiwanese fishing vessels. Later, smaller, more delicately shaped specimens were discovered in deeper waters of the Philippines. It is the largest of the webbed murex. Its webbed varices support the shell in soft sediments and also protect the mollusk's body from predators. Like other muricids, *Siratus alabaster* is a predator of mollusks.

RELATED SPECIES

Siratus tenuivaricosus (Dautzenberg, 1927), from Brazil, resembles *S. alabaster*, but is smaller, shorter, with recurved spines, and has narrower, webbed varices. *Chicoreus palmarosae* (Lamarck, 1822), which ranges from Sri Lanka to the southwestern Pacific, has a brown shell with spines ending in delicate fronds tinted in pink. Specimens from Sri Lanka usually have more frondose and frilly shells than those found elsewhere.

Actual size

The shell of the Alabaster Murex is large, thin, and lightweight. Its spire is tall and pointed, with about 7 to 9 whorls. The whorls have 3 varices and 6 knobs per whorl, and a long, straight, or slightly upturned spine. Fine spiral lines expand to form long and delicate webbed varices that give the shell its striking appearance. Its aperture is large, rounded, and triangular in profile. The columella is smooth, and the columella and lip are flared. The siphonal canal is long and slightly curved. The shell color varies from pure white to cream.

FAMILY	Muricidae
SHELL SIZE RANGE	2⅜ to 8¼ in (60 to 213 mm)
DISTRIBUTION	North Carolina to eastern Mexico
ABUNDANCE	Common
DEPTH	Intertidal 260 ft (80 m)
HABITAT	Rocks and reefs
FEEDING HABIT	Carnivore
OPERCULUM	Corneous, brown

SHELL SIZE RANGE
2⅜ to 8¼ in
(60 to 213 mm)

PHOTOGRAPHED SHELL
7 in
(173 mm)

HEXAPLEX FULVESCENS

GIANT EASTERN MUREX

(SOWERBY II, 1834)

465

Hexaplex fulvescens is the largest species in the genus *Hexaplex* in the Atlantic Ocean. It has a massive shell with many, relatively short, non-branching spines. The genus name, *Hexaplex*, refers to the six varices per whorl that are present on the shell. *Hexaplex fulvescens* is a common species, found both as a Pleistocene fossil, as well as living, from North Carolina to Texas and Mexico. It is a voracious predator of oysters.

RELATED SPECIES

Hexaplex duplex (Röding, 1798) is of similar size and inhabits comparable depths off western Africa. It has short open spines on about eight varices per whorl. The open siphonal canal, columella, and the edges of the open labial spines (longest at the shoulder) are fleshy pink to tan; the rest of the shell is off-white to cream.

The shell of the Giant Eastern Murex is white and bulbous, with a short spire and a moderately long, curving, nearly closed siphonal canal. There are widely spaced, raised spiral cords of red-brown, and 6 to 10 varices bearing fairly short, blunt, open spines that are longer on the shoulders. Its siphonal canal and lip are also lined with spines, which are largest and most open toward the shoulder. The off-white aperture is circular.

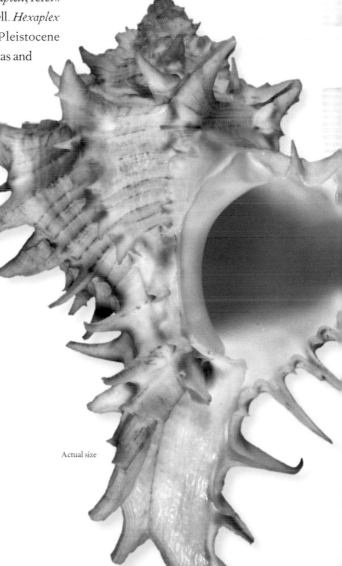

Actual size

FAMILY	Muricidae
SHELL SIZE RANGE	1¾ to 12⅞ in long (45 to 327 mm)
DISTRIBUTION	Red Sea to Indo-West Pacific
ABUNDANCE	Common
DEPTH	Shallow offshore
HABITAT	Coral reefs
FEEDING HABIT	Carnivore
OPERCULUM	Corneous

SHELL SIZE RANGE
1¾ to 12⅞ in long
(45 to 327 mm)

PHOTOGRAPHED SHELL
9½ in
(243 mm)

466

CHICOREUS RAMOSUS
RAMOSE MUREX
(LINNAEUS, 1758)

Chicoreus ramosus is the largest species in the genus *Chicoreus*, a spectacular and solidly heavy shell much admired for its decorative qualities. The species is voraciously carnivorous. It devours other mollusks by boring holes in their shells. Some muricid species are known to be cannibalistic. There is evidence that this species has been collected for food by humans during the late stone age.

RELATED SPECIES

Chicoreus virgineus (Röding, 1798) has a similar sculpture to *C. ramosus* but it bears fewer, shorter, blunt spines instead of fronds. The siphonal canal is completely open, the spire is slightly taller, and there are infrequent, pink-brown spiral bands. It is endemic to the Red Sea.

Actual size

The shell of the Ramose Murex is thick and bulbous, with a short spire and a long, wide, recurving siphonal canal. It is white, with fine spiral cords tinged chestnut brown in places. There are 3 varices on each whorl, separated by a nodulose axial rib, with 10 fronds from the shoulder to the end of the canal; the fronds are much extended at the shoulder and along the upper outer lip. The aperture is circular and large.

FAMILY	Coralliophilidae
SHELL SIZE RANGE	¾ to 1⅝ in (20 to 42 mm)
DISTRIBUTION	Southern Japan to South China Sea to Fiji
ABUNDANCE	Uncommon
DEPTH	230 to 1,900 ft (70 to 580 m)
HABITAT	Rocky bottoms
FEEDING HABIT	Parasitic on corals
OPERCULUM	Corneous, oval

SHELL SIZE RANGE
¾ to 1⅝ in
(20 to 42 mm)

PHOTOGRAPHED SHELL
1⅝ in
(42 mm)

BABELOMUREX DIADEMA
DIADEM LATIAXIS
(ADAMS, 1854)

467

Babelomurex diadema ranges from southern Japan to the South China Sea to Fiji, and lives on rocky bottoms in deep water. The adult shell is variable, and the larval shell is important in species identification. *Babelomurex diadema* may comprise a group of species with very similar shells. The diversity in the genus *Babelomurex* is the result of the high degree of specificity to their coral hosts. The larva only settles when it finds the right coral species.

RELATED SPECIES
Babelomurex princeps (Melvill, 1912), which ranges from the Arabian Gulf to the Philippines, has a similar shell, but has longer spines and more spiral cords. *Babelomurex spinaerosae* (Shikama, 1970), from Japan to the Philippines, has a small shell with long, fine spines. Some shells are almost entirely pink, while others are white, with the spines tinged in pink or purple.

The shell of the Diadem Latiaxis is medium in size, rough, spinose, and fusiform. Its spire is tall, with angulated whorls, a well-marked suture, and a pointed apex. The sculpture consists of axial ridges and 2 to 3 raised, sharp spiral cords, which bear spines; the shoulder has large, recurved spines. The aperture is wide, and the outer lip bears a large spine and several smaller ones. The siphonal canal is short and twisted, and the columella smooth. The shell color ranges from white to pinkish, often with pink or brown blotches near the spiral cords.

Actual size

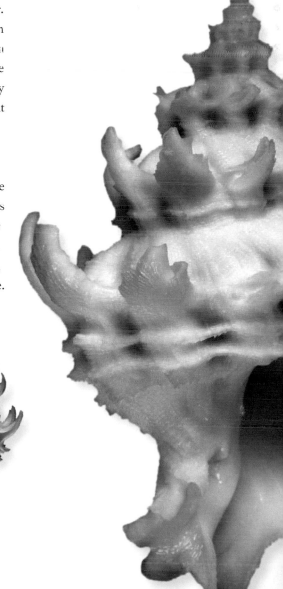

FAMILY	Coralliophilidae
SHELL SIZE RANGE	½ to 2½ in (10 to 64 mm)
DISTRIBUTION	Florida to Brazil
ABUNDANCE	Common
DEPTH	Intertidal to 75 ft (23 m)
HABITAT	Rocky bottoms
FEEDING HABIT	Parasitic on corals
OPERCULUM	Corneous, oval

SHELL SIZE RANGE
½ to 2½ in
(10 to 64 mm)

PHOTOGRAPHED SHELL
2¼ in
(58 mm)

468

CORALLIOPHILA ABBREVIATA
SHORT CORAL SHELL
(LAMARCK, 1816)

The Short Coral Shell is a common shallow-water coralliophilid. It is parasitic on at least eight species of corals, and like other coralliophilids, it is sedentary and lives on its host for months. Groups of up to 20 snails can be found, although most often there are fewer snails per coral colony. Corallivorous snails can affect the population structure of corals.

RELATED SPECIES

Coralliophila violacea Kiener, 1836, from the Indo-Pacific, somewhat resembles *C. abbreviata*, but the sculpture consists of smaller spiral cords, which are usually heavily encrusted. The aperture is deep violet. *Babelomurex diadema* (Adams, 1854), which ranges from southern Japan to the South China Sea to Fiji, has a shell with a tall spire and recurved spines.

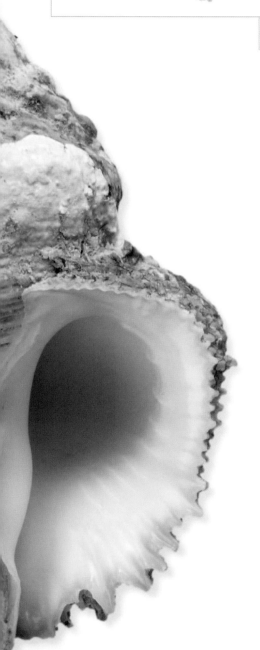

The shell of the Short Coral Shell is larger than most coralliophilids in the Caribbean. It is thick, rough, and has variable shape. Its spire is short, with rounded or angulated shoulders; the suture can be well marked or not. The sculpture consists of many spiral cords covered with small scales and rounded axial ridges, but many shells (such as the one illustrated here) are heavily encrusted. The aperture is oval, the outer lip thick and denticulate within, and the columella smooth. The shell color is usually white, but it can be pinkish or yellowish. The aperture is white.

Actual size

FAMILY	Coralliophilidae
SHELL SIZE RANGE	¼ to 2¾ in (19 to 70 mm)
DISTRIBUTION	Japan to Northeastern Australia
ABUNDANCE	Locally common
DEPTH	165 to 650 ft (50 to 200 m)
HABITAT	Coral reefs and sandy mud bottoms
FEEDING HABIT	Parasitic, feeds on hard corals
OPERCULUM	Corneous, concentric

SHELL SIZE RANGE
¼ to 2¾ in
(19 to 70 mm)

PHOTOGRAPHED SHELL
2¾ in
(69 mm)

469

LATIAXIS MAWAE

MAWE'S LATIAXIS

(GRIFFITH AND PIDGEON, 1834)

Latiaxis mawae has a distinctive shell with a flattened apex and a body whorl that uncoils. It is locally common in deep waters, and is found on sandy mud bottoms. Like other coralliophilids, it lacks a radula, and is parasitic on hard corals, sucking fluids from coral polyps. Coral shells are popular among collectors, and *L. mawae*, with its strange shape, is a particular favorite. There are more than 200 species of coralliophilids worldwide, with the highest diversity in tropical waters.

RELATED SPECIES

Latiaxis pilsbryi Hirase, 1908, from Japan to Vietnam, has a shell that resembles a young *L. mawae*, with a flat spire, but with longer shoulder spines and the body whorl does not separate from previous whorls. *Babelomurex echinatus* (Azuma, 1960), from Japan to the Philippines, is one of the spiniest species of coral shells, with as many as 70 spines on the body whorl.

Actual size

The shell of the Mawe's Latiaxis is thick, with a low or flat apex, and uniquely shaped, with a partially uncoiled body whorl. The early whorls are flattened but gradually become raised and convex. The shoulder has a row of strong, triangular spines, which are curved toward the spire. The sculpture consists of fine spiral lines crossed by axial growth marks. The lower part of the whorl is rounded, with a recurved siphonal canal, forming a wide and deep umbilicus bordered by a row of siphonal canals. The shell color is commonly white to cream, but may be pink, orange, or even purple. The aperture is white.

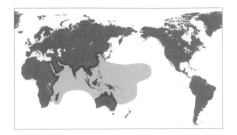

FAMILY	Coralliophilidae
SHELL SIZE RANGE	1½ to 3½ in (40 to 90 mm)
DISTRIBUTION	Indo-Pacific
ABUNDANCE	Common
DEPTH	Intertidal to 1,000 ft (300 m)
HABITAT	Buried in soft coral
FEEDING HABIT	Parasite, feeds on soft corals
OPERCULUM	Corneous, thin

SHELL SIZE RANGE
1½ to 3½ in
(40 to 90 mm)

PHOTOGRAPHED SHELL
2⅞ in
(72 mm)

470

RAPA RAPA

RAPA SNAIL

(LINNAEUS, 1758)

Rapa rapa has the largest shell in the family Coralliophillidae, if the long calcareous tube of *Magilus antiquatus* (which can be longer than *R. rapa*'s shell) is discounted. The shell is thin and fragile, with a spherical shape. *Rapa rapa* is a parasitic snail, specialized in leather corals (Alcyonacea), such as *Sarcophyton* and *Sinularia*. It enters the coral through the stalk and leaves a small hole as the only external sign. It then grows inside the coral, eating it from the inside. *Rapa rapa* lacks a radula, but has a long proboscis. It produces an enzyme that liquefies the host's tissues, which are then ingested by the snail.

RELATED SPECIES

Rapa incurva (Dunker, 1853), from Indo-West Pacific, has a similar but smaller shell, with a short spire. It also grows embedded in leather corals. *Magilus antiquatus* Montfort, 1810, from the Indo-West Pacific, lives and grows into massive hard corals. Its shell is relatively small, but as it grows, it builds a long calcareous tube.

Actual size

The shell of the Rapa Snail is large for the family, thin, fragile, translucent, and globose, with a turnip shape. Its spire is very short to flat, with a well-marked suture. The body whorl is large and globose. The aperture is wide and long, the outer lip is thin, and the sculpture of strong spiral ribs gives the outer lip a saw-tooth edge. The columella is smooth, with a columellar shield and a long siphonal canal. The shell color is uniformly white or cream.

FAMILY	Turbinellidae
SHELL SIZE RANGE	1 to 2 in (25 to 50 mm)
DISTRIBUTION	Western Australia and Queensland
ABUNDANCE	Uncommon
DEPTH	65 to 650 ft (20 to 200 m)
HABITAT	Fine sand bottoms
FEEDING HABIT	Carnivore, feeds on polychaete worms
OPERCULUM	Corneous, thin

SHELL SIZE RANGE
1 to 2 in
(25 to 50 mm)

PHOTOGRAPHED SHELL
1⅛ in
(27 mm)

471

TUDIVASUM SPINOSUM

SPINY HAMMER VASE

(H. AND A. ADAMS, 1863)

Although closely related to *Vasum*, the genus *Tudivasum* has a more restricted distribution, being limited to the coasts of Australia, with a single species known from Zanzibar. The shells tend to be smaller and more globular, with a long and distinct siphonal canal. *Tudivasum spinosum* is an uncommon offshore species. The animal feeds on polychaete worms, which it swallows whole.

RELATED SPECIES

Tudivasum inermis (Angas, 1878) most closely resembles *T. spinosum* in size and shape, but can be distinguished by its smooth surface, darker color, and lack of spines. The larger *T. armigera* has a taller spire, and much longer, radial spines along the shoulder and siphonal canal. *Tudivasum kurzi* (Macpherson, 1963) resembles *T. armigera*, but has numerous short spines along the spiral cords that run between the larger rows of spines.

Actual size

The shell of the Spiny Hammer Vase has a small shell with a short triangular spire, a globular body, and a long, sinuous, axial siphonal canal. The protoconch is large. The shell has low, axial folds and is lined with numerous spiral cords. There is a row of short spines along the shoulder, with each spine on an axial rib. The aperture is ovate, with a short columella that has 3 folds. The shell is cream colored, with bands of reddish brown and flecks or bands of darker color. The aperture and the tip of the siphonal canal are white.

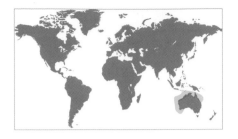

FAMILY	Turbinellidae
SHELL SIZE RANGE	2 to 3 in (50 to 75 mm)
DISTRIBUTION	Northern and Western Australia, and Queensland
ABUNDANCE	Moderately common
DEPTH	Subtidal to 120 ft (40 m)
HABITAT	Sandy bottoms
FEEDING HABIT	Carnivore, feeds on polychaete worms
OPERCULUM	Corneous, thick

SHELL SIZE RANGE
2 to 3 in
(50 to 75 mm)

PHOTOGRAPHED SHELL
2 in
(51 mm)

472

TUDIVASUM ARMIGERA

ARMORED HAMMER VASE

(A. ADAMS, 1855)

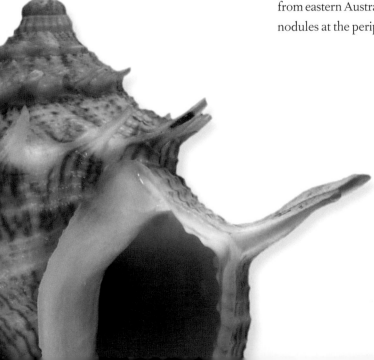

Actual size

Tudivasum armigera is a moderately common species from subtidal sand and rubble bottoms along the tropical shores of northern and Western Australia and Queensland. Shells can be variable, especially in color and in the number and length of spines. The long, axially oriented spines indicate that the animals crawl over the substrate rather than burrow into it.

RELATED SPECIES

Five of the six known species of *Tudivasum* are restricted to the tropical waters of Australia. *Tudivasum spinosum* (H. and A. Adams, 1863) and *T. inermis* (Angas, 1878) are smaller than *T. armigera. Tudivasum inermis* has a smooth, rounded shell, and lacks spines while *T. spinosum* has a single row of short, sharp spines along the shoulder. *Tudivasum rasilistoma* (Abbott, 1959) from eastern Australia, has a larger, heavier shell with prominent nodules at the periphery.

The shell of the Armored Hammer Vase is thick and stout, with a pear-shaped body and a long, axial siphonal canal. The protoconch is large, and the spire weakly stepped. The aperture is ovate, with denticles along the outer lip and folds on the columella. There are long, open, radially oriented spines along the shell periphery, as well as along the base of the siphonal canal. Additional, weaker spines may be present along the basal keel and on the siphonal canal. The color may be whitish to purplish brown, with thin, irregular, darker brown axial markings, and a lighter, spiral bank along the anterior keel.

FAMILY	Turbinellidae
SHELL SIZE RANGE	2 to 3 in (50 to 75 mm)
DISTRIBUTION	Japan to South China Sea
ABUNDANCE	Uncommon
DEPTH	Subtidal to 650 ft (200 m)
HABITAT	Sandy bottoms
FEEDING HABIT	Carnivore, feeds on polychaete worms
OPERCULUM	Corneous, thin

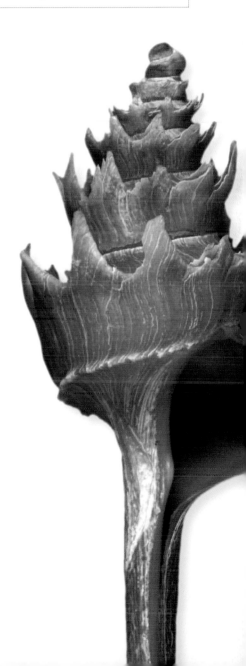

SHELL SIZE RANGE
2 to 3 in
(50 to 75 mm)

PHOTOGRAPHED SHELL
2½ in
(62 mm)

COLUMBARIUM PAGODA

FIRST PAGODA SHELL

(LESSON, 1831)

473

Like all members of the deepwater subfamily Columbariinae, *Columbarium pagoda* inhabits sandy and muddy bottoms along the outer continental shelf and continental slope. The animals have an extremely long proboscis used to feed on tube-dwelling polychaetes. In some regions, several closely related species have similar geographic ranges, but live at different depths.

RELATED SPECIES

Columbarium pagodoides (Watson, 1882) from off eastern Australia, is similar, but has a pronounced keel along the shoulder. *Columbarium harrisae* Harasewych, 1983, also from eastern Australia, has a much larger white shell, with very short spines along the periphery. *Coluzea juliae* Harasewych, 1987, from off southeastern Africa, has a larger, white shell with axial ribs, spiral cords, and spines originating below the shoulder.

The shell of the First Pagoda Shell is very long and narrow with a high, stepped spire and a long, sinuous, axial siphonal canal. The protoconch is large, bulbous, and shiny. The aperture is small and nearly circular. A sharply angled shoulder has numerous, open spines. The basal keel, a second, sharp angle below the shoulder, imparts a rectangular appearance to the whorls. Spiral sculpture is limited to several widely spaced scabrous cords along the upper portion of the siphonal canal. The shell color is a uniform tan, with a darker brown periostracum.

Actual size

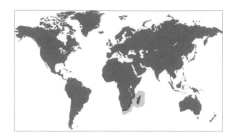

FAMILY	Turbinellidae
SHELL SIZE RANGE	2¾ to 4 in (70 to 100 mm)
DISTRIBUTION	Southeastern Africa
ABUNDANCE	Rare
DEPTH	Deep water
HABITAT	Sand and rubble bottoms
FEEDING HABIT	Carnivore, feeds on polychaete worms
OPERCULUM	Corneous, concentric, with terminal nucleus

SHELL SIZE RANGE
2¾ to 4 in
(70 to 100 mm)

PHOTOGRAPHED SHELL
3½ in
(89 mm)

474

COLUZEA JULIAE
JULIA'S PAGODA SHELL
HARASEWYCH, 1989

Coluzea juliae belongs to a group of deepwater gastropods known as pagoda shells. They live on sandy and muddy bottoms, usually in tropical areas, but extending to polar waters. The animals do not burrow into the sediment. They feed on tube-dwelling polychaetes with their long proboscis.

RELATED SPECIES
Coluzea aapta Harasewych, 1986, from Western Australia, has a row of sharp spines along the shoulder. *Columbarium pagoda* (Lesson, 1831), from Japan and Taiwan, has a bulbous protoconch and sharp, broad, triangular spines on the shoulder.

The shell of the Julia's Pagoda Shell is elongately fusiform, with a tall spire and long siphonal canal. Its spire is high, about one quarter of the shell length and the suture is well impressed with a wide spiral channel separating the whorls. The sculpture consists of 3 spiral ridges, the lowest of which forms a strong keel and bears scaly spines that are more developed on the body whorl. The aperture is ovate and the columella smooth, lacking folds. The siphonal canal is long, about half of the shell length, with spiral cords. The shell color is white or gray and the aperture white.

Actual size

FAMILY	Turbinellidae
SHELL SIZE RANGE	2 to 4½ in (50 to 113 mm)
DISTRIBUTION	Endemic to Brazil
ABUNDANCE	Uncommon
DEPTH	Intertidal to 200 ft (60 m)
HABITAT	Muddy bottoms
FEEDING HABIT	Carnivore, feeds on polychaetes and bivalves
OPERCULUM	Corneous, thick, claw-shaped

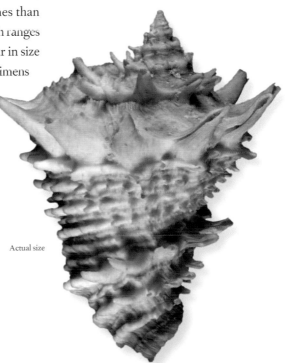

SHELL SIZE RANGE
2 to 4½ in
(50 to 113 mm)

PHOTOGRAPHED SHELL
3⅝ in
(91 mm)

VASUM CASSIFORME

IIELMET VASE

KIENER, 1840

475

Vasum cassiforme is endemic to Brazil, ranging from Rio Grande do Norte to Espirito Santo. It is usually found in shallow water, from the intertidal zone to offshore, on muddy bottoms. Specimens from protected and calm waters tend to develop longer spines than those from open coasts, and shells from deep waters are more elongate than those from shallow waters. *Vasum cassiforme* is very similar to *V. chipolense*, an extinct Miocene fossil from the Chipola Formation in northwestern Florida.

RELATED SPECIES

Vasum muricatum (Born, 1778), from Florida to Venezuela, has a similar but larger shell, usually with fewer spines than *V. cassiforme*. *Vasum turbinellum* (Linnaeus, 1758), which ranges from eastern Africa to Polynesia, has a shell also similar in size and shape, but with fewer, open, large spines. Some specimens have extremely heavy and thick spines. The shell has alternating black and white spiral bands.

The shell of the Helmet Vase is medium-sized, extremely thick, heavy, spiny, and conical. Its spire is short, with a pointed apex, slightly concave whorls, and an impressed suture. The sculpture consists of 12 foliated cords bearing nodules and spines, with the largest spines at the shoulder, and one row of large spines near the anterior end. Its aperture is narrow and long, the outer lip very thick, flared, and heavily denticulate, and the columella has 2 folds. The shell color ranges from white to cream, and the thick glossy shield around the aperture is usually purplish brown.

Actual size

FAMILY	Turbinellidae
SHELL SIZE RANGE	2½ to 5 in (64 to 125 mm)
DISTRIBUTION	Florida to Venezuela, and the Gulf of Mexico
ABUNDANCE	Common
DEPTH	Intertidal to 50 ft (15 m)
HABITAT	Sandy bottoms
FEEDING HABIT	Carnivore, feeds on polychaetes and bivalves
OPERCULUM	Corneous, concentric, unguiculate

SHELL SIZE RANGE
2½ to 5 in
(64 to 125 mm)

PHOTOGRAPHED SHELL
4 in
(99 mm)

VASUM MURICATUM
CARIBBEAN VASE
(BORN, 1778)

The shell of the Caribbean Vase is heavy, extremely thick, and conical. Its spire is short and pointed, with a well-marked suture. The body whorl is large, and sculptured with strong and fine spiral cords that form blunt tubercles at the shoulder and near the base. The aperture is broad and long, tapering toward its anterior end. The columella is thick and callous, with 4 or 5 folds. Underneath a thick brown periostracum, the shell color ranges from creamy to bright white. The aperture is white, and it is sometimes tinged with purple.

Vasum muricatum is a common species, usually found in shallow water near seagrass beds, buried in sand and coral rubble. The juveniles may burrow in the sand, but the adults live on the surface of the sediment. It ranges from Florida to Venezuela, and the Gulf of Mexico. It is a carnivore, preying mostly on polychaete worms and bivalves with its long proboscis. It is active at night and hides during the day. It is sometimes found in groups. The operculum is corneous, claw-shaped, and has a terminal nucleus.

RELATED SPECIES

Vasum cassiforme (Kiener, 1840), which is endemic to north and northeastern Brazil, has a thick callus and parietal shield over the columella. Specimens from calmer waters have more developed spines than those from higher-energy habitats. *Altivasum flindersi* (Verco, 1914), from western and southern Australia, is the largest of the vase shells; it has a tall spire that is longer than half of the shell length.

Actual size

FAMILY	Turbinellidae
SHELL SIZE RANGE	4 to 8½ in (100 to 220 mm)
DISTRIBUTION	Southeast India and Sri Lanka
ABUNDANCE	Abundant offshore
DEPTH	Shallow subtidal to offshore
HABITAT	Sandy bottoms
FEEDING HABIT	Carnivore, feeds on worms
OPERCULUM	Corneous, concentric, elongate

SHELL SIZE RANGE
4 to 8½ in
(100 to 220 mm)

PHOTOGRAPHED SHELL
4½ in
(120 mm)

TURBINELLA PYRUM

INDIAN CHANK

(LINNAEUS, 1758)

477

Turbinella pyrum has the distinction of being one of few shells that are considered sacred; the word "chank" means "the divine conch" in Sanskrit. According to Hindu mythology, a trumpet made from an Indian Chank is carried by the god Krishna as a symbol of his victory over the evil demon Panchajana. The trumpet is used ceremonially in India before combat and in religious ceremonies. Rare sinistral specimens of *T. pyrum* are particularly sought after because they have even greater religious significance. *Turbinella pyrum* is a highly variable species, and several subspecies have been named.

RELATED SPECIES

Turbinella laevigata (Anton, 1839), which is endemic to northeastern Brazil, has a shell that is similar in size, but more slender and with a taller spire than that of *T. pyrum*. *Turbinella angulata* (Lightfoot, 1786), from the West Indies, has a larger and more angular profile, with strong, pointed tubercles on the shoulder.

The shell of the Indian Chank is large, massive, heavy, and club-shaped. Its spire is short, and can be turreted in some shells. The whorls are inflated, and convex, with spiral cords that are strongest in the lower part of the whorl (not evident in the shell illustrated here). The shoulder has tubercles that vary from weak to smooth. The aperture is long, and lanceolate, and the lip is thick and smooth. The columella has 3 or 4 strong folds, and a thick callus posteriorly. The siphonal canal is long. Beneath the thick, dark brown periostracum the shell is white or apricot, and the aperture is yellowish.

Actual size

FAMILY	Turbinellidae
SHELL SIZE RANGE	4¾ to 14¼ in (127 to 365 mm)
DISTRIBUTION	Eastern Mexico to Panama; Bahamas
ABUNDANCE	Common
DEPTH	Subtidal to 150 ft (45 m)
HABITAT	Sandy bottoms
FEEDING HABIT	Carnivore
OPERCULUM	Corneous, ovate

SHELL SIZE RANGE
4¾ to 14¼ in
(127 to 365 mm)

PHOTOGRAPHED SHELL
13¾ in
(345 mm)

478

TURBINELLA ANGULATA
WEST INDIAN CHANK
(LIGHTFOOT, 1786)

The shell of the West Indian Chank is very dense. A high spire is made up of around 7 whorls, each with prominent knobs that point posteriorly. The whorls are separated by a clearly defined suture, below which spiral ridges run down to the shoulder. The ridges usually become weaker at the nodulose periphery of each whorl, and obsolete in the middle of the body whorl. A large ovate aperture is bounded by a columella with 3 distinctive folds and an outer lip that is notched posteriorly. A brown periostracum covers a pale off-white to light orange shell.

Chanks are among the heaviest shells for their size, and the West Indian Chank, being the largest of them all, is one of the largest gastropods found in the Atlantic. It is an important food source in the Bahamas, where it is known as the Pepper Conch. Various chanks have been used to make implements and horns, not least the West Indian Chank, which was used as such by the Classic Maya civilization.

RELATED SPECIES
Pleuroploca gigantea (Kiener, 1840), found from North Carolina to the Gulf of Mexico, although not related is often misidentified as *Turbinella angulata*. *Pleuroploca gigantea* is generally larger and has clearly defined, regularly spaced spiral cords across the entire body whorl (which are absent in the middle of the whorls of *T. angulata*), and a smooth columella compared with *T. angulata*'s plicate one. *Turbinella pyrum* (Linnaeus, 1758) from India to Sri Lanka, has a much shorter spire and less nodulose shoulders, both of which give the shell a much rounder appearance.

Actual size

FAMILY	Turbinellidae
SHELL SIZE RANGE	12 to 39 in (300 to 1,000 mm)
DISTRIBUTION	Northern Australia and Papua New Guinea
ABUNDANCE	Common
DEPTH	Intertidal to 130 ft (40 m)
HABITAT	Intertidal sand flats
FEEDING HABIT	Carnivore, feeds on large polychaete worms
OPERCULUM	Corneous, with terminal nucleus

SHELL SIZE RANGE
12 to 40 in
(300 to 1,000 mm)

PHOTOGRAPHED SHELL
23 in
(578 mm)

SYRINX ARUANUS

AUSTRALIAN TRUMPET
(LINNAEUS, 1758)

479

Syrinx aruanus is the world's largest shelled gastropod. It is a fairly common species in shallow tidal flats, but also occurs to depths of 130 ft (40 m) off northern Australia and Papua New Guinea. Like other turbinellids, it feeds on tube-dwelling polychaetes. The shell is heavy and voluminous. It is covered with a thick brown skin, the periostracum, that flakes off in empty shells. This gastropod is fished for its meat and shell, which is used to carry water or as a trumpet. Because of the ease in collecting this species, some local populations have declined, causing concern about its conservation.

RELATED SPECIES

There is a single species in the genus *Syrinx*. Other species in the family Turbinellidae include species with varied shell shapes, such as the vase shells, for example, *Vasum muricatum* (Born, 1778) from southern Florida and the Caribbean, or the heavy-shelled chanks, such as the Indian chank, *Turbinella pyrum* (Linnaeus, 1758) from the Indian Ocean, which is the closest relative to *Syrinx aruanus*.

The shell of the Australian Trumpet is very large and fusiform with a long and straight anterior canal. The aperture and columella are smooth, and the umbilicus deep and elongated, covered partially by a columellar shield. Shells from northern Australia usually are strongly keeled, while those from Western Australia have rounded shoulders. The embryonic shell is quite long, with many whorls and persists in some juvenile shells, although it often erodes away in larger specimens. The shell color is apricot or cream and the aperture is pale yellow to orange.

Actual size

FAMILY	Ptychatractidae
SHELL SIZE RANGE	1 to 1½ in (27 to 38 mm)
DISTRIBUTION	Off Costa Rica
ABUNDANCE	Rare
DEPTH	6,400 ft (2,000 m)
HABITAT	Ooze bottoms
FEEDING HABIT	Carnivore
OPERCULUM	Absent

SHELL SIZE RANGE
1 to 1½ in
(27 to 38 mm)

PHOTOGRAPHED SHELL
1 in
(26 mm)

EXILIA BLANDA
SMOOTH EXILIA
(DALL, 1908)

Actual size

The shell of the Smooth Exilia is small, thin, fragile, narrow, and fusiform with a tall spire and a long and broad axial siphonal canal. The first few whorls of the spire have axial ribs, with fine spiral cords becoming dominant in later whorls. The aperture is ovate, with a thin parietal callus. There are no folds on the columella. The shell color is yellowish tan.

Exilia blanda is an extremely rare species that is known only from a single specimen collected by a research expedition more than a century ago. This species lives on muddy ooze on the edge of the abyssal plain off the Pacific coast of Costa Rica. It is probably a predator, but may also feed on carrion. Comparisons with other species of *Exilia* suggest that the species may reach a slightly larger size.

RELATED SPECIES

Exilia kiwi (Kantor and Bouchet, 2001) from slightly shallower depths of 4,547 to 5,499 ft (1,386 to 1,676 m) off New Zealand, has a slightly larger and broader shell in which the spiral cords are weak or absent on the final whorl. The shell of *E. hilgendorfi* (Martens, 1897), from Japan and the Philippines, can exceed 3 in (76 mm), and is much thicker. It has axial ribs on the spire and incised spiral furrows on the body whorl, as well as three folds on the columella.

FAMILY	Ptychatractidae
SHELL SIZE RANGE	1¼ to 2 in (30 to 52 mm)
DISTRIBUTION	Taiwan, Philippines, and Indonesia
ABUNDANCE	Rare
DEPTH	820 to 3,280 ft (250 to 1,000 m)
HABITAT	Sandy bottoms
FEEDING HABIT	Carnivore
OPERCULUM	Corneous, thin, reduced

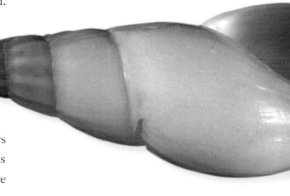

LATIROMITRA BARTHELOWI

BARTHELOW'S LATIROMITRA

(BARTSCH, 1942)

481

Like all members of the family Ptychatrachtidae, *Latiromitra barthelowi* is a deepwater species that is infrequently collected. It has a broad geographic range and appears to be restricted to sandy bottoms along the upper continental slope. The long narrow shells indicate that the animals may burrow in the sand.

RELATED SPECIES

Latiromitra meekiana (Dall, 1889) from comparable depths off Cuba, has a smaller, wider shell, with weaker axial sculpture on the spire and more pronounced columellar folds. *Latiromitra cryptodon* (Fischer, 1882), from deeper waters of the Caribbean, Azores, and Morocco, is similar in size, but has a broader, heavier shell, with more pronounced axial sculpture and a shorter siphonal canal.

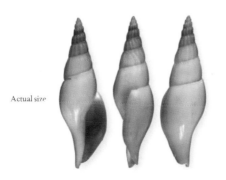

Actual size

The shell of the Barthelow's Latiromitra is comparatively large for the genus, thin, and narrowly fusiform. The spire is tall, the aperture narrow, and the siphonal canal shorter than the aperture and deflected slightly to the left. Axial ribs are prominent on the spire, but the last shell whorl is smooth, or with fine spiral incised lines. The columella has 3 weak folds. The shell color is white, with an olive brown periostracum.

FAMILY	Volutidae
SHELL SIZE RANGE	½ to ¾ in (11 to 19 mm)
DISTRIBUTION	Caribbean to northern Brazil
ABUNDANCE	Rare
DEPTH	Offshore to 100 ft (30 m)
HABITAT	Sandy bottoms
FEEDING HABIT	Carnivore
OPERCULUM	Corneous, elongate

SHELL SIZE RANGE
½ to ¾ in
(11 to 19 mm)

PHOTOGRAPHED SHELL
½ in
(14 mm)

482

ENAETA GUILDINGII
GUILDING'S LYRIA
(SOWERBY I, 1844)

Enaeta guildingii is one of the smallest volutes, and resembles more a columbellid or buccinid than a volute. It is a rare species from the Caribbean and northern Brazil, and not much is known about its biology. There are only about seven species of *Enaeta*, most of them uncommon to rare and all with small shells. There are about 250 species in the family Volutidae worldwide, with the highest diversity in Australia.

RELATED SPECIES

Enaeta reevei (Dall, 1907), from Honduras to Brazil, has a shell similar in size and shape to *E. guildingii*, but the axial ribs on the spire become obsolete on the body whorl. *Enaeta cumingii* (Broderip, 1832), which ranges from Baja California to Peru, is the only common species in the genus, and has a larger and broader shell than *E. guildingii*.

Actual size

The shell of the Guilding's Lyria is very small for the family, solid, elongate-oblong, with a tall spire, that is blunt at the apex. The protoconch is smooth but the teleoconch is sculptured with axial ribs, crossed by spiral striae. Its spire whorls are convex and the suture is impressed. The aperture is narrow and lanceolate. The outer lip is thickened and has a blunt tooth near the posterior end. The columella is concave and has 5 or 6 folds. The shell color is orange to brown, with light-colored lines.

FAMILY	Volutidae
SHELL SIZE RANGE	1 to 2¾ in (25 to 70 mm)
DISTRIBUTION	Australia, from Queensland to New South Wales
ABUNDANCE	Common
DEPTH	Intertidal to 180 ft (55 m)
HABITAT	Sandy bottoms
FEEDING HABIT	Carnivore
OPERCULUM	Absent

SHELL SIZE RANGE
1 to 2¾ in
(25 to 70 mm)

PHOTOGRAPHED SHELL
1¾ in
(44 mm)

AMORIA ZEBRA

ZEBRA VOLUTE

(LEACH, 1814)

483

Amoria zebra, like all species in the genus, is endemic to Australia, ranging from Queensland to New South Wales. It lives on sandy bottoms, from the intertidal zone to offshore. It is a carnivore like other volutes, feeding on other mollusks. This species can sometimes be found in large colonies. Although considered common, *A. zebra* and other *Amoria* species were included on a list of Australian species potentially vulnerable to the shell trade because they can be easily collected in large numbers.

RELATED SPECIES

Amoria dampieria Weaver, 1960, from western and northern Australia, has a shell similar in shape and color pattern to that of *A. zebra* but with thicker axial lines. *Amoria damonii* Gray, 1864, from northwestern Australia to Queensland, has a larger, elongate-ovate shell with variable coloration that may resemble the shell form of the genus *Oliva*.

The shell of the Zebra Volute is small for the family, glossy, oblong-ovate, and with a short spire. The protoconch is blunt and rounded, with slightly concave spire whorls and an indented suture. The spire whorls may have axial ribs but the rest of the shell is smooth and glossy. Its body whorl is large and inflated at the shoulder. The outer lip is thickened and the columella has 4 strong folds. The shell color is white to light brown, sometimes golden, with brown axial lines. The aperture and columella are white.

Actual size

FAMILY	Volutidae
SHELL SIZE RANGE	1⅛ to 4⅜ in (28 to 111 mm)
DISTRIBUTION	Dominican Republic to Venezuela
ABUNDANCE	Common
DEPTH	Shallow subtidal to 33 ft (10 m)
HABITAT	Sandy bottoms
FEEDING HABIT	Carnivore, feeds on other mollusks
OPERCULUM	Corneous, concentric, elongate

SHELL SIZE RANGE
1⅛ to 4⅜ in
(28 to 111 mm)

PHOTOGRAPHED SHELL
2 in
(52 mm)

484

VOLUTA MUSICA
COMMON MUSIC VOLUTE
LINNAEUS, 1758

The shell of the Common Music Volute is thick, heavy, and very variable in shape and color. The shape ranges from ovate to almost triangular. The protoconch is blunt and smooth, and the suture is uneven. The whorls have blunt or sharp knobs on the shoulder. Its aperture is long and moderately wide, and the outer lip reflected. The columella is glazed and has about 10 folds. The shell color is ivory to pinkish white with spiral red-brown lines, blotches, and dots—a pattern that resembles written music.

Voluta musica is a good name for this shell, which is decorated with spiral lines, vertical bars, and blotches reminiscent of sheet music. This common and variable species from the Caribbean lives on sandy bottoms in shallow waters. Some studies suggest that there is sexual dimorphism in shell size, with males usually having smaller and less nodulose shells, and females larger and more strongly knobbed ones. Both shell shape and color vary greatly, and several populations have been erroneously named as separate species. The operculum is corneous, elongate, and relatively small.

RELATED SPECIES

Voluta ebraea Linnaeus, 1758, from northeastern Brazil, has a larger, elongated, and heavy shell with sharp spines at the shoulder, resembling some shells of *V. musica*. *Voluta virescens* Lightfoot, 1786, from Belize to Colombia, has a smaller, more elongated, and usually less knobbed shell than *V. musica*.

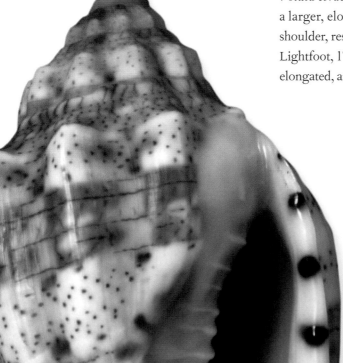

Actual size

FAMILY	Volutidae
SHELL SIZE RANGE	2½ to 4 in (60 to 105 mm)
DISTRIBUTION	South India and Sri Lanka
ABUNDANCE	Uncommon
DEPTH	Intertidal to 80 ft (25 m)
HABITAT	Sandy bottoms
FEEDING HABIT	Carnivore, feeds on other mollusks
OPERCULUM	Corneous, elongate, with terminal nucleus

SHELL SIZE RANGE
2½ to 4 in
(60 to 105 mm)

PHOTOGRAPHED SHELL
2¾ in
(70 mm)

HARPULINA ARAUSIACA

VEXILLATE VOLUTE

(LIGHTFOOT, 1786)

485

Harpulina arausiaca is a beautiful volute with a very restricted range, being endemic to southern India and Sri Lanka. Although it lives in shallow water, sometimes in the intertidal zone, shells are scarce and much sought-after by collectors. Its biology is not well known. There are four species in the genus *Harpulina*, and all are restricted to the same general area.

RELATED SPECIES

Harpulina loroisi (Valenciennes, 1863), which is also endemic to southern India and Sri Lanka, is similar but has dark brown axial bands. It lives in deeper water but it is more commonly collected. *Lyria lyraeformis* (Swainson, 1821), from Kenya to Mozambique, has a handsome, elongate fusiform shell with a tall spire and many axial ribs.

The shell of the Vexillate Volute is medium-sized, solid, heavy, and ovate-fusiform in outline. Its spire is moderately high, with a pointed protoconch. The early whorls are axially ribbed, but the sculpture becomes smooth in later whorls. There is a spiral row of blunt tubercles at the shoulder. Its aperture is semi-ovate and long, the outer lip smooth and sharp, and the columella has 6 to 8 folds. The shell color is off-white to pale pink with bright red-orange axial bands; the aperture is white.

Actual size

FAMILY	Volutidae
SHELL SIZE RANGE	1¾ to 6¼ in (45 to 160 mm)
DISTRIBUTION	Tropical West Pacific, from the Philippines to N. Australia
ABUNDANCE	Common
DEPTH	3 to 65 ft (1 to 20 m)
HABITAT	Sandy and muddy bottoms
FEEDING HABIT	Carnivore, feeds on other mollusks
OPERCULUM	Absent

SHELL SIZE RANGE
1¾ to 6¼ in
(45 to 160 mm)

PHOTOGRAPHED SHELL
2⅞ in
(74 mm)

CYMBIOLA VESPERTILIO
BAT VOLUTE
(LINNAEUS, 1758)

Cymbiola vespertilio is a common shallow-water gastropod with a restricted distribution, ranging from the Philippines to Indonesia and Papua New Guinea, the region known as the "Coral Triangle," and in the Northern Territory, Australia. This species has a shell that is very variable in shape and color, therefore many subspecies and forms have been named. It is collected and sold locally as food, but is not a commercially important species. The specimen in the photograph is sinistral, the result of a genetic mutation; such shells are uncommon in nature, and rare in collections. Normal shells are dextral, with the aperture to the right of the columella.

RELATED SPECIES

There are many named species of *Cymbiola*, including *C. aulica* (Sowerby I, 1825), from the Philippines, which has a similarly shaped shell but with a more reddish tint. *Cymbiola pulchra* (Sowerby I, 1825), from Queensland, Australia, is another variable species with handsomely colored shells, usually with broad spiral bands and white and brown mottlings.

Actual size

The shell of the Bat Volute is heavy, elongate-ovate in general shape but very variable in shape and color. The spire is short, and the shoulder has blunt to sharp spines. The outer surface is smooth and glossy, with fine axial growth lines. The aperture is wide, long, and smooth, and the columella has 4 oblique folds. Its outer lip is smooth and thickened, and the siphonal canal is wide with a deep notch. The shell coloration is quite variable, but is usually cream to olive, with zigzag or tented brown lines, or blotches; the interior is gray to cream, with an orangish columella and lip margin.

FAMILY	Volutidae
SHELL SIZE RANGE	1⅜ to 3¼ in (35 to 81 mm)
DISTRIBUTION	Southernmost Argentina to Antarctica
ABUNDANCE	Rare
DEPTH	Abyssal, about 10,000 ft (3,000 m)
HABITAT	Mud bottoms
FEEDING HABIT	Carnivore, feeds on other mollusks
OPERCULUM	Absent

SHELL SIZE RANGE
1⅜ to 3¼ in
(35 to 81 mm)

PHOTOGRAPHED SHELL
3¼ in
(81 mm)

TRACTOLIRA GERMONAE

GERMON'S VOLUTE

HARASEWYCH, 1987

487

Tractolira germonae is an abyssal gastropod from very deep and cold waters off Antarctica. It lives below the calcium carbonate compensation depth, where, because of the low temperature, and the high pressure and concentration of dissolved CO_2, the calcium carbonate in the shell dissolves in seawater. The shell of *T. germonae* has a thick, dark, protein layer, the periostracum, which protects the shell from being dissolved. In areas where the periostracum is abraded or damaged, the shell dissolves, and the animal must deposit additional shell material from within. The apex is usually eroded.

RELATED SPECIES

The genus *Tractolira* currently has only four species, all from abyssal waters: *T. sparta* Dall, 1896, from the Gulf of Panama to western Mexico, has an elongate shell with a smaller aperture than *T. germonae; T. tenebrosa* Leal and Bouchet, 1989, from off Brazil, has a light periostracum and small knobs on the whorl shoulder; and *T. delli* Leal and Harasewych, 2005, from the Ross Sea, Antarctica, has a large aperture, and fine spiral and axial lines that form a cancellate sculpture.

Actual size

The shell of the Germon's Volute is extremely thin, translucent, and elongate-fusiform. The apex is usually eroded but some shells still show signs of a calcarella at the apex, a projection on the protoconch. The whorls are moderately convex and the suture is impressed. Its sculpture consists of very fine spiral threads and fine growth striae. The aperture is ovate, the outer lip smooth and flared, and the columella smooth. The shell color is white, and the thick periostracum brown.

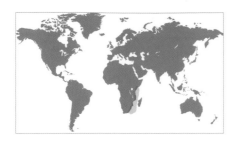

FAMILY	Volutidae
SHELL SIZE RANGE	3 to 5⅝ in (74 to 145 mm)
DISTRIBUTION	East Africa, from Kenya to northern Mozambique
ABUNDANCE	Uncommon
DEPTH	Offshore, to 820 ft (250 m)
HABITAT	Sandy bottoms
FEEDING HABIT	Carnivore, feeds on other mollusks
OPERCULUM	If present, corneous and elongate

SHELL SIZE RANGE
3 to 5⅝ in
(74 to 145 mm)

PHOTOGRAPHED SHELL
3⅜ in
(87 mm)

488

LYRIA LYRAEFORMIS
LYRE-FORMED LYRIA
(SWAINSON, 1821)

The shell of the Lyre-formed Lyria is solid, and elongate-fusiform, with a high spire. The protoconch is bulbous and smooth, and the tall spire whorls are convex, with about 18 axial ribs per whorl. Its aperture is relatively small, oblong and wider anteriorly; the outer lip is smooth and thickened. The columella has 2 or 3 plaits and lyrations centrally, and the siphonal notch is small. The shell color is creamy to pinkish with red-brown spiral bands broken between axial ribs.

Lyria lyraeformis is an uncommon deepwater volute from East Africa. It has seldom been collected live, thus much of its biology is not well known, including its feeding preferences. All volutes are believed to undergo direct development within the egg capsule, and do not have a free-swimming larval stage, which accounts for the narrow distribution of many species. Direct-developers usually have a large larval shell, as seen in many volutes, while gastropods that have a long larval life usually possess small larval shells, such as many species of *Cymatium*.

RELATED SPECIES
Lyria doutei Bouchet and Bail, 1991, from the Saya de Malha Bank, East Africa, has a shell similar to *L. lyraeformis* but with more axial ribs and white coloring. *Lyria beauii* (Fischer and Bernardi, 1857), from the Lesser Antilles, has a smaller shell with shorter spire than *L. lyraeformis*, and brown spots on the inner lip.

Actual size

FAMILY	Volutidae
SHELL SIZE RANGE	1¾ to 4⅜ in (45 to 112 mm)
DISTRIBUTION	Endemic to South Africa
ABUNDANCE	Uncommon
DEPTH	360 to 1,800 ft (110 to 550 m)
HABITAT	On iron-ore and shell rubble bottoms
FEEDING HABIT	Carnivore, feeds on other mollusks
OPERCULUM	Absent

SHELL SIZE RANGE
1¾ to 4⅜ in
(45 to 112 mm)

PHOTOGRAPHED SHELL
3½ in
(91 mm)

VOLUTOCORBIS ABYSSICOLA

DEEPSEA VOLUTE

(ADAMS AND REEVE, 1848)

489

Volutocorbis abyssicola is considered a "living fossil" because of its resemblance to the extinct genus, *Volutilithes*, which flourished form the Cretaceous to the Miocene Period. Its discovery in 1848 caused surprise because it was the first living species of this ancient lineage. Since then, an additional nine living species have been discovered, all in deep water, and most from around South Africa. *Volutocorbis abyssicola* is the largest living species in the genus *Volutocorbis*. It appears to be relatively common in its deepwater habitat. The animal has a long siphon, and the eyes are along the thickened outer edges of the cephalic tentacles.

The shell of the Deepsea Volute is medium to large, lightweight, and elongate-pyriform. Its spire is short, the whorls slightly convex, the apex often eroded, and the suture well marked. The sculpture consists of both axial and spiral ribs forming a cancellate pattern; the interior is smooth. Its aperture is relatively narrow and long, the outer lip thickened and denticulate; the columella has several strong white folds. The shell color is beige or brown, and the aperture is beige.

RELATED SPECIES

Volutocorbis lutosa Koch, 1948, from Angola to South Africa, has a smaller and broader shell, with fewer and less prominent columellar folds. It is found in shallower waters and is locally common. *Neptuneopsis gilchristi* (Sowerby III, 1898), which is also endemic to South Africa, has a large, elongate and slender fusiform shell with a large, bulbous apex.

Actual size

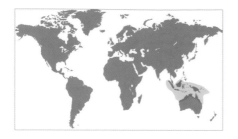

FAMILY	Volutidae
SHELL SIZE RANGE	3⅜ to 6¼ in (70 to 160 mm)
DISTRIBUTION	Northern Australia to Papua New Guinea and Indonesia
ABUNDANCE	Formerly rare; now uncommon
DEPTH	Intertidal to 330 ft (100 m)
HABITAT	Sand bottoms and seagrass
FEEDING HABIT	Carnivore, feeds on other mollusks
OPERCULUM	Absent

SHELL SIZE RANGE
3⅜ to 6¼ in
(70 to 160 mm)

PHOTOGRAPHED SHELL
3¾ in
(95 mm)

VOLUTOCONUS BEDNALLI

BEDNALL'S VOLUTE

(BRAZIER, 1878)

Volutoconus bednalli is a striking shell with a network of chocolate brown lines on a white background. It was once considered one of the rarest volutes, but now pearl divers bring up increasing numbers of specimens. It lives on sandy bottoms and seagrasses, usually in shallow subtidal to offshore waters, but it is sometimes also found in the intertidal zone. There are at least four species in the genus *Volutoconus* living in northern Australia, and all of them are uncommon. They have a needlelike projection of the larval shell that is added by the mantle later in life.

RELATED SPECIES

Volutoconus hargreavesi (Angas, 1872), from western Australia, has a slightly smaller shell that is morphologically variable; the shell of one population from the west coast has a ribbed shell. *Cymbiola vespertilio* (Linnaeus, 1758), from the Philippines to northern Australia, is a common shallow-water volute that is collected for food.

Actual size

The shell of the Bednall's Volute is large for the genus, solid, glossy, and fusiform. Its spire is moderately high and the larval shell large and rounded, with a short and pointed calcarella. The aperture is narrow and long, the outer lip smooth; the columella has 3 or 4 strong folds. The shell is usually smooth, with axial growth striae. The shell color is cream or pinkish with a network of chocolate-brown zigzag lines crossing spiral lines of the same color.

FAMILY	Volutidae
SHELL SIZE RANGE	3 to 6 in (76 to 146 mm)
DISTRIBUTION	Western Pacific
ABUNDANCE	Uncommon
DEPTH	100 to 1,000 ft (30 to 300 m)
HABITAT	Unknown
FEEDING HABIT	Carnivore, feeds on other mollusks
OPERCULUM	Absent

SHELL SIZE RANGE
3 to 6 in
(76 to 146 mm)

PHOTOGRAPHED SHELL
4⅜ in
(110 mm)

491

FULGORARIA RUPESTRIS
ASIAN FLAME VOLUTE
(GMELIN, 1791)

Fulgoraria rupestris is a deepwater species that was once collected by trawlers working off Taiwan. Since the fishing fleets have moved to different areas, these shells are now rarely collected. The habitat of *F. rupestris* remains unknown, since the trawling fishermen did not keep detailed data. Like other volutids, this species is a carnivore and probably feeds on other mollusks, but its biology is poorly known. There are about 25 species recognized in the genus *Fulgoraria*, with most species occurring off the coasts of Japan and China.

RELATED SPECIES

Fulgoraria hamillei (Crosse, 1869), from Japan and Taiwan, has a similarly shaped and sized shell to *F. rupestris*, but with a more impressed suture; it also lacks crenulations on the lip. *Fulgoraria hirasei* (Sowerby III, 1912), from Japan, has a larger, fusiform shell with strong vertical ribs, thin incised spiral lines, and a large aperture that is about half as long as the shell.

The shell of the Asian Flame Volute is solid, thick, and fusiform. It has a moderately tall spire, with a large mammillate protoconch and angular whorls. The penultimate whorl has about 14 axial ribs that become obsolete on the body whorl, which has incised spiral lines. Its aperture is long and the outer lip is thickened and straight-sided with crenulations. The columella has about 7 to 9 folds. The shell color is creamy white to tan with broad, brown, zigzag flames, and the aperture is white to pinkish.

Actual size

FAMILY	Volutidae
SHELL SIZE RANGE	2½ to 6 in (64 to 155 mm)
DISTRIBUTION	North Carolina to the Florida Keys, Gulf of Mexico
ABUNDANCE	Uncommon; perfect shells rare
DEPTH	65 to 300 ft (20 to 90 m)
HABITAT	Sandy bottoms
FEEDING HABIT	Carnivore, feeds on other mollusks
OPERCULUM	Absent

SHELL SIZE RANGE
2½ to 6 in
(64 to 155 mm)

PHOTOGRAPHED SHELL
5¾ in
(147 mm)

492

SCAPHELLA JUNONIA
JUNONIA
(LAMARCK, 1804)

In the nineteenth century, *Scaphella junonia* was considered one of the rarest of volutes. This species is now encountered as a bycatch of shrimp boats, making it among the most common of *Scaphella* species. Perfect specimens are still rare, as most shells have healed growth scars. This mollusk lives on sand offshore from North Carolina to the Florida Keys and in the Gulf of Mexico. Like all volutes, it is a carnivore and feeds on other mollusks.

RELATED SPECIES

About ten species and several subspecies have been recognized in the genus, although some might be variants of the widely distributed species *Scaphella dohrni* (Sowerby III, 1903); other species from the southeastern U.S.A. and Gulf of Mexico include *S. dubia* (Broderip, 1827), with an elongate shell; and *S. gouldiana* (Dall, 1887), with a slender, golden, or pinkish shell that has lighter spiral bands and lacks blotches.

Actual size

The shell of the Junonia is fusiform, solid, and large. The spire is high, with well-impressed suture and the protoconch is smooth with 1½ to 2 whorls. The teleoconch has 5 whorls, finely sculptured with axial ribs, with the last 2 whorls nearly smooth. The shell color ranges from cream to pale yellow with spiral rows of brown rectangular blotches. The long aperture is pinkish and has 4 columellar folds; the columella and aperture are cream.

FAMILY	Volutidae
SHELL SIZE RANGE	4 to 14¼ in (100 to 360 mm)
DISTRIBUTION	Senegal to the Gulf of Guinea, western Africa
ABUNDANCE	Common
DEPTH	Shallow subtidal
HABITAT	Sandy bottoms
FEEDING HABIT	Carnivore, feeds on other mollusks
OPERCULUM	Absent

SHELL SIZE RANGE
4 to 14¼ in
(100 to 360 mm)

PHOTOGRAPHED SHELL
8⅜ in
(215 mm)

CYMBIUM GLANS
ELEPHANT'S SNOUT VOLUTE
(GMELIN, 1791)

493

Cymbium glans is the largest of about a dozen species from West Africa in the genus *Cymbium*. All have cylindrical shells and a rounded apex, with some species having shells that are more inflated or globose than others. Volutids have direct development and a juvenile crawls out of the egg capsule with a shell. As in other volutids, species of *Cymbium* are carnivores and feed on other mollusks, which they capture and immobilize with their large and muscular foot. They live on sandy and muddy bottoms in shallow water.

Actual size

RELATED SPECIES
Cymbium cucumis Röding, 1798 and *C. pepo* (Lightfoot, 1786), both inhabit the waters off central western Africa; the former has a similar but smaller, more elongate, and cylindrical shell; the latter has a large, short, and very inflated shell with a very wide aperture. *Cymbium olla* (Linnaeus, 1758), from the Mediterranean and northwestern Africa, has a smaller and less inflated shell than *C. pepo*, and a short spire with a smooth apex.

The shell of the Elephant's Snout Volute is thin, lightweight but sturdy, large, and cylindrical-ovate in shape. The spire is sunken and the apex rounded. The body whorl is very large; the aperture is widest at the middle, and as long as the shell length. The shell is smooth, with fine axial growth striae, and occasional warty pustules on some specimens. The outer lip is thin and smooth; the columella has 4 columellar folds. The shell color is cream to orange-brown; the aperture is orange.

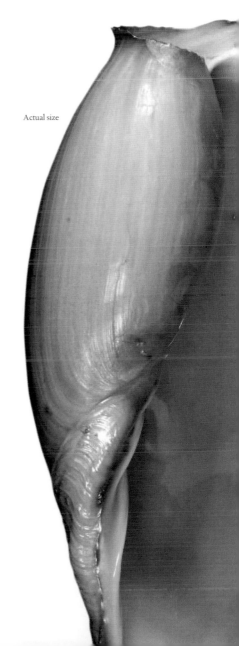

FAMILY	Volutidae
SHELL SIZE RANGE	4 to 10¾ in (100 to 270 mm)
DISTRIBUTION	Rio de Janeiro, Brazil to central Argentina
ABUNDANCE	Common
DEPTH	120 to 240 ft (15 to 200 m)
HABITAT	Sand and mud substrate
FEEDING HABIT	Carnivorous, feeds on mollusks
OPERCULUM	Absent

SHELL SIZE RANGE
4 to 10¾ in
(100 to 270 mm)

PHOTOGRAPHED SHELL
8½ in
(221 mm)

494

ZIDONA DUFRESNEI
ANGULAR VOLUTE
(DONOVAN, 1823)

Zidona dufresnei is a distinctive volute with an angular shoulder, and in some specimens, a long projection added to its apex. Its shell is variable in size and shape, and a few populations have been described as subspecies. The mantle is large and covers the shell, giving it a glazed appearance. The Angular Volute feeds on bivalves. It is often found offshore on beds of the scallop *Chlamys tehuelchus* (d'Orbigny, 1846). It is an edible species, and there is a commercial fishery in the region for *Z. dufresnei*.

RELATED SPECIES

Zidona palliata Kaiser, 1977, from southern Brazil to Antarctica, has a smaller, elongate, and thinner shell. *Voluta musica* Linnaeus, 1758, ranging from the Dominican Republic to Venezuela, has a thick, heavy shell decorated with spiral lines and vertical bars that resemble sheet music.

Actual size

The shell of the Angular Volute is large, thick, glossy, and fusiform with angular whorls. Its spire is tall, and the apex, covered by a callus, can be long and pointed, straight or curved, and less frequently, Y-shaped. Its whorls are angulate, and the sides of the whorls are nearly straight. The aperture is long and subquadrate. The shell is smooth, covered by a glaze, and yellowish orange in color with reddish brown zigzag lines; the aperture is bright or pale orange.

FAMILY	Volutidae
SHELL SIZE RANGE	4 to 9 in (100 to 228 mm)
DISTRIBUTION	South Africa
ABUNDANCE	Common
DEPTH	200 to 1,500 ft (60 to 450 m)
HABITAT	Muddy bottoms
FEEDING HABIT	Carnivore, feeds on other mollusks
OPERCULUM	Corneous, oblong, with a terminal nucleus

SHELL SIZE RANGE
4 to 9 in
(100 to 229 mm)

PHOTOGRAPHED SHELL
9 in
(229 mm)

495

NEPTUNEOPSIS GILCHRISTI
GILCHRIST'S VOLUTE
(SOWERBY III, 1898)

Neptuneopsis gilchristi is a large, common, deepwater species and popular among collectors. It is one of many mollusks endemic to South Africa, including several species of volutes, cowries, and cones. Because of the geographic position of the southern tip of Africa and its upwelling ocean currents, many mollusks and other marine animals and plants are endemic to the region, particularly the deepwater species.

RELATED SPECIES

There is a single species in the genus *Neptuneopsis*. Other related species include *Calliotectum smithi* (Bartsch, 1942), from the Philippines, which has a large, elongated, fusiform shell with axial ribs on the spire whorls. *Zidona dufresnei* (Donovan, 1823), from Brazil to Argentina, has a large glossy shell with an angular body whorl and a projection of the protoconch.

The shell of the Gilchrist's Volute is large, lightweight, and fusiform. It has a high spire, with a large and bulbous protoconch with 2 whorls. The teleoconch has 6 or 7 smooth and convex whorls, with indented suture. Its aperture is wide and semicircular, the columella smooth, and the siphonal canal short. The shell color is pinkish and aperture pinkish to tan. The thin periostracum is olive brown.

Actual size

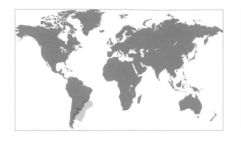

FAMILY	Volutidae
SHELL SIZE RANGE	6 to 20 in (150 to 505 mm)
DISTRIBUTION	Southern Brazil to Argentina
ABUNDANCE	Uncommon
DEPTH	133 to 250 ft (40 to 75 m)
HABITAT	Sandy and muddy bottoms
FEEDING HABIT	Carnivore, feeds on other gastropods
OPERCULUM	Corneous, elongated

SHELL SIZE RANGE
6 to 20 in
(150 to 505 mm)

PHOTOGRAPHED SHELL
13 in
(332 mm)

496

ADELOMELON BECKII
BECK'S VOLUTE
(BRODERIP, 1836)

Beck's Volute is the largest gastropod in the Southern Atlantic, and one of the largest in the world. It is trawled by fishermen from sandy and muddy bottoms from moderately shallow waters. Because it has a variable shell, several forms have been described. Like all volutids, Beck's Volute is a carnivore and feeds on other invertebrates, including other mollusks. Studies of stomach contents have revealed the radular teeth of other volutids, such as *Zidona dufresnei* (Donovan, 1823). The animal has a large foot and it can withdraw completely into its shell.

RELATED SPECIES

Adelomelon riosi Clench and Turner, 1964, also from southern Brazil to southern Argentina has a shell that resembles *A. beckii* but it is smaller, with more inflated and smooth spire whorls. *Zidona dufresnei* (Donovan, 1823), from Brazil to Argentina, has a fusiform, glossy shell with an angular shoulder and a pointed protoconch with calcarella (projection from the protoconch).

Actual size

The shell of the Beck's Volute is very large, fusiform, and slender to inflated, with a tall and pointed spire. The shoulder has knobs, often more pronounced in early whorls and less so in the body whorl. The protoconch is mammillate. Its aperture is large, about half of shell length, lanceolate. The shell is orange to pinkish with brown zigzag longitudinal marks, covered by a thick, brown, deciduous periostracum. The aperture and parietal shield are bright orange to pink, faded to cream in old shells.

FAMILY	Volutidae
SHELL SIZE RANGE	5 to 20¼ in (125 to 515 mm)
DISTRIBUTION	Australia to Papua New Guinea and Indonesia
ABUNDANCE	Common
DEPTH	From shallow subtidal to 33 ft (10 m)
HABITAT	Sand or muddy bottoms
FEEDING HABIT	Carnivore, feeds on other mollusks
OPERCULUM	Absent

SHELL SIZE RANGE
5 to 20¼ in
(125 to 515 mm)

PHOTOGRAPHED SHELL
20¼ in
(514 mm)

497

MELO AMPHORA
GIANT BALER
(LIGHTFOOT, 1786)

Melo amphora is one of the largest gastropods in the world and its shell is voluminous. For that reason, it is often used to carry water, to bail canoes, and for decorative purposes by indigenous peoples. It is also appreciated as food for the quantity and quality of its flesh. The animal has an enormous and muscular foot, which is almost twice as long as its shell when fully extended. The foot color is solid brown or mottled with cream marks. The animal uses its foot to hold its molluscan prey while eating it. It is a voracious carnivore, and sometimes cannibalizes smaller members of the same species.

RELATED SPECIES

The genus *Melo* has few species, such as *M. melo* (Lightfoot, 1786), from the Indian Ocean to the South China Sea, which has a smaller but more globose shell than that of *M. amphora*. The related genus *Cymbiola* has many more species, including *C. vespertilio* (Linnaeus, 1758), from the tropical West Pacific, and *C. imperialis* (Lightfoot, 1786), a common species from the Philippines with a spiny shoulder.

Actual size

The shell of the Giant Baler is very large, globose, and ovate in shape. The spire is short and round, surrounded by a crown of sharp spines at the shoulder. Its body whorl is inflated, with axial growth lines on the outer surface; the lanceolate aperture is broad—almost as long as the shell length. The columella has 3 oblique folds. The shell color ranges from orange to white, decorated with thick, irregular brown to orange axial lines, often with 2 spiral bands of brown blotches. The aperture is glossy, and cream to pink-orange.

FAMILY	Olividae
SHELL SIZE RANGE	½ to 1¼ in (10 to 30 mm)
DISTRIBUTION	Tropical western Pacific
ABUNDANCE	Common
DEPTH	Intertidal to 65 ft (20 m)
HABITAT	Sandy bottoms
FEEDING HABIT	Predatory
OPERCULUM	Absent

SHELL SIZE RANGE
½ to 1¼ in
(10 to 30 mm)

PHOTOGRAPHED SHELL
¾ in
(19 mm)

498

OLIVA CARNEOLA
CARNELIAN OLIVE
(GMELIN, 1791)

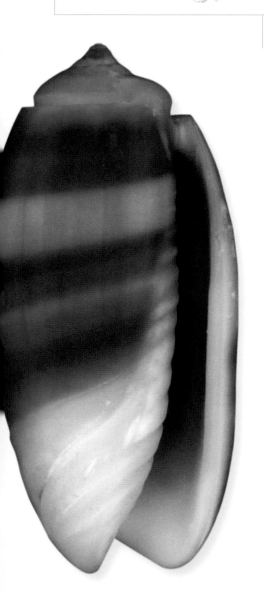

Oliva carneola is a common and small olivid from the tropical western Pacific. It ranges from Japan to Indonesia, and to Melanesia and Polynesia. It is found from the intertidal zone to shallow subtidal waters, buried in sandy bottoms. Its shell is very variable in shape and coloration, but it usually is stout, with a very thick outer lip and orange spiral bands. Like other olivids, it is a predatory snail, and is active at night, when it searches for its invertebrate prey.

RELATED SPECIES

Oliva tessellata Lamarck, 1811, which ranges from the eastern Indian Ocean to New Caledonia, has a shell similar in size and shape to *O. carneola*, but cream or fawn, with purplish brown spots, and a violet aperture. *Oliva bulbosa* (Röding, 1798), from the Red Sea to the western Pacific, also has a similar shell, which can be larger. Its coloration varies widely.

Actual size

The shell of the Carnelian Olive is small, stout, thick, glossy, and elongate-ovate in outline. Its spire is low and pointed, with a thick callus and an incised suture. The surface is smooth and glossy. Its aperture is narrow and long, the outer lip very thick, and the columella has a white callosity with several folds. The shell color is very variable, usually consists of an ivory background with orange or brown spiral bands, or brown zigzag lines, and a white aperture.

FAMILY	Olividae
SHELL SIZE RANGE	¾ to 1¼ in (20 to 32 mm)
DISTRIBUTION	Indo-West Pacific
ABUNDANCE	Uncommon
DEPTH	Intertidal to 65 ft (20 m)
HABITAT	Sandy bottoms near coral reefs
FEEDING HABIT	Carnivore
OPERCULUM	Absent

OLIVA TESSELLATA
TESSELLATE OLIVE
LAMARCK, 1811

SHELL SIZE RANGE
¾ to 1¼ in
(20 to 32 mm)

PHOTOGRAPHED SHELL
1⅛ in
(30 mm)

499

Oliva tessellata is a small but distinctive species, and one of the
few olivids with a spotted shell. Its shell has fairly consistent color
pattern (although rarely some specimens have lines connecting
the spots), but it varies in shape: some shells are slender, while
most are broad and thick. It lives from the intertidal zone to
offshore, on soft bottoms near coral reefs. The animal is white,
and the large, muscular foot has brown spots.

RELATED SPECIES
Oliva maculata Duclos, 1840, from East Africa to the Seychelles,
also has a spotted shell, but it is larger, with a more elongate
profile, and the spots are more numerous and gray. *Oliva
incrassata* Lightfoot, 1786, from Baja California to Peru, has
a very heavy and thick shell, with a thickened shoulder and
angular outer lip.

Actual size

The shell of the Tessellate Olive is small, thick, strong, glossy, inflated, and
oval-cylindrical in outline. Its spire is short, the spire whorls concave, and
the apex pointed, with a thick secondary callus, and an incised groove. The
surface is smooth and glossy. The sides of the body whorl are convex, and
the shell width is about half of the shell length. Its aperture is narrow, the
outer lip thick and smooth, and the columella has several small folds. The
shell color is beige to pale yellow, with spaced-out, purplish brown spots,
and a violet aperture.

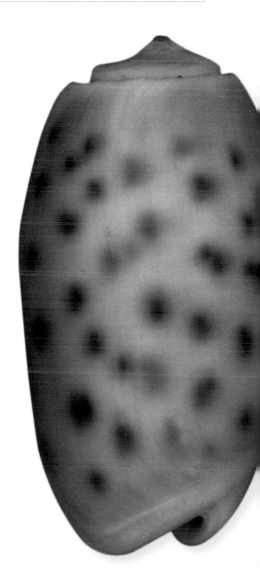

FAMILY	Olividae
SHELL SIZE RANGE	1 to 1½ in (25 to 40 mm)
DISTRIBUTION	Philippines to Melanesia
ABUNDANCE	Common
DEPTH	90 to 150 ft (25 to 45 m)
HABITAT	Sandy bottoms
FEEDING HABIT	Carnivore
OPERCULUM	Absent

SHELL SIZE RANGE
1 to 1½ in
(25 to 40 mm)

PHOTOGRAPHED SHELL
1¼ in
(33 mm)

500

OLIVA RUFULA
RUFULA OLIVE
DUCLOS, 1835

The shell of the Rufula Olive is small-medium in size, thick, glossy, and oval-cylindrical in outline. Its spire is short, with an incised groove and a pointed apex. The sides of the shell are slightly convex, and taper toward the anterior end. The surface is smooth and glossy, but there are shallow furrows parallel to the aperture near the columella. Its aperture is narrow, the outer lip thick and smooth, and the columella has many small folds. The shell color is fawn with darker brown or gray lines forming chevron patterns; the aperture is white.

Oliva rufula is a small olivid with an oval-cylindrical shell and one of the most distinct color patterns in the family, which make it easily recognizable. Some specimens of the variable *O. vidua* may resemble *O. rufula* in shape, but the former is usually larger and the color pattern has dark lines or nearly solid colors. Like *O. vidua*, *O. rufula* has shallow furrows parallel to the aperture. Olivids are sand dwellers, and travel close to the sediment surface with the siphon sticking out, searching for prey. Most species are carnivorous but may also feed on carrion.

RELATED SPECIES

Oliva rufofulgorata Schepman, 1904, with a similar distribution in the western Pacific, has a smaller shell that is oval cylindrical, and vaguely resembles *O. rufula* in color pattern, but the background is cream and the light brown zigzag lines are thin.

Oliva splendidula Sowerby I, 1825, from western Mexico to Peru, has an oval-cylindrical shell decorated with a combination of spiral bands and small tented patterns.

Actual size

FAMILY	Olividae
SHELL SIZE RANGE	1⅛ to 2¼ in (27 to 58 mm)
DISTRIBUTION	Peru to Chile
ABUNDANCE	Common
DEPTH	Shallow subtidal
HABITAT	Sandy bottoms
FEEDING HABIT	Carnivore
OPERCULUM	Absent

SHELL SIZE RANGE
1⅛ to 2¼ in
(27 to 58 mm)

PHOTOGRAPHED SHELL
1½ in
(39 mm)

OLIVA PERUVIANA

PERUVIAN OLIVE

LAMARCK, 1811

501

Oliva peruviana is a well-known species that ranges from Peru to southern Chile. It is easily recognizable by its broad shell with an angled outer lip. It varies widely in shell shape and color, and several subspecies have been named on the basis of these shell characters, for example, *O. peruviana* form *coniformis*, which has a very angulate shell. *Oliva peruviana* lives in shallow waters; it is also common as a Pleistocene fossil in some deposits. Charles Darwin collected some fossil specimens of this olive in Chile during his voyage on the H.M.S. *Beagle*.

The shell of the Peruvian Olive is medium-sized, thick, heavy, inflated, and oval-cylindrical in outline. Its spire is short, with a grooved suture, and a pointed apex. The sides of the shell are convex, and the shell shape varies from oval-cylindrical to conical with an angled shoulder. Its aperture is relatively wide, with a thick and smooth outer lip, and a columella with few small folds. The shell color is variable, usually with a light cream or blue-gray background with red-brown spots or lines; the aperture is white.

RELATED SPECIES

Oliva nitidula sandwicensis Pease, 1860, from the Hawaiian Islands, has a smaller shell that resembles *O. peruviana* in shape, but has a taller spire, usually a smaller shell, and lacks zigzag lines. *Oliva carneola* (Gmelin, 1791), from the tropical western Pacific, has a small, stout shell with a thick outer lip. It is often light colored with spiral bands.

Actual size

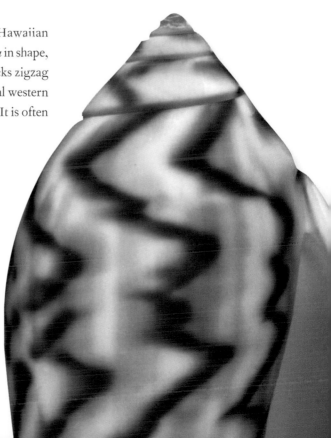

FAMILY	Olividae
SHELL SIZE RANGE	1 to 2¼ in (25 to 55 mm)
DISTRIBUTION	Venezuela to northeastern Brazil
ABUNDANCE	Uncommon
DEPTH	20 to 215 ft (6 to 65 m)
HABITAT	Sandy bottoms
FEEDING HABIT	Carnivore
OPERCULUM	Corneous, thin

SHELL SIZE RANGE
1 to 2¼ in
(25 to 55 mm)

PHOTOGRAPHED SHELL
1⅝ in
(42 mm)

502

ANCILLA LIENARDI
LIENARD'S ANCILLA
(BERNARDI, 1858)

Ancilla lienardi has a very distinctive shell with a bright golden orange color and grooved suture. For a long time it was believed to be endemic to northeastern Brazil, where it is common, but it is also more rarely found as far north as Venezuela. In Brazil it is often found in the stomach of mollusk-eating fishes (*ex pisces*). Many molluscan species are only known *ex pisces* until their habitat is discovered. In the case of *A. lienardi*, it is now known to live offshore on sandy and calcareous algae bottoms.

RELATED SPECIES
Ancillista cingulata (Sowerby I, 1830), endemic to Australia, is probably the largest of the ancillas, reaching 4 in (100 mm) in length. *Ancilla aperta* Sowerby I, 1825, from Somalia to Tanzania, has a much thinner and lighter shell, with a short spire.

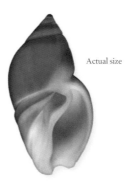

Actual size

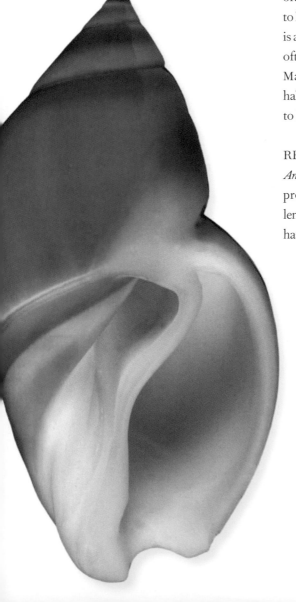

The shell of the Lienard's Ancilla is medium-sized, thick, heavy, glossy, and oval spindle-shaped. Its spire is moderately high, with a well-marked, incised suture, and a rounded apex. The surface is smooth, with a white diagonal spiral groove in the anterior half of the body whorl, and a deep umbilicus. The aperture is wide, the outer lip thick and smooth, the columella concave and smooth. The shell color is usually a solid, bright golden orange, but can be pale yellow or rarely white; the aperture is white.

FAMILY	Olividae
SHELL SIZE RANGE	1¼ to 2¾ in (30 to 70 mm)
DISTRIBUTION	Central Brazil to Argentina
ABUNDANCE	Common
DEPTH	16 to 165 ft (5 to 50 m)
HABITAT	Sandy bottoms
FEEDING HABIT	Carnivore, feeds on other mollusks
OPERCULUM	Absent

SHELL SIZE RANGE
1¼ to 2¾ in
(30 to 70 mm)

PHOTOGRAPHED SHELL
1¾ in
(44 mm)

OLIVANCILLARIA URCEUS
BEAR ANCILLA
(RÖDING, 1798)

503

Olivancillaria urceus is the largest and heaviest species in the genus *Olivancillaria*. There are several species in the genus, and all are endemic to South America, usually ranging from Brazil to Argentina. It is a common species on sandy bottoms, from shallow subtidal to offshore depths. Its flesh is protein-rich, and is harvested commercially. Like other olivids, it feeds on other mollusks.

RELATED SPECIES

Olivancillaria vesica auricularia (Lamarck, 1810), from southern Brazil to Argentina, has a medium-sized shell that is stout and broad. The aperture is very wide, and the outer lip is ear-shaped. *Ancilla lienardi* (Bernardi, 1858), from Venezuela to northeastern Brazil, has a glossy, bright orange or yellow shell.

Actual size

The shell of the Bear Ancilla is medium-sized, thick, heavy, glossy, and triangular-ovate in outline. Its spire is very short, wide, flattened, covered by a thick callus, and a small, pointed apex. The suture is channeled. The body whorl has convex sides, a smooth surface with fine axial lines, and a diagonal spiral line from the columella to the corner of the outer lip. The aperture is wide, the outer lip thick, the columella has small folds, and a very large callus may develop near the posterior end of the columella. The shell color is gray-brown, with a white to orange aperture.

FAMILY	Olividae
SHELL SIZE RANGE	1⅛ to 2¼ in (28 to 55 mm)
DISTRIBUTION	Western Mexico to Peru
ABUNDANCE	Common
DEPTH	Low intertidal to 90 ft (27 m)
HABITAT	Sandy bottoms
FEEDING HABIT	Carnivore
OPERCULUM	Absent

SHELL SIZE RANGE
1⅛ to 2¼ in
(28 to 55 mm)

PHOTOGRAPHED SHELL
1¾ in
(44 mm)

504

OLIVA SPLENDIDULA
SPLENDID OLIVE
SOWERBY I, 1825

Actual size

Oliva splendidula ranges from western Mexico to Peru, and lives in shallow waters. Olive shells are glossy because lateral expansions of the foot cover the shell. As in cowries, pigment is secreted in semitranslucent layers of calcium carbonate that are added onto the dorsum of the shell, resulting in a three-dimensional color pattern (most molluscan shells only have two-dimensional patterns). Unlike all other olives, which have three prismatic layers, *Oliva splendidula* has four. This extra layer explains in part its complex color pattern.

RELATED SPECIES

Oliva sericea (Röding, 1798), from the eastern Indian Ocean to Polynesia, has a variable shell. Some specimens may resemble *O. splendidula*, with spiral bands and tented markings, but their shells are larger, and the spire is shorter. *Oliva porphyria* (Linnaeus, 1758), from the Gulf of California to Panama, has the largest shell in the family. It has a large tented pattern on its shell.

The shell of the Splendid Olive is medium in size, thick, glossy, and oval-cylindrical in outline. Its spire is moderately high, the suture incised, and the apex pointed and tinged in dull pink. The shell sides are slightly convex and taper anteriorly. Its aperture is relatively wide for the genus, the outer lip thick and smooth, and the columella has many small folds. The shell color is complex, with 2 broad brown spiral bands over a cream background, and peppered with fine dots and triangles in cinnamon brown or cream; the lip is white and the aperture yellowish.

FAMILY	Olividae
SHELL SIZE RANGE	⅞ to 2½ in (21 to 60 mm)
DISTRIBUTION	Red Sea to South Africa and to western Pacific
ABUNDANCE	Common
DEPTH	Intertidal to shallow subtidal
HABITAT	Sandy bottoms
FEEDING HABIT	Carnivore
OPERCULUM	Absent

SHELL SIZE RANGE
⅞ to 2½ in
(21 to 60 mm)

PHOTOGRAPHED SHELL
1⅞ in
(49 mm)

OLIVA BULBOSA

INFLATED OLIVE

RÖDING, 1798

505

Oliva bulbosa is aptly named, for its shell is often rather inflated and bulbous. It varies in shell shape and color, and several subspecies have been named on the basis of shell patterns, which seem to be somewhat consistent within populations. The shell can be readily distinguished from other olives by its thickness, the presence of a crestlike fold in the columella, and the elevated callus near the posterior canal. There are approximately 150 species in the genus *Oliva*.

RELATED SPECIES

Oliva bulbiformis Duclos, 1840, from the eastern Indian Ocean to the western Pacific, has a smaller, less inflated shell, but some specimens resemble *O. bulbosa*. It lacks the crestlike fold in the columella and the callus near the posterior canal. *Oliva peruviana* Lamarck, 1811, also has a cylindrical inflated shell, but it is widest at the posterior third of the shell, and the spire is slightly taller.

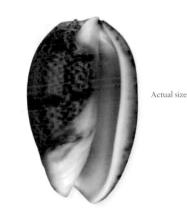

Actual size

The shell of the Inflated Olive is medium in size, thick, heavy, and cylindrical inflated in the middle of the shell. Its spire is short, sometimes shorter than the posterior callus in the aperture; the suture is incised. Its surface is smooth, but the columella has one crestlike bulge on the base, beside a few small folds, inside the aperture. The aperture is moderately wide and long, and the outer lip varies from thick to very thick. The shell color varies widely, from solid white or orange, to small brown spots or irregular bands or blotches. The aperture is white.

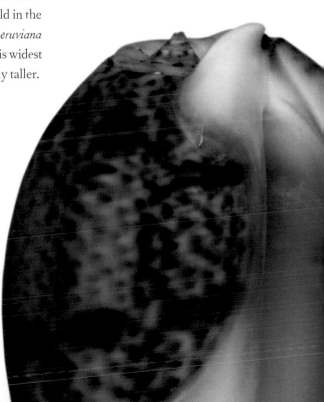

FAMILY	Olividae
SHELL SIZE RANGE	1¼ to 2 in (30 to 50 mm)
DISTRIBUTION	Somalia to Tanzania
ABUNDANCE	Uncommon
DEPTH	Offshore
HABITAT	Sandy bottoms
FEEDING HABIT	Carnivore
OPERCULUM	Corneous, thin

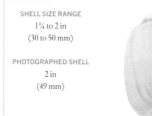

SHELL SIZE RANGE
1¼ to 2 in
(30 to 50 mm)

PHOTOGRAPHED SHELL
2 in
(49 mm)

506

ANCILLA APERTA

GAPING ANCILLA

SOWERBY I, 1825

Ancilla aperta has a very thin shell and a large aperture, which gives it its name. It lives offshore on sandy bottoms and is collected infrequently by trawlers between Somalia and Tanzania. This species was once considered as juveniles of *Ancilla mauritiana*, but *A. aperta* is now considered as a distinct species. There are about 40 living species in the genus *Ancilla* and related genera (*Eburnea* and *Anolacia*) worldwide, although most live in the Indian Ocean.

RELATED SPECIES

Ancilla mauritiana Sowerby I, 1830, from Somalia to Tanzania and Madagascar, has a slightly larger shell, with a taller spire, and fine axial lines. *Ancilla lienardi* (Bernardi, 1858), ranging from Venezuela to northeastern Brazil, has a handsome, solid shell.

Actual size

The shell of the Gaping Ancilla is medium-sized, thin, lightweight, globose, and oval-fusiform in outline. Its spire is short, with concave spire whorls, and a pointed apex; the suture is not well marked. The surface of the shell is smooth, with 1 diagonal line from the mid-columella to the anterior outer lip. The aperture is wide and long, the outer lip thin, and the columella smooth and somewhat twisted. The shell color is reddish brown, with a pale orange aperture and white anterior columella.

FAMILY	Olividae
SHELL SIZE RANGE	1¼ to 3½ in (30 to 90 mm)
DISTRIBUTION	North Carolina to Yucatán, Mexico
ABUNDANCE	Abundant
DEPTH	Intertidal to 425 ft (130 m)
HABITAT	Sandy bottoms
FEEDING HABIT	Carnivore
OPERCULUM	Absent

SHELL SIZE RANGE
1¼ to 3½ in
(30 to 90 mm)

PHOTOGRAPHED SHELL
2⅝ in
(68 mm)

507

OLIVA SAYANA

LETTERED OLIVE

RAVENEL, 1834

Oliva sayana is the state shell of South Carolina, where it is abundant. It ranges from North Carolina to Yucatán, Mexico. It lives from the intertidal zone to offshore depths, on sandy bottoms. It buries in the sand during the day and is active at night. Its popular name comes from the markings on its shell, some of which may be reminiscent of letters. It is known as a fossil since at least the Pliocene Epoch.

RELATED SPECIES

Oliva circinata Marrat, 1871, from eastern Mexico to Brazil, has a shell similar in shape and color pattern, but is smaller and slightly broader than *O. sayana*. *Oliva tessellata* Lamarck, 1811, from the Indo-West Pacific, has a small, oval-cylindrical shell with a cream background and purplish brown spots.

The shell of the Lettered Olive is medium-large for the family, thick, solid, slender, and cylindrical in outline. Its spire is moderately tall, with a pointed apex, and a sharp ridge bordering the grooved suture. The surface of the shell is smooth and glossy, with a diagonal line from the anterior third of the columella to the anterior end of the outer lip. The aperture is narrow, the outer lip thick and the columella has several small folds. The shell color is variable, usually gray or cream background with reddish brown zigzag markings, and a pale purplish aperture.

Actual size

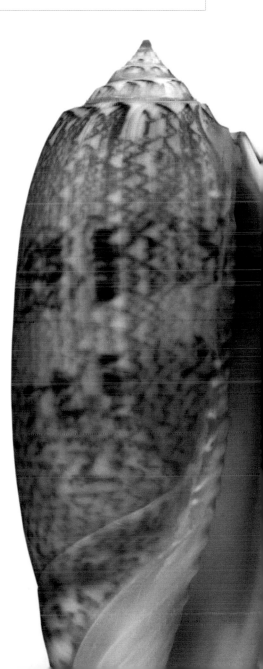

FAMILY	Olividae
SHELL SIZE RANGE	1¼ to 3¼ in (30 to 80 mm)
DISTRIBUTION	West Africa, from Mauritania to Angola
ABUNDANCE	Common
DEPTH	6 to 33 ft (2 to 10 m)
HABITAT	Sandy bottoms
FEEDING HABIT	Carnivore
OPERCULUM	Absent

SHELL SIZE RANGE
1¼ to 3¼ in
(30 to 80 mm)

PHOTOGRAPHED SHELL
2⅞ in
(73 mm)

AGARONIA ACUMINATA
POINTED ANCILLA
(LAMARCK, 1811)

Agaronia acuminata is the largest species in the genus *Agaronia*. Its name refers to its pointed apex. It is a common species in shallow waters in western Africa, ranging from Mauritania to Angola. Its shell looks like that of a slender *Oliva* with a tall spire. The animal cruises just below the surface of the sand looking for its prey, with only the siphon above the surface; it leaves a distinct trail in the sand. There are about 17 living species in the genus *Agaronia* worldwide; the highest diversity is in western Africa.

RELATED SPECIES

Agaronia travassosi Morretes, 1938, which is endemic to Brazil, and ranges from Espirito Santo to Santa Catarina, has a shell that resembles *A. acuminata* in size and shape. It is slightly smaller, and broader, with a wider aperture. *Ancillista cingulata* (Sowerby I, 1830), from western and northern Australia, has a large, very thin, and oval-cylindrical shell. It has a tall spire and a relatively large and rounded apex.

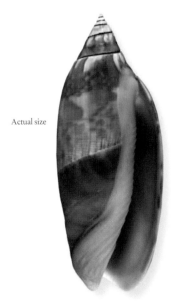

Actual size

The shell of the Pointed Ancilla is medium-large for the family, smooth, lightweight, and oval-fusiform in outline. Its spire is moderately tall, with a pointed apex, straight-sided whorls, and a channeled suture. The surface of the shell is smooth and glossy, with a diagonal spiral line from the mid-columella to the anterior part of the outer lip. The aperture is narrow, the outer lip sharp and convex, and the columella has a white callus with many small folds. The shell color has a gray to fawn background, with 2 broad spiral bands, irregular markings, and a white or beige aperture.

FAMILY	Olividae
SHELL SIZE RANGE	1¼ to 3¾ in (32 to 95 mm)
DISTRIBUTION	Gulf of California to Peru
ABUNDANCE	Common
DEPTH	Intertidal to 33 ft (10 m)
HABITAT	Sandy bottoms
FEEDING HABIT	Carnivore
OPERCULUM	Absent

SHELL SIZE RANGE
1¼ to 3¾ in
(32 to 95 mm)

PHOTOGRAPHED SHELL
3⅛ in
(79 mm)

509

OLIVA INCRASSATA

ANGLED OLIVE

(LIGHTFOOT, 1786)

Oliva incrassata is one of the largest and heaviest olivids. The adult shell often has a thickened shoulder, giving it a distinctive angular outline above the middle of the shell. Juveniles do not have the angular callus. The shell color is usually gray to brown, peppered with brown markings or fine zigzag lines, but occasionally it can be solid yellow or very dark. *Oliva incrassata* is found on the outer side of sandbars at extremely low tides, or at low subtidal depths. It is found as a Pleistocene fossil in some deposits, such as on Magdalena Island, Lower California.

RELATED SPECIES

Oliva polpasta Duclos, 1833, which ranges from Baja California to Peru, has a small to medium shell that can be thick and heavy, and have a somewhat angular body whorl. The body whorl typically has small yellow triangles just below the suture, a pattern that some authors call "cogwheel." *Oliva peruviana* Lamarck, 1811, from Peru to Chile, also has an angular shell, but it is smaller and much thinner than *O. incrassata*.

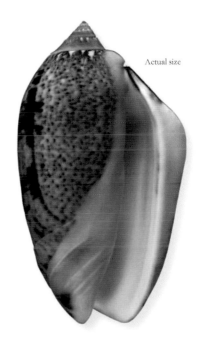

Actual size

The shell of the Angled Olive is medium to large, very thick, heavy, and angularly swollen above the middle. Its spire is short and pointed, with a grooved suture. The surface is smooth and glossy, with a diagonal raised line from the columella to the anterior edge of the outer lip. The aperture is moderately wide and the outer lip is very thick, with a callus at the shoulder. The columella has several small folds and a white callus. The shell color of the shell is gray or brown, occasionally yellow, with fine zigzag patterns, and the aperture is white.

FAMILY	Olividae
SHELL SIZE RANGE	2⅛ to 4 in (55 to 100 mm)
DISTRIBUTION	Endemic to the northern half of Australia
ABUNDANCE	Common
DEPTH	Intertidal to 260 ft (80 m)
HABITAT	Sand flats
FEEDING HABIT	Carnivore
OPERCULUM	Corneous, thin, small, oval

SHELL SIZE RANGE
2⅛ to 4 in
(55 to 100 mm)

PHOTOGRAPHED SHELL
3⅝ in
(92 mm)

ANCILLISTA CINGULATA
CINGULATE ANCILLA
(SOWERBY I, 1830)

The shell of the Cingulate Ancilla is large, very thin, lightweight, elongate, inflated, and oval-fusiform in outline. Its spire is tall, with a relatively large and rounded white apex, slightly convex whorls, and a shallow suture. The surface of the shell is smooth, with a spiral band below the suture. The aperture is large, about half of the shell length, the outer lip thin, and the columella smooth and curved. The shell ranges from creamy gray to fawn, with spiral brown bands anteriorly, and a white spiral band below the suture; the aperture is the same color as the exterior.

Ancillista cingulata is a thin-shelled, large olivid that is endemic to Australia. It ranges from Western Australia to northern Australia and Queensland. Records from deep water off Indonesia need confirmation. This species can be found on intertidal sand flats but also offshore. Like other olivids, *Ancillista cingulata* has a large and wide muscular foot with an anterior shield-shaped portion (the propodium) and two side lobes (the metapodium) that cover the shell. When disturbed, the animal is able to swim by flapping the propodium. The posterior of the foot is bifurcated. The foot is mottled in brown, fawn, and white.

RELATED SPECIES
Ancillista velesiana Iredale, 1930, from southern Queensland to New South Wales, Australia, has a very similar shell in size and coloration. It is considered by some authors a subspecies of *A. cingulata*. *Agaronia acuminata* (Lamarck, 1811), from Mauritania to Angola, western Africa, has an elongated, slender shell with a pointed apex. The coloration is variable, usually with two broad spiral bands and irregular brown markings.

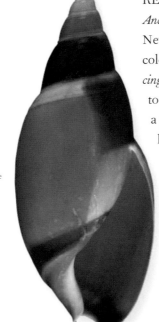

Actual size

FAMILY	Olividae
SHELL SIZE RANGE	2 to 5 in (50 to 130 mm)
DISTRIBUTION	Baja California, Mexico to Peru
ABUNDANCE	Uncommon
DEPTH	Intertidal to 80 ft (25 m)
HABITAT	Sandy bottoms
FEEDING HABIT	Carnivore, feeds on other mollusks
OPERCULUM	Absent

SHELL SIZE RANGE
2 to 5 in
(50 to 130 mm)

PHOTOGRAPHED SHELL
4⅛ in
(106 mm)

OLIVA PORPHYRIA

TENT OLIVE

(LINNAEUS, 1758)

511

Oliva porphyria has the largest shell in the family Olividae, and one of the most distinctive. It is an uncommon species living in tropical west America, from the intertidal zone to shallow subtidal sandy bottoms. The shell is highly polished by the lobes of its large foot, which envelopes the entire shell while the animal cruises buried in sand. It is a predator of other mollusks, usually bivalves or gastropods, and is active at night, burying in the sand during the day. It uses its large muscular foot to hold its prey. There are hundreds of species of Olividae worldwide.

RELATED SPECIES

Oliva incrassata Lightfoot, 1786, which ranges from California to Peru, has the thickest and heaviest, but not the largest, shell in the genus *Oliva*. *Olivancillaria urceus* (Röding, 1798), from Brazil and Argentina, has a distinctive triangular shell with a wide aperture. Typically there is a large callus posterior to the aperture.

The shell of the Tent Olive is heavy, solid, cylindrical, and inflated. Its spire is short with a sharp protoconch and a narrow, channeled suture. Its body whorl is large and inflated, with a long, narrow aperture. The surface is smooth and glossy. The lip is thick, slightly concave in the middle, and smooth. The columella is thickly callused. The shell has a pale violet-pinkish background with rich, brown tent markings; the aperture is orange to pale yellow.

Actual size

FAMILY	Olivellidae
SHELL SIZE RANGE	½ to 1⅛ in (12 to 28 mm)
DISTRIBUTION	Panama to northern Peru
ABUNDANCE	Common
DEPTH	Intertidal
HABITAT	Sandy shores and mudflats
FEEDING HABIT	Carnivore
OPERCULUM	Corneous, thin

SHELL SIZE RANGE
½ to 1⅛ in
(12 to 28 mm)

PHOTOGRAPHED SHELL
⅝ in
(17 mm)

OLIVELLA VOLUTELLA
VOLUTE-SHAPED DWARF OLIVE
(LAMARCK, 1811)

Olivella volutella is one of around 20 species of *Olivella* found on the tropical west American coast. It is particularly abundant on the mud flats of Panama. It travels relatively quickly, using its propodium as a plow to cut through the sand and mud. A small, sandy mound marks the point where it buries itself. From beneath the surface it senses the presence of its prey through its snorkel-like siphon. It feeds at night on tiny bivalves, crustaceans, and other invertebrates.

RELATED SPECIES

Olivella gracilis (Broderip and Sowerby I, 1829) has a delicate, pale, all-white shell and shares the same distribution as *O. volutella*. *Olivella dama* (Wood, 1828) occurs slightly farther north, on sand spits in the Gulf of California. It has a light gray-brown shell, cream fasciole, and dark-streaked suture.

Actual size

The shell of the Volute-Shaped Dwarf Olive is moderately thin, quite hard, and highly glossy. It has a fairly high spire with a fine suture. The aperture is narrow and slightly triangular. It is distinguished by the fine grooves on the columella and fasciole, which, however, do not extend into the body whorl. The shell color is most commonly a uniform purple-brown, sometimes with white or cream bands, with a white fasciole.

FAMILY	Olivellidae
SHELL SIZE RANGE	½ to 1⅛ in (13 to 28 mm)
DISTRIBUTION	Vancouver Island to Baja California
ABUNDANCE	Common
DEPTH	Low intertidal to 165 ft (50 m)
HABITAT	Sandy bottoms, lagoons, and bays
FEEDING HABIT	Carnivore
OPERCULUM	Absent

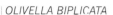

SHELL SIZE RANGE
½ to 1⅛ in
(13 to 28 mm)

PHOTOGRAPHED SHELL
⅞ in
(22 mm)

OLIVELLA BIPLICATA

PURPLE DWARF OLIVE

(SOWERBY I, 1825)

513

Olivella biplicata is one of the largest species in the genus *Olivella*. It is common in lagoons and protected bays, ranging from the lower intertidal zone to offshore, on sandy bottoms. It is active mostly at night, and larger animals occur higher on the beach than smaller ones. The species is often found in large aggregations. The male locates the female by following her track, and temporarily attaches himself to her shell; mating can last as long as three days. The female lays small egg capsules attached individually to hard objects such as stones or shells.

RELATED SPECIES

Olivella volutella (Lamarck, 1811), from Panama to northern Peru, has a smaller, narrower shell than *O. biplicata*. Its shell is thin, and the columella is nearly straight until it curves anteriorly, with several folds. *Olivella nivea* (Gmelin, 1791), which ranges from North Carolina to central Brazil, has a narrow shell that resembles *O. volutella* in shape, but has a wider aperture and smoother columella. It has a quite variable color pattern.

The shell of the Purple Dwarf Olive is small, robust, smooth, and oval in outline. Its spire is short, with a pointed apex, a fine impressed suture, and slightly convex spire whorls. The surface is smooth, with a white and purple diagonal line from the mid-columella to the anterior edge of the outer lip. The aperture is relatively wide, the outer lip thin, the columella has 2 to 3 folds anteriorly. The shell color is typically gray, but can be whitish or brownish, with a white columellar callus and a purplish interior.

Actual size

FAMILY	Pseudolividae
SHELL SIZE RANGE	¾ to 1 in (19 to 25 mm)
DISTRIBUTION	Eastern Australia
ABUNDANCE	Uncommon
DEPTH	250 to 500 ft (80 to 150 m)
HABITAT	Fine sand and mud bottoms
FEEDING HABIT	Carnivore
OPERCULUM	Large, oval

SHELL SIZE RANGE
¾ to 1 in
(19 to 25 mm)

PHOTOGRAPHED SHELL
¾ in
(21 mm)

514

ZEMIRA AUSTRALIS
SOUTHERN ZEMIRA
(SOWERBY I, 1833)

Zemira is one of 14 genera, many now extinct, in the family Pseudolividae. This family may be recognized by having a tooth on the outer lip that produces a spiral groove along the lower half of the shell. The Pseudolividae have declined in diversity and geographic range since the Paleogene Period. Less than 20 species survive today. Most live in deeper water, some along the outer continental shelf, others at bathyal to abyssal depths.

RELATED SPECIES

Zemira bodalla Garrard, 1966, the only other living species of *Zemira*, inhabits the outer continental shelf off Queensland. It has a slightly larger, thicker, and broader shell with coarser, more homogeneous spiral cords, a less conspicuous furrow behind the labial tooth, and more broadly and evenly distributed blotches of dark brown pigment.

The shell of the Southern Zemira is small, ovate, and relatively thick, with a characteristic channel between the suture and the shoulder. The spire is high and stepped, and the anterior is rounded. The aperture is oval, with a thick, smooth columella, a short labial tooth along the outer lip, and a siphonal notch. The shell exterior is smooth, with shallow spiral grooves. The surface of the labial tooth is concave, forming a deeper furrow around the shell. The shell color is white to tan, with irregular darker brown spots and blotches, particularly along the shoulder.

Actual size

FAMILY	Pseudolividae
SHELL SIZE RANGE	1⅜ to 2 in (35 to 52 mm)
DISTRIBUTION	El Salvador to Ecuador
ABUNDANCE	Common
DEPTH	Intertidal to 17 ft (5 m)
HABITAT	Rocks in mud flats
FEEDING HABIT	Generalized carnivore and scavenger
OPERCULUM	Corneous, claw-shaped

SHELL SIZE RANGE
1⅜ to 2 in
(35 to 52 mm)

PHOTOGRAPHED SHELL
1½ in
(39 mm)

TRIUMPHIS DISTORTA

DISTORTED TRIUMPHIS

(WOOD, 1828)

515

Triumphis distorta has a curiously shaped aperture with a strong posterior canal that forms an earlike projection, which gives the aperture a distorted appearance. The thickened outer lip and narrow aperture are responses against predators. Many mollusks respond to predation by making thicker shells, reinforcing certain parts of the shell, adding spines, and other features. *Triumphis distorta* is a common pseudolivid found on rocks in mud flats in the intertidal zone and shallow subtidal waters.

RELATED SPECIES

Triumphis subrostrata (Wood, 1828), from western Mexico to Colombia, has a small but solid, thick fusiform shell. It has one row of short spines on the periphery of the spire whorls, with the rest of the shell smooth. *Cantharus undosus* Linnaeus, 1758, from the Indo-West Pacific, has a small and attractive, white or light brown shell ornamented with brown spiral ridges.

The shell of the Distorted Triumphis is medium in size for the family, strong, thick, and barrel-shaped in outline. It has a short spire with cancellate sculpture, which fades with growth; the suture is well marked. The aperture is narrow and spindle-shaped, with an earlike extension of the posterior canal. The outer lip is thickened, and lirated within. The siphonal canal is short, and the columella smooth. The shell color is white mottled with brown, and the aperture is white.

Actual size

FAMILY	Strepsiduridae
SHELL SIZE RANGE	¾ to 3 in (20 to 73 mm)
DISTRIBUTION	South Africa to Mozambique
ABUNDANCE	Uncommon
DEPTH	65 to 500 ft (20 to 150 m)
HABITAT	Sandy mud bottoms
FEEDING HABIT	Carnivore
OPERCULUM	Absent

SHELL SIZE RANGE
¾ to 3 in
(20 to 73 mm)

PHOTOGRAPHED SHELL
⅞ in
(22 mm)

516

MELAPIUM ELATUM
ELATED ONION SHELL
(SCHUBERT AND WAGNER, 1829)

The genus *Melapium* has a very limited geographic distribution. It has been placed in the family Strepsiduridae based on the similarity of its shell to the more widespread fossil genus *Strepsidura*, which ranged from the Cretaceous to the Eocene Period. The animal of *Melapium* crawls along the bottom on a broad, rounded foot with bright concentric bands of color (blue, yellow-red, blue, and red) near its edge. Females attach leathery egg capsules directly to their shell on the columellar part of the siphonal canal.

RELATED SPECIES

Melapium lineatum (Lamarck, 1822) occurs in the same geographic area, but has a more evenly rounded shell with a shorter, less pronounced spire and a siphonal canal that lacks the pronounced siphonal fasciole. *Melapium lineatum* lacks purple pigment on the columella, and has axial bands that are lighter in color and may be more wavy and diffuse. Some specimens have a series of dark brown blotches along the shell periphery.

Actual size

The shell of the Elated Onion Shell is nearly spherical, with a glossy, porcelaneous surface, a small, short spire, and a distinctive siphonal canal with a strong, sharp oblique fasciole. The aperture is broad, with a weak shoulder and rounded, smooth outer lip. The columella, interrupted by a sharp fold, is often suffused with a dark purple. The shell color is cream to orange-tan with numerous, irregular, dark brown axial lines.

FAMILY	Babyloniidae
SHELL SIZE RANGE	1¼ to 2 in (33 to 50 mm)
DISTRIBUTION	Endemic to South Africa
ABUNDANCE	Uncommon
DEPTH	80 to 330 ft (25 to 100 m)
HABITAT	Sandy and muddy bottoms
FEEDING HABIT	Primarily scavenger
OPERCULUM	Corneous, thin and flexible, with terminal nucleus

SHELL SIZE RANGE
1¼ to 2 in
(33 to 50 mm)

PHOTOGRAPHED SHELL
1½ in
(38 mm)

BABYLONIA PAPILLARIS

SPOTTED BABYLON

(SOWERBY I, 1825)

517

Babylonia papillaris is an attractive shell that is endemic to South Africa, and distinguished from other species in the family by its finely spotted shell. It lives offshore in shallow to moderately deep waters, on sandy and muddy bottoms. It has been found, along with other shells, in the stomachs of large molluscivorous fishes such as the musselcracker. A rare live-collected specimen had a bright orange-red body peppered with white blotches.

RELATED SPECIES

Babylonia zeylanica (Bruguière, 1789), from India and Sri Lanka, has a shell with large irregular blotches and an umbilicus tinted in purple. *Babylonia lani* Gittenberger and Gould 2003, from the South China Sea and Gulf of Thailand, has a shell that resembles *B. papillaris*, but has three spiral rows of large blotches.

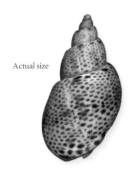

Actual size

The shell of the Spotted Babylon is small, slender, and elongated, with a tall spire and well-marked suture. The larval shell is rounded and white. The surface of the shell is smooth and glossy, as are the columella, aperture, and outer lip. The aperture is lanceolate, and there is a broad, thick, white parietal shield over the columella. Adult shells lack an umbilicus. The shell color is white with spiral rows of rounded brown spots.

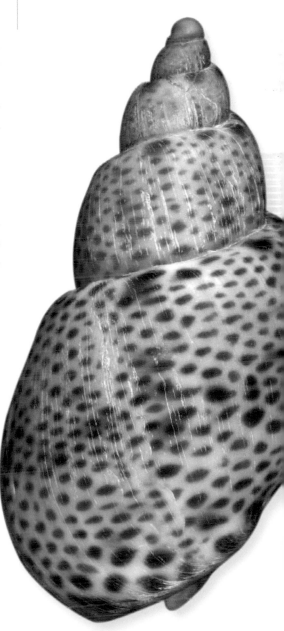

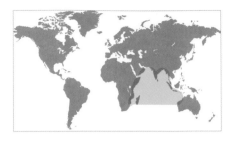

FAMILY	Babyloniidae
SHELL SIZE RANGE	1½ to 3 in (40 to 75 mm)
DISTRIBUTION	Indian Ocean
ABUNDANCE	Locally abundant
DEPTH	17 to 200 ft (5 to 60 m)
HABITAT	Sandy and muddy bottoms
FEEDING HABIT	Primarily scavenger
OPERCULUM	Corneous, thin and flexible, with terminal nucleus

SHELL SIZE RANGE
1½ to 3 in
(40 to 75 mm)

PHOTOGRAPHED SHELL
2⅛ in
(53 mm)

518

BABYLONIA SPIRATA
SPIRAL BABYLON
(LINNAEUS, 1758)

The shell of the Spiral Babylon is heavy, solid, and broadly ovate with a tall spire. Its whorls are separated by a channeled suture with a sharp edge. The shell, outer lip, and columella are smooth, and there is a heavy callus on the columella. The umbilicus is open in adult shells. Its shell color is white, with irregular light brown blotches and spots; the aperture is white, and the apex purple.

Babylonia spirata is the most common of the babyloniids that have a deep channel between whorls. It is an abundant gastropod in the Indian Ocean. It usually lives in shallow subtidal waters but can also be found in deeper waters, on sandy and muddy bottoms. In India, it is found primarily as a bycatch of trawl and push nets, but is also collected by skin divers. There is a good potential for the aquaculture of *B. spirata*, because it can be raised in captivity, and there is a market for its meat and shell.

RELATED SPECIES
Babylonia canaliculata Schumacher, 1817, from the Arabian Sea, resembles *B. spirata*, but with a shorter and broader shell. *Babylonia magnifica* Fraussen and Stratmann, 2005, from Japan to Thailand, has three broad dark brown bands on the body whorl.

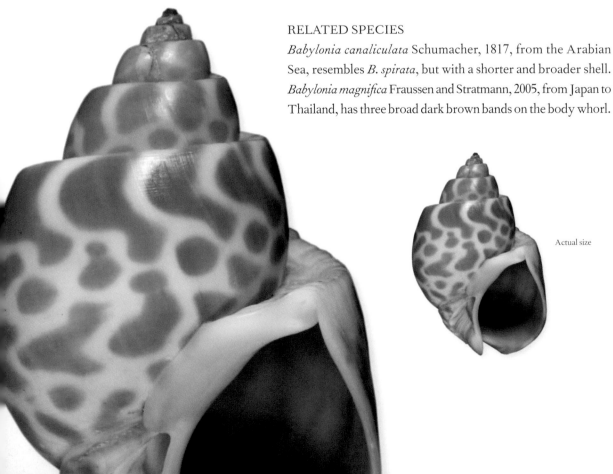

Actual size

FAMILY	Babyloniidae
SHELL SIZE RANGE	2 to 3½ in (50 to 85 mm)
DISTRIBUTION	India and Sri Lanka
ABUNDANCE	Abundant
DEPTH	Intertidal to 65 ft (20 m)
HABITAT	Sandy and muddy bottoms
FEEDING HABIT	Primarily scavenger
OPERCULUM	Corneous, thin and flexible, with terminal nucleus

SHELL SIZE RANGE
2 to 3½ in
(50 to 85 mm)

PHOTOGRAPHED SHELL
2¼ in
(56 mm)

BABYLONIA ZEYLANICA

INDIAN BABYLON

(BRUGUIÈRE, 1789)

Babylonia zeylanica is abundant in southern India and Sri Lanka, where it occurs on intertidal to shallow water, sandy and muddy bottoms. It is primarily a scavenger, feeding on carrion. Like other species of *Babylonia*, it is edible and sold throughout Asia for food. This species is also used in the aquarium trade, and its shell can be found in pet stores as homes for hermit crabs. There are about 15 species in the genus *Babylonia*, all restricted to the Indian or Pacific oceans.

The shell of the Indian Babylon is slender and smooth, with a well-marked suture and a high spire. Its body whorl is large, with fine axial lines, and a lanceolate aperture with a short siphonal notch. The columella is smooth, with a single fold near the top of the aperture, a white columellar callus, and a deep umbilicus. The base color of the shell is white with spiral rows of brown to light brown blotches. The umbilicus and apex are tinged in violet.

RELATED SPECIES

Babylonia spirata (Linnaeus, 1758), from the Indian Ocean to the west Pacific, has a deep suture; *B. areolata* (Link, 1807), from Taiwan to Sri Lanka, has a channeled suture; *Babylonia japonica* (Reeve, 1842), from Japan and Taiwan, lacks a violet fasciole.

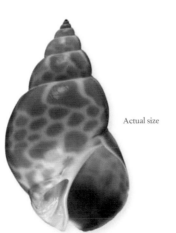

Actual size

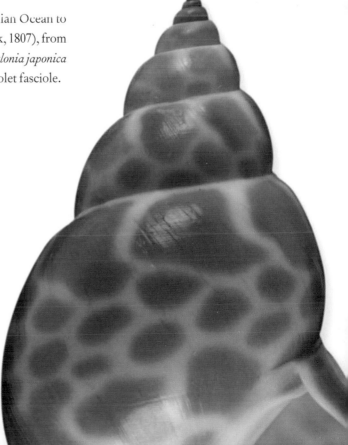

FAMILY	Babyloniidae
SHELL SIZE RANGE	2 to 3⅜ in (50 to 85 mm)
DISTRIBUTION	Taiwan to Thailand
ABUNDANCE	Rare
DEPTH	Offshore to 100 ft (30 m)
HABITAT	Sandy and muddy bottoms
FEEDING HABIT	Primarily scavenger
OPERCULUM	Corneous, thin and flexible, with terminal nucleus

SHELL SIZE RANGE
2 to 3⅜ in
(50 to 85 mm)

PHOTOGRAPHED SHELL
3⅛ in
(78 mm)

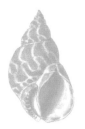

520

BABYLONIA PERFORATA
PERFORATE BABYLON
(SOWERBY II, 1870)

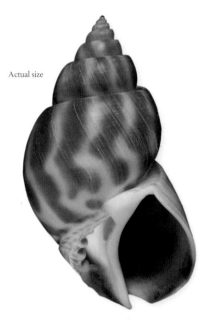

Actual size

Babylonia perforata is a rare representative of the family Babyloniidae. It has an elongated shell with a broad, channeled suture. Its umbilicus is surrounded by denticles. It is most commonly found around Taiwan. Babyloniids have separate sexes, and in some species, such as *B. spirata*, the females have a slightly larger and heavier shell than males. They congregate for mating and spawning, and groups of about 15 to 20 females lay eggs in the same region.

RELATED SPECIES

Babylonia japonica (Reeve, 1842), from the Japonic region, is one of the larger species in the family Babyloniidae. It has an elongated shell with a tall, pointed spire. *Babylonia kirana* Habe, 1959, from Okinawa, Japan, is the only living species in the family lacking color markings; it has a plain chestnut-colored shell.

The shell of the Perforate Babylon is solid, large for the family, and elongated with a tall, pointed spire. It has a channeled suture. Its columella and outer lip are smooth, and there is a strong callus on the posterior columella. The umbilicus is open in adult shells and surrounded by denticles. The surface of the shell is smooth, except near the umbilicus, which has a strong spiral ridge. The shell color is cream with hazy pale brown blotches.

FAMILY	Harpidae
SHELL SIZE RANGE	½ to 1¾ in (12 to 43 mm)
DISTRIBUTION	Lower Gulf of California to Peru, Galápagos Islands
ABUNDANCE	Uncommon
DEPTH	Intertidal
HABITAT	Under rocks
FEEDING HABIT	Carnivore
OPERCULUM	Corneous, thin

MORUM TUBERCULOSUM

LUMPY MORUM

(REEVE, 1842)

SHELL SIZE RANGE
½ to 1¾ in
(12 to 43 mm)

PHOTOGRAPHED SHELL
1⅛ in
(27 mm)

There are more than two dozen known living species of the genus *Morum*. Based on anatomical studies *Morum* has recently been reclassified as a member of the family Harpidae, having previously been regarded as belonging to the Cassidae or helmet shells. Like other Harpidae they have the facility of autotomy, shedding part of their foot if disturbed (as some lizards do their tails). Although carnivorous, members of the subfamily Moruminae have few teeth on the radula, and are believed to suck the fluids from their crustacean prey.

RELATED SPECIES

Morum oniscus (Linnaeus, 1767), in the Caribbean, is similar to *M. tuberculosum* and the two species probably share a common ancestor. It has three rows of spiral knobs instead of five, and a pustulate parietal shield; it is white with brown patches rather than vice versa.

The shell of the Lumpy Morum is conical, and tapers convexly. The spire is flat apart from the apex. The narrow aperture extends the full length of the shell, which is dark brown with 3 broad spiral bands of irregular white splashes, and 5 spiral rows of shallow blunt tubercles. The outer lip is highly dentate, and the interior and columella are pale to orange yellow. A smooth undeveloped parietal wall is rather transparent.

Actual size

FAMILY	Harpidae
SHELL SIZE RANGE	1¼ to 2⅜ in (30 to 66 mm)
DISTRIBUTION	North Carolina to Brazil
ABUNDANCE	Uncommon
DEPTH	100 to 300 ft (30 to 90 m)
HABITAT	On and under rocks
FEEDING HABIT	Carnivore
OPERCULUM	Corneous, thin

SHELL SIZE RANGE
1¼ to 2⅜ in
(30 to 66 mm)

PHOTOGRAPHED SHELL
1⅞ in
(47 mm)

522

MORUM DENNISONI

DENNISON'S MORUM

(REEVE, 1842)

The shell of Dennison's Morum is distinguished by its extended parietal shield, which is orange with dense, tiny, white pustulate tubercles. The orange lip is also lined with strong, fine white folds. Its spire is short and the body whorl is highly sculpted with narrow, cancellate axial ribs. The shell is white with indistinct spiral bands of orange.

Like other morums, *Morum dennisoni* feeds on crustaceans, which it captures and envelops in its large foot. The minute radular teeth are likely used to rasp a hole between the joints of the crabshell, through which digestive enzymes are injected and through which the liquified tissues are consumed. Until dredging off Barbados in the 1980s brought more specimens to light, there were fewer than 40 known examples of this deepwater species. It is still uncommon in collections and highly prized by collectors.

RELATED SPECIES

Morum veleroae (Emerson, 1968) in the eastern Pacific is a cognate species, a sort of non-identical twin, of *M. dennisoni*. The sculpture on the body whorl is less pronounced, and the spiral banding more so. The two developed separately after common ancestors were separated when the Panamanian landmass separated the Caribbean from the Pacific, about 3 million years ago.

Actual size

FAMILY	Harpidae
SHELL SIZE RANGE	¾ to 2½ in (20 to 65 mm)
DISTRIBUTION	Red Sea to Indo-Pacific
ABUNDANCE	Common
DEPTH	Intertidal to 100 ft (30 m)
HABITAT	Sandy bottoms
FEEDING HABIT	Carnivore, feeds on small crabs
OPERCULUM	Absent

SHELL SIZE RANGE
¾ to 2½ in
(20 to 65 mm)

PHOTOGRAPHED SHELL
1⅞ in
(47 mm)

HARPA AMOURETTA

MINOR HARP

(RÖDING, 1798)

525

Harpa amouretta is easily recognized by its elongate shape and what is probably the tallest spire among the harps, although its shell shape is variable. It is also one of the most widely distributed of *Harpa* species, ranging from the Red Sea, south to South Africa and throughout the Indian Ocean to Hawaii and the Marquesas Islands. The form of this species that occurs in the Red Sea and western Indian Ocean is stout, rather heavy, and pale in color. Another, more elongate and slender form with a darker shell is more common in the eastern Indian and western Pacific oceans.

The shell of the Minor Harp is small to medium, narrow, cylindrical, and elongate, with straight sides. Its spire is relatively high (for the family), with a well-marked suture. The body whorl is large, with about 12 to 14 slightly axial ribs. The aperture is long, with a thickened outer lip, that forms a sharp angle at the shoulder. The siphonal notch is short, and the columella is smooth. The shell color is cream with pale brown spiral bands and lines, and the aperture is white or cream.

RELATED SPECIES

Harpa gracilis (Röding, 1798), from South Africa to the Indo-West Pacific, has a similar but smaller shell, with more axial ribs per whorl. *Harpa ventricosa* Lamarck, 1801, from the Red Sea, Arabian Gulf, and Indian Ocean, has a large, broader shell, with about 15 axial ribs per whorl.

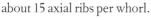

Actual size

FAMILY	Harpidae
SHELL SIZE RANGE	2 to 4 in (50 to 100 mm)
DISTRIBUTION	Baja California to Peru
ABUNDANCE	Uncommon
DEPTH	Intertidal to 200 ft (60 m)
HABITAT	Sandy and muddy bottoms
FEEDING HABIT	Carnivore, feeds on crustaceans
OPERCULUM	Absent

SHELL SIZE RANGE
2 to 4 in
(50 to 100 mm)

PHOTOGRAPHED SHELL
3¼ in
(82 mm)

HARPA CRENATA

PANAMA HARP

SWAINSON, 1822

Harpa crenata is the only harp in the Panamic region. It is uncommon, occurring at depths from the intertidal zone to about 200 ft (60 m), generally on muddy bottoms. It is a carnivore, feeding on crabs and other crustaceans that it captures with its large and muscular foot, which produces profuse amounts of mucus. Its tiny radula pierces the thin membrane between the crab's segments and the harp injects saliva with digestive fluids through the opening. The harp then feeds on the liquefied tissues. Harps are more active at night, hiding during the day. There are about 40 species in the family Harpidae, most in the tropical Indo-Pacific.

RELATED SPECIES

Harpa amouretta (Röding, 1798) is a common species from the Indo-West Pacific. It has a small and slender shell with about 13 axial ribs per whorl and a relatively tall spire. *Harpa goodwini* Rehder, 1993, a rare species endemic to Hawaii, has about 14 to 16 axial ribs crossed by four broad orange or pink bands.

Actual size

The shell of the Panama Harp is medium-sized, inflated, and ventricose. The spire is short and pointed, with a smooth and glossy protoconch, and a well-impressed suture. The whorls are angular, sharply keeled at the shoulder, with pointed spines at each of the axial ribs. The body whorl has 12 to 15 narrow and low axial ridges reflected backward. The columella is smooth and curved near the anterior end. The shell color is tan to pinkish with fine, irregular brown lines between ribs, and brown blotches on the columella.

FAMILY	Harpidae
SHELL SIZE RANGE	2 to 5¼ in (50 to 133 mm)
DISTRIBUTION	Red Sea and Arabian Gulf, Indian Ocean
ABUNDANCE	Common
DEPTH	Intertidal
HABITAT	Sandy bottoms
FEEDING HABIT	Carnivore, feeds on crustaceans
OPERCULUM	Absent

SHELL SIZE RANGE
2 to 5¼ in
(50 to 133 mm)

PHOTOGRAPHED SHELL
3⅜ in
(86 mm)

HARPA VENTRICOSA
VENTRAL HARP
LAMARCK, 1801

525

Harpa ventricosa is one of the largest harps, and a common species in the intertidal sandy bottoms of the Red Sea, Arabian Gulf, and Indian Ocean. It is a predatory gastropod, feeding on small crabs and shrimp, which it captures with its large foot. Harps have the ability to shed the posterior part of the foot when molested, leaving the still-moving piece of tissue behind while the snail escapes. Like many sand-dwelling snails, harps lack an operculum. They have such a small radula that for a long time scientists believed they lacked one.

RELATED SPECIES
Harpa major (Röding, 1798), from the Indo-Pacific, including Hawaii, has a large, slightly more elongated shell, with axial ribs more spaced out. *Harpa costata* (Linnaeus, 1758), which is endemic to the Mascarene Islands off eastern Africa, is the rarest of the harps. It has a large shell with 30 to 40 axial ribs per whorl.

sized, thick, heavy, and globose-ovate. The spire is short, with a smooth and glossy violet protoconch and well-impressed suture. The whorls are angular, sharply keeled at the shoulder, with pointed spines at each of the axial ribs. The body whorl is inflated and large, with about 15 strong axial ribs, sharply reflected backward. The columella is smooth. The shell color is tan to pinkish with fine, crescent brown lines between ribs, 3 broad darker spiral bands, and 2 large brown blotches on the columella.

Actual size

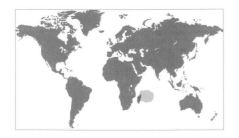

FAMILY	Harpidae
SHELL SIZE RANGE	2¼ to 4¼ in (60 to 110 mm)
DISTRIBUTION	Mascarenes Islands and northeast Madagascar
ABUNDANCE	Rare
DEPTH	Shallow subtidal to 40 ft (12 m)
HABITAT	Sandy bottoms
FEEDING HABIT	Carnivore, feeds on crustaceans
OPERCULUM	Absent

SHELL SIZE RANGE
2¼ to 4¼ in
(60 to 110 mm)

PHOTOGRAPHED SHELL
3⅝ in
(92 mm)

526

HARPA COSTATA
IMPERIAL HARP
(LINNAEUS, 1758)

The shell of the Imperial Harp is medium-sized, inflated, and ventricose. The spire is short and pointed, with a smooth and glossy rosy protoconch, and well-impressed suture. The whorls are angular, sharply keeled at the shoulder. Its body whorl is inflated and very large; with 30 to 40 closely spaced axial ridges, sharply reflected backward. The aperture is large, with ribs showing through the aperture. The outer lip and columella are smooth. The shell color is creamy yellow with light brown and pink spiral bands crossing the radial ribs, and brown blotches on the columella.

Harpa costata, which has a beautiful shell, is the rarest species in the genus *Harpa* and highly prized among collectors. It is endemic to the Mascarene Islands (Mauritius and Réunion) and northeastern Madagascar. It is easily distinguished from all other harps by the large number of axial ribs, about 30 to 40 per whorl. The animal is about twice the size of the shell when extended. It lives on sandy bottoms in shallow depths. Like other harps, *H. costata* is a carnivore, feeding on crustaceans, especially crabs.

RELATED SPECIES
Harpa crenata Swainson, 1822, which ranges from Baja California to Peru, has an inflated shell with more broadly spaced axial ribs than *H. costata*, with 12 to 15 axial ridges per whorl. *Harpa doris* Röding, 1798, from Cape Verde to Angola, has a more elongate and less inflated shell than most harps. It has two or three broad, spiral reddish bands with white and brown markings.

Actual size

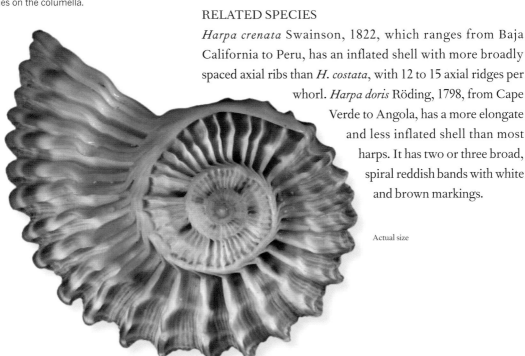

FAMILY	Cystiscidae
SHELL SIZE RANGE	up to ⅛ in (1 to 4 mm)
DISTRIBUTION	Florida to northern Brazil
ABUNDANCE	Common
DEPTH	Intertidal to 240 ft (75 m)
HABITAT	Muddy and coralline sandy bottoms
FEEDING HABIT	Carnivore
OPERCULUM	Absent

GIBBERULA LAVALLEANA
SNOWFLAKE MARGINELLA
(D'ORBIGNY, 1842)

SHELL SIZE RANGE
up to ⅛ in
(1 to 4 mm)

PHOTOGRAPHED SHELL
less than ⅛ in
(2 mm)

529

Gibberula lavalleana is a tiny gastropod with a glossy, domed shell. It is usually found in shallow waters in muddy bottoms or associated with algae, but in Texas, it is found on calcareous sediments near offshore coral reefs. Juvenile shells are translucent, but they become less so as the shell thickens. Even in adults, however, the multicolored internal organs and mantle of the animal can be seen through the shell. There are many living species in the family Cystiscidae, with more than 100 species in the western Atlantic alone.

RELATED SPECIES

Persicula moscatellii (Boyer, 2004), from northeastern Brazil, has a small, elongated, beige-colored shell decorated with thin brown lines that are reminiscent of a shark's teeth. *Persicula cingulata* (Dillwyn, 1817), from northwestern Africa, has a larger shell that is similar in shape to *G. lavalleana*. The shell is opaque, pale yellow with reddish spiral lines.

Actual size

The shell of the Snowflake Marginella is very small, glossy, translucent, globose, and dome-shaped, and tapering anteriorly. Its spire is very short, with slightly convex whorls, and an impressed suture best observed when seen from the posterior (apical) view. Its surface is smooth and glossy, with weak axial striae. The aperture is narrow and long, the outer lip thickened and denticulate in adults. The columella has 3 to 4 folds, and a siphonal notch is present. The shell color is translucent in juveniles, and white in adults. The interior is the same color as the exterior.

FAMILY	Cystiscidae
SHELL SIZE RANGE	½ to 1⅛ in (14 to 28 mm)
DISTRIBUTION	Western Sahara to Senegal, Canary Islands
ABUNDANCE	Common
DEPTH	Offshore
HABITAT	Muddy bottoms
FEEDING HABIT	Carnivore
OPERCULUM	Absent

SHELL SIZE RANGE
½ to 1⅛ in
(14 to 28 mm)

PHOTOGRAPHED SHELL
⅝ in
(16 mm)

528

PERSICULA CINGULATA

GIRDLED MARGINELLA

(DILLWYN, 1817)

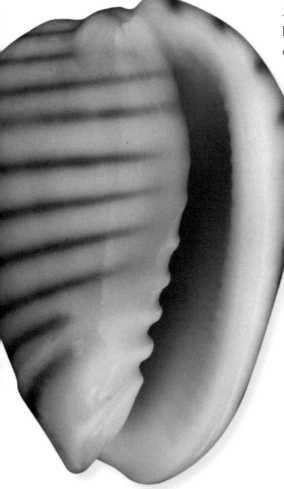

Persicula cingulata is a common species from northwestern Africa, found in sandy and muddy bottoms offshore. It has a beautiful shell, shiny and glossy, with spiral reddish lines. Like other cystiscids, its biology is poorly known. Cystiscids had been classified as marginellids, but certain characters, such as radular morphology and internal shell whorls that are partially resorbed, suggest they are indeed a separate family, probably closer to Olividae than to the Marginellidae. Molecular and anatomical studies are needed to better elucidate the relationships in this group.

RELATED SPECIES

Persicula canariensis Clover, 1972, from the Canary Islands, has an elongated shell with a low spire and light brown to gray coloration. *Gibberula caelata* (Monterosato, 1877), which ranges from Spain to Algeria, has a tiny ovate shell with a narrow aperture. The shell color can be reddish pink to pale yellow.

Actual size

The shell of the Girdled Marginella is small, glossy, and inflated-ovate, with a sunken spire covered by a callus. Its surface is smooth and glossy. The sides of body whorl are convex, tapering toward the anterior end. The aperture is narrow and long, and slightly curved; the outer lip raised, thickened, and finely denticulate. The columella has 7 folds and a callus at the posterior end, and a notched siphonal canal. The shell color is pale yellow to white, with reddish spiral lines, and the aperture is white.

FAMILY	Cystiscidae
SHELL SIZE RANGE	½ to 1 in (13 to 25 mm)
DISTRIBUTION	Mauritania to Guinea
ABUNDANCE	Uncommon
DEPTH	Offshore
HABITAT	Sandy bottoms
FEEDING HABIT	Carnivore
OPERCULUM	Absent

SHELL SIZE RANGE
½ to 1 in
(13 to 25 mm)

PHOTOGRAPHED SHELL
⅞ in
(22 mm)

PERSICULA PERSICULA

SPOTTED MARGINELLA

(LINNAEUS, 1758)

Persicula persicula can be separated from other cystiscids by its broadly ovate and spotted shell. It is an uncommon species from western Africa, found offshore on sandy bottoms. Like other sand-dwelling gastropods, it lacks an operculum. It is a carnivore, and preys on other mollusks. The cystiscid animal is often brightly colored and has a broad mantle that envelopes the entire shell, resulting in a glossy shell. Many species in the family are small. Although they occur worldwide, cystiscids are more common in the tropics, and are particularly diverse in western and southern Africa.

RELATED SPECIES

Persicula cingulata (Dillwyn, 1817), from northwestern Africa, has an inflated, ovate shell with reddish spiral lines. *Gibberula lavalleana* (d'Orbigny, 1842), ranging from North Carolina to Brazil, has a tiny translucent white shell that looks like a miniature white *Persicula*.

Actual size

The shell of the Spotted Marginella is small, thick, glossy, and ovate, with a sunken spire covered by a callus. It superficially resembles a cowrie shell, but its smooth, thickened outer lip and columella with 6 to 9 folds readily distinguish it from a cowrie. Its surface is smooth and glossy. The narrow aperture is as long as the shell. The shell varies from white or pale yellow to brown, and is peppered with brown spots, sometimes arranged in spiral streaks; the aperture is white.

FAMILY	Marginellidae
SHELL SIZE RANGE	⅜ to ⅝ in (9 to 16 mm)
DISTRIBUTION	North Carolina to eastern Brazil
ABUNDANCE	Abundant
DEPTH	Intertidal to 15 ft (5 m)
HABITAT	Sand under rocks
FEEDING HABIT	Carnivore
OPERCULUM	Absent

SHELL SIZE RANGE
⅜ to ⅝ in
(9 to 16 mm)

PHOTOGRAPHED SHELL
½ in
(14 mm)

530

VOLVARINA AVENA

ORANGE-BANDED MARGINELLA

(KIENER, 1834)

Like most marginellids, *Volvarina avena* prefers shallow, warm seas, where it hides below the surface of the sand floor to prey on other mollusks. Throughout a large portion of its range, from Florida to Brazil, this small shell is often found together with the marginellid, *Prunum guttatum*, the White-Spotted Marginella.

RELATED SPECIES

Marginella philippinarum (Redfield, 1848), is almost identical in shape, color, and habitat, but occurs only in the Philippines and western Australia. It is slightly larger than *V. avena*. Otherwise it is almost indistinguishable, even sharing the white, unimpressed suture on the spire and a faint concavity of the outer lip.

Actual size

The shell of the Orange-banded Marginella is narrow with a short spire. Only a white, unincised suture defines the spire whorls, and only a thickening of the upper outer lip separates the spire from the body whorl. It is a highly glossy shell, cream to deep pink with 3 broad, indistinct, usually orange spiral bands. The columella displays the defining marginellid characteristic of 4 folds on its lower half.

FAMILY	Marginellidae
SHELL SIZE RANGE	⅜ to ⅝ in (10 to 16 mm)
DISTRIBUTION	Southeastern Australia and Tasmania
ABUNDANCE	Common
DEPTH	Subtidal to 30 ft (10 m)
HABITAT	Sand
FEEDING HABIT	Carnivore
OPERCULUM	Absent

SHELL SIZE RANGE
⅜ to ⅝ in
(10 to 16 mm)

PHOTOGRAPHED SHELL
⅝ in
(15 mm)

AUSTROGINELLA MUSCARIA
FLY MARGINELLA
(LAMARCK, 1822)

531

The Fly Marginella, with its relatively narrow range of distribution, is one of the southernmost marginellids, a long way from the tropical and subtropical waters preferred by most of its related species. The family Marginellidae comprise a large and global family, of more than 600 species, with the largest diversity present in the tropical seas off western Africa. Tropical eastern America also hosts a diverse population.

RELATED SPECIES

Marginella fischeri Bavay, 1902, is another small, off-white marginella with a limited range, in this case the Philippines. It has a short spire, with little more than the flattened apex showing. There are faint spiral grooves on the whorl, and the lip is much thickened and calloused. There are about six columellar folds, occurring along its whole length, not just the lower half as in *Austroginella muscaria.*

Actual size

The shell of the Fly Marginella is off-white and bulbous, tapering concavely. Its spire is moderately tall with convex whorls and a lightly impressed suture. The lip and columella are calloused, forming a thickened belt around the base of the shell. The inner face of the columella shows the 4 folds that are typical of most marginellas.

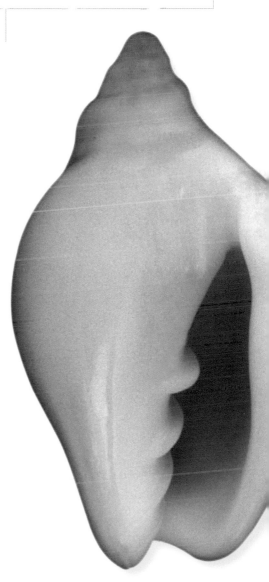

FAMILY	Marginellidae
SHELL SIZE RANGE	½ to ¾ in (10 to 20 mm)
DISTRIBUTION	Florida, Gulf of Mexico to Venezuela
ABUNDANCE	Moderately common
DEPTH	Offshore to 60 ft (20 m)
HABITAT	Grasses
FEEDING HABIT	Carnivore
OPERCULUM	Absent

SHELL SIZE RANGE
½ to ¾ in
(10 to 20 mm)

PHOTOGRAPHED SHELL
¾ in
(19 mm)

532

PRUNUM CARNEUM
ORANGE MARGINELLA
(STORER, 1837)

Many marginellas have evolved and diversified rapidly, with small populations adapting in response to specific local conditions. These have resulted in a great many species, each with relatively restricted distributions. *Prunum carneum* is one such example, having moved from the shallow sandy waters occupied by many marginellas to exploit the predatory opportunities of sea grass fields deeper offshore in the western Caribbean.

Actual size

RELATED SPECIES
Marginella prunum (Gmelin, 1791), the Plum Marginella, is distributed farther east than the Orange Marginella, from the lower Caribbean to Brazil. It is of a similar shape and size, but has a wider aperture particularly toward the anterior canal. The suture is marked by a fine white band; the columella is orange like the rest of the whorl, and streaked with fine white axial lines.

The shell of the Orange Marginella is thick and highly polished, with a very narrow aperture. Its lip is white. The columella has 4 folds on its lower half and is a very pale orange across the parietal wall. The rest of the shell is orange, with faint, paler spiral bands on the shoulders of the short spire and in the lower half of the body whorl.

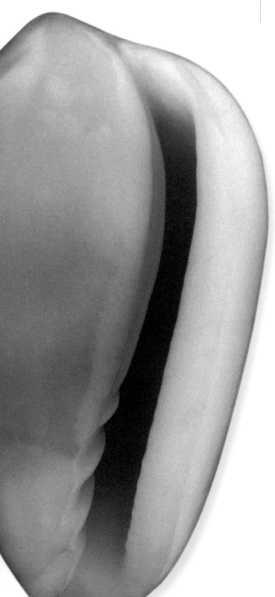

FAMILY	Marginellidae
SHELL SIZE RANGE	¾ to 1¾ in (20 to 44 mm)
DISTRIBUTION	West Africa, from Mauritania to Guinea
ABUNDANCE	Uncommon
DEPTH	Offshore
HABITAT	Sandy bottoms
FEEDING HABIT	Carnivore
OPERCULUM	Absent

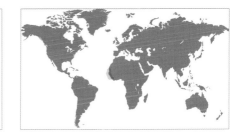

SHELL SIZE RANGE
¾ to 1¾ in
(20 to 44 mm)

PHOTOGRAPHED SHELL
1⅛ in
(28 mm)

GLABELLA PSEUDOFABA

QUEEN MARGINELLA

(SOWERBY II, 1846)

533

The rare *Glabella pseudofaba* has a beautiful spotted shell. It lives in West Africa, on offshore sandy bottoms. There are hundreds of marginellid species worldwide, although the family is more common in tropical waters. The life history is poorly known, but some species are reportedly carnivores, feeding on foraminifera, other gastropods, or sometimes on carrion. Like cowries and olives, the animal envelopes the entire glossy shell with its mantle.

Actual size

RELATED SPECIES

Glabella faba (Linnaeus, 1758), from Senegal, has a very similarly spotted pattern, but a more slender and smaller shell than *G. pseudofaba*. *Glabella mirabilis* (H. Adams, 1869), from the Red Sea and Gulf of Aden, has numerous axial ribs and a broader shell. The lip is thick and spotted in red or brown.

The shell of the Queen Marginella is medium-sized, glossy, and solid, with a moderately tall spire and pointed apex. The whorls are smooth and have an angular profile, with about 15 rounded knobs at the shoulder. The shell is widest at about the middle, tapering anteriorly. The aperture is narrow and long. The outer lip is thickened and denticulate and the columella has 4 strong folds. The shell color is mottled white and brown with spiral rows of squarish black or brown spots, and the aperture is white.

FAMILY	Marginellidae
SHELL SIZE RANGE	¾ to 1⅜ in (19 to 35 mm)
DISTRIBUTION	Mauritania to Sierra Leone
ABUNDANCE	Uncommon
DEPTH	Offshore, moderately deep
HABITAT	Sand
FEEDING HABIT	Carnivore
OPERCULUM	Absent

SHELL SIZE RANGE
¾ to 1⅜ in
(19 to 35 mm)

PHOTOGRAPHED SHELL
1⅜ in
(34 mm)

534

MARGINELLA PETITII

PETIT'S MARGINELLA

DUVAL, 1841

Marginella petitii has a restricted distribution, ranging from Mauritania to Sierra Leone. It lives offshore, on sandy bottoms. Marginellas are present in all oceans at all depths and even (in one case in Thailand) in fresh water. Western Africa has the greatest concentration of species, thanks perhaps to its own great variety of habitats, from sand to mud to coral, in waters of all depths.

RELATED SPECIES

Marginella petitii shares its geographic range with several other species that have color patterns of spiral dark spots. *Marginella faba* (Linnaeus, 1758) has small to medium dots, mostly distributed in spiral lines of vertical pairs. Among the species of *Glabella* (with well-defined axial ribs), *G. harpaeformis* (Sowerby II, 1846) has tiny spots occurring sparsely, and those of *G. pseudofaba* (Sowerby II, 1846) are no more numerous but slightly larger.

The shell of Petit's Marginella is smooth and glossy. It is white, with a triangular, reticulate, orange to chestnut brown pattern. There are regular spiral bands of small, dark brown dots, and also 3 broad bands of dark smudges on the body whorl. Its spire is short, with convex whorls that are defined by shallow sutures. The lip is much thickened; the columella has 4 folds.

Actual size

FAMILY	Marginellidae
SHELL SIZE RANGE	⅞ to 2⅛ in (22 to 53 mm)
DISTRIBUTION	Malay peninsula, Burma to Thailand
ABUNDANCE	Uncommon
DEPTH	Infratidal, shallow
HABITAT	Sand
FEEDING HABIT	Carnivore
OPERCULUM	Absent

SHELL SIZE RANGE
⅞ to 2⅛ in
(22 to 53 mm)

PHOTOGRAPHED SHELL
1⅜ in
(34 mm)

CRYPTOSPIRA ELEGANS
ELEGANT MARGINELLA
(GMELIN, 1791)

535

Marginellas get their family name from their distinctive
feature, a thickened apertural margin or lip. There are examples
throughout the world of extravagantly lipped marginellids,
especially in western Africa. Nowhere, however, is that lip
more elegantly displayed than on this Asian species, *Cryptospira
elegans*, where the aperture takes on the appearance of a brightly
painted mouth, or perhaps the orange-tasseled hem of a light
gray, tweed skirt.

RELATED SPECIES
Cryptospira strigata (Dillwyn, 1817) is identical in range, shape,
size, and color, but here it is the fine axial stripes that are the
darker shade, interrupted by broken, pale gray spiral bands. Both
species have a finely dentate lip and typically pleated marginellid
columella (here with more than four folds).

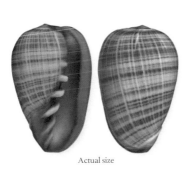

Actual size

The shell of the Elegant Marginella is ovate and
very polished. Its spire is depressed and callused
over. The body whorl, which is swollen at the
posterior, is pale blue-gray with darker spiral
bands of variable width, crossed by closely
packed, fine, pale gray axial stripes, giving the
appearance of woven cloth. The anterior edge
of the columella and the much thickened lip
are all bright orange, completely encircling
the narrow aperture.

FAMILY	Marginellidae
SHELL SIZE RANGE	¾ to 2¼ in (18 to 56 mm)
DISTRIBUTION	Morocco to Senegal, Cape Verde, and the Canary Islands
ABUNDANCE	Moderately common
DEPTH	Offshore to 250 ft (80 m)
HABITAT	Sand
FEEDING HABIT	Carnivore
OPERCULUM	Absent

SHELL SIZE RANGE
¾ to 2¼ in
(18 to 56 mm)

PHOTOGRAPHED SHELL
1½ in
(37 mm)

MARGINELLA GLABELLA
SHINY MARGINELLA
(LINNAEUS, 1758)

Marginella glabella is one of the many Marginellidae found on the western coasts of Africa, a region especially rich in marginellid diversity. Spotted patterns are often a distinguishing feature in the variation between species. In the case of *M. glabella*, for example, several other shells in the region share its general shape and color, differing primarily in the size of their spots, shell, or spire.

RELATED SPECIES

Marginella irrorata (Menke, 1828) is a smaller but very similar form. Its spire and recurved lip are off-white, and the pale spots showing through on the pink body whorl are much finer so that dorsally it almost resembles a strawberry in a pool of cream.

Actual size

The shell of the Shiny Marginella is ovate with a slight concavity anteriorly. Color is variable, but is typically cream overlaid with pale brown to strawberry pink with 3 slightly darker spiral bands on the whorl, through which the cream base shows as a speckle of small to medium spots. The thickened and finely dentate lip recurves as an orange fossa, and there are 4 columellar folds.

FAMILY	Marginellidae
SHELL SIZE RANGE	1⅜ to 4 in (34 to 101 mm)
DISTRIBUTION	Endemic to Brazil
ABUNDANCE	Uncommon
DEPTH	10 to 80 ft (3 to 25 m)
HABITAT	Sand
FEEDING HABIT	Carnivore
OPERCULUM	Absent

SHELL SIZE RANGE
1⅜ to 4 in
(34 to 101 mm)

PHOTOGRAPHED SHELL
2½ in
(63 mm)

BULLATA BULLATA

BUBBLE MARGINELLA

(BORN, 1778)

537

Few of the many hundreds of the species in the family Marginellidae are much more than 1 in (25 mm) long, and *Bullata bullata* is among the largest living species. It is endemic to northeastern Brazil, ranging from Bahia to Espirito Santo. Like all the family, *B. bullata* is a carnivorous predator, feeding on bivalves and crustaceans, which it locates through its sensitive siphon while buried in sand.

RELATED SPECIES

Bullata matthewsi (Van Mol and Tursch, 1967) is a smaller, less elongated species. It too occurs off the shores of Brazil. Its outer lip is a paler yellow, and the overall color is pale to mid orange.

The shell of the Bubble Marginella is elongated and ovate. It is beige to peach pink, with white to pink growth lines and indistinct, slightly darker spiral bands. Its spire is completely sunken. The lip is swollen at the top and rises just beyond the spiral suture. There is a broadening of the aperture toward the siphonal notch. The columella and inner lip are white. The outer lip is recurved and apricot to orange; there are 4 anterior columellar folds.

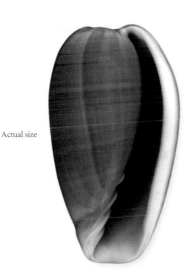

Actual size

FAMILY	Marginellidae
SHELL SIZE RANGE	2¾ to 5 in (70 to 125 mm)
DISTRIBUTION	Endemic to South Africa
ABUNDANCE	Uncommon
DEPTH	230 to 1,650 ft (70 to 500 m)
HABITAT	Offshore soft bottoms
FEEDING HABIT	Carnivore
OPERCULUM	Absent

SHELL SIZE RANGE
2¾ to 5 in
(70 to 125 mm)

PHOTOGRAPHED SHELL
4¾ in
(118 mm)

538

AFRIVOLUTA PRINGLEI
PRINGLE'S MARGINELLA
TOMLIN, 1947

The shell of Pringle's Marginella is lightweight and delicate, distinguished by a large, creamy white round callus between the spire and the aperture, and by 4 large orange columellar folds. Its spire is short, with a blunt, rounded apex and a moderately impressed suture. The long body whorl is pale pink-brown marked with many white growth lines and with 1 broad, paler spiral band rising from the lower outer lip.

Despite the presence of the typical bladelike marginellid columellar folds, *Afrivoluta pringlei* was classified as a volute for many years after its discovery. Closer anatomical study has now confirmed that it is the one of the largest of the Marginellidae, although even within that family it is quite unique in form. Like most of its family it has a specific and localized distribution, endemic to the deep waters off South Africa.

RELATED SPECIES
Marginellona gigas (Martens, 1904), is an even larger species from the eastern Indian Ocean to the South China Sea. It has an evenly colored glossy shell, but unlike *A. pringlei*, it does not have a callus beneath the spire or pronounced columellar folds. Some specimens had been confused for volutes until their anatomy was better studied.

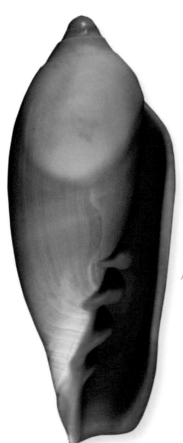

Actual size

FAMILY	Mitridae
SHELL SIZE RANGE	⅜ to 1 in (11 to 25 mm)
DISTRIBUTION	Indo-Pacific
ABUNDANCE	Uncommon
DEPTH	Offshore
HABITAT	Sand
FEEDING HABIT	Carnivore, scavenger
OPERCULUM	Absent

SHELL SIZE RANGE
⅜ to 1 in
(11 to 25 mm)

PHOTOGRAPHED SHELL
¾ in
(19 mm)

IMBRICARIA PUNCTATA
BONELIKE MITER
(SWAINSON, 1821)

539

There are around 500 species of miter shells, evenly split between the Mitridae and Costellariidae families. A simple rule of thumb is that in Mitridae it is the spiral sculpture that dominates, whereas members of the family Costellariidae are distinguished by axial sculpture. In addition there are fine spiral lirations inside the aperture of Costellariidae that are absent from Mitridae. The highest diversity of both families is in the Indo-Pacific region, although they are found in all the world's tropical and temperate seas.

RELATED SPECIES
Imbricaria carbonacea (Hinds, 1844) is a fairly common, slightly smaller shell of similar shape, found in subtidal sands off western and southwestern Africa. It has a more open aperture and the interior is creamy white. The exterior is variable but typically a rich marmelade orange.

The shell of the Bonelike Miter is conical with rounded shoulders, tapering convexly to the anterior. The spire ranges from short to very short, with a lightly incised suture. There are light, spiral punctuated grooves on the body. The color is very pale orange, paler at the lip and spire. Its lip is finely dentate, there are around 6 shallow folds on the columella. The fasciole is callused.

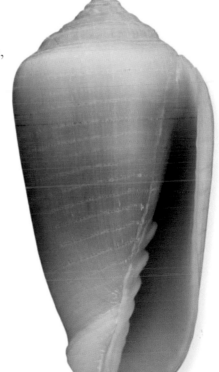

Actual size

FAMILY	Mitridae
SHELL SIZE RANGE	½ to 1 in (13 to 26 mm)
DISTRIBUTION	Western and central Pacific
ABUNDANCE	Moderately common
DEPTH	Subtidal
HABITAT	Coral sands
FEEDING HABIT	Carnivore, scavenger
OPERCULUM	Absent

SHELL SIZE RANGE
½ to 1 in
(13 to 26 mm)

PHOTOGRAPHED SHELL
1 in
(25 mm)

540

IMBRICARIA CONULARIS
CONE MITER
(LAMARCK, 1811)

Imbricaria is a genus in the *family Mitridae,* occurring only in the warmth of tropical waters, where they are found in coral sands. Like all miters they are scavenging carnivores, preying on sipunculans in shallow water at or below the low-tide level. *Imbricaria conularis* has a shell with distinctive shape and coloration, which makes it easily recognized. Its conic shell resembles those of cone shells, but the columellar folds reveal that it is a mitrid.

RELATED SPECIES
Imbricaria olivaeformis (Swainson, 1821), also from the tropical Pacific, is pale green-yellow, elongatedly ovate instead of conical, its spire rounded and convex instead of pointed and concave. The apex and the anterior margin are purple, adding to its fruitlike appearance.

Actual size

The shell of the Cone Miter is conical with a moderately short, concave spire. The apex is purple, the outer lip white, and the rest of the shell a pale milky and purple, with irregular white blotches between mid-brown, wavy, faintly punctured spiral cords. Its aperture is narrow, and the interior is dark orange-brown. There are very shallow axial ribs on all shoulders, and shallow folds on the columella.

FAMILY	Mitridae
SHELL SIZE RANGE	¾ to 1⅜ in (19 to 35 mm)
DISTRIBUTION	Red Sea to Indo-Pacific
ABUNDANCE	Uncommon
DEPTH	Intertidal to 260 ft (80 m)
HABITAT	Sandy bottoms and seaweed beds
FEEDING HABIT	Carnivore
OPERCULUM	Absent

ZIBA ANNULATA

RINGED MITER

(REEVE, 1844)

SHELL SIZE RANGE
¾ to 1⅜ in
(19 to 35 mm)

PHOTOGRAPHED SHELL
1¼ in
(30 mm)

541

Ziba annulata has a wide distribution, ranging from the Red Sea to Mozambique, and to the Marquesas Islands. It is found from the intertidal zone to offshore depths, on sand or in seaweed beds. Like many sand-dwelling gastropods, miters lack an operculum, but the aperture is usually narrow, and the presence of columellar folds may deter some predators. Some miters lay eggs in vase-shaped egg capsules. The number of eggs in each capsule varies from about 100 to 500.

RELATED SPECIES

Ziba maui (Kay, 1979), from Hawaii to the Philippines, has a shell similar in shape and color pattern, but can reach a slightly larger size. *Scabricola fissurata* (Lamarck, 1811), from the Red Sea and Indian Ocean, has a narrow, torpedo-shaped shell, with pale brown spiral bands and white reticulations.

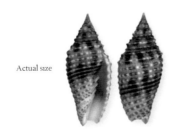

Actual size

The shell of the Ringed Miter is medium in size, solid, sculptured, and fusiform. Its spire is tall, with a pointed apex, well-marked suture, and convex whorls. The sculpture consists of rounded, or sometimes keeled, spiral cords, and 2 rows of small punctations in between cords. The aperture is narrow and long, the outer lip is dentate, and the columella has 4 to 6 folds. The shell color is white or pinkish, with brown dashes on cords. The aperture is stained in brown.

FAMILY	Mitridae
SHELL SIZE RANGE	¾ to 2⅝ in (21 to 65 mm)
DISTRIBUTION	Red Sea to Indo-West Pacific
ABUNDANCE	Uncommon
DEPTH	Subtidal
HABITAT	Coral sand and rubble
FEEDING HABIT	Carnivore, scavenger
OPERCULUM	Absent

SHELL SIZE RANGE
¾ to 2⅝ in
(21 to 65 mm)

PHOTOGRAPHED SHELL
1½ in
(37 mm)

542

SCABRICOLA FISSURATA

RETICULATE MITER

(LAMARCK, 1811)

Actual size

Species in the genus *Scabricola* are an uncommon to rare, and found mostly in the Indian and western and central Pacific regions. They are characterized by the smooth graceful lines that taper at either end. *Scabricola fissurata* is found at the western end of the range, but other species occur as far east as Hawaii and the Marquesas Islands. All miters studied to date have been found to feed exclusively on sipunculans, which also inhabit coral rubble and sandy bottoms.

RELATED SPECIES

Scabricola variegata (Gmelin, 1791) from the central and western Pacific, is unusual. It has an elongated spiral, scalelike sculpture in bands of white and orange-brown, which gives rise to its common name, the Snake Miter. The suture is deeper than usual, the edge of the lip is roundly notched, and there are shallow columellar folds.

The shell of the Reticulate Miter is fusiform with a tall spire whose whorls follow the lines of the body, defined only by a fine suture. The apex is very sharp, and the spire is covered with fine spiral punctuate lines that become obsolete on the body whorl. It is off-white to pale chestnut with darker gray-brown bands below the shoulder, all with a fine, white, reticulate triangular pattern.

FAMILY	Mitridae
SHELL SIZE RANGE	1 to 2 in (25 to 50 mm)
DISTRIBUTION	Gulf of California to Ecuador
ABUNDANCE	Uncommon
DEPTH	30 to 300 ft (9 to 90 m)
HABITAT	Sand and mud flats
FEEDING HABIT	Carnivore, scavenger
OPERCULUM	Absent

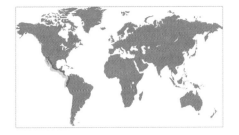

SHELL SIZE RANGE
1 to 2 in
(25 to 50 mm)

PHOTOGRAPHED SHELL
1⅞ in
(48 mm)

543

SUBCANCILLA ATTENUATA

SLENDER MITER

(BRODERIP, 1836)

It is an illustration of the pitfalls of early shell discovery and classification that this species has been given about half a dozen names, most of them to variations in the early nineteenth century. It has been placed in at least three different genera over the years, but it is now classified in the genus *Subcancilla*. These are characterized by their elegant bullet shape, the moderately impressed suture, and the predominance of fine regular spiral cords on body and spire.

RELATED SPECIES

Subcancilla hindsii (Reeve, 1844) shares the size, range, and habitat of *S. attenuata*, but lives farther offshore in deeper water. It varies from light brown to pale purple, with white spiral bands on the shoulders, and spiral cords are a striking dark to red brown. The periphery of the aperture is tinted pale orange.

The shell of the Slender Miter is a slender elongated bullet with a tall spire and a moderately incised suture. It is all white, with fine, sharp, raised spiral cords that are more prominent on the slightly rounded shoulders. The aperture is very finely lined in orange along its outer edge. There are 2 or 3 small folds on the columella.

Actual size

FAMILY	Mitridae
SHELL SIZE RANGE	¾ to 2⅝ in (19 to 65 mm)
DISTRIBUTION	Red Sea to western Pacific
ABUNDANCE	Uncommon
DEPTH	Shallow subtidal to 100 ft (30 m)
HABITAT	Sand
FEEDING HABIT	Carnivore
OPERCULUM	Absent

SHELL SIZE RANGE
¾ to 2⅝ in
(19 to 65 mm)

PHOTOGRAPHED SHELL
2⅛ in
(53 mm)

544

NEOCANCILLA PAPILIO
BUTTERFLY MITER
(LINK, 1807)

Neocancilla papilio has an elegant shell with both spiral cords and axial ribs that form a cancellate texture. It lives on sand and gravel bottoms, ranging from the intertidal zone to offshore. The animal has a pale orange foot spotted in white. It ranges from the Red Sea to the Indo-West Pacific. A smaller form of the species lives in Hawaii.

RELATED SPECIES

Neocancilla clathrus (Gmelin, 1791) also from the Indo-Pacific, is slightly smaller and with more rounded spiral whorls. It is cream to beige-orange, with dark brown spiral bands and moderately large white blotches. The interior is pale pink.

Actual size

The shell of the Butterfly Miter is bullet-shaped with moderately rounded whorls. It is covered in spiral ridges, steeper on the posterior side, separated by 1 to 2 very fine cords. The spiral sculpture is interrupted by closely spaced axial grooves. There are occasional purple spots on the ridges, tending to elongate spirally within 2 broad, brown spiral bands on the body whorl. The aperture is orange to mid brown.

FAMILY	Mitridae
SHELL SIZE RANGE	1⅝ to 3 in (40 to 74 mm)
DISTRIBUTION	Eastern Florida to Yucatán and Honduras
ABUNDANCE	Rare
DEPTH	6 to 100 ft (2 to 30 m)
HABITAT	Sand
FEEDING HABIT	Carnivore, scavenger
OPERCULUM	Absent

MITRA FLORIDA

FLORIDA MITER

GOULD, 1856

SHELL SIZE RANGE
1⅝ to 3 in
(40 to 74 mm)

PHOTOGRAPHED SHELL
2⅛ in
(54 mm)

545

Mitra florida is one of the largest and rarest miters in the tropical western Atlantic. It lives on and under rubble at the base of reefs, and is occasionally collected by scuba divers, particularly during night dives. This species has nine columellar folds, of which the posteriormost two are prominent and the others weak.

RELATED SPECIES

Mitra floridula Sowerby II, 1874, from the Indo-Pacific is reminiscent in appearance of *M. florida*. The spiral spotting is white and less frequent, and the overall color is mid brown with cream bands along the shoulders intersecting large, irregular, white blotches.

Actual size

The shell of the Florida Miter is fusiform with well-rounded whorls, a bulbous body, and a moderate spire. There are very fine spiral grooves near the apex. The aperture is white to pale brown. The outer lip is thin, and the columella has up to 9 closely spaced folds, all but the posteriormost recessed within the aperture. The shell is white to pale pink with a few large, pale to mid brown blotches and a sparse spiral pattern of slightly elongated chestnut spots.

FAMILY	Mitridae
SHELL SIZE RANGE	1¾ to 6⅜ in (44 to 160 mm)
DISTRIBUTION	Red Sea to Indo-Pacific
ABUNDANCE	Rare
DEPTH	Shallow to 120 ft (40 m)
HABITAT	Reef flats
FEEDING HABIT	Carnivore, scavenger
OPERCULUM	Absent

SHELL SIZE RANGE
1¾ to 6⅜ in
(44 to 160 mm)

PHOTOGRAPHED SHELL
4 in
(103 mm)

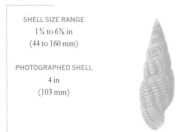

546

MITRA INCOMPTA
TESSELLATE MITER
(LIGHTFOOT, 1786)

The diversity of miters is highest in the shallow waters of the tropical Indo-Pacific Ocean, where the various co-occurring species partition their habitat by substrate type. Some inhabit sandy bottoms while others are restricted to rocky rubble at various depths. Like many of the larger species, *Mitra incompta* inhabits subtidal reef flats.

RELATED SPECIES

Mitra puncticulata (Lamarck, 1811) is another miter inhabiting the reefs of the Indo-Pacific. At about 2 in (55 mm) it looks like a stout version of *M. incompta*—it has the same punctate spiral grooves, coronated suture, and crenulated lip. The shell is dark orange, with a fine cream spiral band below the suture and a broader one on the body whorl.

The shell of the Tesssellate Miter is slender and elongate, the spire longer than the body whorl, and the suture coronated. Shallow axial ribs cross more deeply incised spiral grooves that terminate at a crenulated outer lip. The aperture is tan, with 5 or 6 bold columellar folds and a weak fasciole. The shell is cream to pale orange, with pale and dark brown axial streaks, and an indistinct broad, pale spiral band on the lower body.

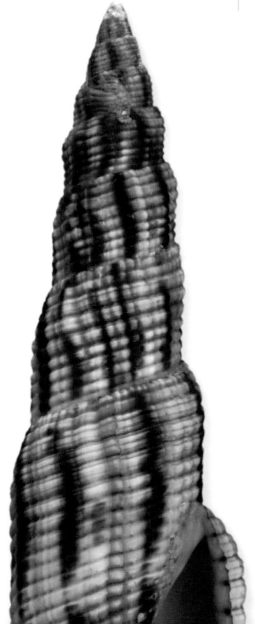

Actual size

FAMILY	Mitridae
SHELL SIZE RANGE	1½ to 7 in (40 to 180 mm)
DISTRIBUTION	Red Sea to Indo-Pacific; Galápagos Islands
ABUNDANCE	Common
DEPTH	Intertidal to 260 ft (80 m)
HABITAT	Sandy bottoms
FEEDING HABIT	Carnivore, feeds on sipunculan worms
OPERCULUM	Absent

SHELL SIZE RANGE
1½ to 7 in
(40 to 180 mm)

PHOTOGRAPHED SHELL
5 in
(131 mm)

MITRA MITRA

EPISCOPAL MITER

(LINNAEUS, 1758)

547

Mitra mitra is the largest species in the family Mitridae, and one of the most widely distributed, ranging from the Red Sea, throughout the Indo-Pacific, to the Galápagos Islands. During the day it stays buried in sand, becoming active at night, when it crawls out of the sand to forage. Like other mitrids, it is believed to feed exclusively on sipunculan worms (peanut worms). *Mitra mitra* has a very long and slender proboscis. Large specimens are used by Pacific islanders as a chiseling tool.

RELATED SPECIES

After *Mitra mitra*, the next largest miters are *M. papalis* (Linnaeus, 1758), from the Indo-West Pacific, and *M. swainsoni* Broderip, 1836, from Baja California to Peru. The former has a suture with short spines, and a white shell peppered with reddish brown blotches. The latter has a smooth shell with a tall, stepped spire, and is uniformly cream-colored.

Actual size

The shell of the Episcopal Miter is large, solid, heavy, smooth, and elongate-ovate in outline. Its spire is tall, with a shallow suture; the spire whorls have spiral grooves, but they fade in later whorls, which are smooth. The aperture is about the same length as the spire. The outer lip is thick, and the lower margin has a finely serrated edge; the columella has 4 to 5 folds. The shell color is white, with spiral rows of squarish orange or reddish blotches, and a white or pale yellow aperture.

FAMILY	Pleioptygmatidae
SHELL SIZE RANGE	2¾ to 5 in (69 to 125 mm)
DISTRIBUTION	Endemic to Honduras
ABUNDANCE	Uncommon
DEPTH	120 to 500 ft (35 to 150 m)
HABITAT	Rocky bottoms
FEEDING HABIT	Carnivore
OPERCULUM	Absent

SHELL SIZE RANGE
2¾ to 5 in
(69 to 125 mm)

PHOTOGRAPHED SHELL
3¾ in
(94 mm)

548

PLEIOPTYGMA HELENAE
HELEN'S MITER
(RADWIN AND BIBBEY, 1972)

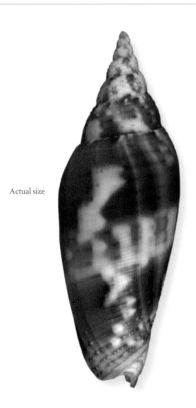

Actual size

Pleioptygma helenae is a unique species that for many years puzzled malacologists, because some of its shell features resemble those from several unrelated gastropod families. Based on those shell characters, it was placed by different authors in the families Volutidae, Mitridae, and Costellariidae. It is an uncommon, deepwater gastropod endemic to Honduras. The first live-collected material finally became available for study in 1989; its anatomy proved to be different enough from all known gastropods to be placed in its own family, Pleioptygmatidae. Judging from its anatomy, it probably feeds on soft-bodied prey such as polychaetes and sipunculan worms.

RELATED SPECIES
Pleioptygma helenae is the only known living species in the family Pleioptygmatidae. There are three named (and possibly two unnamed) Miocene and Pliocene fossil species from the Carolinas and Florida. The family is probably most closely related to the Mitridae.

The shell of the Helen's Miter is medium-large, solid, moderately lightweight, elongate, and fusiform in outline. Its spire is tall, with slightly convex whorls, a pointed apex, and impressed suture. The sculpture consists of moderate to very sharp spiral cords anteriorly, which become obsolete posteriorly. The aperture is long and moderately wide, the outer lip smooth and thin, and the columella has 6 to 9 folds. The shell is marbled with orange-brown and white, with about 20 dashed spiral lines, and the aperture is white.

FAMILY	Volutomitridae
SHELL SIZE RANGE	1⅛ in (28 mm)
DISTRIBUTION	Bellinghausen Abyssal Plain
ABUNDANCE	Rare
DEPTH	14,500 to 15,775 ft (4,419 to 4,808 m)
HABITAT	Abyssal plain
FEEDING HABIT	Carnivore
OPERCULUM	Absent

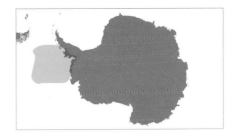

SHELL SIZE RANGE
1⅛ in
(28 mm)

PHOTOGRAPHED SHELL
1⅛ in
(28 mm)

DAFFYMITRA LINDAE

LINDA'S MITER-VOLUTE

HARASEWYCH AND KANTOR, 2005

549

Daffymitra lindae lives at depths of 15,000 ft (4,600 m) on the abyssal plains off Antarctica, and is the only member of the family Volutomitridae to inhabit the deep sea. At present, this species is known only from the single specimen shown here. It is a predator that crawls on mud and sand bottoms. Like all deepsea mollusks that live below the calcium carbonate compensation depth, the shell is very thin, and covered by an outer protein layer, the periostracum, which protects it from dissolving into the surrounding seawater.

RELATED SPECIES

Daffymitra lindae may be easily distinguished from *Paradmete fragillima* (Watson, 1882), a close relative from shallower waters off Antarctica, by its larger, thinner, more inflated shell with a shorter spire, and longer siphonal canal. The Arctic *Volutomitra alaskana*, as well as tropical members of the family all have much thicker shells without pronounced sculpture and with taller spires and thicker columellar folds.

Actual size

The shell of Linda's Miter-volute is small, very thin, and fragile. It has a spire of moderate height, inflated whorls, a large, oval aperture, and a long, broad, siphonal canal. The surface has many fine, sharp, flaring riblets, and less prominent spiral cords. The columella has 3 closely spaced, sharp spiral folds recessed within the aperture. The shell is white, and covered by a thin, olive brown periostracum.

FAMILY	Volutomitridae
SHELL SIZE RANGE	1 to 2 in (25 to 50 mm)
DISTRIBUTION	Northern Pacific, Japan to Alaska, California
ABUNDANCE	Uncommon
DEPTH	Offshore to 500 ft (150 m)
HABITAT	Sand and mud bottoms
FEEDING HABIT	Carnivore
OPERCULUM	Absent in adults

SHELL SIZE RANGE
1 to 2 in
(25 to 50 mm)

PHOTOGRAPHED SHELL
1½ in
(40 mm)

VOLUTOMITRA ALASKANA

ALASKA MITER-VOLUTE

DALL, 1902

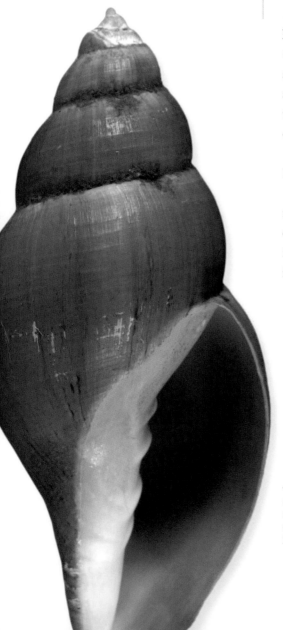

Volutomitra alaskana is an uncommon species that lives in sandy bottoms on the outer continental shelf of the northern Pacific Ocean, along the coasts of northern Japan, Siberia, Alaska, and northern California. Little is known of its ecology, but the anatomy of its jaw and radula indicates that it probably consumes the blood or body fluids of its prey. Larvae and juveniles have an operculum, but it is lost in adults.

RELATED SPECIES

Volutomitra alaskana has a larger and thicker shell than that of most miter-volutes, and lacks the pronounced axial ribs and well-demarcated shoulder common in the tropical members of the family. It resembles *Paradmete fragillima* (Watson, 1882), which is smaller, thinner, and has a proportionally larger aperture.

Actual size

The shell of the Alaska Miter-Volute is narrow and evenly fusiform, with a high conical spire and a narrow, ovate aperture. Its surface is smooth, with numerous, very fine, and closely spaced spiral threads. The long, straight columella has 3 to 4 strong, rounded spiral folds. The shell color is white to yellowish ivory. The periostracum is dark chestnut brown.

FAMILY	Costellariidae
SHELL SIZE RANGE	¾ to 1⅜ in (20 to 35 mm)
DISTRIBUTION	Japan to Indonesia and Tonga
ABUNDANCE	Common
DEPTH	Intertidal to shallow subtidal
HABITAT	Rocky and boulder bottoms
FEEDING HABIT	Carnivore
OPERCULUM	Absent

SHELL SIZE RANGE
¾ to 1⅜ in
(20 to 35 mm)

PHOTOGRAPHED SHELL
1 in
(26 mm)

551

ZIERLIANA ZIERVOGELI

ZIERVOGEL'S MITER

(GMELIN, 1791)

Zierliana ziervogeli has one of the thickest and most robust shells among the costellariids. There are few species in the genus *Zierliana*, and they are the only costellariids that have apertural teeth on both the outer lip and columella. Thickening of the shell and the development of heavy apertural teeth are some of the adaptations of gastropods in response to crabs and other predators.

RELATED SPECIES

Zierliana woldemarii (Kiener, 1838), which ranges from the Andaman Sea to the Solomon Islands, resembles *Z. ziervogeli*, but it is more elongated and has a taller spire. *Vexillum crocatum* (Lamarck, 1812), from the Red Sea and Indo-West Pacific, is fusiform with a cancellated sculpture and angulated whorls.

Actual size

The shell of the Ziervogel's Miter is small, thick, solid, stout, glossy, and pyriform. Its spire is short for the family, the suture is well marked, and the spire whorls are slightly concave. The sculpture consists of a smooth and glossy surface with several rounded spiral cords. The aperture is narrow and elongated, the outer lip thick and dentate, and the columella has 3 to 4 heavy folds. The shell color is dark brown or black, with the apex white and the following whorls light brown. The aperture and teeth are white.

FAMILY	Costellariidae
SHELL SIZE RANGE	1½ to 1⅛ in (13 to 28 mm)
DISTRIBUTION	Indo-West Pacific
ABUNDANCE	Uncommon
DEPTH	Intertidal to 30 ft (9 m)
HABITAT	Sandy bottoms near coral reefs
FEEDING HABIT	Carnivore
OPERCULUM	Absent

SHELL SIZE RANGE
1½ to 1⅛ in
(13 to 28 mm)

PHOTOGRAPHED SHELL
1⅛ in
(28 mm)

VEXILLUM CADAVEROSUM
GHASTLY MITER
(REEVE, 1844)

Vexillum cadaverosum is a small costellariid. Its scientific name means "like a cadaver," referring to its mostly white shell. Costellariids are carnivores, and feed on other mollusks, although some species feed on ascidians. They kill their prey using a venom produced by the salivary gland, which is different from the venom gland of the unrelated cone shells. Costellariid shells vary in shape, but often have predominantly axial sculpture.

RELATED SPECIES

Vexillum pagodula (Hervier, 1898), from the Marianas Islands to Papua New Guinea and Fiji, has a shell that resembles *V. cadaverosum*, but it is smaller, has fewer axial ribs and a broad olive brown spiral band. *Vexillum rugosum* (Gmelin, 1792), from the Indo-Pacific, has a larger, broadly fusiform shell.

Actual size

The shell of the Ghastly Miter is small, thick, solid, and ovately elongate. Its spire is tall, with an impressed suture and angulated whorls. The sculpture consists of 10 to 12 keeled axial ribs, with a sharp knob at the shoulder, crossed by several spiral grooves, forming a cancellated pattern. The aperture is narrow, the outer lip lirated within, and the columella has 4 folds. The shell color is mostly white or off-white, with a narrow, pale brown spiral band just above the suture. The aperture is white.

FAMILY	Costellariidae
SHELL SIZE RANGE	¾ to 1½ in (17 to 36 mm)
DISTRIBUTION	Red Sea to Indo-West Pacific
ABUNDANCE	Uncommon
DEPTH	Shallow subtidal
HABITAT	Coral sand and rubble
FEEDING HABIT	Carnivore, scavenger
OPERCULUM	Absent

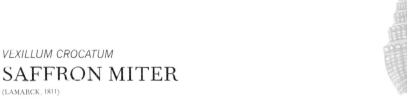

SHELL SIZE RANGE
¾ to 1½ in
(17 to 36 mm)

PHOTOGRAPHED SHELL
1⅛ in
(28 mm)

VEXILLUM CROCATUM

SAFFRON MITER

(LAMARCK, 1811)

553

In classifying *Vexillum crocatum* it would be easy to be distracted
from the axial sculpture by the shell's very pretty colors. It has
distinct spiral cords, but, like all Costellariidae, it is the axial ribs
that dominate the sculpture of the shell. It is an inhabitant of the
sand and rubble of tropical reefs where it scavenges for living
and fallen food. Specimens found beached often have a damaged
or missing apex.

RELATED SPECIES

Vexillum plicarium (Linnaeus, 1758) is another orange Indo-
Pacific sand-dweller, slightly larger than *V. crocatum*. Its spiral
color bands are broader and the cords narrower, giving a less
stepped effect at the shoulders. The shell itself is slightly larger,
with stronger fasciole and a dark brown aperture.

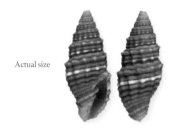

Actual size

The shell of the Saffron Miter is pale to mid orange with a
moderately tall spire. The whorls are angulated, and the whole
shell is covered in deep axial ribs and low spiral cords, slightly
raised on the shoulders. There is fine spiral banding of chestnut
brown between the cords, and 1 cord is white. The aperture is
pale tan, and has about 4 columellar folds.

FAMILY	Costellariidae
SHELL SIZE RANGE	1¼ to 2½ in (30 to 63 mm)
DISTRIBUTION	East Africa to New Caledonia
ABUNDANCE	Uncommon
DEPTH	Intertidal to 165 ft (50 m)
HABITAT	Sandy bottoms near coral reefs
FEEDING HABIT	Carnivore
OPERCULUM	Absent

SHELL SIZE RANGE
1¼ to 2½ in
(30 to 63 mm)

PHOTOGRAPHED SHELL
1⅞ in
(46 mm)

554

COSTATE MITER
(GMELIN, 1791)

Vexillum costatum has an elegant and slender shell, with a smooth surface and incised axial and spiral lines that form a cancellate pattern. The proboscis and the foot are magenta, peppered with white blotches. The cephalic tentacles are small and slender, and each eye is located at a swelling close to the tentacle base. Several gastropod groups lack an operculum as adults, but usually the juveniles have one; costellariids lack an operculum even as juveniles.

RELATED SPECIES

Vexillum politum Reeve, 1844, from the Philippines, has a similar shell, but it is smaller and the surface is mostly smooth. The shell color varies from tan to bright orange, with a thin beige spiral line above the suture. *Vexillum citrinum* (Gmelin, 1791), from the Indo-West Pacific, has a slender shell with a tall spire and a more typical sculpture of sharp axial ridges and weaker spiral cords.

The shell of the Costate Miter is medium in size, solid, glossy, slender, and fusiform. Its spire is very tall, and accounts for about half of the shell length; the suture is impressed, the sides slightly convex, and the apex pointed (partially eroded in the shell illustrated here). Its sculpture consists of evenly spaced, incised spiral and axial lines, forming a lattice pattern. The aperture is narrow and elongate, the outer lip thick and lirate within, and the columella bears 4 to 5 folds. The shell color is white, with variegated orange blotches. The aperture is orange.

Actual size

FAMILY	Costellariidae
SHELL SIZE RANGE	1 to 2½ in (24 to 64 mm)
DISTRIBUTION	Indo-Pacific
ABUNDANCE	Moderately common
DEPTH	Intertidal to 65 ft (20 m)
HABITAT	Sand and mud
FEEDING HABIT	Carnivore, scavenger
OPERCULUM	Absent

SHELL SIZE RANGE
1 to 2½ in
(24 to 64 mm)

PHOTOGRAPHED SHELL
1⅞ in
(46 mm)

VEXILLUM RUGOSUM

RUGOSE MITER

(GMELIN, 1791)

555

Although many *costellariids* carry impressive spiral coloring, it is the strong axial ribs that distinguish them most visibly from mitrids. They differ in other details, too—the costellarid aperture is lirate, and their radular teeth are curved and monocuspid. The mitrid proboscis can be extended and retracted, while that of the Ribbed Miters is shorter, and often does not extend much beyond the tentacles. Both the Mitridae and the Costellariidae have been successful, each with around 250 species spread throughout the world's tropical and temperate oceans.

RELATED SPECIES

Vexillum vulpecula (Linnaeus, 1758) in the Indo-Pacific, and its northern Australian variant *V. v. jukesii* resemble colorful versions of *V. rugosum*. The former has orange and mid-brown banding on a white or occasionally yellow shell. The latter has broader, darker banding with less white, and the aperture is dark brown.

The shell of the Rugose Miter is roughly sculpted, with low spiral cords and strong axial ribs on both the body whorl and the tall spire. It is white to cream, and has bold, very dark spiral bands around the suture and on the body, between which are other, finer, dark and yellow bands. The aperture is narrow, with a convex lower outer lip, and 4 columellar folds.

Actual size

FAMILY	Costellariidae
SHELL SIZE RANGE	2 to 3⅜ in (50 to 86 mm)
DISTRIBUTION	Indo-West Pacific
ABUNDANCE	Uncommon
DEPTH	Shallow subtidal to 165 ft (50 m)
HABITAT	Sandy bottoms near coral reefs
FEEDING HABIT	Carnivore
OPERCULUM	Absent

SHELL SIZE RANGE
2 to 3⅜ in
(50 to 86 mm)

PHOTOGRAPHED SHELL
2½ in
(62 mm)

556

VEXILLUM CITRINUM

QUEEN VEXILLUM

(GMELIN, 1791)

Vexillum citrinum has one of the most colorful shells in the family. It is a variable species, and some forms have been described on the basis of coloration. It lives on sandy bottoms, feeding on other gastropods, usually in shallow, subtidal tropical waters. Vexillums were originally classified as Mitridae because of overall shell resemblance. However, differences such as apertural lirations, which the mitrids lack, separate vexillums from miters.

RELATED SPECIES

Vexillum taeniatum (Lamarck, 1811), from the Indo-West Pacific, is shorter and broader, and has a larger aperture than *V. citrinum*. *Vexillum stainforthii* (Reeve, 1844), ranging from Japan to Australia, has a shell with 10 to 11 strong axial ribs and six scarlet spiral bands per whorl that appear only on the axial bands.

The shell of the Queen Vexillum is slender, thick and fusiform, with a tall spire. The whorls are angulated and the suture is impressed. The sculpture consists of many broad axial, sharply ridged ribs, crossed by smaller spiral riblets. The aperture is narrow and long, the outer lip straight-sided, and the columella has 5 folds. The siphonal canal is curved dorsally. The shell color varies widely, but consists of combinations of brown, orange, yellow, or white spiral bands, with a white or yellowish aperture.

Actual size

FAMILY	Cancellariidae
SHELL SIZE RANGE	⅜ to 1⅛ in (9 to 27 mm)
DISTRIBUTION	Circumarctic
ABUNDANCE	Uncommon
DEPTH	13 to 4,600 ft (4 to 1,400 m)
HABITAT	Sandy and muddy bottoms
FEEDING HABIT	Suctorial feeder
OPERCULUM	Absent

SHELL SIZE RANGE
⅜ to 1⅛ in
(9 to 27 mm)

PHOTOGRAPHED SHELL
¾ in
(18 mm)

ADMETE VIRIDULA

GREENISH ADMETE

(FABRICIUS, 1780)

Admete viridula is a small cancellariid with a wide circumarctic distribution: it ranges from northern Europe, to Russia, and Greenland. Its shell has mostly spiral ridges. There are several hundred living species in the family Cancellariidae, with the highest diversity in the Indo-Pacific and the eastern Pacific. The oldest cancellariid fossils date from the Cretaceous Period. Most cancellariids live in tropical and subtropical waters, but admetines are mostly limited to polar regions and the deep sea.

The shell of the Greenish Admete is small, rough, and globose-fusiform. Its spire is moderately short, with rounded whorls and a well-impressed suture. The sculpture consists of predominantly axial ribs in the earlier whorls; gradually the axial ribs fade and spiral cords dominate. Some shells still have axial sculpture on the body whorl. The large body whorl has mostly spiral ridges. The aperture is oval, the outer lip thin, and the columella has only a small siphonal fold. The shell is cream in color and covered with a thin brown periostracum. The aperture is white.

RELATED SPECIES

Admete unalaskhensis (Dall, 1873), from eastern Russia and northern Japan to Alaska and Oregon, has a small, elongated shell with a stepped, tall spire. The surface has strong spiral ridges and weaker axial ribs, forming a reticulate pattern. *Cancellaria mitriformis* Sowerby I, 1832, from Nicaragua to Peru, has an elongated fusiform shell with a delicate lattice sculpture.

Actual size

FAMILY	Cancellariidae
SHELL SIZE RANGE	¾ to 1 in (19 to 27 mm)
DISTRIBUTION	Costa Rica to Ecuador, Galápagos Islands
ABUNDANCE	Uncommon
DEPTH	Offshore to 100 ft (30 m)
HABITAT	Sandy bottoms
FEEDING HABIT	Suctorial feeder
OPERCULUM	Absent

SHELL SIZE RANGE
¾ to 1 in
(19 to 27 mm)

PHOTOGRAPHED SHELL
¾ in
(20 mm)

558

TRIGONOSTOMA MILLERI

MILLER'S NUTMEG

BURCH, 1949

Trigonostoma milleri has a distinctive shell, with uncoiled whorls like a corkscrew, making it one of the more popular species of nutmegs. As the genus name suggests, it has a triangular aperture. Most species of Cancellariidae have small shells, growing up to about 1 in (25 mm) or less in length. As the family name suggests, the cancellate sculpture on the shell is characteristic, with spiral cords and axial ribs intersecting. Like many sand-dwelling snails, nutmegs lack an operculum. There are well over 150 species of cancellariids worldwide, with the fauna of the eastern Pacific being particularly diverse.

The shell of the Miller's Nutmeg is small, scalariform, and uncoiled, with a triangular whorl and aperture profile. The whorls are sharply keeled at the shoulder and in the lower part of the whorl, bearing short spines that are bent toward the coiling axis. The lower spines and those close to the aperture are larger than those on the shoulder. The whorls are so loosely coiled that succeeding whorls do not touch one another. The shell color ranges from cream to light brown.

RELATED SPECIES

Trigonostoma elegantulum Smith, 1947, also from the eastern Pacific and the Galápagos Islands, has a low spire and angular whorls. *Trigonostoma thysthlon* Petit and Harasewych, 1987, which ranges from Japan to Australia, is a small nutmeg with sharply keeled shoulders with short spines and a wide umbilicus.

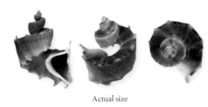

Actual size

FAMILY	Cancellariidae
SHELL SIZE RANGE	¾ to 1⅜ in (19 to 35 mm)
DISTRIBUTION	Nicaragua to Peru
ABUNDANCE	Uncommon
DEPTH	Subtidal to 120 ft (35 m)
HABITAT	Muddy bottoms
FEEDING HABIT	Suctorial feeder
OPERCULUM	Absent

SHELL SIZE RANGE
¾ to 1⅜ in
(19 to 35 mm)

PHOTOGRAPHED SHELL
1⅛ in
(27 mm)

CANCELLARIA MITRIFORMIS
MITER-SHAPED NUTMEG
SOWERBY I, 1832

Cancellaria mitriformis has a small but elegant fusiform shell that somewhat resembles a miter shell, hence the name. Its purplish brown and very cancellated shell is easily recognized by its elongated spindle shape. Crabbed specimens may be found in the intertidal zone, but the live cancellariid lives offshore, on muddy bottoms. The feeding habits of cancellariids are not well known; anatomical studies cannot find food remains in the stomach, other than some sediment. Some authors have suggested that the group may be specialized ectoparasites of large gastropods or fishes. At least a few cancellariids are known to suck the blood of sleeping fishes.

RELATED SPECIES
Cancellaria nodulifera Sowerby I, 1825, from the Japonic region, has a larger and broader shell, with a wide aperture and a sculpture of strong axial folds and spiral ribs, with nodules. *Scalptia mercadoi* Old, 1968, from the Philippines, has an attractive shell with very strong, inclined axial varices crossed by fine spiral threads, and brown spiral bands. The varices make the shell somewhat resemble an epitoniid, but the aperture with a siphonal canal immediately separates it from epitoniids.

The shell of the Miter-shaped Nutmeg is medium-sized, slender, elongate, and fusiform. Its spire is tall and stepped, with a well-marked suture. The sculpture consists of evenly spaced axial and spiral cords, with the spiral ones slightly heavier, forming a lattice pattern. The aperture is elongate-oval, the outer lip thick and crenulated with the spiral cords, and the columella has 1 strong and several smaller folds. The shell color is purplish brown, with a slightly lighter interior.

Actual size

FAMILY	Cancellariidae
SHELL SIZE RANGE	⅞ to 1½ in (23 to 40 mm)
DISTRIBUTION	Sri Lanka to the Philippines and Australia
ABUNDANCE	Uncommon
DEPTH	Offshore to 330 ft (100 m)
HABITAT	Sandy gravel bottoms
FEEDING HABIT	Suctorial feeder
OPERCULUM	Absent

SHELL SIZE RANGE
⅞ to 1½ in
(23 to 40 mm)

PHOTOGRAPHED SHELL
1⅜ in
(34 mm)

TRIGONOSTOMA SCALARE
TRIANGULAR NUTMEG
(GMELIN, 1791)

Trigonostoma scalare has one of the largest shells in the genus. Its distinctive shape separates it from most species in this diverse family: it has a pagoda-shaped shell with sharply keeled whorls that are flat on top. It is an uncommon species found offshore, and like many cancellariids, it lives in sandy bottoms. The diet of cancellariids is not well known because no gut contents can be readily identified. Adaptations in the alimentary system suggest that they are suctorial feeders. Some cancellariids have a trend toward the loss of the radula.

RELATED SPECIES

Trigonostoma milleri Burch, 1949, from the Panamic region, has a small, scalariform shell that is nearly completely uncoiled. It has a sharp shoulder with short spines. *Axelella smithii* (Dall, 1888), from North Carolina to Colombia, has a small, ovate shell. It is one of the few species of which the anatomy has been studied in detail.

The shell of the Triangular Nutmeg is large for the genus, spiny, and pagoda-shaped in outline. Its spire is tall and stepped, and the whorls barely touch the preceding whorl. The aperture has the triangular shape that is characteristic of the genus, and the lip is thickened. The umbilicus is deep and wide, with a sharp keel around it. The sculpture consists of axial ribs that become spinose at the shoulder, and spiral riblets. The shell color is white or beige, and the interior of the aperture is light brown.

Actual size

FAMILY	Cancellariidae
SHELL SIZE RANGE	⅞ to 2 in (22 to 50 mm)
DISTRIBUTION	Philippines and Indonesia
ABUNDANCE	Rare
DEPTH	780 to 1,100 ft (240 to 335 m)
HABITAT	Soft bottoms
FEEDING HABIT	Suctorial feeder
OPERCULUM	Absent

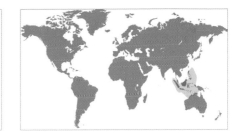

PLESIOTRITON VIVUS
(HABE AND OKUTANI, 1981)

SHELL SIZE RANGE
⅞ to 2 in
(22 to 50 mm)

PHOTOGRAPHED SHELL
1⅝ in
(42 mm)

561

Plesiotriton vivus is a rare deepwater cancellariid restricted to the Philippines and Indonesia. It has a tall spire and the spire whorls bulge over the suture in some places. It resembles the shell of some species of the unrelated *Colubraria* genus, but its distinct radula shows that it is a cancellariid. Cancellariids have a radula with a single row of very long teeth, with interlocking cusps at the anterior end. It is believed that the radula and other anatomical features of the alimentary system, such as a long proboscis, are specializations for its suctorial feeding.

RELATED SPECIES

Tritonoharpa siphonata (Reeve, 1844), from Baja California to Panama, has one of the more elongated shells in the family. It superficially resembles a stretched out *Plesiotriton vivus*, with a cancellate texture, and a smooth outer lip. *Cancellaria cooperi* Gabb, 1865, has a fusiform shell with an angular shoulder and a tall spire. The shell is yellow-brown or orange-brown in color.

Actual size

The shell of *Plesiotriton vivus* is medium-sized for the family, rough, elongate, and fusiform. Its spire is tall, with a pointed apex, convex whorls, and a well-impressed and uneven suture. The sculpture consists of rounded axial folds crossed by spiral threads which form nodules. There are unevenly spaced varices, but less than 2 per whorl. The aperture is lanceolate, the outer lip thickened and dentate, and the columella has 2 to 3 small folds. The shell color is cream with a light brown axial band, and the aperture is white.

FAMILY	Cancellariidae
SHELL SIZE RANGE	1½ to 2½ in (40 to 60 mm)
DISTRIBUTION	Japan to East China Sea
ABUNDANCE	Uncommon
DEPTH	16 to 165 ft (5 to 50 m)
HABITAT	Sandy and muddy bottoms
FEEDING HABIT	Suctorial feeder
OPERCULUM	Absent

SHELL SIZE RANGE
1½ to 2½ in
(40 to 60 mm)

PHOTOGRAPHED SHELL
1¾ in
(43 mm)

CANCELLARIA NODULIFERA
KNOBBED NUTMEG
SOWERBY I, 1825

Cancellaria nodulifera is found in shallow subtidal waters in the Japonic region, occurring from northern Japan to the East China Sea. It has a broad, barrel-shaped shell with a wide oval aperture and strong axial and spiral ribs with nodules, especially at the shoulder, where the nodules can be pointed, making it easily recognized. It is known also as a Pleistocene fossil from deposits near Tokyo and other locales in Japan. Cancellariids are distributed worldwide; while most species are tropical or subtropical, some are limited to temperate regions.

RELATED SPECIES

Sveltia gladiator Petit, 1976, is endemic to the Galápagos Islands. Its delicate shell has a tall spire, and where a spiral cord crosses a varix, it forms spines; it is one of the few cancellariids with a spiny shell. *Cancellaria crawfordiana* (Dall, 1892), from California, has a fusiform shell with a tall spire. Its surface has the typical cancellate sculpture common in the family.

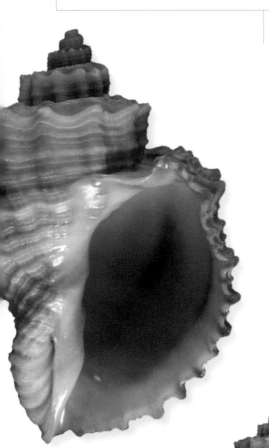

Actual size

The shell of the Knobbed Nutmeg is medium in size, thick, broad, and barrel-shaped. Its spire is moderately short, with whorls angular at the shoulder, and a well-marked suture. The surface has strong axial ribs crossed by spiral threads, forming nodules at the intersections. The aperture is large and wide, the outer lip thin and crenulated, with expansions corresponding to the spiral threads. The columella has 3 small folds, and the fasciole is partially covered by a columellar callus. The shell color is apricot and the interior cream.

FAMILY	Cancellariidae
SHELL SIZE RANGE	¾ to 2 in (20 to 50 mm)
DISTRIBUTION	Spain to Angola; Mediterranean
ABUNDANCE	Common
DEPTH	35 to 135 ft (10 to 40 m)
HABITAT	Muddy and shell hash bottoms
FEEDING HABIT	Suctorial feeder
OPERCULUM	Absent

SHELL SIZE RANGE
¾ to 2 in
(20 to 50 mm)

PHOTOGRAPHED SHELL
1¾ in
(44 mm)

563

CANCELLARIA CANCELLATA

CANCELLATE NUTMEG

(LINNAEUS, 1767)

Cancellaria cancellata is the type species of the genus *Cancellaria*. It is a common species that ranges from shallow subtidal to offshore depths, often on muddy bottoms. Its center of distribution is probably the western African coast; it is rare in the Mediterranean. It is believed to parasitize animals such as fish, sucking blood or other bodily fluids. *Cancellaria cancellata* also occurs in fossil deposits of the Miocene age, as well as in Pliocene and Pleistocene strata.

RELATED SPECIES

Cancellaria reticulata (Linnaeus, 1767), which ranges from North Carolina to Brazil, is a common species found in shallow water. Its shell is thick and oval, with a pointed apex, and a cancellate surface sculpture. *Cancellaria cooperi* Gabb, 1865, has a larger, more elongated shell with angular shoulders and a tall spire. Its shell is lined with thin, brown spiral bands, and the sculpture is dominated by axial ribs, which form nodules along the shoulder.

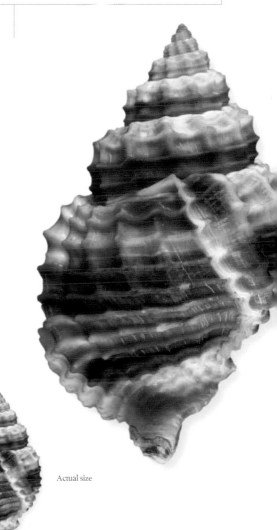

The shell of the Cancellate Nutmeg is medium in size, thick, and oval-globose. Its spire is moderately tall, with a pointed apex, well-marked suture, and whorls with an angular shoulder. The sculpture consists of strong axial and spiral ribs, forming a lattice pattern. The aperture is oval and small, the outer lip thickened and denticulate, and the columella has strong folds. The shell color is white or tan with brown spiral bands, and the aperture is white.

Actual size

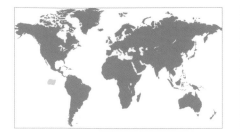

FAMILY	Cancellariidae
SHELL SIZE RANGE	1½ to 2 in (40 to 50 mm)
DISTRIBUTION	Galápagos Islands
ABUNDANCE	Rare
DEPTH	Deep water, to 650 ft (200 m)
HABITAT	Hard substrates
FEEDING HABIT	Suctorial feeder
OPERCULUM	Absent

SHELL SIZE RANGE
1½ to 2 in
(40 to 50 mm)

PHOTOGRAPHED SHELL
2 in
(51 mm)

564

SVELTIA GLADIATOR
GLADIATOR NUTMEG
PETIT, 1976

Sveltia gladiator is one of very few cancellariids with a spiny shell. Cancellariids occur worldwide, but are more common in the tropics, being particularly diverse in the eastern Pacific. Little is known about the life history or ecology of cancellariids. Most species, including *S. gladiator*, have a radula with a single row of unusually long and flexible teeth. A few species are known to be suctorial feeders, sucking blood or fluids from tissues of living prey such as fish and other mollusks.

RELATED SPECIES

Sveltia centrota (Dall, 1896), from Baja California to Peru, has a shell that is similar to that of *S. gladiator* in shape but with shorter spines. *Cancellaria gemmulata* (Sowerby I, 1832), from the eastern Pacific and Galápagos, has a more typical cancellariid shell, albeit short, with cancellate sculpture.

The shell of the Gladiator Nutmeg is medium-sized, high-spired, shouldered, and spiny. The spire comprises half the shell length. The suture is impressed. A long spine is present on each varix where it intersects the strong shoulder. The whorls have 7 or 8 strong spiral cords, crossed by 8 to 10 axial varices per whorl. The aperture is large and wide, broadly ovate, and the outer lip reflected, with small spines. The columella has 3 folds. The shell color is cream.

Actual size

FAMILY	Cancellariidae
SHELL SIZE RANGE	2 to 2¾ in (50 to 69 mm)
DISTRIBUTION	Washington, U.S.A., to Baja California
ABUNDANCE	Uncommon
DEPTH	Offshore to 2,000 ft (600 m)
HABITAT	Sandy and muddy bottoms
FEEDING HABIT	Suctorial feeder
OPERCULUM	Absent

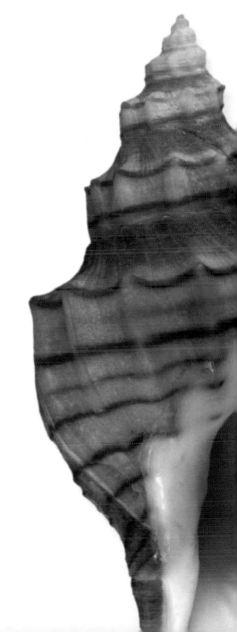

SHELL SIZE RANGE
2 to 2¾ in
(50 to 69 mm)

PHOTOGRAPHED SHELL
2½ in
(62 mm)

CANCELLARIA COOPERI

COOPER'S NUTMEG

GABB, 1865

565

Cancellaria cooperi is a large cancellariid that lives offshore on soft bottoms. Females lay long-stalked egg capsules; the larvae have an operculum, but all adult cancellariids lack one. Studies of the ecology of this species revealed that is parasitizes fish. *Cancellaria cooperi* is active at night, and searches for its prey using chemical cues. When it finds a sleeping electric ray, it extends its long proboscis and uses its radula to penetrate soft tissues such as gills, in order to feed on the blood of its host.

RELATED SPECIES

Cancellaria nodulifera Sowerby I, 1825, has a broad shell with a wide aperture and short spire. The shell has strong axial folds and spiral ribs. *Plesiotriton vivus* (Habe and Okutani, 1981), from the Philippines and Indonesia, has an elongate, fusiform shell with irregular spire whorls and unevenly spaced varices.

The shell of the Cooper's Nutmeg is large for the family, thick, heavy, and elongate-fusiform. Its spire is tall, with a pointed apex and spire whorls with an angular shoulder. The sculpture consists of strong axial ribs with sharp tubercles at the shoulder. The body whorl is large and the siphonal canal short and strong. The aperture is lanceolate, the outer lip thick and lirate within, and the columella has 2 folds. The shell color is yellow brown or orange-brown, with brown spiral lines. The aperture is white.

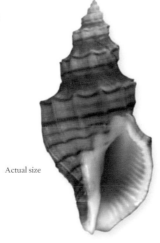

Actual size

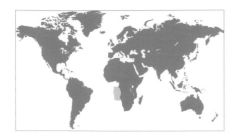

FAMILY	Clavatulidae
SHELL SIZE RANGE	1½ to 2¼ in (38 to 58 mm)
DISTRIBUTION	Western Africa
ABUNDANCE	Common
DEPTH	Subtidal to 100 ft (30 m)
HABITAT	Sandy bottoms
FEEDING HABIT	Carnivore
OPERCULUM	Corneous, thin

SHELL SIZE RANGE
1½ to 2¼ in
(38 to 58 mm)

PHOTOGRAPHED SHELL
2 in
(49 mm)

566

CLAVATULA IMPERIALIS
IMPERIAL TURRID
(LAMARCK, 1816)

Actual size

Clavatula imperialis belongs to the superfamily Conoidea, whose members have evolved a venom apparatus for hunting their prey. There are many thousands of species in this superfamily, and the relationships among the various families included within it are still poorly understood. Most of the adaptations associated with the venom apparatus are biochemical and anatomical, including the development of a large venom gland and bulb, which produce and inject a "cocktail" of toxins into prey.

RELATED SPECIES
Clavatula kraepelini (Strebel, 1912), which ranges from Senegal to Angola, is a slightly smaller species easily recognized by its smooth, glossy, narrowly fusiform shell. *Clavatula bimarginata* (Lamarck, 1822) from Mauritania to South Africa has a strongly biconical shell, with a siphonal canal as long as the aperture.

The shell of the Imperial Turrid is moderately large and broadly fusiform, with a stepped, conical spire and a broadly ovate aperture with a narrow, spiny shoulder. The posterior notch is below the shoulder. It is narrow, deep, and slit-like, and produces a slit band (selenizone) that runs below the base of the spines. The siphonal canal is short, broad, and slightly deflected to the right. Other than the shoulder spines, the sculpture is limited to weak spiral cords and occasionally stronger growth lines. The shell color is white with the slit band being brownish interrupted with white. The aperture is white.

FAMILY	Clavatulidae
SHELL SIZE RANGE	1¼ to 3¼ in (32 to 83 mm)
DISTRIBUTION	Mauritania to Senegal, Canary Islands
ABUNDANCE	Moderately common
DEPTH	Offshore, to 250 ft (75 m)
HABITAT	Sandy bottoms
FEEDING HABIT	Carnivore
OPERCULUM	Corneous, thin

SHELL SIZE RANGE
1¼ to 3¼ in
(32 to 83 mm)

PHOTOGRAPHED SHELL
2¼ in
(56 mm)

PUSIONELLA NIFAT
NIFAT TURRID
(BRUGUIÈRE, 1789)

567

Like most members of the Conoidea superfamily, *Pusionella nifat* is a predator that hunts polychaete worms inhabiting offshore sandy bottoms. The smooth, streamlined shell indicates that this species burrows into the sand. The radular ribbon of members of the family Clavatulidae contains three teeth per row, with the central tooth reduced and the outer teeth long, sharply pointed, and furrowed, possibly to conduct toxins. Other conoidean families have lost the central tooth and modified their outer teeth into hollow tubes with a sharp, barbed distal end.

RELATED SPECIES

Pusionella vulpina (Born, 1780), with a narrow range limited to the coast of Senegal, has a smaller, tapered, ivory-colored shell. *Pusionella milleti* (Petit, 1851), also from Senegal, has a slightly smaller, fusiform shell that is white to chestnut brown.

The shell of the Nifat Turrid is large, smooth, and roundly fusiform. Its spire is tall and conical, with evenly rounded whorls. The aperture is oval, without a conspicuous posterior notch, and with a short, broad siphonal canal that crosses the coiling axis. Surface sculpture, when distinguishable, is limited to very fine spiral threads and axial growth lines. The shell is white or ivory, with 5 rows of tan to dark brown rectangular spots. The 2 rows between the white bands at the suture and shell periphery sometimes fuse to produce axial blotches. The aperture is white within.

Actual size

FAMILY	Clavatulidae
SHELL SIZE RANGE	2 to 3 in (50 to 75 mm)
DISTRIBUTION	Red Sea to Indo-West Pacific
ABUNDANCE	Uncommon
DEPTH	Offshore to 100 ft (35 m)
HABITAT	Muddy bottoms
FEEDING HABIT	Carnivore
OPERCULUM	Corneous, thin

SHELL SIZE RANGE
2 to 3 in
(50 to 75 mm)

PHOTOGRAPHED SHELL
3 in
(75 mm)

TURRICULA TORNATA

TURNED TURRID

(DILLWYN, 1817)

Turricula tornata has a fairly large, smooth shell, suggesting that it is a burrower in subtidal sandy bottoms, a habitat it shares with many other members of the family. When large numbers of species with similar feeding adaptations co-occur, each may specialize on a different prey. In contrast, some species of *Turricula* have a generalized diet, feeding on more than a dozen different polychaete worms.

RELATED SPECIES

Turricula javana (Linnaeus, 1767), another, wide-ranging Indo-West Pacific species, is similar in size and general shape, but uniform tan in color. *Turricula nelliae* (Smith, 1877), from Indonesia, has a smaller shell with a taller spire. The shell color consists of brown and white axial bands.

The shell of the Turned Turrid is large and spindle-shaped, with a tall spire, small, oval aperture, and a long, narrow, axial siphonal canal. The spire comprises almost half the shell length. The rounded shoulder forms an obtuse angle along the middle of the whorls. The aperture is narrowly ovate, with a broad, triangular posterior notch that extends from the suture to the shoulder. The outer lip is thin and the columella is smooth, with a conspicuous siphonal fold. The shell surface is fairly smooth, with very fine growth lines and occasionally spiral threads. The base color is ivory, with brownish blotches along the shoulder, and narrow, parallel, brown lines that run at a diagonal angle to the outer lip.

Actual size

FAMILY	Clavatulidae
SHELL SIZE RANGE	2 to 3 in (50 to 77 mm)
DISTRIBUTION	Indo-West Pacific; Japan
ABUNDANCE	Common
DEPTH	30 to 260 ft (10 to 80 m)
HABITAT	Offshore muddy bottoms
FEEDING HABIT	Carnivore
OPERCULUM	Corneous, leaf-shaped

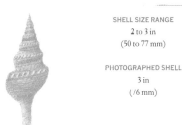

SHELL SIZE RANGE
2 to 3 in
(50 to 77 mm)

PHOTOGRAPHED SHELL
3 in
(76 mm)

TURRICULA JAVANA

JAVA TURRID

(LINNAEUS, 1767)

509

One of the largest species of clavatulids commonly found, the Java Turrid occurs at a wide range of depths, up to around 260 ft (80 m). It feeds on polychaete worms, which it finds on muddy substrates, using its poisonous, arrowlike tooth to overcome its prey. The species is commonly caught in trawler nets, but is not a significant fisheries species.

RELATED SPECIES

Turricula tornata (Dillwyn, 1817), found in the Red Sea, Thailand, and West Pacific, is narrower with a longer spire and siphonal canal, and lacks knobs on the periphery of the whorls. *Toxiclionella haliplex* (Bartsch, 1915), which is endemic to South Africa, has a smaller shell with a tall spire, a rounded apex, and a cream spiral band alternated with a brown band.

The shell of the Java Turrid has a relatively long, high spire. The periphery of the whorls carries oblique, pale nodules. It has a deep suture. There are 2 spiral ridges between the suture and the shoulder. Distinctive spiral ridges run from the shoulder of the body whorl to the end of the siphonal canal, which can occasionally be twisted. There is a sinuous columella and notch at the posterior end of outer lip. The uniform color generally ranges from pale to dark brown.

Actual size

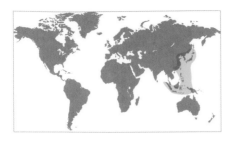

FAMILY	Drilliidae
SHELL SIZE RANGE	½ to ⅝ in (13 to 17 mm)
DISTRIBUTION	Japan to Indonesia, New Guinea and the Philippines
ABUNDANCE	Rare
DEPTH	200 to 500 ft (60 to 150 m)
HABITAT	Sandy bottoms
FEEDING HABIT	Carnivore
OPERCULUM	Corneous, thin

SHELL SIZE RANGE
½ to ⅝ in
(13 to 17 mm)

PHOTOGRAPHED SHELL
⅝ in
(15 mm)

570

CONOPLEURA STRIATA

STRIATED TURRID

(HINDS, 1844)

The shell of the Striated Turrid is small and biconical, with a broad spire, and a narrowly ovate aperture. At the anterior end of the aperture, a short siphonal canal flanks a siphonal fasciole. The posterior margin of the aperture forms a deep, narrow posterior notch. The outer margins of successive notches are joined to form a raised outer wall, giving the shell a conical appearance. The shell color ranges from white to golden tan.

Conopleura striata has a small, fragile shell and the scarcity of repaired breaks on the shell indicates that it does not often encounter predators. Like most other conoideans it feeds primarily or exclusively on polychaete worms. It is a member of the family Drilliidae, a group characterized by having an unusual radula with five teeth per row, including a reduced central tooth, which is flanked by rakelike teeth and long, narrow, sharply pointed teeth. The rake-like teeth may assist with swallowing the polychaete prey.

RELATED SPECIES

Clavus canicularis (Röding, 1798), which has a similar distribution, has a shell that is twice the size and much heavier; the margins of its posterior notches form spires that are not joined. *Guraleus kamakuranus* (Pilsbry, 1904) from Japan and Korea, is a minute species, about one-third the size of *C. striata*, yet has a similar shell shape.

Actual size

FAMILY	Drilliidae
SHELL SIZE RANGE	1 to 1¼ in (25 to 32 mm)
DISTRIBUTION	Tropical western Pacific
ABUNDANCE	Uncommon
DEPTH	Shallow water to 65 ft (20 m)
HABITAT	Sandy gravel
FEEDING HABIT	Carnivore
OPERCULUM	Corneous, thin

SHELL SIZE RANGE
1 to 1¼ in
(25 to 32 mm)

PHOTOGRAPHED SHELL
1⅛ in
(28 mm)

CLAVUS CANICULARIS

LITTLE DOG TURRID

(RÖDING, 1798)

571

Clavus canicularis lives on gravel bottoms at subtidal depths. Its broad shell, with long, outwardly directed spines, suggests that the animal crawls over rather than burrows into the substrate. Most snails that are active predators have a long siphon; in *C. canicularis*, the siphon is extended far beyond the edge of a short, broad siphonal canal. When hunting, the animal moves the siphon from side to side, and locates its prey by "smelling" the water brought through the siphon.

RELATED SPECIES

Clavus exasperatus (Reeve, 1843), from the central Pacific, is similar in size, general proportions, and coloration, yet lacks the broad, flaring spines that are formed by the expanded margins of the posterior notch. *Clavus enna* (Dall, 1918), from the southwestern Pacific, grows to twice the size. The much smaller *C. lamberti* (Montrouzier, 1860), from the western Pacific, has a glossy, colorful shell, with prominent axial ribs, and spiral bands of white, yellow, and brown.

The shell of the Little Dog Turrid is small and rhombic in outline, with a high, conical spire, and an elongate aperture. The posterior notch in the aperture is about as broad as the short siphonal canal. The sculpture consists of 8 to 10 prominent axial ribs that, in most specimens, have a broad open spine at the shoulder. Spiral sculpture is limited to fine spiral threads. The shell color is white, with a darker, tan to brown band across the middle or base of the shell. The aperture is white.

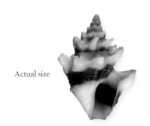

Actual size

FAMILY	Drilliidae
SHELL SIZE RANGE	1 to 2 in (25 to 52 mm)
DISTRIBUTION	Indo-West Pacific
ABUNDANCE	Uncommon
DEPTH	Subtidal to 60 ft (20 m)
HABITAT	Sand and rubble bottom
FEEDING HABIT	Carnivore
OPERCULUM	Corneous, thin

SHELL SIZE RANGE
1 to 2 in
(25 to 52 mm)

PHOTOGRAPHED SHELL
1¾ in
(44 mm)

572

CLAVUS FLAMMULATUS
FLAMING TURRID
(MONTFORT, 1810)

Clavus flammulatus is an uncommon species with a broad geographic range throughout the shallow waters of the tropical Indian and Pacific oceans. A closer inspection of the illustrated shell will reveal that the animal that made it survived an attack by a crushing predator, probably a crab. When attacked, a snail will withdraw deep within its shell, and seal the opening with its operculum. However, the crab will hold the shell and try to peal back the outer lip by breaking off pieces with its claws.

RELATED SPECIES

Clavus bilineatus (Reeve, 1845), from the Indo-Pacific, is a small turrid that has two narrow lines, one white and the other dark brown, that run along the shoulder over a reddish brown shell. *Clavus opalus* (Reeve, 1845) from the Philippines and Fiji, is half the size, white to light tan, and has prominent axial ribs.

The shell of the Flaming Turrid is of moderate size and biconic, with a tall spire that is longer than half the shell length, and an elongated aperture with nearly straight sides. The siphonal canal is broad, short, and barely distinguishable. The posterior notch is as broad, with a dorsally reflected flared edge. The shoulder is sharply angled, and has 10 to 12 axial nodes, each with a short spine. Spiral sculpture is limited to fine threads near the anterior margin of the shell. The base color is white with 5 or 6 rows of dark brown spots that are irregular in size and shape. Spots in adjacent rows above and below the shoulder often fuse. The aperture is white within, but may reflect external color patterns near the thin outer lip.

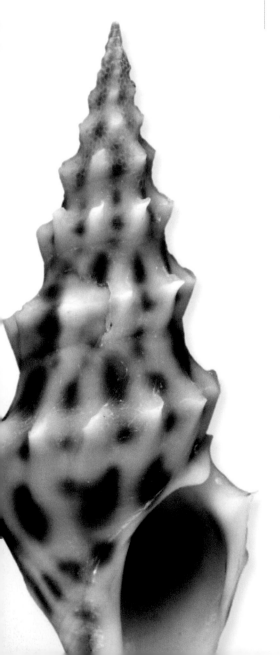

Actual size

FAMILY	Drilliidae
SHELL SIZE RANGE	1¼ to 2 in (30 to 52 mm)
DISTRIBUTION	Northern coast of South America
ABUNDANCE	Uncommon
DEPTH	20 to 65 ft (6 to 20 m)
HABITAT	Sandy bottoms
FEEDING HABIT	Carnivore
OPERCULUM	Corneous, thin

SHELL SIZE RANGE
1¼ to 2 in
(30 to 52 mm)

PHOTOGRAPHED SHELL
1¾ in
(46 mm)

DRILLIA GIBBOSA
HUMPED TURRID
(BORN, 1778)

573

Drillia gibbosa has a very narrow range along the coasts of Colombia and Venezuela. The limited geographic distribution of this species is a result of its biology. Like other species of *Drillia*, females deposit their eggs in small, round, flat, leathery capsules that are attached to rocks, shell fragments, or other hard objects. Each capsule contains less than a dozen eggs. The larvae undergo all developmental stages within the capsule, and hatch as small, crawling juveniles. The distribution of these animals is thus limited to how far they can crawl after hatching.

RELATED SPECIES

Drillia clavata (Sowerby I, 1834), from Costa Rica to Ecuador, is similar in shape, but has a smooth, glossy surface. *Drillia albicostata* (Sowerby I, 1834), endemic to the Galápagos Islands, is also similar, but has smooth, sinuous, axial bands.

The shell of the Humped Turrid is moderate in size, with a very tall, conical, slightly stepped spire. The aperture is elongate and oval, with a distinct, long, siphonal canal and a narrowed posterior notch. Whorls are strongly shouldered, with the posterior fold placed between the suture and shoulder. The sculpture consists of numerous, closely spaced, axial ribs, slightly oblique to the coiling axis, which are crossed by spiral cords, producing weak reticulate sculpture. The base color is white or ivory, with brownish spriral bands interrupted by a band of white at mid-whorl. The aperture is whitish, but translucent, showing pigmentation on the outer surface of shell.

Actual size

FAMILY	Pseudomelatomidae
SHELL SIZE RANGE	1¼ to 2 in (32 to 50 mm)
DISTRIBUTION	Sea of Cortéz to Ecuador, Galápagos Islands
ABUNDANCE	Common
DEPTH	Intertidal to 100 ft (30 m)
HABITAT	Sandy bottoms
FEEDING HABIT	Carnivore
OPERCULUM	Corneous, leaf-shaped, reduced

SHELL SIZE RANGE
1¼ to 2 in
(32 to 50 mm)

PHOTOGRAPHED SHELL
1⅞ in
(47 mm)

574

HORMOSPIRA MACULOSA
BLOTCHY TURRID
(SOWERBY I, 1834)

Like other members of the family Pseudomelatomidae, species in the genus *Hormospira* have a distinctive radula with three rows of teeth, each with a single cusp, and a poison gland. They live in relatively shallow waters along the inner continental shelf off the Pacific coast of central and South America. They inhabit sandy bottoms and feed on small invertebrates, such as polychaetes.

RELATED SPECIES

There is only a single species in the genus *Hormospira*. The related species *Tiariturris libya* (Dall, 1919) and *T. spectrabilis* Berry, 1958, resemble *H. maculosa*, but differ in having a larger shell with narrow axial ribs below the knobs along the shoulder, and a thicker, darker periostracum. *Pseudomelatoma penicillata* (Carpenter, 1864) can be easily distinguished by its slightly smaller size and shorter siphonal canal, as well as by its lack of a distinctive shoulder and by the presence of axial ribs.

The shell of the Blotchy Turrid is large, thin, narrow, and fusiform. The spire is very tall and conical. A raised spiral band with nodules runs along the shoulder. The aperture is narrowly ovate, with a moderately deep posterior sinus and a broad siphonal canal that is nearly as long as the aperture. Sculpture is limited to fine spiral striae. The shell is white, with irregular blotches of darker brown that may coalesce to form axial bands. The periostracum is thin and light brown.

Actual size

FAMILY	Turridae
SHELL SIZE RANGE	¼ to ½ in (7 to 13 mm)
DISTRIBUTION	Caribbean, Honduras to Brazil
ABUNDANCE	Common
DEPTH	Subtidal
HABITAT	Rocks
FEEDING HABIT	Carnivore
OPERCULUM	Corneous

SHELL SIZE RANGE
¼ to ½ in
(7 to 13 mm)

PHOTOGRAPHED SHELL
⅜ in
(10 mm)

MONILISPIRA QUADRIFASCIATA

FOUR-BANDED TURRID

REEVE, 1845

575

The genus *Monilispira* was formerly considered a subgenus of
the very diverse genus *Crassispira*. *Monilispira quadrifasciata*
has a chocolate brown shell with a distinctive color pattern of
white spiral bands and axial sculpture that make it one of the
most recognizable species among the Crassispirinae, a subfamily
of the family Turridae.

RELATED SPECIES

There is a particular concentration of crassispirines around the
Panamanian isthmus—one authority lists more than 60 species
on the west coast alone, including the striking *Monilispira
ochsneri* (Hertlein and Strong, 1949). It has a yellowish spire
and a dark brown body with a broad yellow spiral band on the
shoulders that is divided in three by two fine red-brown ones.

Actual size

The shell of the Four-banded Turrid is fusiform,
with a tall spire of about 7 whorls. All except
the protoconch bear a close sculpture of well-
defined axial ribs, which may show as white
through the dark brown background color.
There is a wide, shallow, white spiral cord flanked
by 2 much finer cords. The shell has a short
anterior canal and a deep, narrow posterior slit.

FAMILY	Turridae
SHELL SIZE RANGE	½ to 1 in (11 to 25 mm)
DISTRIBUTION	East Africa to Hawaii
ABUNDANCE	Uncommon
DEPTH	Intertidal to 165 ft (50 m)
HABITAT	Rocky bottoms
FEEDING HABIT	Carnivore
OPERCULUM	Corneous

SHELL SIZE RANGE
½ to 1 in
(11 to 25 mm)

PHOTOGRAPHED SHELL
¾ in
(20 mm)

576

TURRIDRUPA BIJUBATA
CRESTED TURRID
(REEVE, 1843)

Turridrupa bijubata is one of the many small turrids from the Indo-Pacific region. The family Turridae is the most diverse of all mollusks, and its members occur worldwide, in all latitudes and depths. Intensive field studies in several Indo-Pacific locations have revealed that species diversity is very high, but species densities are very low, and most are known from only a few specimens. *Turridrupa bijubata* is not well represented in museum collections. Recent molecular studies suggest that some forms that superficially look alike actually represent many distinct species. Since most turrids are small or minute, certainly thousands of new species await discovery.

The shell of the Crested Turrid is small, solid, stout, sculptured, and fusiform, with a tall spire and short siphonal canal. Its spire has many whorls, the suture is not very clear in early whorls, and the apex is small and rounded (eroded in the shell illustrated here). The sculpture consists of several raised, sharp, spiral cords, with the second cord below the suture being the strongest. The aperture is elongate, the outer lip lirate, with a deep, U-shaped anal notch and a truncated siphonal canal. The shell color is dark brown, with the spiral cords ranging from white to yellowish.

RELATED SPECIES

Turridrupa weaveri Powell, 1967, from the Hawaiian Islands, has a shell similar in shape and size, but with rounded spiral cords. Most of the shell is red-brown with white blotches. *Monilispira quadrifasciata* (Reeve, 1845), from Honduras to Brazil, has a shell with similar shape, but with different sculpture: axial ribs on the spire whorls, a spiral cord below the suture, and spiral cords in the lower part of the body whorl.

Actual size

FAMILY	Turridae
SHELL SIZE RANGE	1¾ to 2⅜ in (45 to 60 mm)
DISTRIBUTION	Florida to Colombia
ABUNDANCE	Very rare
DEPTH	180 to 4,850 ft (55 to 1,480 m)
HABITAT	Sandy and muddy bottoms
FEEDING HABIT	Carnivore
OPERCULUM	Corneous, elongate

SHELL SIZE RANGE
1¾ to 2⅜ in
(45 to 60 mm)

PHOTOGRAPHED SHELL
2¼ in
(57 mm)

COCHLESPIRA ELEGANS

ELEGANT STAR TURRID

(DALL, 1881)

Cochlespira elegans has an elegant, pagoda-shaped shell. The whorls are very angulated, and the shoulder has a sharp keel, which bears two spiral rows of short spines pointed posteriorly. It lives on sandy muddy bottoms in deep water, and is rarely collected. Species in the subfamily Cochlespirinae have pagoda-shaped shells with a long siphonal canal, and the anal notch has a rounded triangular shape. Most species live in deep water and are rare. Cochlespirines are among the oldest turrids, dating back from the Cretaceous Period.

The shell of the Elegant Star Turrid is medium in size, thin-walled, elongate, and pagoda-shaped. Its spire is tall, the suture well marked, the apex pointed, and the whorls sharply keeled at the shoulder. The sculpture consists of numerous beaded spiral cords, 2 spiral rows of short spines; the area above the shoulder is smooth. Its aperture is triangular and elongated, continuing into the long siphonal canal; the anal notch is located above the shoulder. The outer lip is thin and the columella smooth. The shell color is off-white, with a white aperture.

RELATED SPECIES

Cochlespira pulchella (Schepman, 1913), from the Philippines and Indonesia, has a smaller shell, with larger spines at the shoulder. The slope between the spines and the suture is concave. *Polystira albida* (Perry, 1811), from Florida to northern Brazil, is the largest turrid in the Atlantic. Its shell is elongately fusiform, with a tall spire, long siphonal canal, and many sharp spiral cords.

Actual size

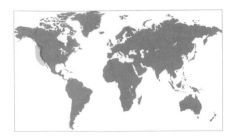

FAMILY	Turridae
SHELL SIZE RANGE	⅜ to 2⅞ in (44 to 71 mm)
DISTRIBUTION	Vancouver Island to lower California, U.S.A.
ABUNDANCE	Common
DEPTH	250 to 2,000 ft (80 to 600 m)
HABITAT	Mud
FEEDING HABIT	Carnivore
OPERCULUM	Corneous

SHELL SIZE RANGE
⅜ to 2⅞ in
(10 to 71 mm)

PHOTOGRAPHED SHELL
2¼ in
(68 mm)

578

ANTIPLANES PERVERSA

LARGE PERVERSE TURRID

(GABB, 1865)

The genus *Antiplanes* includes both sinistral and dextral species. The type species of the genus *Antiplanes* is *A. perversa*, which is sinistral. Most species live in deep water, usually have a relatively short and rounded posterior notch, and a smooth surface. The siphonal canal varies in length. Fossils of *A. perversa* have been found in Pleiocene deposits around the northern Pacific rim. It ranges from Alaska to lower California, U.S.A.

RELATED SPECIES

Antiplanes sanctiioannis (Smith, 1875) is a dextral species from deep water off eastern Russia and northern Japan. Its shell is chalky like that of many antiplanes, and white with no other color markings, although it often retains the remains of its thick brown periostracum.

The shell of the Large Perverse Turrid is smooth and auger-like, with a tall spire of convex whorls. It is predominantly white with a spiral band of pink tan immediately below the deeply impressed suture, with pink blushing on the body whorl and inside the aperture, but the apex and anterior columella remain white. The anterior canal varies in length from short to moderately long; the posterior slit is short and rounded.

Actual size

FAMILY	Turridae
SHELL SIZE RANGE	1⅝ to 3⅜ in (40 to 85 mm)
DISTRIBUTION	Bering Sea, Japan to Alaska
ABUNDANCE	Uncommon
DEPTH	Deep water
HABITAT	Sand
FEEDING HABIT	Carnivore
OPERCULUM	Corneous

SHELL SIZE RANGE
1⅝ to 3⅜ in
(40 to 85 mm)

PHOTOGRAPHED SHELL
2⅝ in
(68 mm)

AFORIA CIRCINATA

RIDGED TURRID

(DALL, 1873)

579

Species in the genus *Aforia* are mostly found in deep waters. The genus is particularly diverse in Antarctica and the Bering Sea. *Aforia circinata* ranges from Japan, north to the Bering Sea, to Alaska. Mature females of the Ridged Turrid exhibit an unusual but not unique feature, a third notch in the aperture which is associated with breeding activity, possibly copulation or oviposition (egg laying).

RELATED SPECIES

Aforia magnifica (Strebel, 1908) is very similar in appearance to *A. circinata*, but is found on the opposite side of the globe, in deep waters around the Antarctic islands and peninsula. Like *A. circinata*, it is matte white, with a sculpture of fine spiral cords and a ridge on the shoulder of each whorl.

The shell of the Ridged Turrid is matte white and fusiform, with a sculpture of fine, low spiral cords. Along the shoulders of the body and tall spire there is a sharp ridge, whose profile shows in a deviation of its thin outer lip. The columella has a white parietal callus. The very long siphonal canal is shallow, open, recurved, and slightly twisted.

Actual size

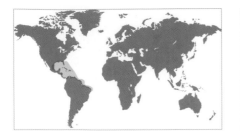

FAMILY	Turridae
SHELL SIZE RANGE	1¾ to 5 in (44 to 127 mm)
DISTRIBUTION	Florida to the Gulf of Mexico and Brazil
ABUNDANCE	Common
DEPTH	50 to 750 ft (15 to 230 m)
HABITAT	Sandy and muddy bottoms
FEEDING HABIT	Carnivore
OPERCULUM	Corneous, elongate

SHELL SIZE RANGE
1¾ to 5 in
(44 to 127 mm)

PHOTOGRAPHED SHELL
2¾ in
(69 mm)

580

POLYSTIRA ALBIDA

WHITE GIANT TURRID

(PERRY, 1811)

Polystira albida is indeed a white giant in the family; it is one of the largest turrids, and the largest turrid in the western Atlantic. It is a common species that ranges from Florida to the Gulf of Mexico and to northern Brazil. It lives offshore on sandy and muddy bottoms. The shell often has healed scars, which represent unsuccessful attacks by predators such as crabs. Although shell collectors do not like shells that bear scars, those shells are important and informative to biologists. The study of scars provides an insight into the ecology of the species and its predators.

RELATED SPECIES

Polystira coltrorum Petuch, 1993, from northeastern Brazil, has a shell similar to *P. albida*, but it is smaller, and the spiral cords are smoother. Some shells are tinged in tan. *Turris crispa* (Lamarck, 1816), from the Indo-Pacific, is the largest species in the family Turridae. Its shell is also fusiform with a tall spire, although the siphonal canal is proportionally shorter than in *P. albida*. The shell is white, peppered with brown dashes or blotches on the spiral cords.

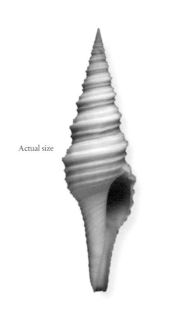

Actual size

The shell of the White Giant Turrid is large for the family, thick, solid, and elongately fusiform, with a tall spire and long siphonal canal. Its spire has a pointed apex, well-marked suture, and angular whorls. The sculpture consists of many strong, sharp spiral cords, with the strongest one at the shoulder, adjacent to the anal notch. The aperture is narrow and elongate, the outer lip crenulated with the spiral cords, and the columella smooth. The shell color is pure white, occasionally cream, the early whorls are often tinged in tan or brown. The operculum is elongated and brown.

FAMILY	Turridae
SHELL SIZE RANGE	1⅜ to 3½ in (35 to 90 mm)
DISTRIBUTION	India to Japan and Australia
ABUNDANCE	Uncommon
DEPTH	Shallow subtidal to 165 ft (50 m)
HABITAT	Sandy bottoms
FEEDING HABIT	Carnivore
OPERCULUM	Corneous, elongate

SHELL SIZE RANGE
1⅜ to 3½ in
(35 to 90 mm)

PHOTOGRAPHED SHELL
3½ in
(88 mm)

LOPHIOTOMA INDICA
INDIAN TURRID
(RÖDING, 1798)

581

Lophiotoma indica is the largest shallow-water species in the genus *Lophiotoma*, although it also occurs offshore. Shell shape varies with depth; shallow-water shells are usually more robust than those from deep water. Although its name indicates it lives in India, *L. indica* is more common around the Philippines and Australia. Several subspecies have been described on the basis of shell features. The features used to identify turrids include the shape, size, sculpture, and number of whorls of the larval shell, the number and type of radula teeth, and position of the anal notch. However, the apex is often eroded or missing, thus making identification more difficult in this very large family.

RELATED SPECIES

Lophiotoma millepunctata (Sowerby III, 1908), which ranges from Japan to New Zealand, has a smaller shell with a tall spire but a shorter siphonal canal. The shell is aptly named: it is decorated with seemingly a thousand spots. *Turris babylonia* (Linnaeus, 1758), from the Philippines to Indonesia and the Solomon Islands, is a common, large turrid with a shell that resembles *L. indica*, but has a slightly shorter siphonal canal. The shell is white, with squarish black spots arranged in spiral lines.

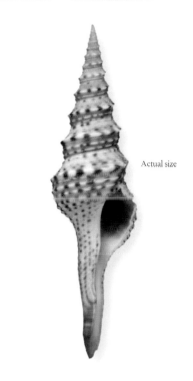

Actual size

The shell of the Indian Turrid is medium in size, solid, and elongately fusiform, with a tall spire and a long siphonal canal. The spire whorls are angulated, the suture well marked, and the apex pointed. Its sculpture consists of 1 strong spiral cord at the shoulder and many smaller cords below (2 on the spire whorls). The aperture is oval, and continues anteriorly into the long, open siphonal canal; the anal notch is on the shoulder of the body whorl. The shell color comprises a white background with brown maculations.

FAMILY	Turridae
SHELL SIZE RANGE	2½ to 6 ¼ in (60 to 160 mm)
DISTRIBUTION	Indo–West Pacific
ABUNDANCE	Uncommon
DEPTH	30 to 100 ft (10 to 30 m)
HABITAT	Offshore muddy bottoms
FEEDING HABIT	Carnivore
OPERCULUM	Corneous

SHELL SIZE RANGE
2 ½ to 6 ¼ in
(60 to 160 mm)

PHOTOGRAPHED SHELL
5 in
(127 mm)

TURRIS CRISPA

SUPREME TURRID

(LAMARCK, 1816)

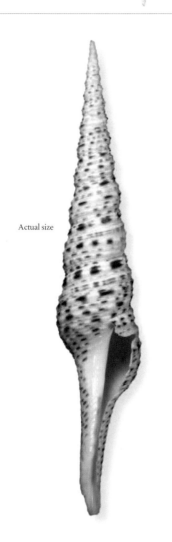

Actual size

The Supreme Turrid is the largest of the turrids. The size of the long, narrow shell, sculptured with variously sized spiral ridges and regularly spaced, chocolate-brown blotches makes this a popular species with collectors. The shell varies considerably, and this has led to the identification of a number of subspecies, including *Turris crispa variegata*, *T. c. yeddoensia*, and *T. c. intricata*.

RELATED SPECIES

Lophiotoma acuta (Perry, 1817), found in Indo–West Pacific, is shorter than *Turris crispa*, with deeper sutures. The shoulders of each whorl feature two distinctive ridges. Generally speckled in appearance, *L. acuta* is not as heavily marked as *T. crispa*. *Turris babylonia* (Linnaeus, 1758) from the Philippines, Indonesia, and the Solomon Islands is also shorter. Its deep sutures feature relatively thick spiral ridges, regularly marked with dark brown squares, which are notably larger than other shell markings.

The shell of the Supreme Turrid has a tall, sharply pointed spire and a shorter, but generally straight siphonal canal. Flattened whorls are covered in spiral cords and ridges of varying heights and thicknesses, all covered in regularly spaced brown blotches, which are often aligned on the siphonal canal and body whorl. The aperture is white, with a smooth columella, and a notch at the posterior end of the outer lip.

FAMILY	Conidae
SHELL SIZE RANGE	½ to ⅞ in (12 to 23 mm)
DISTRIBUTION	Netherland Antilles, West Indies
ABUNDANCE	Uncommon
DEPTH	Shallow subtidal to 20 ft (6 m)
HABITAT	Rocky bottoms
FEEDING HABIT	Carnivore, feeds on polychaetes
OPERCULUM	Corneous, elongate

SHELL SIZE RANGE
½ to ⅞ in
(12 to 23 mm)

PHOTOGRAPHED SHELL
⅝ in
(16 mm)

CONUS HIEROGLYPHUS

HIEROGLYPHIC CONE

DUCLOS, 1833

Conus hieroglyphus is a small cone with a narrow distribution, apparently endemic to the ABC Islands (Aruba, Bonaire, and Curaçao), Netherland Antilles. Specimens from the west coast of Aruba are usually the largest, with a spotted reddish brown color, while shells from the east coast are smaller and darker. *Conus hieroglyphus* lives under or on rocks in shallow waters.

RELATED SPECIES

Conus selenae Van Mol, Tursch, and Kempf, 1967, from northern and nothwestern Brazil, has a similar, but smaller shell, which can be smooth, have spiral incised lines, or have a somewhat cancellate sculpture. The color ranges from white to orange or light brown. *Conus genuanus* Linnacus, 1758, from west Africa, has spiral rows of dark squarish dots and dashes, interspaced with white bars, and alternating with light brown bands.

Actual size

The shell of the Hieroglyphic Cone is small, lightweight, glossy, and conical, with a moderately high spire. Its spire is stepped, with a well-marked suture, and the apex is pointed but rounded. The body whorl is slightly convex and the shoulder rounded. The sculpture consists of spiral rows of raised beads, and several spiral cords anteriorly. The aperture is wide and the outer lip is thin. The shell color is usually dark reddish brown with 3 spiral bands of whitish irregular blotches. The aperture is pale violet.

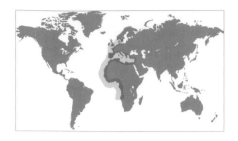

FAMILY	Conidae
SHELL SIZE RANGE	½ to ⅞ in (10 to 23 mm)
DISTRIBUTION	Northern Europe to Angola; Mediterranean
ABUNDANCE	Uncommon
DEPTH	33 to 330 ft (10 to 100 m)
HABITAT	Sand, rubble, rock, and near seagrasses
FEEDING HABIT	Carnivore
OPERCULUM	Corneous

SHELL SIZE RANGE
½ to ⅞ in
(10 to 23 mm)

PHOTOGRAPHED SHELL
⅞ in
(23 mm)

584

RAPHITOMA PURPUREA
PURPLE RAPHITOMA
(MONTAGU, 1803)

Raphitoma purpurea is the largest species of the genus *Raphitoma* in European waters. It is less common than some of the other species in its genus, and is more common in southern Britain than in the north. It is easily recognized by its elegant, slender shell with a reticulated sculpture, and by its distinctive thickened and denticulate outer lip. *Raphitoma purpurea* is found offshore among eelgrass, *Zostera*, and oarweed, *Laminaria*, or under stones and in crevices.

RELATED SPECIES

Raphitoma linearis (Montagu, 1803), from Norway to the Canary Islands and the Mediterranean, is similar but differs by having a smaller, stouter shell, with stronger axial ribs, and less numerous spiral cords. *Gymnobela edgariana* (Dall, 1889), a deepwater species from Louisiana to Curaçao, has a fusiform, smooth shell with angulated whorls and an aperture that is about half the shell length.

Actual size

The shell of the Purple Raphitoma is small, lightweight but solid, slender, and fusiform with a tall spire. The suture is well marked, the apex narrow; there are about 8 to 12 whorls. Its spire whorls have about 7 spiral ridges, while the body whorl has about 25, crossed by 18 to 20 axial ribs, forming beads at the intersections. The outer lip is thickened and denticulate, and the aperture lanceolate. The shell color is usually reddish brown, and the lip white.

FAMILY	Conidae
SHELL SIZE RANGE	¾ to 2¾ in (20 to 69 mm)
DISTRIBUTION	Indo-West Pacific
ABUNDANCE	Uncommon
DEPTH	Intertidal to 400 ft (120 m)
HABITAT	Sandy bottoms and under corals
FEEDING HABIT	Carnivore, feeds on other mollusks
OPERCULUM	Corneous, elongate

SHELL SIZE RANGE
¾ to 2¾ in
(20 to 69 mm)

PHOTOGRAPHED SHELL
1⅝ in
(43 mm)

CONUS PERTUSUS

PERTUSUS CONE

HWASS *IN* BRUGUIÈRE, 1792

585

Conus pertusus is usually easily recognized by its orange to pink shell with three spiral rows of white blotches, and a convex spire with a small and pointed apex. However, the coloration is variable, and some shells are mostly white or pale yellow, with pale brown blotches. It is widely distributed throughout the Indo-Pacific, ranging from eastern Africa to Hawaii. It lives in shallow waters as well as offshore, on sandy bottoms and under corals. It is uncommon in most places.

RELATED SPECIES

Conus vexillum Gmelin, 1791, also from the Indo-West Pacific, has a much larger shell, but small specimens may resemble *C. pertusus*. Large shells can vary from pale yellow to dark brown. *Conus cedonulli* Linnaeus, 1767, from the Lesser Antilles, has a beautiful orange-brown shell decorated with spiral rows of white irregular blotches and spots.

Actual size

The shell of the Pertusus Cone is medium in size, moderately heavy, somewhat glossy, and conical. Its spire is short, convex, with a shallowly incised suture, and a small, pointed apex. The body whorl can have straight or convex sides. The sculpture varies from smooth to corded, with strong spiral ribs. The aperture is narrow, wider anteriorly, and the outer lip thin. The shell color ranges from bright orange-red, to deep pink to whitish, usually with 3 spiral rows of irregular white blotches.

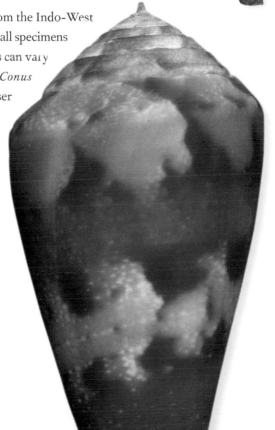

FAMILY	Conidae
SHELL SIZE RANGE	1½ to 3 in (40 to 78 mm)
DISTRIBUTION	Lesser Antilles
ABUNDANCE	Uncommon
DEPTH	6 to 165 ft (2 to 50 m)
HABITAT	Rocky bottoms
FEEDING HABIT	Carnivore, feeds on polychaete worms
OPERCULUM	Corneous, with terminal nucleus, rather small

SHELL SIZE RANGE
1½ to 3 in
(40 to 78 mm)

PHOTOGRAPHED SHELL
1⅞ in
(48 mm)

586

CONUS CEDONULLI
MATCHLESS CONE
LINNAEUS, 1767

Conus cedonulli was one of the rarest shells in the eighteenth century. Indeed, in 1796 a specimen brought more than six times as much as a painting by Vermeer sold at the same auction. It is still considered rare to uncommon and it is prized by collectors for its beautiful pattern. However, with the advent of scuba diving, it is now found more often. All cone shells are venomous and should be handled with care when alive. The venom of *C. cedonulli* is not fatal to humans but its sting may still be painful. There are more than 500 living species in the genus *Conus*.

RELATED SPECIES

The *C. cedonulli* complex includes closely related species such as *C. mappa* Lighfoot, 1786; *C. aurantius* Hwass, 1792, and *C. pseudaurantius* Vink and Cosel, 1985, all from the southern Caribbean, and the more widely ranging *C. regius* Gmelin, 1791, which occurs from Georgia, U.S.A., to southern Brazil. All species have variable shells.

The shell of the Matchless Cone is thick and conical, and has a long and narrow aperture, in which the lip is nearly parallel to the columella. The spire is short and stepped, and the body whorl is straight-sided, with sculpture consisting of fine spiral lines that are strongest near the base. The shell is white and handsomely decorated with irregular spiral lines, beads, and blotches that may range in color from yellow to orange to brown. The shell pattern is extremely variable, and several subspecies have been named.

Actual size

FAMILY	Conidae
SHELL SIZE RANGE	1⅜ to 2½ in (35 to 64 mm)
DISTRIBUTION	Gulf of Mexico and Caribbean
ABUNDANCE	Rare
DEPTH	Deep water to 1,600 ft (500 m)
HABITAT	Sand and mud
FEEDING HABIT	Carnivore
OPERCULUM	Corneous

SHELL SIZE RANGE
1⅜ to 2½ in
(35 to 64 mm)

PHOTOGRAPHED SHELL
1⅞ in
(18 mm)

PLEUROTOMELLA EDGARIANA

EDGAR'S PLEUROTOMELLA

(DALL, 1889)

587

Pleurotomella is a deepwater genus of the Conidae. It is characterized by a tall and usually highly sculpted spire, and by a long siphonal canal. Its species are found in temperate and cold waters as well as the tropical environments typical of most Conidae. Like others of that family, it can deliver a powerful sting to its prey through the poison glands connected to its hollow teeth. *Pleurotomella edgariana* has a thick, tough periostracum and a clawlike operculum.

RELATED SPECIES

Pleurotomella packardi (Verrill, 1872) is a rough, sculpturally more typical member of the genus. It is off-white, with fine widely spaced spiral cords that cross high spiraling axial ribs on well-rounded whorls. The spire is tall, the aperture ovate, and the anterior canal short and open. It is about ¾ in (18 mm) long, and occurs in the North Sea and north-western Atlantic from Norway to Gibraltar.

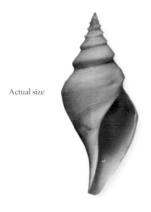

Actual size

The shell of Edgar's Pleurotomella is shiny and fusiform, and a white to pale tan conical body bearing a tall spire with low sloping shoulders, nodular on the upper whorls, and a deeply impressed suture. There are faint fine spiral cords, occuring especially lower on the body. The aperture is narrowly ovate, with a thin outer lip and moderately long open anterior canal. The interior is white and glossy.

FAMILY	Conidae
SHELL SIZE RANGE	1¼ to 3 in (33 to 75 mm)
DISTRIBUTION	Senegal to Angola, Cape Verde
ABUNDANCE	Uncommon
DEPTH	Shallow subtidal to 330 ft (100 m)
HABITAT	Rocky bottoms
FEEDING HABIT	Carnivore, feeds on polychaete worms
OPERCULUM	Corneous, elongate

SHELL SIZE RANGE
1¼ to 3 in
(33 to 75 mm)

PHOTOGRAPHED SHELL
2 in
(52 mm)

588

CONUS GENUANUS
GARTER CONE
LINNAEUS, 1758

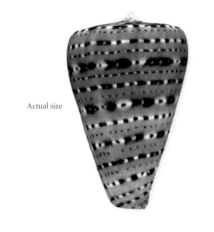

Actual size

Conus genuanus is an unmistakable and handsome cone with a shell decorated with light and pale brown spiral bands, and spiral rows of dark rectangular blotches alternated with dots. Because of the dots and blotches pattern, it is also known as the Morse Code Cone. It usually lives in shallow waters, although it can also occur in greater depths. It preys on polychaete worms, especially the fire worm, *Hermodice carunculata*. Specimens kept in an aquarium refused to eat anything but fire worms. It is found as a subfossil in the Canary Islands. The color pattern tends to fade in older specimens in collections.

RELATED SPECIES

Conus betulinus Linnaeus, 1758, which ranges from southeastern Africa to Polynesia, has a color pattern that may resemble *C. genuanus*, but its shell is generally larger. The shell color is orange and the spiral patterns have dots of uniform size. *Conus pulcher* Lightfoot, 1786, from western Sahara to Angola, is the largest species in the family. Juveniles have bright patterns that become pale in large shells.

The shell of the Garter Cone is medium in size, solid, heavy, glossy, with flat sides, and conic in outline. Its spire is short and pointed, and the suture is barely visible. The aperture is long, narrow, and nearly parallel to the body whorl; the outer lip is thin and sharp. The body whorl is smooth except for growth lines. Its shell has a creamy white background with pale pink or orange spiral bands alternated with darker bands, covered with spiral rows of dark brown or black blotches and white, sometimes with a dot in between. The aperture is white.

FAMILY	Conidae
SHELL SIZE RANGE	1¼ to 5¼ in (30 to 135 mm)
DISTRIBUTION	Indo-West Pacific
ABUNDANCE	Common
DEPTH	Intertidal to 230 ft (70 m)
HABITAT	Sandy bottoms
FEEDING HABIT	Carnivore, feeds on polychaete worms
OPERCULUM	Corneous, elongate

SHELL SIZE RANGE
1¼ to 5¼ in
(30 to 135 mm)

PHOTOGRAPHED SHELL
2¼ in
(57 mm)

CONUS FIGULINUS

FIG CONE

LINNAEUS, 1758

589

Conus figulinus has a heavy shell with a short spire and many brown spiral lines. It is a common cone from shallow waters, and is found on sandy bottoms, in which it lives semi-buried. The females lay large, pink egg capsules in the summer. Unlike most cones, *C. figulinus* always lays its egg capsules in sand, with the first four or five capsules, which lack eggs, used to anchor the egg mass in the sand. The eggs develop into planktonic veliger larvae.

RELATED SPECIES

Conus betulinus Linnaeus, 1758, which ranges from eastern Africa to Polynesia, has a similar shell, but is larger and heavier, with spiral rows of dots or blotches, instead of continuous lines. *Conus geographus* Linnaeus, 1758, from the Indo-West Pacific, has a medium to large shell that is relatively lightweight and thin. The aperture is wide anteriorly, and the whorl shoulder has tubercles. It is a fish eater, and one of the most poisonous cones.

The shell of the Fig Cone is medium-sized, very heavy, with some gloss, and a triangular conical shape. Its spire is moderately to very short, the suture is incised and not conspicuous, and the apex is sharp. The body whorl is large, with rounded shoulders, a smooth surface, some spiral threads anteriorly, and the occasional growth line. The aperture is long, relatively wide, of uniform width, and the outer lip is strong and sharp. Its outer color is gray to dark tan, with numerous dark brown spiral lines, and a white aperture.

Actual size

FAMILY	Conidae
SHELL SIZE RANGE	1⅝ to 4¼ in (42 to 109 mm)
DISTRIBUTION	Indo-West Pacific
ABUNDANCE	Common
DEPTH	Intertidal to 820 ft (250 m)
HABITAT	Sandy and sandy mud bottoms
FEEDING HABIT	Carnivore, feeds on other mollusks
OPERCULUM	Corneous, elongate

SHELL SIZE RANGE
1⅝ to 4¼ in
(42 to 109 mm)

PHOTOGRAPHED SHELL
2⅜ in
(61 mm)

590

CONUS AMMIRALIS
ADMIRAL CONE
LINNAEUS, 1758

Conus ammiralis has a distinctive pattern of brown spiral bands, alternated with caramel, narrower spiral bands, and decorated with irregular white triangles. Some subspecies have been named on the basis of color patterns and shell sculpture. *Conus ammiralis* lives from the intertidal zone to deep waters, on sand, mud, and coral rubble bottoms. It feeds on other gastropods using its harpoonlike teeth to inject them with a paralytic toxin.

RELATED SPECIES

Conus milneedwardsii Jousseaume, 1894, from the Red Sea and Indian Ocean to the South China Sea, has a large, off-white shell with triangular markings. It has one of the tallest spires in the genus. *Conus figulinus* Linnaeus, 1758, from the Indo-West Pacific, has a heavy, broad shell with a short spire.

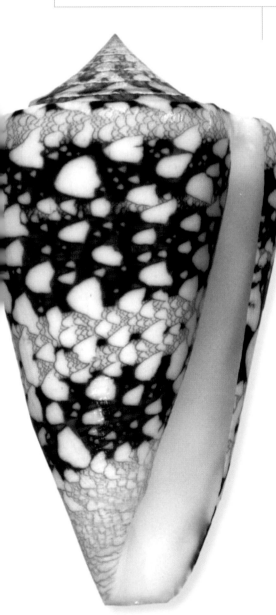

The shell of the Admiral Cone is medium-sized, somewhat glossy, and conical in outline. Its spire is moderately high, with a poorly marked suture, concave sides, and a pointed apex. The body whorl has nearly straight sides, angulate shoulders, and a mostly smooth surface. The aperture is relatively wide, broader anteriorly, and the outer lip is thin and sharp. The shell color pattern is variable, usually with 2 broad, brown spiral bands, 3 caramel bands, and white tented marking of varying sizes. The aperture is white.

Actual size

FAMILY	Conidae
SHELL SIZE RANGE	1¼ to 3⅜ in (32 to 87 mm)
DISTRIBUTION	Indo-Pacific
ABUNDANCE	Common
DEPTH	165 to 1,400 ft (50 to 425 m)
HABITAT	Rocky bottoms
FEEDING HABIT	Carnivore, feeds on polychaete worms
OPERCULUM	Corneous, elongate

CONUS ORBIGNYI
ORBIGNY'S CONE
AUDOUIN, 1831

SHELL SIZE RANGE
1¼ to 3⅜ in
(32 to 87 mm)

PHOTOGRAPHED SHELL
2½ in
(63 mm)

591

Conus orbignyi is a deepwater cone shell with an easily recognized elongate and slender shell, with a tall spire. It has a very wide distribution: it ranges from East Africa, throughout the Indo-Pacific. It has not been found in all places in between, but where it is found, it is usually common. Cone snails have modified radular teeth, each transformed into a little harpoon that carries a powerful venom.

RELATED SPECIES

Conus sauros García, 2006, from Texas and Louisiana, has a similar but apparently thicker shell. It is also a deepwater species. *Conus pertusus* Hwass in Bruguière, 1792, from the Indo-West Pacific, has a rounded, convex spire with a small and pointed apex, and usually brightly colored in orange-red or deep pink.

Actual size

The shell of Orbigny's Cone is medium-sized, lightweight, and elongate-conical. Its spire is moderately high, with a row of rounded tubercles at the shoulder, and a sharp, pointed apex. The body whorl is slightly concave and elongate, with a long siphonal canal. The aperture is long and narrow, and the outer lip is thin and fragile. The sculpture consists of many, low, flat spiral ridges alternated with shallow grooves, which have fine axial lines. The shell color is white or cream with brown spiral bands, and sometimes blotches.

FAMILY	Conidae
SHELL SIZE RANGE	1⅛ to 4¼ in (42 to 110 mm)
DISTRIBUTION	Indo-Pacific
ABUNDANCE	Common
DEPTH	Intertidal to 260 ft (80 m)
HABITAT	Sandy bottoms among rocks and corals
FEEDING HABIT	Carnivore, feeds on polychaete worms
OPERCULUM	Corneous, elongate

SHELL SIZE RANGE
1½ to 4¼ in
(42 to 110 mm)

PHOTOGRAPHED SHELL
3⅛ in
(79 mm)

592

CONUS IMPERIALIS
IMPERIAL CONE
LINNAEUS, 1758

Conus imperialis is one of the most easily recognizable cones. Although its shell varies in shape and coloration, its coronate shoulders, short spire, and spiral rows of black dots and dashes are highly distinctive. It is a common species that occurs from the intertidal zone to offshore, on coralline sandy bottoms. It is a specialized carnivore feeding only on the amphinomid polychaete worm *Eurythoe complanata*. Its shell often has growth marks or healed scars from attempted predation, and may be covered with coralline algae deposits.

RELATED SPECIES

Conus dorreensis Péron, 1807, endemic to Western Australia, has a unique periostracum decorated with a broad, tan spiral band lined with black. *Conus ammiralis* Linnaeus, 1758, from the Indo-West Pacific, has a beautiful pattern of two broad, brown spiral bands and three caramel bands, and white triangular markings.

Actual size

The shell of the Imperial Cone is medium-sized, thick, heavy, and conical in outline. Its spire is very low or flat, with sharp tubercles at the shoulders. The body whorl may have spiral ridges anteriorly, but is mostly smooth, with the occasional growth mark. The aperture is narrow and long, widening anteriorly, and the outer lip is sharp. The shell color is creamy white, covered by 2 broad, light brown or greenish brown spiral bands and many wavy spiral rows of black dashes and spots. The whitish aperture is stained purple anteriorly.

FAMILY	Conidae
SHELL SIZE RANGE	2¾ to 4½ in (70 to 120 mm)
DISTRIBUTION	Japan to northwestern Australia
ABUNDANCE	Uncommon
DEPTH	200 to 2,000 ft (60 to 600 m)
HABITAT	Muddy bottoms
FEEDING HABIT	Carnivore, feeds on polychaete worms
OPERCULUM	Absent

SHELL SIZE RANGE
2¾ to 4½ in
(70 to 120 mm)

PHOTOGRAPHED SHELL
3¼ in
(83 mm)

593

THATCHERIA MIRABILIS

JAPANESE WONDER SHELL

ANGAS, 1877

Thatcheria mirabilis is one of the most distinctive shells in the world. In fact, it is so different from any other shell, that when the single specimen was discovered and described as a new species, many scientists believed it to be a malformed specimen. It took more than half a century before other specimens became available. The shape of this shell is said to have inspired Frank Lloyd Wright in designing the Guggenheim Museum in New York. *Thatcheria mirabilis* is the single species in its genus.

RELATED SPECIES

Raphitoma purpurea (Montagu, 1803), from northern Europe to Angola and the Mediterranean, has an elongate fusiform shell with a reticulated sculpture. *Tritonoturris poppei* Vera-Peláez and Vega-Luz, 1999, from the Philippines, has a fusiform shell with a tall spire and smooth surface. Both species were formerly classified as a "turrids," but are now members of the conid subfamily Raphitomidae.

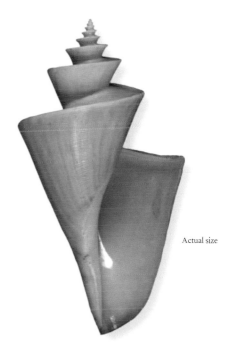

Actual size

The shell of the Japanese Wonder Shell is thin, lightweight, and angular. Its spire is high and stepped. The protoconch is sculptured with criss-crossing diagonal lines forming diamonds. Its spire whorls are nearly flat, with a deep suture and a sharp keel at the shoulder. Its body whorl is large, narrowing sharply toward the wide siphonal canal. The turrid notch is a wide groove about one-quarter of a whorl long. The aperture is large, about half of shell length, with a very angular profile and a smooth columella. The shell color is dull yellow, and the aperture and columella are white and glossy.

FAMILY	Conidae
SHELL SIZE RANGE	1¾ to 5 in (44 to 129 mm)
DISTRIBUTION	Red Sea to Indo-West Pacific
ABUNDANCE	Common
DEPTH	Shallow subtidal
HABITAT	Sandy bottoms near coral reefs
FEEDING HABIT	Carnivore, feeds on fishes
OPERCULUM	Corneous, concentric, elongate

SHELL SIZE RANGE
1¾ to 5 in
(44 to 129 mm)

PHOTOGRAPHED SHELL
3½ in
(88 mm)

594

CONUS STRIATUS
STRIATE CONE
LINNAEUS, 1758

The shell of the Striate Cone is medium-sized, solid, and cylindrical. The spire is low, the protoconch pointed, and the spire whorls concave, with a sharply keeled shoulder. The body whorl is slightly convex, with the widest point below the shoulder. Its aperture is very long and narrow, gradually widening anteriorly. The sculpture consists of fine spiral ridges crossed by axial growth marks. The color is variable, with a white to pinkish background mottled with irregular, broad spiral bands and light brown to black zigzag axial markings. The aperture is white.

Conus striatus is a common fish-eating cone with a wide geographic distribution. It inhabits shallow sandy bottoms, especially near coral reefs, from the Red Sea to the South Pacific Ocean. Because of its potent venom it should be handled carefully, more so than most other cones. This handsome shell is polymorphic, varying greatly in color and shape. Although the outer portions of the shell are thick and stout, the animal resorbs almost all of the walls of earlier whorls within the shell, to the point that they are paper thin and nearly transparent.

RELATED SPECIES

Closely related to *Conus striatus* from the Indo-West Pacific include *C. consors* Sowerby I, 1833, which has a slender shell with a short but pointed spire; *C. magus* Linnaeus, 1758, has a polychromic shell, usually with dark blotches on the spire; and *C. stercumuscarum* Linnaeus, 1758, which has a white shell covered with small brown spots in irregular axial rows.

Actual size

FAMILY	Conidae
SHELL SIZE RANGE	1¼ to 6 in (31 to 150 mm)
DISTRIBUTION	Indo-West Pacific
ABUNDANCE	Common
DEPTH	Intertidal to 165 ft (50 m)
HABITAT	Rocky bottoms
FEEDING HABIT	Carnivore, feeds on other mollusks
OPERCULUM	Corneous, elongate

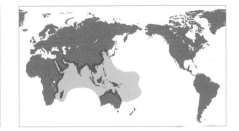

SHELL SIZE RANGE
1¼ to 6 in
(31 to 150 mm)

PHOTOGRAPHED SHELL
3½ in
(88 mm)

CONUS MARMOREUS

MARBLE CONE

LINNAEUS, 1758

Conus marmoreus is the type species of the genus *Conus*. It is a well-known species, with a striking black and white shell that makes it easily recognized. It is very variable in pattern, ranging from large white tents (triangles) arranged in diagonal spiral rows, to small and very dense tents; a variation from New Caledonia can have caramel-colored or even pure white shells. It is found from the intertidal zone to offshore, on rocky bottoms and under corals. *Conus marmoreus* feeds on other mollusks. Unlike most cones, it is active during the day.

The shell of the Marble Cone is medium to large in size, heavy, with some gloss, and conical in outline. Its spire is short, with a well-marked suture, and large, rounded tubercles at the shoulder. The body whorl has nearly straight sides, and is only slightly convex posteriorly. The aperture is long, widest anteriorly, with a thick or thin outer lip. The surface is usually smooth, but may have spiral lines anteriorly. The shell color is black with white triangular marks arranged in diagonal spiral rows, and a white aperture.

RELATED SPECIES

Conus nobilis victor Broderip, 1842, endemic to Bali and Flores Islands, in Indonesia, has a shell with a tented pattern resembling *C. marmoreus*. Its background is caramel, with a few broad, brown spiral bands, and white tents. *Conus orbignyi* Audouin, 1831, spans all of the Indo-Pacific. Its shell is slender and elongate, with a tall spire.

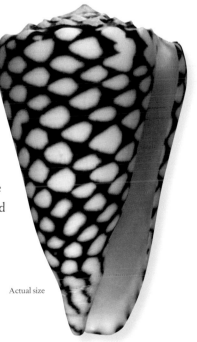

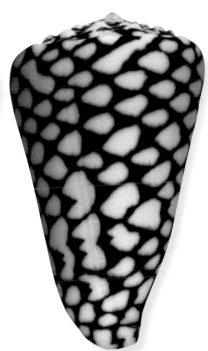

Actual size

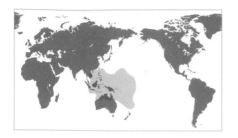

FAMILY	Conidae
SHELL SIZE RANGE	2¾ to 6¼ in (70 to 162 mm)
DISTRIBUTION	Tropical west Pacific
ABUNDANCE	Uncommon
DEPTH	Intertidal (rare) to 260 ft (80 m)
HABITAT	Fine sand bottoms
FEEDING HABIT	Carnivore, feeds on other gastropods
OPERCULUM	Corneous, concentric, elongate, small

SHELL SIZE RANGE
2¾ to 6¼ in
(70 to 162 mm)

PHOTOGRAPHED SHELL
4 in
(101 mm)

596

CONUS GLORIAMARIS

GLORY-OF-THE-SEA CONE

CHEMINITZ, 1777

Conus gloriamaris was once considered one of the world's rarest shells—only about a dozen shells were known by 1900. However, since the advent of scuba diving and the discovery of its habitat in Papua New Guinea and the Philippines a few decades ago, it is now collected often, although most specimens are collected using tangle nets in deeper waters. *Conus gloriamaris* is one of many cone species that feed on other mollusks, especially gastropods, by injecting a venomous cocktail of peptides using its highly modified radula.

RELATED SPECIES

There are many species of tented cones, including *Conus milneedwardsi* Jousseaume, 1854, from the Indian Ocean and the South China Sea, with a tall and stepped spire; *C. textile* Linnaeus, 1758, from the Indo-Pacific, has a short spire and ovate shell; and *C. marmoreus* Linnaeus, 1758, from the Indo-West Pacific, has a depressed spire and coronated shoulder, and typically has a black and white shell.

Actual size

The shell of the Glory-of-the-sea Cone is relatively large, tall-spired, slender, and solid. The spire is sharp, slightly stepped, and accounts for about one-quarter of the shell length. The body whorl is nearly straight-sided, smooth, and glossy. The aperture is long, becoming slightly broader anteriorly. The lip is thin and sharp, and the columella smooth. The shell color is cream to bluish white, with about 5 broad, darker spiral bands, packed with small tented or triangular markings white within, and 1 brown band on spire whorls. The aperture is white.

FAMILY	Conidae
SHELL SIZE RANGE	1¾ to 6½ in (13 to 166 mm)
DISTRIBUTION	Indo-West Pacific
ABUNDANCE	Common
DEPTH	Shallow subtidal
HABITAT	Sandy bottoms near coral reefs
FEEDING HABIT	Carnivore, feeds on fishes
OPERCULUM	Corneous, concentric, elongate

CONUS GEOGRAPHUS
GEOGRAPHY CONE
LINNAEUS, 1758

SHELL SIZE RANGE
1¾ to 6½ in
(43 to 166 mm)

PHOTOGRAPHED SHELL
4¼ in
(106 mm)

597

Conus geographus is the most venomous of all mollusks. Most species of *Conus* feed on polychaete worms. Their venoms are specialized for worms and are not especially dangerous to humans. Many other cones are molluscivores, feeding on other mollusks. Their venoms are potentially more harmful to humans. A relatively small group of *Conus* species, including *C. geographus*, feed on fish. Their venoms are adapted to vertebrates, and can cause human fatalities. At least 30 human deaths have been attributed to stings of cone shells, most to *C. geographus*.

RELATED SPECIES

Recent molecular studies indicate that piscivory has evolved independently at least three times in Conidae. Thus, not all fish-eating cones are closely related. *Conus radiatus* Gmelin, 1791, also from the Indo-West Pacific, is from a different lineage than *C. geographus*, as is *C. purpurascens* Sowerby I, 1833, a species from the tropical eastern Pacific. *Conus ermineus* Born, 1778, the only fish-eating cone presently known from the western Atlantic, is closely related to *C. purpurascens*.

Actual size

The shell of the Geography Cone is relatively large, inflated, thin, lightweight, and glossy. The spire is very low, with short tubercles at the shoulder. The sculpture of the body whorl is smooth, with fine, axial growth lines. The periostracum is thick, brown, and may have raised ridges. The aperture is broad anteriorly and the smooth columella ends abruptly. The base color is white, cream, or pinkish, with 2 or more chestnut brown spiral bands, and with numerous, fine, tented marking adjacent to the bands. The aperture is white.

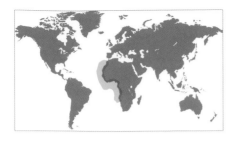

FAMILY	Conidae
SHELL SIZE RANGE	2 to 9 in (50 to 226 mm)
DISTRIBUTION	West Africa, from western Sahara to Angola
ABUNDANCE	Common; very large shells rare
DEPTH	Shallow subtidal
HABITAT	Sandy bottoms
FEEDING HABIT	Carnivore, feeds on other mollusks
OPERCULUM	Corneous, concentric, elongate

SHELL SIZE RANGE
2 to 9 in
(50 to 226 mm)

PHOTOGRAPHED SHELL
5¼ in
(135 mm)

598

CONUS PULCHER
BUTTERFLY CONE
LIGHTFOOT, 1786

The Butterfly Cone is the largest of the cone shells living today, reaching up to 9 in (226 mm) in length, although most specimens in collections are less than half that size. It is often found with growth scars, the result of unsuccessful attacks from predators. Its shell is variable in shape and coloration. Like other cones, it is a predatory snail, usually active at night. *Conus pulcher* can be easily distinguished from the many species of western African cones by its exceptionally large size and distinctive color pattern.

RELATED SPECIES

Most of the cones with which *C. pulcher* occurs, such as *C. mercator* Linnaeus, 1758 and *C. genuanus* Linnaeus, 1758, are much smaller. The Indo-Pacific species, *C. leopardus* (Röding, 1798) approaches *C. pulcher* in size, but has a far heavier shell, marked with simple, spiral rows of dots.

Actual size

The shell of the Butterfly Cone is very large, but relatively light and thin, considering its size. The spire is low, with an angled shoulder, and slightly concave spire whorls that form a well-marked suture. The aperture is long and wide, becoming slightly broader anteriorly. The lip is thin and straight, and the columella smooth. The sculpture of low spiral ridges near the base is more evident in young specimens. The base color is white or cream, covered with multiple spiral bands of brownish squares and dots of varying sizes and shades. Color patterns are less pronounced in large individuals.

FAMILY	Terebridae
SHELL SIZE RANGE	½ to 1½ in (13 to 40 mm)
DISTRIBUTION	Florida to central Brazil
ABUNDANCE	Common
DEPTH	Intertidal
HABITAT	Sandy beaches
FEEDING HABIT	Carnivore, feeds on polychaete worms
OPERCULUM	Corneous, small, oval

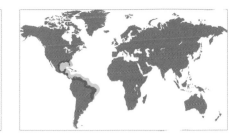

SHELL SIZE RANGE
½ to 1½ in
(13 to 40 mm)

PHOTOGRAPHED SHELL
¾ in
(21 mm)

HASTULA SALLEANA
SALLÉ'S AUGER
(DESHAYES, 1859)

599

Hastula salleana is a common terebrid found in the surf zone
on sandy beaches. It is a fast burrower and can quickly bury
itself in sand when stranded above the waterline. It can also use
its foot as a water sail to ride back with receding waves. It is a
specialized carnivore that feeds on polychaete worms, which
live on sandy beaches. There are about 270 living species in the
family Terebridae worldwide, with the highest diversity found
in the Indo-Pacific. The fossil record of the family dates back to
the late Cretaceous period.

RELATED SPECIES
Hastula strigilata (Linnaeus, 1758), which ranges from the
Red Sea, throughout the Indo-Pacific, and reaches Hawaii and
French Polynesia, has a similar shell that can be larger and more
colorful. It can be easily separated by the spiral row of brown
spots or blotches. *Impages hectica* (Linnaeus, 1758), also from
the Red Sea and Indo-Pacific, has a larger, slightly broader shell.

Actual size

The shell of the Sallé's Auger is medium in
size, thin-walled but strong, glossy, and elongately
conical. Its spire is very tall, with over 12 whorls,
a pointed apex, and nearly flat, tapering sides.
The sculpture consists of 1 spiral row of short
axial ribs below the suture; the rest of the shell
is smooth. The aperture is rectangular, the outer
lip thin, and the columella smooth. The shell
color ranges from bluish to brownish gray, with
a brown spiral band below the suture, and a thin
white band anteriorly. The aperture is tan.

FAMILY	Terebridae
SHELL SIZE RANGE	¾ to 1½ in (19 to 37 mm)
DISTRIBUTION	Southeastern Australia
ABUNDANCE	Uncommon
DEPTH	Subtidal to 500 ft (150 m)
HABITAT	Sandy bottoms
FEEDING HABIT	Carnivore
OPERCULUM	Corneous, oval

SHELL SIZE RANGE
¾ to 1½ in
(19 to 37 mm)

PHOTOGRAPHED SHELL
1 in
(24 mm)

600

DUPLICARIA USTULATA
SCORCHED AUGER
(DESHAYES, 1857)

Duplicaria ustulata has a distinctive shell that is perhaps the broadest in the genus *Duplicaria*. It is easily recognized by a deep spiral groove below the suture of the whorls, which have strong axial ribs. It is endemic to Australia, and distributed from New South Wales to Victoria and Tasmania. It seems to be more common in Tasmania than elsewhere. It is usually found offshore, but occasionally occurs subtidally. Some researchers classify it in the genus *Pervicacia* because its columella is smoother than the typically twisted columella found in species in the genus *Duplicaria*, for example, in *D. duplicata*.

RELATED SPECIES
Duplicaria gemmulata (Kiener, 1839), from southeastern Brazil to Chile, has a larger, more slender shell with longer axial ribs, two white spiral bands, and two spiral beaded rows per whorl. *Duplicaria duplicata* (Linnaeus, 1758), which ranges from the Red Sea to South Africa to the western Pacific, has a large, elegant shell with a deep spiral groove below the suture, and many flat, wide axial ribs per whorl.

Actual size

The shell of the Scorched Auger is medium in size, glossy, and conical in outline. Its spire is tall, with a pointed apex, a well-marked suture, and spiral groove below the suture. The sculpture consists of strong axial ribs below, and sometimes also above, the spiral groove, with about 20 to 25 ribs on the penultimate whorl. The aperture is wide, the outer lip thin, and the columella smooth and curved. The shell color is tan or beige, with the aperture of similar color.

FAMILY	Terebridae
SHELL SIZE RANGE	1¼ to 3¼ in (30 to 80 mm)
DISTRIBUTION	Red Sea to Indo-West Pacific
ABUNDANCE	Common
DEPTH	Intertidal
HABITAT	Sandy bottoms
FEEDING HABIT	Carnivore
OPERCULUM	Corneous, oval

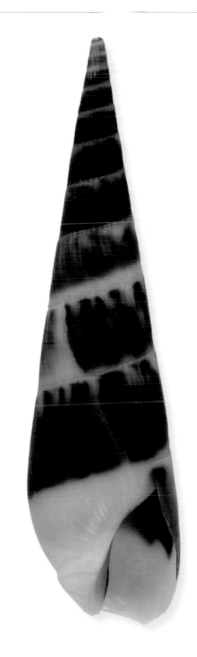

SHELL SIZE RANGE
1¼ to 3¼ in
(30 to 80 mm)

PHOTOGRAPHED SHELL
1⅜ in
(36 mm)

601

IMPAGES HECTICA

SANDBEACH AUGER

(LINNAEUS, 1758)

Impages hectica is a common auger found on open, surf-swept sandy beaches, from the Red Sea to the Indo-West Pacific. All terebrids are carnivores, although there are three basic types of feeding in the family. *Impages hectica* belongs to a group that possesses specialized radular teeth that impale their polychaete worm prey, and deliver a powerful venom that paralyzes it. The terebrid then engulfs the prey. Many terebrids, however, have lost the venom gland, and ingest their prey whole. A third group of terebrids has a special organ that grasps and ingests the tentacles of cirratulid polychaetes.

RELATED SPECIES

Hastula solida (Deshayes, 1857), from the tropical Indo-West Pacific, has a smaller, cylindrical-ovate shell with a reticulated pattern. Although its shell looks different from that of *I. hectica*, molecular studies show they are closely related. *Terebra maculata* (Linnaeus, 1758), from the Red Sea to the eastern Pacific, is the largest of the augers. It looks like a giant *Impages hectica*, but is broader and heavier, and with a more elongated aperture.

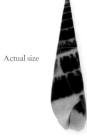

Actual size

The shell of the Sandbeach Auger is medium in size, glossy, and conical in shape. Its spire is tall, the suture is incised, the apex is pointed, and the sides of the whorls are almost straight. The surface is smooth, with fine growth marks. The aperture is leaf-shaped, the outer lip thin, and the columella smooth and curved. The background color is cream, with a black spiral band that ranges from broad to thin. The aperture is white.

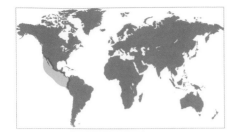

FAMILY	Terebridae
SHELL SIZE RANGE	¾ to 2¾ in (20 to 70 mm)
DISTRIBUTION	Southern California to Peru and Galápagos Islands
ABUNDANCE	Abundant
DEPTH	Intertidal to 360 ft (110 m)
HABITAT	Sandy and muddy bottoms
FEEDING HABIT	Carnivore
OPERCULUM	Corneous, oval

SHELL SIZE RANGE
¾ to 2¾ in
(20 to 70 mm)

PHOTOGRAPHED SHELL
1⅝ in
(42 mm)

602

TEREBRA ARMILLATA
COLLAR AUGER
HINDS, 1844

Terebra armillata has a distinctive shell that bears a strong, beaded spiral cord just below the suture, reminiscent of a collar or bracelet wrapped around a person's arm. It also has incised spiral lines in early whorls, but the sculpture becomes dominated by curved axial ribs toward the body whorl. The shell is variable in shape and color. *Terebra armillata* is an abundant terebrid, ranging from the intertidal zone to offshore, on sandy and muddy bottoms. It also occurs as Pleistocene fossils in Magdalena Bay, California.

RELATED SPECIES

Terebra dislocata (Say, 1822), which ranges from Maryland to Brazil, and in the eastern Pacific, from California to Panama, has a shell similar in size and sculpture. *Terebra triseriata* Gray, 1834, from the Indo-Pacific, including Hawaii, has one of the narrowest and most elongate shells in the Terebridae. Large shells may have more than 40 whorls.

The shell of the Collar Auger is medium in size, sculptured, and conical in shape. Its spire is tall, with an incised suture and a pointed apex. The sculpture is variable; in some shells, the early whorls have incised spiral lines, while others have axial ribs; in later whorls, axial ribs dominate in some shells, while others have beaded spiral cords. The beaded spiral cord also varies in weight. The aperture is elongate, the outer lip thin, and the columella twisted. The shell color varies from a cream to brown background, with gray, white, or brown spiral bands.

Actual size

FAMILY	Terebridae
SHELL SIZE RANGE	1 to 3 in (25 to 76 mm)
DISTRIBUTION	Indo-West Pacific
ABUNDANCE	Common
DEPTH	Intertidal to 400 ft (120 m)
HABITAT	Sandy bottoms
FEEDING HABIT	Carnivore
OPERCULUM	Corneous, oval

SHELL SIZE RANGE
1 to 3 in
(25 to 76 mm)

PHOTOGRAPHED SHELL
2¼ in
(56 mm)

HASTULA LANCEATA

LANCE AUGER
(LINNAEUS, 1767)

603

Hastula lanceata, with its long narrow shell and elegant markings, is popular with shell collectors. Its architecture and markings vary little, making this one of the easiest species to identify with confidence. As with all augers, the Lance Auger is carnivorous, feeding on polychaete worms. Members of the *Terebra* and *Hastula* genera have poisonous mouthparts. They inject the prey with a paralyzing poisonous toxin through a harpoonlike, hollow tooth before consuming it.

RELATED SPECIES

Hastula penicillata (Hinds, 1844), shares the same range, but is shorter with a less pointed spire and smoother sutures. Although the markings of *H. penicillata* occasionally resemble those of *H. lanceata*, they vary widely throughout the species. *Terebra areolata* (Link, 1807), also from Indo-West Pacific, is generally longer, and spotted rather than axially lined. The blotches just above the suture are the largest.

The shell of the Lance Auger has a tall, sharply pointing spire, although the tip is often absent. Axial ribs on the small posterior whorls fade on the larger whorls. Each whorl is decorated with regularly spaced, red-brown axial lines, which begin just below the shallow but distinctive suture. Axial markings on the body whorl are less regular, and can be interrupted by a white band. The aperture is narrow and small, with a concave posterior columella that recurves at the lower half.

Actual size

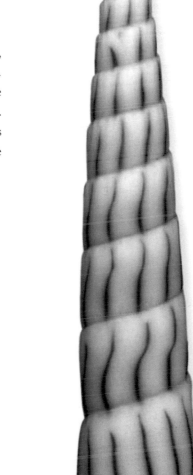

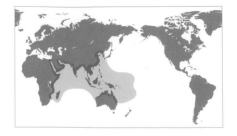

FAMILY	Terebridae
SHELL SIZE RANGE	¾ to 3⅝ in (20 to 93 mm)
DISTRIBUTION	Red Sea to Indo-West Pacific
ABUNDANCE	Common
DEPTH	Intertidal to 200 ft (60 m)
HABITAT	Sandy bottoms
FEEDING HABIT	Carnivore
OPERCULUM	Corneous, oval

SHELL SIZE RANGE
¾ to 3⅝ in
(20 to 93 mm)

PHOTOGRAPHED SHELL
3⅛ in
(78 mm)

604

DUPLICARIA DUPLICATA
DUPLICATE AUGER
(LINNAEUS, 1758)

Actual size

Duplicaria duplicata has an elegant shell with many broad axial ribs and one spiral groove, thus resulting in two zones, which often have different colors. Its shell varies in sculpture and color. It lives on clean sandy bottoms, but not near seagrasses, and is more common offshore. Despite the fact that there are three main feeding types within the family Terebridae, recent molecular studies suggest that the family is monophyletic, meaning that all species share a common ancestor.

RELATED SPECIES
Duplicaria australis (Smith, 1873), from northern and western Australia and Fiji, has a shell that resembles *D. duplicaria*, but it is more slender, the axial ribs are raised throughout the shell, and the spiral band has a different sculpture than the rest of the shell. *Terebra dimidiata* Linnaeus, 1758, which ranges from the Red Sea to Hawaii, has a large, smooth shell, with a smooth spiral band. The shell color is orange, with a thin, spiral white band and irregular axial white lines.

The shell of the Duplicate Auger is medium in size, solid, glossy, and elongately conical. Its spire is tall, with a pointed apex, a grooved suture, and nearly straight sides. The sculpture consists of many broad, but sometimes thin, axial ribs, crossed by a spiral groove below the suture. The axial ribs are elevated in early whorls, but become flat toward the body whorl. The aperture is elongate, the outer lip thin, and the columella has 1 diagonal fold. The shell color can be white, beige, orange, gray, or brown, and may be solid or mottled.

FAMILY	Terebridae
SHELL SIZE RANGE	2 to 5⅜ in (50 to 136 mm)
DISTRIBUTION	Indo-West Pacific to Hawaii
ABUNDANCE	Uncommon
DEPTH	Subtidal to 500 ft (150 m)
HABITAT	Sandy and muddy bottoms
FEEDING HABIT	Carnivore
OPERCULUM	Corneous, oval

TEREBRA TRISERIATA

TRISERIATE AUGER

GRAY, 1834

SHELL SIZE RANGE
2 to 5⅜ in
(50 to 136 mm)

PHOTOGRAPHED SHELL
3¼ in
(84 mm)

605

Terebra triseriata is the most elongate of all coiled gastropods. It has a shell that is very narrow and long, and large shells can have more than 40 whorls. Carrying a shell that is very long and narrow presents physical challenges to the mollusk; the aperture is relatively small, and the ratio between shell weight and foot muscle mass may be close to the functional limit to mobility. *Terebra triseriata* ranges from Japan to the Philippines, but is more common around Japan. It lives in sandy and muddy bottoms, ranging from shallow subtidal waters to offshore

RELATED SPECIES

Terebra pretiosa Reeve, 1842, from Japan to the Philippines, also has a very elongate shell, but it is slightly broader, the sculpture consists of wavy axial ribs, and there is one brown spiral band per whorl. *Terebra armillata* Hinds, 1844, from California to Peru, and the Galápagos Islands, has a smaller, much shorter shell, with a single beaded spiral cord and curved axial ribs.

Actual size

The shell of the Triseriate Auger is large for the family, thin, beaded, extremely narrow, and elongately conical. Its spire is very tall, with a pointed apex, well-marked suture, and nearly flat sides. Its sculpture consists of 2 beaded spiral cords just below the suture, and 3 to 4 thinner spiral cords, crossed by axial threads, forming a cancellate pattern. The aperture is rectangular, the outer lip thin, and the columella smooth. The shell color is cream or pale orange; sometimes the 2 stronger cords are lighter in color.

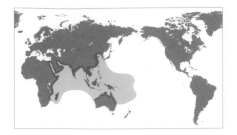

FAMILY	Terebridae
SHELL SIZE RANGE	2¼ to 6½ in (55 to 166 mm)
DISTRIBUTION	Indo-West Pacific
ABUNDANCE	Common
DEPTH	Shallow water
HABITAT	Sandy bottoms
FEEDING HABIT	Carnivore
OPERCULUM	Corneous

SHELL SIZE RANGE
2¼ to 6½ in
(55 to 166 mm)

PHOTOGRAPHED SHELL
4¼ in
(107 mm)

606

TEREBRA DIMIDIATA
DIVIDED AUGER
LINNAEUS, 1758

Another venomous snail of the *Terebra* genus, the Divided Auger—also known as the Orange or Dimidiate Auger—burrows in the sandy substrate that it inhabits, leaving a distinctive trail. The Divided Auger gets its common name from the thin paler shelf at the posterior end of each whorl, which creates the impression that each whorl is divided in two. The shell is sufficiently thin to allow sunlight to pass through it.

RELATED SPECIES
Terebra crenulata (Linnaeus, 1758), from Indo-West Pacific is more solid and features shoulders with tubercles, interspersed with short red-brown lines; in some individuals tubercles can grow to more prominent knobs. The color is paler and less orange than *T. diamidiata*. *Terebra babylonia* (Lamarck, 1822), also from Indo-West Pacific, is shorter with distinctive, wavy, regularly spaced axial ridges running the length of the shell; they intersect spiral ridges giving the shell a nodulose appearance.

Actual size

The shell of the Divided Auger is delicate with a high, pointed spire. Mature individuals may have up to 20 flattish whorls, each of which appears divided due to a narrow platform that sits beneath the suture. White, wavy axial streaks are distinctive on the lower two-thirds of each whorl, and become less distinctive on the upper third. The aperture is less elongate than in other augers, with a weakly folded columella. The shell has an overall orange-red color, with wavy white axial lines.

FAMILY	Terebridae
SHELL SIZE RANGE	6 to 10 in (150 to 250 mm)
DISTRIBUTION	Red Sea to Indo-Pacific
ABUNDANCE	Locally common
DEPTH	Subtidal to 650 ft (200 m)
HABITAT	Sandy bottoms
FEEDING HABIT	Carnivore, feeds on polychaete worms
OPERCULUM	Corneous, thin

SHELL SIZE RANGE
6 to 10 in
(150 to 250 mm)

PHOTOGRAPHED SHELL
7⅛ in
(184 mm)

TEREBRA MACULATA

MARLINSPIKE AUGER

LINNAEUS, 1758

607

Terebra maculata, with its strong and heavy shell, is the largest of the *Terebra* genus. It moves just below the surface of the sand—usually leaving a clearly defined trail—in search of worms, which it immobilizes with its venomous barb. Tracking to the end of such a trail and carefully taking up a scoop of sand will often reveal an individual. The Terebridae resemble closely the Cerithiidae, but the former feature smaller, less regular apertures, with one or two folds on the columella, and generally exhibit more flattened whorls.

RELATED SPECIES

Terebra areolata (Link 1807), also from the Indo-Pacific, has four spiral rows of medium brown blotches on the body whorl, compared with the two less regular, purple-brown blotches found on the body whorl of *T. maculata*. *Terebra crenulata* (Linnaeus, 1758), has more pronounced shoulders with small nodules, it also lacks the purple-brown blotches of *T. maculata*.

Actual size

The shell of the Marlinspike Auger is long and narrow with a high, pointed spire. The aperture is relatively wide, with the parietal wall set at an angle of around 120 degrees to the folded columella. The fasciole, although small, is clearly defined with a strong central groove. The posterior half of each whorl has distinctive, purple-brown blotches that vary in regularity, and in front of which are found a line of much smaller but similarly colored blotches. The shell is generally pale tan to light brown in color.

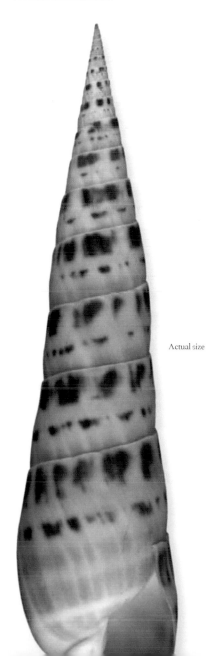

FAMILY	Architectonicidae
SHELL SIZE RANGE	⅜ to ¾ in (8 to 18 mm)
DISTRIBUTION	Red Sea to Indo-West Pacific
ABUNDANCE	Common
DEPTH	Intertidal to 65 ft (20 m)
HABITAT	Sand
FEEDING HABIT	Carnivore
OPERCULUM	Corneous, spiral

SHELL SIZE RANGE
⅜ to ¾ in
(8 to 18 mm)

PHOTOGRAPHED SHELL
⅝ in
(15 mm)

HELIACUS AREOLA
VARIEGATED SUNDIAL
(GMELIN, 1791)

Species in the family Architectonicidae are found in the various depths of tropical and sub-tropical seas. Most are in the Indo-Pacific, with others in the western Atlantic and eastern Pacific. Sundials are a relatively small family, of about 130 species in several genera including *Discotectona*, *Architectonica*, and the more sculptural *Heliacus*. One of the smallest and most highly decorated is *Heliacus areola*. It occurs from eastern Africa to the Central Pacific, and ranges from the intertidal zone to offshore. Its spire varies from short to tall.

RELATED SPECIES

Architectonica perspectiva (Linnaeus, 1758), is a larger, more common shell from deep water in the same region that has a superficially similar beaded sculpture. It has a striking sequence of spiral decoration: above the suture there is a beaded cream cord with chestnut splashes, then a broad cream cord with axial grooves and two thin bands of chestnut then a white band at the top.

Actual size

The shell of the Variegated Sundial is small, generally white to cream, with more or less frequent dark or red-brown axial streaks and spots that sometimes overwhelm the white. Its spire is relatively short, with a rounded apex on a flattened body. An open umbilicus is lined by a spiral ridge of white or off-white beads. The aperture is circular.

FAMILY	Architectonicidae
SHELL SIZE RANGE	1 to 2 in (24 to 50 mm)
DISTRIBUTION	Japan to northern Australia
ABUNDANCE	Rare
DEPTH	150 to 700 ft (50 to 200 m)
HABITAT	Sand
FEEDING HABIT	Carnivore
OPERCULUM	Corneous

SHELL SIZE RANGE
1 to 2 in
(24 to 50 mm)

PHOTOGRAPHED SHELL
2 in
(50 mm)

DISCOTECTONICA ACUTISSIMA

SHARP-EDGED SUNDIAL

(SOWERBY III, 1914)

609

Discotectonica acutissima has a circular shell with a depressed spire that gives it a discoidal shape. It occupies a zone defined not by latitudes of common warmth but by a narrow, longitudinal band that includes temperate as well as equatorial seas. The species is tolerant to wide variations in water temperature. All architectonicids have larvae that can spend long periods of time in the plankton and disperse over great distances. Larval shells are sinistral, but the coiling direction is reversed to dextral in the adult shell.

RELATED SPECIES

Philippia radiata (Röding, 1798), the Radial Sundial from the shallows of the Indo-Pacific is similar in overall color, although it is smaller and relatively much taller. The orange streaks have here coalesced into flamelike trickles, running down from solid orange spiral bands below the suture.

The shell of the Sharp-edged Sundial is flat, with a short to very short spire on a flattened discoidal shell with a sharp keel along its periphery. It is off-white to beige, with a paler spiral band above the suture. The whorls are marked with very close spiral grooves and oblique axial orange streaks. The umbilicus is open and deep, lined with a beaded spiral cord.

Actual size

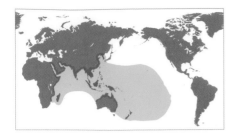

FAMILY	Architectonicidae
SHELL SIZE RANGE	¾ to 3¼ in (19 to 82 mm)
DISTRIBUTION	Indo-Pacific, including Hawaii
ABUNDANCE	Common
DEPTH	30 to 150 ft (10 to 50 m)
HABITAT	Sand
FEEDING HABIT	Carnivore
OPERCULUM	Corneous

SHELL SIZE RANGE
¾ to 3¼ in
(19 to 82 mm)

PHOTOGRAPHED SHELL
2½ in
(62 mm)

610

ARCHITECTONICA MAXIMA
GIANT SUNDIAL
(PHILIPPI, 1849)

The largest of the sundials, *Architectonica maxima* has a suitably broad range. It is found from New Zealand to Hawaii and from South Africa to Polynesia. Like many architectonids, it lives on sand near coral reefs, where it feeds on a diet of anemones and coral polyps. The shell is porcelaneous, and specimens are often damaged, particularly around the aperture.

RELATED SPECIES

Architectonica laevigata (Lamarck, 1816) is, in contrast to other beaded sundials, quite smooth and glossy. It has a distinct suture on a tall spire, with around four spiral grooves on the whorls. It is cream to pale violet, with blotches of pale chestnut brown. *Architectonica laevigata* is fairly common in the shallow waters of the Indian Ocean.

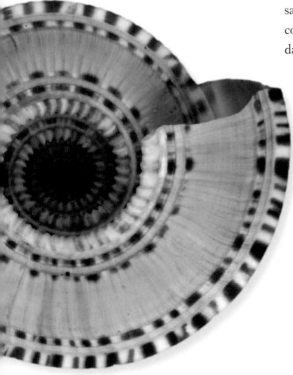

Actual size

The shell of the Giant Sundial is slightly convex on the base, with a short wide spire. It is distinguished by the spiral sequence on its whorls: above the suture, there are 2 ridges of white and brown blotches separated by a wide channel, then a wider channel in beige or pink then 2 more ridges, axially grooved and separated by a thin channel. The base has 2 pairs of white and brown blotched ridges, separated by a wide band of shallow axial pleats. At the center the open umbilicus is lined with a denticulate spiral ridge.

FAMILY	Rissoellidae
SHELL SIZE RANGE	less than ⅛ in (1 to 2 mm)
DISTRIBUTION	Florida to Northern Brazil
ABUNDANCE	Common
DEPTH	Intertidal to 80 ft (25 m)
HABITAT	On algae near coral reefs
FEEDING HABIT	Grazer
OPERCULUM	Calcareous, semicircular, with a peg

SHELL SIZE RANGE
less than ⅛ in
(1 to 2 mm)

PHOTOGRAPHED SHELL
less than ⅛ in
(1 mm)

RISSOELLA CARIBAEA

CARIBBEAN RISSO

REHDER, 1943

611

Rissoella caribaea is a common microgastropod from the tropical Western Atlantic. It is found crawling on macroalgae near coral reefs, and feeds on detritus, algal filaments, and diatoms. The shells in the family are usually smooth, clear, and translucent when fresh, and whitish when empty and dry. The color of the animal can be seen through the shell and it may help in the identification of the species, especially since many species are similar in shape and size. For example, *R. caribaea* has a black animal, while the animal of *R. galba* is yellow. There are about 40 species of Rissoellidae worldwide, all in tropical and temperate waters.

RELATED SPECIES

Rissoella galba Robertson, 1961, from the Gulf of Mexico to the Bahamas, has a shell similar in shape and color but about half of the size of *R. caribaea*, and with a shorter apex. *Rissoela longispira* Kay, 1979, from Hawaii, also has a similar shaped shell but with a longer spire than *R. caribaea*. The animal is rose colored with gray spots.

Actual size

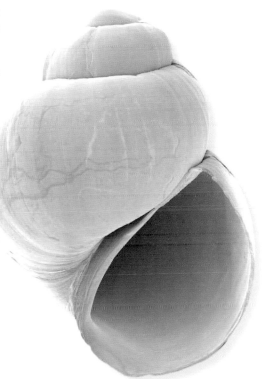

The shell of the Caribbean Risso is very small, thin, fragile, translucent, and ovately conical. Its spire is short, with convex whorls and an impressed suture. The body whorl is large and inflated, the aperture semicircular, and the outer lip thin and simple. The columella is smooth, and produces a narrow, chink-like umbilicus. The operculum is semicircular and not spirally coiled. It has a short internal peg that keeps it in place against the columella. The empty shell color is translucent and whitish.

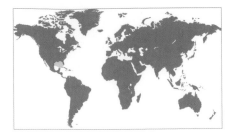

FAMILY	Omalogyridae
SHELL SIZE RANGE	less than ⅛ in (0.7 mm)
DISTRIBUTION	Northern Cuba to Texas
ABUNDANCE	Uncommon
DEPTH	13 to 160 ft (4 to 50 m)
HABITAT	On algae near coral reefs
FEEDING HABIT	Herbivore, feeds on macroalgae
OPERCULUM	Corneous, with central nucleus

SHELL SIZE RANGE
less than ⅛ in
(0.7 mm)

PHOTOGRAPHED SHELL
less than ⅛ in
(0.7 mm)

612

OMALOGYRA ZEBRINA

ZEBRINA OMALOGYRA

ROLÁN, 1992

Omalogyra zebrina is one of the world's smallest shells, with the adult shell growing to only about ½₂ in (0.7 mm) in diameter. It is locally common, but because of its microscopic size, it is seldom collected. It is usually found in shallow waters, on macroalgae, on which it feeds by piercing the algal cell with its radula and sucking out the contents. The operculum is corneous and has a central nucleus. Omalogyrids are sequential hermaphrodites, and begin life as males. Omalogyrids are found worldwide, and all have tiny, planispiral shells that are usually translucent.

RELATED SPECIES

Ammonicera lineofuscata Rolán, 1992, and *A. minortalis* Rolán, 1992, are both from the Gulf of Mexico to the Canary Islands; the former has a minute white shell with a reddish brown spiral band on both sides, while the latter is even smaller, at almost half the size of *O. zebrina*, and has a brown, translucent shell.

Actual size

The shell of the Zebrina Omalogyra is microscopic, planispiral, and has a sunken spire. The shell is a flattened disk, and nearly bilaterally symmetric. The protoconch has very fine axial threads. The aperture is circular and the whorl sides are rounded. The shell is sculpted with numerous fine axial cords. The shell color is white and translucent, with axial rows of reddish brown blotches on both sides of the shell.

FAMILY	Omalogyridae
SHELL SIZE RANGE	less than ⅛ in (0.5 to 0.7 mm)
DISTRIBUTION	Cuba to Texas, Canary Islands
ABUNDANCE	Common
DEPTH	10 to 80 ft (3 to 24 m)
HABITAT	On algae near coral reefs
FEEDING HABIT	Herbivore, feeds on macroalgae
OPERCULUM	Corneous, with central nucleus

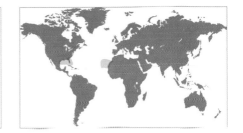

SHELL SIZE RANGE
less than ⅛ in
(0.5 to 0.7 mm)

PHOTOGRAPHED SHELL
less than ⅛ in
(0.7 mm)

AMMONICERA LINEOFUSCATA
BROWN-LINED AMMONICERA
ROLÁN, 1992

613

Ammonicera lineofuscata is one of the smallest known gastropods, as are most species in the family Omalogyridae. The animal is nearly transparent and carries its shell in a position close to vertical when crawling. It feeds on macroalgae and lives in shallow subtidal waters. Many omalogyrids are sequential hermaphrodites, although some seem to be simultaneous hermaphrodites. There are currently fewer than 40 species of Omalogyridae worldwide.

RELATED SPECIES

Ammonicera minortalis Rolán, 1992, from the Gulf of Mexico and Canary Islands, and *Omalogyra japonica* (Habe, 1972), from the western Pacific, have both been considered the world's smallest gastropods. Reaching only half of the size of *A. lineofuscata*, they both have a brown shell with about 18 axial ribs.

Actual size

The shell of the Brown-lined Ammonicera is minute, planispiral, and has a sunken spire. The shell is a flattened disk, and nearly bilaterally symmetrical. The protoconch has very fine axial threads. The aperture is circular and the whorl sides are rounded. The shell is sculpted with numerous fine axial cords. The shell color is white and translucent, with a thin reddish brown spiral band on both sides of the shell.

FAMILY	Pyramidellidae
SHELL SIZE RANGE	⅜ to 1¼ in (10 to 30 mm)
DISTRIBUTION	Indo-West Pacific
ABUNDANCE	Uncommon
DEPTH	Intertidal to 65 ft (20 m)
HABITAT	Sand
FEEDING HABIT	Parasitic carnivore
OPERCULUM	Corneous, oval, paucispiral

SHELL SIZE RANGE
⅜ to 1¼ in
(10 to 30 mm)

PHOTOGRAPHED SHELL
¾ in
(20 mm)

614

OTOPLEURA AURISCATI
CAT'S EAR PYRAM
<small>(HOLTEN, 1802)</small>

There are more than 6,000 species of Pyramidellidae, a family found at every depth of every sea on the globe. Most of them are less than ½ in (13 mm) long, and many are microscopic. Although some actively prey on bivalves or worms, most are ectoparasites, with each species specializing in a particular host. They have no radula, but a long stiletto-like proboscis for piercing their victim and sucking out fluid and tissue.

RELATED SPECIES

Otopleura mitralis (Adams, 1855) is a predatory pyram with a similar sculpture of fine axial ribs. It is white, with more or fewer brown to purple splashes inside and out, sometimes in spiral bands. The aperture is slightly more open toward the siphonal notch, and there are three folds on the columella.

Actual size

The shell of the Cat's Ear Pyram is bulbous, and the aperture accounts for about half of the shell length. The suture is impressed, making a sloping step at the top of each whorl. It is off-white, and its sculpture of narrow, close axial ribs is decorated in broken spiral bands of blurred tan to dark brown. The outer lip is thin, and there are 3 dental pleats on the columella.

FAMILY	Pyramidellidae
SHELL SIZE RANGE	½ to 2 in (14 to 50 mm)
DISTRIBUTION	Indo-West Pacific
ABUNDANCE	Uncommon
DEPTH	Intertidal to shallow subtidal
HABITAT	Sand bays
FEEDING HABIT	Parasitic carnivore
OPERCULUM	Corneous, oval, paucispiral

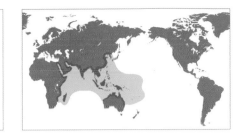

SHELL SIZE RANGE
½ to 2 in
(14 to 50 mm)

PHOTOGRAPHED SHELL
1 in
(27 mm)

PYRAMIDELLA TEREBELLUM

TEREBRA PYRAM

(MÜLLER, 1774)

615

All species of Pyramidellidae are hermaphrodites, and some produce spermatophores, capsules of spermatozoa transferred from one individual to another. Most are parasites, and their eggs are laid in large masses of jelly outside the shells of their chosen hosts. The larval shells are sinistral—wound counterclockwise—but adults are dextral; this results in heterostrophic nuclear whorls, with the axis of the larval shell forming a considerable angle with the axis of the adult shell.

RELATED SPECIES

Pyramidella dolabrata (Linnaeus, 1758) occurs both in the Indo-Pacific and the Caribbean, and *P. terebellum* was for a time considered to be merely a Pacific-only color-variant of it. *Pyramidella dolabrata* has fewer and paler spiral bands, generally tan to mid chestnut in color; and there is a deeper groove behind the columellar folds than is present on *P. terebellum*.

The shell of the Terebra Pyram is smooth and auger-like, with a very tall spire on a rounded body whorl. Its suture is moderately deep, and the slightly convex spire whorls bear 3 dark brown spiral bands on a white to cream background. There are 4 more brown bands on the body whorl, which are visible through the narrow aperture. The columella shows faint folds.

Actual size

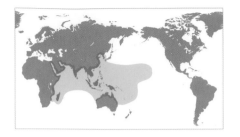

FAMILY	Pyramidellidae
SHELL SIZE RANGE	¾ to 1⅝ in (20 to 40 mm)
DISTRIBUTION	Indo-Pacific
ABUNDANCE	Common
DEPTH	Intertidal to shallow subtidal
HABITAT	Sand
FEEDING HABIT	Parasitic carnivore
OPERCULUM	Corneous, oval, paucispiral

SHELL SIZE RANGE
¾ to 1⅝ in
(20 to 40 mm)

PHOTOGRAPHED SHELL
1¼ in
(30 mm)

616

PYRAMIDELLA TESSELLATA
TESSELLATE PYRAM
(ADAMS, 1854)

Pyrams are predatory parasites—they don't kill their prey, but merely stab it with their sharp proboscis and drain off tissue and fluid. Most pyrams are very small, and *Pyramidella tessellata* is one of the largest in a family that rarely exceeds ½ in (13 mm). It had been considered a variant of *P. sulcata*, which shares its habitat and range; the latter is paler and less patterned, and has a wide groove for a suture, but is very similar in profile and size.

RELATED SPECIES
Pyramidella acus (Gmelin, 1791) is larger still, usually growing to 2 in (50 mm). It has a creamy shell and a deep suture, and is decorated with spiral bands of large, dark brown spots—two or three on each spiral whorl and five on the body whorl. It has the same sequence of columellar folds, but a thinner outer lip.

The shell of the Tessellate Pyram is auger-like. On a background of white to pale brown, it has broad spiral bands of more or less vertical lines that align with adjacent bands to produce broken axial flames of color. The lip is moderately thin, and there are 3 folds on the columella, largest toward the posterior. The apex is often broken in beached specimens.

Actual size

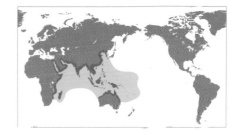

FAMILY	Acteonidae
SHELL SIZE RANGE	⅜ to 1 in (10 to 25 mm)
DISTRIBUTION	Indo-West Pacific
ABUNDANCE	Rare
DEPTH	Intertidal to 300 ft (100 m)
HABITAT	Sand
FEEDING HABIT	Carnivore, feeds on polychaete worms
OPERCULUM	Corneous, elongate

SHELL SIZE RANGE
⅜ to 1 in
(10 to 25 mm)

PHOTOGRAPHED SHELL
⅞ in
(23 mm)

ACTEON VIRGATUS

STRIPED ACTEON

(REEVE, 1842)

617

The family Acteonidae is considered the most primitive among the opistobranch order Cephalaspidea. There is a tendency in other families for the shell to be thin, reduced, or completely lost; however, acteonids have a thick, external shell. The animal of *Acteon virgatus* can withdraw completely into its shell, and there is an operculum. There are about 50 living species of acteonids worldwide, ranging from the intertidal zone to deep water. The earliest acteonids date from the Cretaceous Period.

RELATED SPECIES

Acteon eloisae (Abbott, 1973), endemic to Oman, is rare and very popular with collectors because of its striking spirals of large, orange axial blotches outlined in dark brown. The spiral cords are wider and flatter toward the middle of the body whorl.

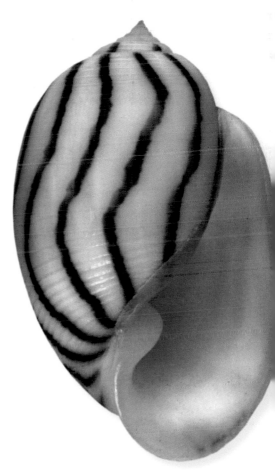

The shell of the Striped Acteon
is ovate. The short spire has a moderately impressed suture. The aperture broadens anteriorly. There is 1 twisted fold on the columella. Fine spiral cords become wider but almost smooth at the middle of each whorl. The shell is white, with dark brown, fine, irregular axial lines.

Actual size

FAMILY	Acteonidae
SHELL SIZE RANGE	½ to 1⅛ in (12 to 27 mm)
DISTRIBUTION	Indo-Pacific
ABUNDANCE	Common
DEPTH	Intertidal to 100 ft (30 m)
HABITAT	Sand
FEEDING HABIT	Carnivore, feeds on polychaete worms
OPERCULUM	Corneous, elongate

SHELL SIZE RANGE
½ to 1⅛ in
(12 to 27 mm)

PHOTOGRAPHED SHELL
1 in
(24 mm)

618

PUPA SOLIDULA
SOLID PUPA
(LINNAEUS, 1758)

The shell of the Solid Pupa is ovate with a moderately short spire, encircled by fairly narrow rounded spiral cords. It is white, and the cords are marked with elongated spots of black, tan, or, more commonly, red; one or two cords may lack spots altogether. Apex and aperture are all white. There is 1 double fold on the lower columella and a single fold farther up.

Actual size

Pupa is one of the several genera that come under the general heading of bubble shells. Broadly similar to the acteons, they differ in having a second, smaller columellar fold higher up, and in exhibiting broader, more rounded spiral cords. *Pupa solidula* occurs widely throughout its range and there is a rarer variation, *P. s. fumata* with paler gray or beige markings, which is confined between the Philippines and western Australia.

RELATED SPECIES

Pupa sulcata (Gmelin, 1791), which is less common within the same range of distribution, has a slightly more swollen body than *P. solidula*. The spiral cords are more flattened, and the black or tan spots are smudged. The spire and anterior margin may be stained tan.

FAMILY	Aplustridae
SHELL SIZE RANGE	½ to 1¼ in (12 to 30 mm)
DISTRIBUTION	Indo-West Pacific
ABUNDANCE	Moderately common
DEPTH	Intertidal to 6 ft (2 m)
HABITAT	Sand and mud bottoms
FEEDING HABIT	Carnivore, feeds on polychaete worms
OPERCULUM	Absent

SHELL SIZE RANGE
½ to 1¼ in
(12 to 30 mm)

PHOTOGRAPHED SHELL
½ in
(12 mm)

HYDATINA AMPLUSTRE
ROYAL PAPER BUBBLE
LINNAEUS, 1758

619

Hydatina amplustre has one of the smallest and more heavily calcified shells in the family Aplustridae. Unlike other aplustrids, which usually have an animal that is more colorful than the shell. *H. amplustre* has a beautifully colored shell and a translucent grayish animal that envelopes the shell. It has two pairs of tentacles on the cephalic shield. The animal relies on acid glands on its skin to produce a distasteful chemical for its defense. It can also seek refuge by burrowing into the sand. There are about a dozen living species in the family Aplustridae worldwide, in tropical and subtropical waters, with the highest diversity in the Indo-Pacific.

RELATED SPECIES

Hydatina physis (Linnaeus, 1758), from the Red Sea to the Indo-Pacific, has a large, globose shell with thin spiral brown lines. *Micromelo undata* (Bruguière, 1792), which is a circumtropical species, has a small, white shell with a network of thin, dark red lines. The animal is translucent gray with opaque white spots and a colorful bluish edge with a bright yellow band and a thin red line.

The shell of the Royal Paper Bubble is small, thin, translucent and glossy, with an oval-elongate outline. Its spire is short and blunt, and the suture is well marked. The shell is smooth, with only weak growth lines. The aperture is elongate and moderately narrow, and the outer lip and columella are smooth. The shell color is white with 2 broad rosy bands bordered by thin black lines. The bands of color are visible within the aperture.

Actual size

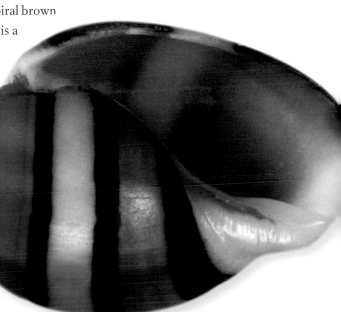

FAMILY	Hydatinidae
SHELL SIZE RANGE	⅝ to 2½ in (15 to 65 mm)
DISTRIBUTION	Red Sea to Indo-Pacific
ABUNDANCE	Common
DEPTH	Intertidal to 90 ft (28 m)
HABITAT	Silty sand bottoms
FEEDING HABIT	Carnivore, feeds on polychaete worms
OPERCULUM	Absent

SHELL SIZE RANGE
⅝ to 2½ in
(15 to 65 mm)

PHOTOGRAPHED SHELL
1⅛ in
(28 mm)

620

HYDATINA PHYSIS

GREEN-LINED PAPERBUBBLE

(LINNAEUS, 1758)

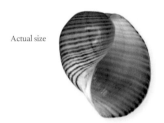

Actual size

Hydatina physis is a colorful, shelled sea slug with a wide distribution in the Indo-Pacific, occurring from the Red Sea to South Africa, and throughout the Indo-Pacific. It has a paper-thin and fragile shell that offers little protection against predators. The large, frilly pink body has white edges and evokes rose petals; it cannot retract completely into its shell. *Hydatina physis* and other aplutrids lack an operculum, and the eyes are two small black dots directly on the body, not at the base of cephalic tentacles, which are also lacking; instead, it has four flat and broad tentacular processes. It is a specialized predator of polychaete worms.

RELATED SPECIES

Hydatina vesicaria (Lightfoot, 1786), from Florida to Brazil, is very similar, but it differs by having a smaller and more slender shell with a short spire. *Hydatina amplustre* (Linnaeus, 1758), from the Indo-Pacific, has a small, very distinctive shell with two broad spiral rosy bands and three white bands, separated by thin black lines.

The shell of the Green-lined Paperbubble is large for the family, very thin, lightweight, smooth, and globular in outline. It has a sunken spire and a large body whorl. The aperture is wide and the outer lip is thin and smooth, as is the entire shell, except for weak growth striae. The shell color is creamy white with many fine, wavy, closely spaced, brown lines of varying thickness. The columella and aperture are white; the inner edge of the outer lip is dark.

FAMILY	Cylichnidae
SHELL SIZE RANGE	1¼ to 3 in (32 to 75 mm)
DISTRIBUTION	Iceland to the Canaries, and the Mediterranean
ABUNDANCE	Locally common
DEPTH	Shallow subtidal to 2,300 ft (700 m)
HABITAT	Sandy bottoms
FEEDING HABIT	Carnivore, feeds on bivalves and worms
OPERCULUM	Absent

SCAPHANDER LIGNARIUS

WOODY CANOEBUBBLE

(LINNAEUS, 1767)

SHELL SIZE RANGE
1¼ to 3 in
(32 to 75 mm)

PHOTOGRAPHED SHELL
2⅛ in
(56 mm)

621

Scaphander lignarius is one of the largest representatives of the family Cylichnidae. The animal burrows in sand as deep as 2 in (50 mm), searching for its prey: bivalves, polychaetes, foraminiferans, and small crustaceans. Like other cylichnids, the animal is too large to withdraw completely into its shell. The flattened head bears no tentacles and has a cephalic shield. The large foot has parapodial lobes used for swimming. The animal has three large, calcified gastric or gizzard plates that are used to grind its food, aided by strong gizzard muscles. Like many other sand-dwellers, there is no operculum. There are about 50 species in the family Cylichnidae worldwide.

RELATED SPECIES

Scaphander watsoni Dall, 1869, which ranges from North Carolina to Venezuela, has a shell similar in shape to *S. lignarius*, but smaller, with more widely spaced spiral lines and white or cream in color. *Akera bullata* (Müller, 1776), from the Mediterranean and northeastern Atlantic, has a more cylindrical and fragile shell that is translucent.

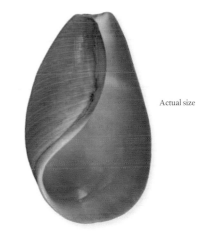

Actual size

The shell of the Woody Canoebubble is thin but sturdy, ovate, and has a sunken spire. The body whorl is greatly expanded anteriorly, narrowing toward the apex. The aperture is as long as the shell and widest anteriorly. The outer lip is thin and extends above the spire. The columella is smooth and curved, with a white parietal shield. The sculpture consists of fine incised spiral lines, crossed by fine growth lines. The shell is tan with a darker periostracum, and white inside.

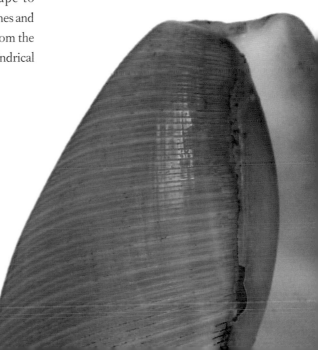

FAMILY	Haminoeidae
SHELL SIZE RANGE	⅝ to 1¼ in (15 to 32 mm)
DISTRIBUTION	Ireland to the Mediterranean
ABUNDANCE	Common
DEPTH	Shallow subtidal to 16 ft (5 m)
HABITAT	Sandy and muddy bottoms
FEEDING HABIT	Herbivore, feeds on diatoms
OPERCULUM	Absent

SHELL SIZE RANGE
⅝ to 1¼ in
(15 to 32 mm)

PHOTOGRAPHED SHELL
1⅛ in
(29 mm)

622

HAMINOEA NAVICULA
NAVICULA PAPERBUBBLE
(DA COSTA, 1778)

The shell of the Navicula Paperbubble is small, thin, fragile, globose, and involute. The spire is sunken, and the large and inflated body whorl envelops the entire shell. The sculpture consists of thin spiral lines crossed by growth lines. The aperture is large, and the outer lip is thin and extends beyond the spire. The columella is smooth and sigmoid, and there is a parietal shield. The shell color is white to pale yellow, with a thin periostracum.

Haminoea navicula is the largest haminoeid from European shallow waters, and one of the largest species in the family. It is found on sandy and muddy bottoms. It is a herbivore, feeding mostly on diatoms but also on vegetal detritus. Haminoeids resorb the internal whorls so that the large animal, which is more than double of the size of the shell, can withdraw completely into its shell. It has a large cephalic shield and parapodial lobes. The head does not have tentacles, and the foot lacks an operculum. Some haminoeids burrow into the sediment during the day and are active at night.

RELATED SPECIES

Haminoea antillarum (d'Orbigny, 1841) is a common species that ranges from Florida to Brazil. It resembles *H. navicula* but has a more globose and smaller shell, with an aperture that is wider anteriorly. *Atys naucum* (Linnaeus, 1758), from the west Pacific to Hawaii, has a thick, involute, and rounded shell.

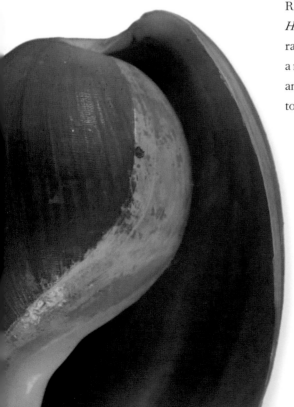

Actual size

FAMILY	Haminoeidae
SHELL SIZE RANGE	⅝ to 2 in (15 to 50 mm)
DISTRIBUTION	Indo Pacific, reaching Hawaii
ABUNDANCE	Common
DEPTH	Intertidal to 90 ft (27 m)
HABITAT	Sandy bottoms
FEEDING HABIT	Herbivore
OPERCULUM	Absent

SHELL SIZE RANGE
⅝ to 2 in
(15 to 50 mm)

PHOTOGRAPHED SHELL
2 in
(51 mm)

ATYS NAUCUM

WHITE PACIFIC ATYS

(LINNAEUS, 1758)

623

Atys naucum is one of the largest species in the family Haminoeidae, a group of herbivore snails with thin, translucent, and inflated shells. The shell of *A. naucum* is widest at the middle, and involute, with the spire completely enveloped by the later whorls of the shell. The animal has a broad foot with lateral extensions, called parapodia, that protect the shell. Some species are able to swim short distances by flapping their parapodia. Haminoeids are usually found in shallow waters on soft sediments, and occur worldwide, both in tropical and temperate waters.

RELATED SPECIES

Atys cylindricum (Helbling, 1779), from the Indo-West Pacific, has an elongated cylindrical shell. The shell color is white to cream, and the sculpture consists of fine spiral lines near the extremities. *Haminoea navicula* (da Costa, 1778), from Ireland to the Mediterranean, has a small, inflated, thin, and fragile shell.

The shell of the White Pacific Atys is lightweight, globose, rather solid, and large for the family. The shell is involute, with a small depression in the posterior end, above the immersed spire. The body whorl is large and inflated, with fine spiral grooves that are deeper toward the extremities, and crossed by weak growth lines. The aperture is wide and long, spanning the length of the shell. The columella is smooth and bent anteriorly. The shell color is pure white, often covered by an orange-brown, deciduous periostracum.

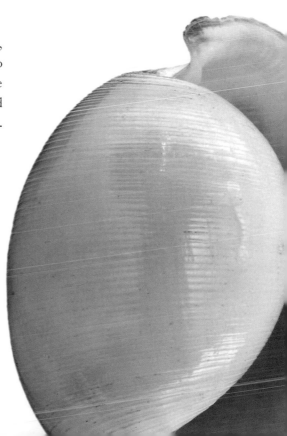

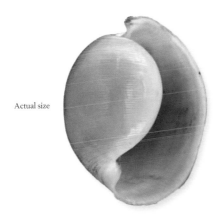

Actual size

FAMILY	Smaragdinellidae
SHELL SIZE RANGE	⅜ to ⅝ in (8 to 15 mm)
DISTRIBUTION	Indo-Pacific to Hawaii
ABUNDANCE	Abundant
DEPTH	Intertidal zone
HABITAT	Rocky shores or among algae
FEEDING HABIT	Herbivore
OPERCULUM	Absent

SHELL SIZE RANGE
⅜ to ⅝ in
(8 to 15 mm)

PHOTOGRAPHED SHELL
½ in
(11 mm)

624

SMARAGDINELLA CALYCULATA

SMARAGDINELLA CALYCULATA

(BRODERIP & SOWERBY I, 1829)

Smaragdinella calyculata is a small sea slug that is part of a group known as bubble shells. Its shell is reduced to basically just the body whorl, with a wide aperture. The animal is dark green, and much larger than the shell; the shell protects vital organs. It is the equivalent of a limpet among the bubble shells, and lives in intertidal areas, attached to bare rock or algae.

RELATED SPECIES

Smaragdinella sieboldi Adams, 1864, from the Indo-West Pacific, is smaller and has a translucent white shell with a spoon-shaped outline. The animal is similar to *S. calyculata*, but paler green, speckled with white. *Phanerophthalmus smaragdinus* (Rüpell and Leuckart, 1828), also from the Indo-West Pacific, has a more elongate, green body, and a rudimentary white shell.

Actual size

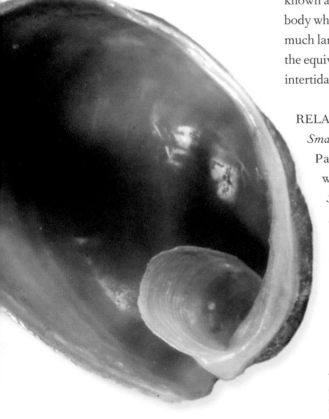

The shell of **Smaragdinella calyculata** is small, reduced, solid, and cap-shaped. Its spire is hidden by the body whorl; the previous whorls are reduced to a spoon-shaped projection on the inner lip. The surface is smooth, with growth marks. The aperture is wide, the outer lip thin, and the columella smooth. The color is olive-green (yellow in live animals), and the inner lip is white.

FAMILY	Bullidae
SHELL SIZE RANGE	¼ to 2½ in (20 to 65 mm)
DISTRIBUTION	Indo-West Pacific
ABUNDANCE	Very common
DEPTH	Intertidal to shallow subtidal
HABITAT	Grass beds
FEEDING HABIT	Herbivore
OPERCULUM	Absent

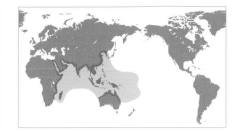

SHELL SIZE RANGE
¼ to 2½ in
(20 to 65 mm)

PHOTOGRAPHED SHELL
2⅛ in
(54 mm)

BULLA AMPULLA

AMPULLE BUBBLE

LINNAEUS, 1758

Species in the genus *Bulla* usually have a deeply sunken apex, which gives the impression of a long, narrow umbilicus. Although they have no operculum, all *Bulla* species are able to withdraw completely into their shells. The family is found in shallow waters of all the world's seas. *Bulla ampulla* is a nocturnal herbivore; it spends its days buried in the sandy substrate, but is easily seen by flashlight emerging at night to feed on sea weeds and grasses.

RELATED SPECIES

Bulla striata (Bruguière, 1792) is a smaller Atlantic and Mediterranean version of *B. ampulla*. It is narrower, slightly broader anteriorly, and ovate in outline. There are fine spiral lines near both extremities. It has a similar range of shell colors as *B. ampulla*.

Actual size

The shell of the Ampulle Bubble is a egg-shaped, with a sunken spire. It is smooth except for growth lines, varying from cream-pink to tan-gray with mid to dark brown blotching all over. The aperture is white, longer than the body whorl at both ends and broadening markedly to the anterior. The outer lip extends past the sunken spire.

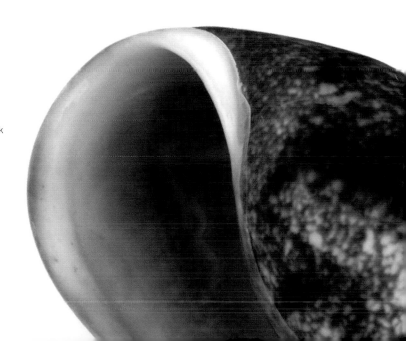

FAMILY	Akeridae
SHELL SIZE RANGE	1 to 1½ in (25 to 40 mm)
DISTRIBUTION	Mediterranean, northeastern Atlantic
ABUNDANCE	Uncommon
DEPTH	Intertidal to shallow subtidal
HABITAT	Soft bottoms, especially mud
FEEDING HABIT	Herbivore, grazes on algae
OPERCULUM	Absent

SHELL SIZE RANGE
1 to 1½ in
(25 to 40 mm)

PHOTOGRAPHED SHELL
⅝ in
(16 mm)

AKERA BULLATA
BUBBLE AKERA
(MÜLLER, 1776)

Akera bullata is a member of the Akeridae, a small family of primitive sea hares that have a thin, inflated shell and an elongated neck-head region. The foot has parapodia that fold over the body. If disturbed, the animal can swim by flapping its parapodia, in jerky movements that can last up to half an hour. The animal cannot withdraw completely into its shell, and therefore lacks an operculum. It spends most of the time buried in muddy substrates, with only the head emerging. It is sometimes observed in large swimming swarms. There are only four living species recognized in the Akeridae, all with shells similar to that of *Akera bullata*.

RELATED SPECIES

Akera soluta Gmelin, 1791 ranges from East Africa to the central Pacific. The Akeridae are related to sea hares such as *Aplysia morio* (Verrill, 1901), from Rhode Island to Texas and Bermuda. It is a large sea hare that can grow up to 16 in (400 mm); its reduced, internal shell can reach about 2½ in (60 mm) in length.

Actual size

The shell of the Bubble Akera is thin, fragile, translucent, convolute, and ovate. The spire is either flat or very low, with a well-marked channeled suture. It is sculptured with many thin spiral lines and axial growth lines, but is generally smooth. The aperture is large and elongate, almost as long as the shell, and widest at the anterior end; the columella is smooth. The shell color is tan posteriorly, and grayish anteriorly.

FAMILY	Aplysiidae
SHELL SIZE RANGE	1 to 2⅝ in (25 to 67 mm)
DISTRIBUTION	Indo-West Pacific
ABUNDANCE	Common
DEPTH	Shallow subtidal to 40 ft (12 m)
HABITAT	Grass beds or on soft bottoms
FEEDING HABIT	Herbivore, feeds on macroalgae
OPERCULUM	Absent

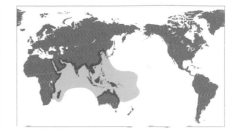

SHELL SIZE RANGE
1 to 2⅝ in
(25 to 67 mm)

PHOTOGRAPHED SHELL
1¾ in
(44 mm)

DOLABELLA AURICULARIA

SHOULDERBLADE SEA CAT

(LIGHTFOOT, 1786)

627

Dolabella auricularia is a large sea hare, a member of the family
Aplysiidae, a group of gastropods which have a reduced,
internal shell. The head of sea hares has two projections, called
rhinophores, which make the head resemble a hare or rabbit,
hence the name. *Dolabella auricularia* is a herbivore and feeds
on brown and green macroalgae. It is commonly found in bays
and lagoons, usually on seagrass beds, but also in tide pools. Like
other aplysiids, it releases a purple ink when disturbed as a means
of defense, similar to the ink of cephalopods. The internal shell is
the remnant of the external shell of other gastropods. It is located
on the dorsum, under the mantle, and over the gill and viscera.

Actual size

RELATED SPECIES

Dolabella gigas (Rang, 1828), from the Indian Ocean, has a
slightly larger shell with a rounded and larger saucer-shaped
expansion of the apex than that of *D. auricularia*. *Aplysia
vaccaria* Winkler, 1955, from California, is the largest
species in the family Aplysiidae, and a contender for the
title of the world's largest gastropod, because it can
grow to about 40 in (1 m), although it has a reduced,
internal shell.

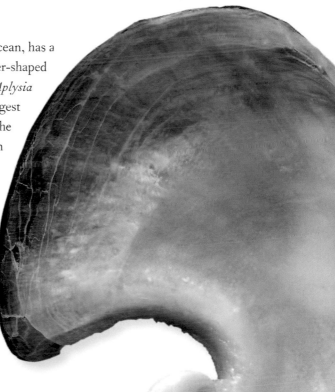

The internal shell of the Shoulderblade Sea Cat is
reduced, flattened and platelike. It is highly calcified
for the family; juvenile shells become brittle or
deformed from drying. It is ear-shaped (hence the
species name), with a saucer-shaped expansion
around the apex in large shells. The uncoiled shell has
a rapid growth rate, and is sculptured with thin axial
growth lines. The shell color is white, with a light
brown periostracum covering the dorsal side of the
shell, but it peels easily in juvenile shells.

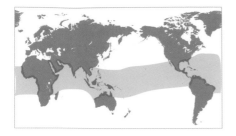

FAMILY	Umbraculidae
SHELL SIZE RANGE	3 to 4 in (75 to 100 mm); body: up to 11 in (280 mm)
DISTRIBUTION	Worldwide in warm waters
ABUNDANCE	Uncommon
DEPTH	Intertidal to 900 ft (275 m)
HABITAT	Soft bottoms near coral reefs
FEEDING HABIT	Carnivore, feeds on sponges
OPERCULUM	Absent

SHELL SIZE RANGE
3 to 4 in
(75 to 100 mm)

PHOTOGRAPHED SHELL
2⅜ in
(60 mm)

628

UMBRACULUM UMBRACULUM
UMBRELLA SHELL
(LIGHTFOOT, 1786)

Umbraculum umbraculum is a large, opistobranch gastropod that has a flattened, limpetlike shell that covers less than half of its large body. The animal carries its shell like a little umbrella on its back, hence the name. It is usually found in shallow waters and tide pools but is sometimes dredged in deep waters. Its yellow or orange body is covered with large pustules that mimic some of its prey sponges both in color and texture. It has side gills and eyes at the base of the tentacles, and a broad and long radula with an estimated 800,000 teeth. There is a single species in the family Umbraculidae with worldwide distribution.

Actual size

RELATED SPECIES

The family's closest relatives are the Tylonidae, with includes *Tylodina corticalis* (Tate, 1889), from Australia, and *T. americana* Dall, 1890, from the Gulf of Mexico and the Caribbean. Both species have yellow shells similar to *U. umbraculum*, but the former grows to 1 in (25 mm) in length, while the latter has axial rays and reaches about half of that size.

The shell of the Umbrella Shell is flattened, plate-like, conical, and elliptical. The only coiling is seen in the protoconch, which has one whorl, but it is often eroded in larger shells. The sculpture consists of fine concentric growth lines, and sometimes fine radial lines or undulations on the internal side of the shell. The external side is covered by a brown periostracum. The shell color is white to yellow, with a yellow, white, or brown off-centered apex.

FAMILY	Cavolinidae
SHELL SIZE RANGE	¼ to ¾ in (5 to 20 mm)
DISTRIBUTION	Worldwide
ABUNDANCE	Common
DEPTH	0 to 100 ft (30 m)
HABITAT	Pelagic
FEEDING HABIT	Generalist
OPERCULUM	Absent

SHELL SIZE RANGE
¼ to ¾ in
(5 to 20 mm)

PHOTOGRAPHED SHELL
¼ in
(18 mm)

CAVOLINIA TRIDENTATA

THREE-TOOTHED CAVOLINE

NIEBUHR, 1775

Cavolinia tridentata is a large and common cavolinid; cavolinids are a group of pelagic gastropods also known as sea butterflies, because of the large foot lobes that flap like the wings of a butterfly. It can swim at about 5½ in (140 mm) per second. Like all cavolinids, it has a shell that is glassy, thin, translucent, and bilaterally symmetrical. The family includes shells with very different shapes. They are generalist feeders, and produce a mucous net that traps plankton. There are about 30 living species in the family Cavolinidae worldwide, and all are pelagic. The earliest known cavolinid fossil dates from the Eocene Epoch.

RELATED SPECIES

The following cosmopolitan species represent the main different shell shapes in the family: *Clio pyramidata* Linnaeus, 1758, has a very fragile pyramidal shell with a pointed end that is seldom collected whole. *Creseis acicula* (Rang, 1828), has a long, narrow, straight, tapering shell with a single opening. *Diacria trispinosa* (Lesueur, 1821), has a broad and flattened shell with three spines, with the middle one very long. *Cuvierina columnella* (Rang, 1827) has a small, bottle-shaped shell.

Actual size

The shell of the Three-toothed Cavoline is small, thin, fragile, lightweight, glassy, and globular. There are no traces of spiral coiling; instead, its shell is bilaterally symmetrical. The surface is smooth and glossy, with fine growth lines. The aperture is narrow and curved, with a thickened outer lip. Opposite from the aperture there are 3 spines, the middle one the longest. The shell color is golden brown.

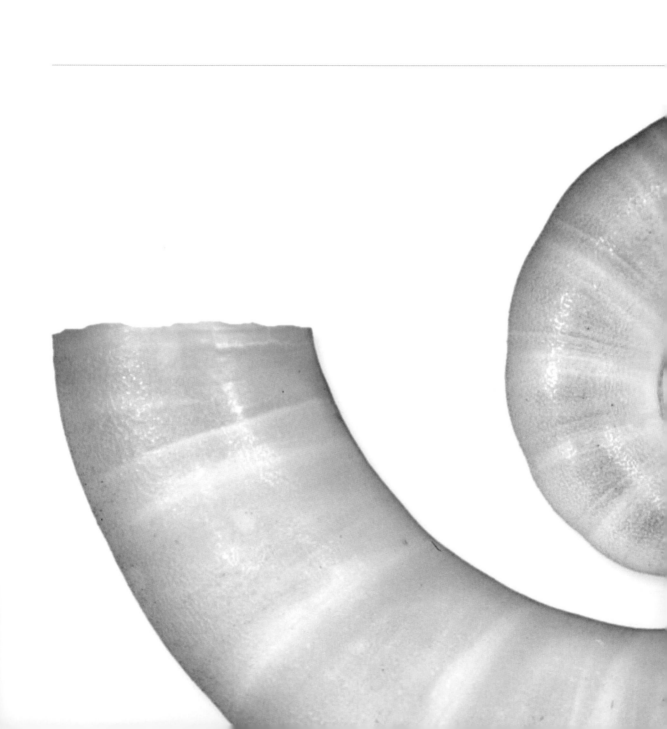

CEPHALOPODS

Of the approximately 900 species of cephalopods living today, only six, all in the primitive genus *Nautilus*, still have external shells. These animals occupy only the last chamber of the bilaterally symmetrical shell; the other chambers are filled with gas and used to control buoyancy. Nautilus species inhabit the deeper waters along coral reefs during the day, but rise to shallower waters at night to feed, using the 80 to 90 tentacles around their head. Sepia, cuttlefish, and squid have internal shells that are greatly reduced. They swim rapidly and use their eight arms and two tentacles in order to capture prey. Octopuses lack any shell at all, and have eight arms.

Female Paper Nautilus produce an egg case that looks like a shell. Like the shell of the Nautilus, it is bilaterally symmetrical, but much thinner and is not subdivided into chambers, nor does it float when empty.

FAMILY	Nautilidae
SHELL SIZE RANGE	6 to 10½ in (150 to 268 mm)
DISTRIBUTION	Indo-West Pacific
ABUNDANCE	Common
DEPTH	650 to 1,500 ft (200 to 450 m), shallower at night
HABITAT	Water column where coral reefs drop off into deep water
FEEDING HABIT	Carnivore, feeds on hermit crabs and fish
OPERCULUM	Absent, but its leathery hood serves a similar function

SHELL SIZE RANGE
6 to 10½ in
(150 to 268 mm)

PHOTOGRAPHED SHELL
6⅜ in
(166 mm)

632

NAUTILUS POMPILIUS
CHAMBERED NAUTILUS
LINNAEUS, 1758

The shell of the Chambered Nautilus is large, involute, and planispiral, lightweight and thin-shelled, with a wide aperture. The shell is bilaterally symmetrical. Unlike some other *Nautilus* species, the Chambered Nautilus has a closed umbilicus. The creamy or white shell is sculptured with growth striae, and decorated with brownish to reddish irregular stripes that fade toward the aperture. The part of the shell adjacent to the aperture is stained in brown or black and it represents the part where the hood of the animal touches the shell. The aperture and chambers are pearly white.

Nautilus are the only living cephalopods with a true external shell. They are regarded as living fossils, the only survivors of a long line of shelled cephalopods with a fossil record extending more than 400 million years. Nautilus live in deep waters, as deep as 1,500 ft (450 m) during the day, rising to about 330 ft (100 m) at night. The shell is partitioned into chambers, with the animal occupying the last chamber; the others are filled with gas and liquid. All the chambers are connected by a hollow tube, the siphuncle, which the Nautilus use to regulate buoyancy.

RELATED SPECIES

Nautilus macromphalus Sowerby II, 1849, from off New Caledonia and northeastern Australia, has a large umbilicus on both sides of the shell; *Nautilus belauensis* Sanders, 1981 from Palau, has a shell similar to *N. pompilius*, while *N. stenomphalus* Sowerby II, 1849, from the Great Barrier Reef, Australia, has a callus over the umbilicus.

Actual size

FAMILY	Nautilidae
SHELL SIZE RANGE	6 to 8½ in (180 to 215 mm)
DISTRIBUTION	New Guinea and Solomon Islands
ABUNDANCE	Uncommon
DEPTH	330 to 1,000 ft (100 to 300 m)
HABITAT	Water column where coral reefs drop off into deep water
FEEDING HABIT	Carnivore, feed on shrimp, crabs and fishes
OPERCULUM	Absent, but its leathery hood serves a similar function

SHELL SIZE RANGE
6 to 8½ in
(180 to 215 mm)

PHOTOGRAPHED SHELL
6½ in
(168 mm)

633

NAUTILUS SCROBICULATUS

CRUSTY NAUTILUS

LIGHTFOOT, 1786

Nautilus scrobiculatus is a species of *Nautilus* with a narrow distribution, living in deep waters off Papua New Guinea to the Solomon Islands. Like other cephalopods, *Nautilus* species swim by jet propulsion, pumping water through a muscular funnel, the hyponome, which propels the animal backward. The Crusty Nautilus has about 90 tentacles, which it uses to hunt for prey, aided by chemical cues. Like other species of *Nautilus*, its shell has as many as 30 hollow chambers.

The shell of the Crusty Nautilus is large, planispiral, and has a large, open umbilicus on both sides of the shell. The body whorl has sinusoidal radial folds. The external surface of the shell is dull cream, adorned with thin, straight, radial, and brownish to reddish stripes spanning a quarter to a half of the body whorl. The part adjacent to the aperture is stained in black by the animal. The aperture and chambers are pearly white.

RELATED SPECIES

Nautilus perforatus Conrad, 1849, from around Bali, Indonesia, its closest relative, also has an umbilicus. Both species are far less common than *N. pompilius* Linnaeus, 1758, the most wide-spread species. *Nautilus belauensis* Saunders, 1981, from Palau, has the second largest shell in the genus. *Nautilus macromphalus* Sowerby II, 1849, from New Caledonia to northeastern Australia, has the smallest shell.

Actual size

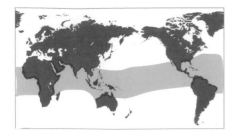

FAMILY	Spirulidae
SHELL SIZE RANGE	¾ to 1¼ in (20 to 30 mm)
DISTRIBUTION	Worldwide in warm waters
ABUNDANCE	Common
DEPTH	330 to 3,300 ft (100 to 1,000 m)
HABITAT	Pelagic
FEEDING HABIT	Carnivore
OPERCULUM	Absent

SHELL SIZE RANGE
¾ to 1¼ in
(20 to 30 mm)

PHOTOGRAPHED SHELL
⅞ in
(23 mm)

634

SPIRULA SPIRULA

SPIRULA

LINNAEUS, 1758

Spirula spirula is a small, deepwater cephalopod that is rarely seen alive, so its biology is not well known. It occurs worldwide in warm waters, living in the water column. They have a daily migration, moving from depths ranging from 1,650 to 3,300 ft (500 to 1,000 m) during the day to shallower waters at night. Like *Nautilus* species, they use their internal, chambered shell to control their buoyancy. Their shell is buoyant, so empty shells float on the surface of the sea and can be carried by currents for long distances. The Spirula is the only extant species in the genus *Spirula*, in the family Spirulidae.

RELATED SPECIES

Its closest relatives are fossil spirulids and an extinct group of cephalopods, the belemnoids, which had an internal, straight, chambered shell, and which resembled modern squids. Modern relatives of the Spirula include cuttlefishes and squids.

Actual size

The shell of the Spirula is small, formed as an open coil in a single plane. Internally, the shell is divided into chambers that are connected by a narrow curved tube, the siphuncle. The aperture is round, and its concave inner surface, or septum, is pearly. The surface of the shell is chalky white, but each septum can be clearly seen as a cream colored band. Although strong, the shell can break more easily at the septum, revealing the internal chambers and the siphuncle.

FAMILY	Sepiidae
SHELL SIZE RANGE	12 to 16 in (300 to 400 mm)
DISTRIBUTION	Tropical Indo-West Pacific
ABUNDANCE	Common
DEPTH	33 to 330 ft (10 to 100 m)
HABITAT	Neritic and demersal
FEEDING HABIT	Carnivore, feeds on fishes and crustaceans
OPERCULUM	Absent

SHELL SIZE RANGE
12 to 16 in
(300 to 400 mm)

PHOTOGRAPHED SHELL
10½ in
(261 mm)

SEPIA PHARAONIS

PHARAOH CUTTLEFISH

EHRENBERG, 1831

635

Cuttlefishes are cephalopods that have a calcareous internal chambered shell. Like the species of *Nautilus*, they control their buoyancy by pumping water and gases in and out of the shell chambers. Cuttlefish are voracious predators, using their ten arms with suckers to prey on crustaceans and fishes. They are able to change their body color rapidly, flashing color patterns and using body posture to communicate with other cuttlefishes. There are more than 100 species of cuttlefishes around the world. They occur in all tropical, subtropical, and temperate waters except the coasts of the Americas.

RELATED SPECIES

Among the more common species are *Sepia officinalis* Linnaeus, 1758, a large species from the Mediterranean and eastern Atlantic. *Sepia apama* Gray, 1849, from southern Australia, is the largest cuttlefish, reaching over 20 in (50 cm) in mantle length.

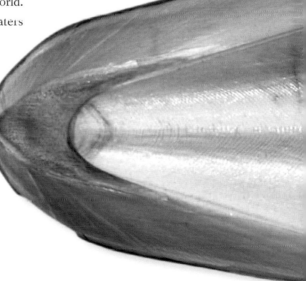

Actual size

The shell of the Pharaoh Cuttlefish is known as cuttlebone. It is a large, thick, calcareous shell that has dozens of chambers, making it lightweight. It is lanceolate in outline, nearly flat, with a broad chitinous margin and a short spine in the posterior end. The dorsal surface is granular; the ventral surface has a shallow longitudinal groove and many shallow ribs that correspond to the chambers. The shell is chalky white and quite brittle; it floats, and is commonly found washed ashore.

FAMILY	Argonautidae
SHELL SIZE RANGE	3¼ to 4 in (80 to 98 mm)
DISTRIBUTION	Baja California to Panama
ABUNDANCE	Uncommon
DEPTH	From near the surface to deep water
HABITAT	Oceanic pelagic
FEEDING HABIT	Carnivore, feeds on crustaceans and other mollusks
OPERCULUM	Absent

SHELL SIZE RANGE
3¼ to 4 in
(80 to 98 mm)

PHOTOGRAPHED SHELL
3¾ in
(95 mm)

636

ARGONAUTA CORNUTA
HORNED PAPER NAUTILUS
CONRAD, 1854

Argonauta cornuta is perhaps the rarest of the argonautids. Although it is reminiscent of *Nautilus* species, its "shell" is not the equivalent of this or any other molluscan shell; it is an evolutionary novelty found only in the family Argonautidae. This egg case is formed by a secretion from the female's two webbed dorsal arms. Like all octopods, it has eight arms. This egg case has two rows of sharp and relatively long tubercles, and hornlike lateral protrusions.

The egg case of the Horned Paper Nautilus is medium-sized, paper-thin, lightweight, fragile, laterally compressed, and discoid in outline. Each side has a pointed and long projection. Its sculpture consists of radial raised ribs; every other rib ends in a sharp tubercle at the periphery of the shell. Internally, the radial ribs show as grooves. The aperture is long and wide, and the outer lip thin. The color is white, and the tubercles are stained in brown in early whorls but fade to white toward the body whorl; the interior is white.

Actual size

RELATED SPECIES
Argonauta nouryi Larois, 1852, from the southwest Pacific and also from Baja California to Panama, has the most elongated egg case in the family, with a long aperture. *Argonauta argo* Linnaeus, 1758, is a cosmopolitan species in warm waters, and the most common and largest argonautid. Sometimes there are mass strandings of this species in southern Australia and South Africa.

FAMILY	Argonautidae
SHELL SIZE RANGE	10 to 12 in (250 to 300 mm)
DISTRIBUTION	Worldwide in warm waters
ABUNDANCE	Can be locally common
DEPTH	3 to 500 ft (1 to 150 m)
HABITAT	Oceanic pelagic
FEEDING HABIT	Carnivore, feeds on small crustaceans, mollusks and jellyfish
OPERCULUM	Absent

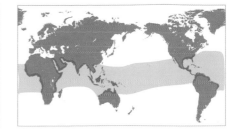

SHELL SIZE RANGE
10 to 12 in
(250 to 300 mm)

PHOTOGRAPHED SHELL
9¾ in
(245 mm)

ARGONAUTA ARGO
PAPER NAUTILUS
LINNAEUS, 1758

The Paper Nautilus or Paper Argonaut is the largest and most common species in the family, and one of several species of pelagic octopuses. Its "shell" is a case produced by the female octopus to protect her eggs. There is an extreme sexual dimorphism, with males growing to ⅝ in (15 mm) and females up to 4 in (100 mm). The shell diameter can reach 12 in (300 mm). There are six species of *Argonauta*, most occurring worldwide.

RELATED SPECIES

Argonauta nodosa Lightfoot, 1786, from the Indo-Pacific, also has a large egg case, similar to that of *A. argo*; *A. hians* Lightfoot, 1786, from tropical waters worldwide, is less common and has a smaller, darker egg case; *A. bottgeri* Maltzan, 1881, from the Indo-Pacific, is the smallest of the argonauts and has an egg case resembling that of *A. hians*.

The egg case of the Paper Nautilus is paper-thin, white, and lacks the chambers present in the shells of true *Nautilus*. It is composed entirely of calcite (without aragonite), and is somewhat flexible but fragile. It is discoidal and laterally compressed, with 2 rows of knobs along the periphery, forming a sharp keel. These knobs are black or dark brown in early whorls but fade to white toward the body whorl. The egg case is sculptured by up to 50 irregular, smooth radial ribs that appear as grooves along the inner surface.

Actual size

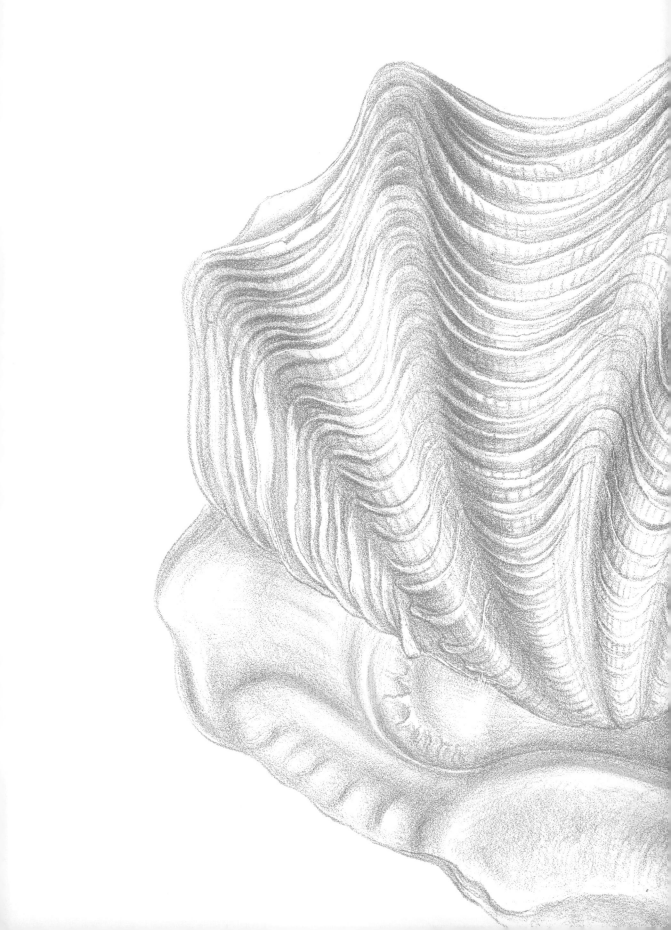

APPENDICES

GLOSSARY

Abyssal Occurring at ocean depths ranging from 13,000 to 20,000 ft (4,000 to 6,000 m).

Adductor muscle Muscle joining the two valves of a bivalve mollusk, and closing the shell when contracted.

Anoxic Without oxygen.

Anterior (prefix: antero-) The forward portion of the animal, near the head.

Aperture The opening through which the animal extends from the shell, and the most recent part of the shell to be formed.

Apex The first portion of the shell to be formed, situated at the tip of the spire.

Apophysis A projecting structure to which muscle is attached.

Aragonite A crystalline form of calcium carbonate. Mother-of-pearl and nacre are composed of aragonite.

Autotomize To shed a body part, usually the posterior part of the foot in snails, when the animal is attacked.

Axial Parallel to the shell axis.

Axis A line passing through the apex around which the whorls of the shell are coiled.

Author The person who first publishes the name of the taxon. The name of the author, and the year in which the taxon description was published follow the name of the taxon. If the author and date are not in parentheses, it means that this author published the species name in the genus in which it is currently cited. If the author and date are enclosed by parentheses, it indicates that the species was originally published in a genus different from the one to which it is currently assigned.

Basalt A volcanic rock.

Bathyal Occurring at ocean depths ranging from 3,300 to 13,000 ft (1,000 to 4,000 m). Generally corresponds to the continental slope and continental rise.

Bathymetric Related to depth in a body of water.

Bead A round raised feature, smaller than a nodule, repeated to form a linear pattern on the surface of a shell resembling a necklace.

Benthic Living on the bottom in an aquatic environment.

Biconic Shaped like two oppositely oriented cones, joined at their bases.

Bifurcate Split or divide into two.

Boreal Relating to the far northern regions.

Bilateral symmetry A type of symmetry in which the two sides of an animal are mirror images of each other.

Body whorl The last whorl (360 degrees) of a gastropod shell to be deposited.

Byssus Proteinaceous fibers secreted by the foot of a bivalve and used to form a temporary attachment to a hard substrate.

Calcarella A type of protoconch or larval shell.

Calcareous Composed of or containing calcium carbonate.

Calcite A crystalline form of calcium carbonate.

Callus A thickened layer of shell along the parietal or columellar region of the shell.

Canal A channel in the shell that is occupied by an organ of the mollusk, for example, the siphonal or anal canal.

Cancellate Having surface sculpture of intersecting axial ribs and spiral cords.

Carina A sharp keel-like ridge.

Carinated Having a carina, especially along the shoulder or shell periphery.

Cephalic Pertaining to the head.

Channel A deep groove, usually following a spiral path, forming part of the interior or exterior surface decoration of a shell.

Chemoautotroph An organism that derives energy from chemical oxidation of inorganic compounds rather than from sunlight.

Chitinous Composed of chitin, a semitransparent, horny substance found in the shells of some mollusks.

Cilia (adj. ciliary) Microscopic, hairlike projection on the surface of some cells.

Clade A group of related organisms that include an ancestor and all of its decendants.

Collar A thin ribbonlike exterior surface rib, often with a frilly outer edge, encircling and more or less perpendicular to the shell.

Columella The central pillar of the shell, formed by the inner lip of the aperture around the axis of coiling.

Conchologist A student of molluscan shells.

Congener An organism belonging to the same genus as another.

Coralline Coral-like.

Cord A thick, rounded, continuous sculptural element.

Corneous Made of a hornlike substance such as conchiolin or scleroprotein; not calcareous.

Coronate Crownlike; having spines, beads, or tubercles along the shoulder of a gastropod shell.

Costa (adj. costate) An element of axial sculpture, which may be rounded and riblike, or flange-like.

Crenulate With alternating furrows and ridges, which are scalloped or corrugated.

Denticle A single, toothlike projection. A shell with such projections is termed dentate or denticulate.

Dentition The pattern of teeth in a mollusk.

Deposit feeder A mollusk that feeds on decomposing organic matter deposited on the floor of its habitat, either on the floor surface or by tunneling deep into it.

Dextral Coiled in a clockwise direction when viewed from the apex of a shell; relating to shells with the aperture to the right of the coiling axis.

Diatom A type of algae common in the phytoplankton, that is, encased in a cell wall made of silica.

Digitation A fingerlike projection.

Dissoconch That portion of the shell of a bivalve produced after the larva undergoes metamorphosis; the adult shell.

Distal Situated farthest from the base or point of attachment; distant.

Dorsum (adj: dorsal) The upper side of a mollusk, or of any animal that moves in a horizontal position.

Epibenthic Living on the surface of sediments in an aquatic environment.

Epifaunal Living on the surface of a substrate (for example, rocks, pilings, or other animals) in an aquatic environment.

Family A group of related genera all of which are more closely related to each other than to any genus outside the family.

Fasciole A raised spiral band formed by successive growth lines along the edges of a canal.

Filter feed To feed on particles or microorganisms filtered from the water.

Flame/flammule An irregular but repeated color marking on the exterior surface.

Fold A spirally wound ridge on the columella of a gastropod.

Foliated Having thin, leaflike layers or plates.

Foliose Leaflike.

Foramen An opening, hole, or passage through the shell.

Foraminifera A phylum of single-celled micro-organisms with a shell formed of calcium carbonate.

Fossa A long, shallow, narrow depression in the surface of a shell.

Frondose Like a frond or leaf, divided into many sections.

Fusiform Spindle-shaped; swollen in the middle, tapering toward the ends.

Gamete A cell (egg or sperm) that fuses with another cell during fertilization to produce an ovum.

Genus (pl. genera) A group of related species and subspecies all of which are more closely related to each other than to any species outside the genus.

Gill A respiratory organ in aquatic organisms.

Girdle A band of leathery or muscular tissue holding the valves of a chiton in place.

Globose Spherical.

Gorgonian A type of soft coral.

Granulose Having a surface covered with granules.

Growth line A line on the shell surface that marked the position of the shell margin during an earlier stage of growth; delineate increments of shell growth.

Hadal Inhabiting depths greater than 20,000 ft (6,000 m), in ocean trenches.

High-energy beach/shore Beach or shore that is pounded by waves.

Hinge Region along the dorsal portion of bivalve shells that allows for limited motion of the two valves relative to each other.

Holdfast A structure that anchors a sessile animal or plant to the substrate. Usually applies to bivalves.

Imbricate With overlapping scales.

Intertidal The region exposed at low tide and submerged at high tide.

Interstice (pl. interstices) The space between two neighboring features such as teeth or ribs.

Involute A type of shell-coiling in which the last whorl completely envelopes the previous whorls.

Inductura A layer of shell material along the inner lip (parietal region and or columella) of a gastropod shell.

Keel A sharp, raised, blade-like spiral sculptural element. Usually present along the shell periphery or shoulder.

Labrum The outer lip of a coiled shell.

Lamella (adj. lamellose) A thin platelike structure that often occurs in multiples.

Lanceolate Leaf-shaped; narrow and tapering to a point.

Lenticular Shaped like a lens, convex on both sides.

Ligament An elastic, structure made of conchiolin, which joins the valves of a bivalve and provides the force to open the valves when the adductor muscles are relaxed.

Lip The margin of the aperture. The inner lip extends from the suture to the base of the columella, and includes the parietal region. The outer lip is the part of the apertural margin farthest from the shell axis. It also extends from the suture to the base of the columella.

641

Lira (pl. lirae) A narrow, linear ridge along the shell, or within its outer lip.

Lunule A heart-shaped, depressed region situated in front of the umbones in many bivalves.

Maculation A pattern of spots.

Malacologist A specialist in the study of mollusks, especially one who deals with the animal as well as the shell.

Mammillate With dome-shaped or nipple-like projections.

Mantle The outer portion of a mollusk's body that secretes the shell.

Margin Edge of the shell.

Monocuspid Having a single cusp or point. Usually refers to teeth.

Multispiral With multiple whorls around a central point.

Muscle scar A region on the interior surface of the shell to which a muscle is or was attached.

Nacre An iridescent inner shell layer composed of thin layers of aragonite. Also called mother-of-pearl.

Nodulose Having small knobs or nodules.

Notch An indentation in the shell margin, usually with a V-shaped or U-shaped profile.

Ocelli Multiple, light-sensing eyes, as in the mantle margin of a scallop.

Operculum A circular or elongated structure produced by some gastropods to block or seal the aperture of the shell when the animal withdraws. May be corneous or calcareous.

Opistobranch A group of gastropods that have their gill posterior to the heart.

Ovate Egg-shaped

Pallial Pertaining to the mantle.

Pallial line A narrow line or muscle scar that marks the line of attachment of the mantle to the shell in some bivalves.

Pallial sinus An indentation in the pallial line marking the attachment of the muscles that retract the siphon in some bivalves.

Palp A broad, flattened appendage near the mouth of bivalves.

Parapodia Lateral extensions of the foot.

Parietal The posterior portion of the inner lip of a gastropod, between the columella and the suture.

Paucispiral Having few whorls around a central point.

Peg A short, cylindrical projection emanating from the interior surface of an operculum.

Pelagic Living in open ocean waters; free-swimming or floating.

Periostracum A thin organic layer of conchiolin, the sometimes fibrous outer coating of the shell during the life of many mollusks.

Phylum A unit by which organisms are classified. Ranks below kingdom, but above class.

Plait A raised fold of shell that occurs on the columella.

Plica A raised fold or ridge.

Planispiral Coiled in a single plane.

Plankton Organisms that drift near the surface of the ocean.

Polychaete A type of worm.

Polymorphic Having multiple forms.

Porcelaneous Having a smooth white surface like fine porcelain.

Posterior (prefix: postero-) The rear portion of the animal; farthest from the head along the axis of the body.

Protoconch Larval shell of a gastropod; that part of the shell formed before the larva undergoes metamorphosis. Situated at the tip of the apex.

Prodisoconch Larval shell of a bivalve. That part of the shell formed before the larva undergoes metamorphosis. Situated at the tip of the umbo.

Punctate With small, needlelike depressions.

Pustulose Having pustules.

Pyriform Pear-shaped.

Quadrate Square.

Radial Extending outward from the center.

Radula A flexible ribbon supporting multiple rows of chitinous teeth. A feeding structure unique to mollusks, but absent in bivalves.

Ramp A shelf-like ledge at the top of a whorl immediately below the suture.

Ray A linear mark of surface color, radiating like a ray of light from a central point such as the apex.

Recurved Curving backward or inward.

Resilifier A recess in the bivalve shell where the ligament is attached.

Reticulated Having a netlike pattern.

Rib A raised, rounded, structure on the shell surface.

Rostrum A tapering structure resembling a bird's beak.

Rugose Having a surface with wrinkles and ridges.

Sagittal Relating to the plane of bilateral symmetry, or any plane parallel to it.

Scalariform Resembling a ladder.

Scar A repaired break in the shell.

Seep An area of the ocean floor where cold gases (cold seep) or hot springs (hydrothermal vent) seep through the earth's crust. The resulting environment supports species not reliant on food and energy produced by photosynthesis.

Selenizone A narrow, conspicuous, parallel-sided band of shell material that runs along the whorls of shells that have a slit. It originates at the back of the slit and has the same width.

Septum A wall dividing a cavity or structure into smaller ones.

Shell hash Substrate of coarsely broken shell.

Shoulder An angulation in the curvature of the whorl, usually fairly close to the suture.

Sigmoid S-shaped.

Siphon A fleshy tube through which water enters the mantle cavity.

Siphonal canal An elongated, semi-tubular extension of the aperture that protects the siphon of the snail.

Siphonal notch A rounded notch at the anterior end of the aperture, through which the animal extends and retracts its siphon.

Sipunculan A phylum of bilaterally symmetrical unsegmented marine worms.

Slit A long, narrow opening in the shell margin of some gastropods.

Species A group of all organisms capable in interbreeding and producing fertile offspring.

Spicule A small spike or needlelike structure.

Spinose Having spines.

Spire The portion of the shell between the apex and the body whorl.

Stria A shallow, incised groove in the surface of the shell.

Stromboid notch A sinuous indentation along the outer lip of the shell, near the siphonal canal. Prevalent in the family Strombidae, but also present in other gastropods.

Subovate Roughly in the form of an egg.

Substrate An underlying layer.

Subtidal Below the low tide line.

Sulcus A grove, furrow, or depression in the surface of a shell.

Superfamily A group of related families all of which are more closely related to each other than to any family outside the superfamily.

Supralittoral The zone above the high tide line, which is regularly splashed by waves, but not normally submerged.

Supratidal The area adjacent to and just above the high tide line.

Suspension feeder Animals that feed on material that is suspended in the water, usually by straining them from the water; also known as a filter feeder.

Suture A line along the shell surface along which adjacent whorls join.

Symbiont An organism that is the beneficiary of a symbiotic relationship with another organism.

Teleoconch That portion of the shell produced after the larva undergoes metamorphosis; the adult shell.

Tenting mark A pattern of triangular color markings on the surface of a shell with the appearance of a field of tents.

Terminal Marking or located at the extremity of a given physical or decorative feature.

Tooth Protuberance inside the lip of a gastropod shell, or on the hinge of a bivalve shell.

Trema (plural: tremata) An opening in the shell that allows excretory products to exit.

Truncate Having a squared-off end.

Tubercle A wartlike projection.

Turbinate Shaped like an inverted cone.

Turriculate In the form of a turret.

Type species The species that serves as the basis for defining a genus or subgenus.

Unguiculate Resembling claws.

Umbilicus A hollow, usually conical opening in the base of the shell. Present in shells in which the entire aperture, including its inner edge, is completely outside the coiling axis.

Umbo (plural: umbones) The first part of a bivalve shell to form; the apex of each valve.

Valve A distinct, calcified structure that forms all or part of the shell.

Varix (plural: varices) A thickening along a lip of the shell, usually indicating an interruption in growth and reinforcement of the shell edge.

Veliger A larval stage of mollusks, possessing a velum.

Velum The membrane of the larvae of some species of mollusk, covered in fine cilia whose waving motions aid the larva's movement.

Ventrum (adj: ventral) The lower side of a mollusk, or any animal that moves in a horizontal position.

Ventricose Distended, inflated, or swollen, especially on one side.

Water column A vertical aquatic zone from surface to sea floor, discussed as a habitat in terms of the chemical, physical, and biological variations within it.

Whorl A complete rotation (360 degrees) of the shell around the coiling axis.

Wing The greatly extended ear on one side of the umbo of some bivalves.

RESOURCES

BOOKS

The following is a small sampling of additional books or other resources available to those interested in learning more about mollusks or shell collecting.

General

Abbott, R. Tucker, *Kingdom of the Seashell* (Crescent Books, 1988)

Abbott, R. Tucker, *Seashells of the World: a guide to the better known species* (St. Martin's Press, 2002)

Abbott, R. T. and S. P. Dance, *Compendium of Seashells* (E. P. Dutton, Inc., New York, 1982)

Dance, S. P., *Shells* (DK Publishing, Inc., New York, 2002)

Harasewych, M. G., *Shells, Jewels from the Sea* (Courage Books, Philadelphia, 1989)

Robin, A., *Encyclopedia of Marine Gastropods* (ConchBooks, Wiesbaden, 2008)

Rosenberg, G., *The Encyclopedia of Seashells* (Dorset Press, New York, 1992)

Stix, H., M. Stix, R. T. Abbott, and H. Landshoff, *The Shell* (Abrams, 1978)

Regional guides

Abbott, R. T., *American Seashells*, 2nd edition (Van Nostrand Reinhold Company, New York, 1974)

Dance, S. P. (ed.) *Seashells of Eastern Arabia* (Motivate Publishing, Dubai, 1998)

Kay, E. A., *Hawaiian Marine Shells* (Bishop Museum Press, Honolulu, 1979)

Keen, A. M., *Sea Shells of Tropical West America*, 2nd edition (Stanford University Press, Stanford, 1971)

Lamprell, K. and T. Whitehead, *Bivalves of Australia. Volume 1* (Crawford House Press Pty Ltd., Bathurst, 1992)

Lamprell, K. and J. Healy, *Bivalves of Australia. Volume 2* (Backhuys Publishers, Leiden, 1998)

Mikkelsen, P. M. and R. Bieler, *Seashells of Southern Florida. Living Marine Mollusks of the Florida Keys and Adjacent Regions. Bivalves* (Princeton University Press, Princeton, 2008)

Okutani, T. (ed.) *Marine Mollusks in Japan* (Tokai University Press, Tokyo, 2000)

Poppe, G. T., *Philippine Marine Mollusks* (ConchBooks, Hackenheim, 2008)

Poppe, G. T. and Y. Goto, *European Seashells* (Verlag Christa Hemmen, Wiesbaden, 1991–1993)

Rios, E. C., *Compendium of Brazilian Sea Shells* (Universidade Federal do Rio Grande and Museu Oceanográfico Prof. Eliézer de Carvalho Rios, Rio Grande, 2009)

Thach, N. N., *Shells of Vietnam* (ConchBooks, Hackenheim, 2005)

Tunnell, J. W., Andrews, J., Barrera, N., and F. Moretzsohn, *Encyclopedia of Texas Seashells* (Texas A&M University Press, Texas, 2010)

Wilson, B., *Australian Marine Shells* (Odyssey Publishing, Kallaroo, 1993–1994)

Zhongyan, Q. (ed.) *Seashells of China* (China Ocean Press, Beijing, 2004)

Techniques and collections

Jacobson, M. K. (ed.), "How to study and collect shells: A symposium." 4th edition (American Malacological Union, Inc. Wrightsville Beach, NC, 1974)

Pisor, D. L., *Pisor's Registry of World Record Size Shells*, 5th edition (ConchBooks, Hackenheim, 2008)

Sturm, C. F., T. A. Pearce, and A. Valdés. *The Mollusks: A Guide to their study, collection, and preservation* (Universal Publishers, 2006)

Identification guides for popular groups

Houart, R., *The genus* Chicoreus *and related genera: Gastropoda (Muricidae) in the Indo-West Pacific* (Editions du Muséum, Paris,1992)

Lorenz, F. and A. Hubert, *A Guide to Worldwide Cowries*, 2nd edition (Conchbooks, Hackenheim, 2000)

Lorenz, F. and D. Fehse, *The Living Ovulidae. A Manual of the Families of Allied Cowries: Ovulidae, Pediculariidae and Eocypraeidae* (ConchBooks, Hackenheim, 2009)

Radwin, G. E. and A. D'Attillio, *Murex Shells of the World* (Stanford University Press, Stanford, 1976)

Röckel, D, W. Korn, and A. J. Kohn, *Manual of the living Conidae. Volume 1. Indo-Pacific Region* (Wiesbaden, 1995)

Rombouts, A., *Guidebook to Pecten Shells. Recent Pectinidae and Propeamussiidae of the World* (Universal Book Services/Dr. W. Backhuys, Leiden, 1991)

Slieker F. J. A. *Chitons of the World, An Illustrated Synopsis of Recent Polyplacophora* (L'Informatore Piceno, 2000)

Weaver, C. S. and J. E. du Pont, *The Living Volutes* (Delaware Museum of Natural History, Greenville, 1970)

NATIONAL AND INTERNATIONAL ORGANIZATIONS

American Malacological Society
http://www.malacological.org/index.php

Conchologists of America, Inc.
http://www.conchologistsofamerica.org/home/

Malacological Society of Australasia
www.malsocaus.org/

Malacological Society of London
http://www.malacsoc.org.uk/

Unitas Malacologica
http://www.unitasmalacologica.org

USEFUL WEB SITES

The following is a small sampling of web sites that are useful and informative. Most feature excellent photographs of shells and living animals that may be helpful for identification. Using a search engine to query a taxon or topic will bring many additional web sites for examination.

The Bailey-Matthews Shell Museum
http://www.shellmuseum.org/
A museum web site with excellent resources and links.

Conchology, Inc.
http://www.conchology.be/
A dealer's price list, but with many useful and informative links.

The Conus Biodiversity Web Site
http://biology.burke.washington.edu/conus/
A comprehensive web site dealing exclusively with the genus *Conus*. Includes a catalog of species and illustrations of many type specimens.

A Database of Western Atlantic Marine Mollusca
http://www.malacolog.org/
A database for research on the systematics, biogeography, and diversity of mollusks

Hardy's Internet Guide to Marine Gastropods
http://www.gastropods.com/
A partly illustrated catalog of recent marine gastropods.

Jacksonville Shells
http://www.jaxshells.org/
An excellent web site providing information about mollusks at many levels. Its emphasis is the fauna of northeastern Florida.

Let's Talk Seashells
http://www.letstalkseashells.com/
Informative lists and links, and a discussion forum.

OBIS Indo-Pacific Molluscan Database
http://clade.ansp.org/obis/find_mollusk.html
A database of marine mollusks in the tropical Indo-West Pacific.

Sea Slug Forum
http://www.seaslugforum.net
This site deals mostly with shell-less gastropods, but includes information on bubble shells and sea hares, which have internal shells.

645

EVOLUTIONARY CLASSIFICATION *of the* MOLLUSCA

The mollusca in phylogenetic (evolutionary) order. Families in **bold** typeface are those represented in ***The Book of Shells***

♦ Taxa that live exclusively freshwater

■ Taxa that are exclusively terrestrial

❍ Taxa that lack a calcified shell

❍ CLASS APLACOPHORA
  ❍ Subclass Chaetodermomorpha
    ❍ Family Chaetodermatidae
    ❍ Family Prochaetodermatidae
  ❍ Subclass Neomeniomorpha
    ❍ Family Dondersiidae
    ❍ Family Lepidomeniidae
    ❍ Family Neomeniidae
    ❍ Family Phyllomeniidae
    ❍ Family Pruvotinidae
    ❍ Family Proneomeniidae
    ❍ Family Epimeniidae

CLASS MONOPLACOPHORA
  Order Tryblidiida
    Family Laevipilinidae
    Family Micropilinidae
    Family Monoplacophoridae
    Family Neopilinidae

CLASS POLYPLACOPHORA
  Order Neoloricata
    Suborder Lepidopleurina
      Family Lepidopleuridae
      Family Hanleyidae
      Family Xylochitonidae
      Family Abyssochitonidae
    Suborder Choriplacina
      Family Choriplacidae
    Suborder Ischnochitonina
      Family Ischnochitonidae
      Family Schizochitonidae
      Family Mopaliidae
      Family Chitonidae
    Suborder Achantochitonina
      Family Achantochitonidae
      Family Cryptoplacidae

CLASS BIVALVIA
  Subclass Protobranchia
    Order Nuculoida
      Superfamily Nuculoidea
        Family Nuculidae
      Superfamily Pristiglomoidea
        Family Pristiglomidae
    Order Solemyoida
      Superfamily Solemyoidea
        Family Solemyidae
      Superfamily Manzanelloidea
        Family Manzanellidae
    Order Nuculanoida
      Superfamily Nuculanoidea
        Family Nuculanidae
        Family Malletiidae
        Family Neilonellidae
        Family Yoldiidae
        Family Siliculidae
        Family Phaseolidae
        Family Tindariidae
  Subclass Pteriomorphia
    Order Arcoida
      Superfamily Arcoidea
        Family Arcidae
        Family Cucullaeidae
        Family Noetiidae
        Family Glycymerididae
      Superfamily Limopsoidea
        Family Limopsidae
        Family Philobryidae
    Order Mytiloida
      Superfamily Mytiloidea
        Family Mytilidae
    Order Pterioida
      Superfamily Pterioidea
        Family Pteriidae
        Family Isognomonidae
        Family Malleidae
        Family Pulvinitidae
      Superfamily Ostreoidea
        Family Ostreidae
        Family Gryphaeidae
      Superfamily Pinnoidea
        Family Pinnidae
    Order Limoida
      Superfamily Limoidea
        Family Limidae
    Order Pectinoida
      Superfamily Pectinoidea

        Family Pectinidae
        Family Entoliidae
        Family Propeamussidae
        Family Spondylidae
      Superfamily Plicatuloidea
        Family Plicatulidae
      Superfamily Anomioidea
        Family Anomiidae
        Family Placunidae
      Superfamily Dimyoidea
        Family Dimyidae
  Subclass Paleoheterodonta
    Order Trigonioida
      Superfamily Trigonioidea
        Family Trigoniidae
    ♦ Order Unionoida
      ♦ Superfamily Unionoidea
        ♦ Family Unionidae
        ♦ Family Margaritiferidae
      ♦ Superfamily Etherioidea
        ♦ Family Etheriidae
        ♦ Family Hyriidae
        ♦ Family Mycetopodidae
        ♦ Family Iridinidae
  Subclass Heterodonta
    Order Carditoida
      Superfamily Crassatelloidea
        Family Crassatellidae
        Family Cardiniidae
        Family Astartidae
        Family Carditidae
        Family Condylocardiidae
    Order Anomalodesmata
        Family Pholadomyidae
        Family Parilimyidae
        Family Pandoridae
        Family Lyonsiidae
        Family Clavagellidae
        Family Laternulidae
        Family Periplomatidae
        Family Spheniopsidae
        Family Thraciidae
        Family Myochamidae
        Family Cleidothaeridae
    Order Septibranchia
        Family Verticordiidae
        Family Poromyidae
        Family Cuspidariidae
    Order Veneroida
      Superfamily Lucinoidea

Family Lucinidae
Family Ungulinidae
Family Thyasiridae
◆Family Cyrenoididae
Superfamily Chamoidea
Family Chamidae
Superfamily Galeommatoidea
Family Lasaeidae
Family Galeommatidae
Family Montacutidae
Superfamily Hiatelloidea
Family Hiatellidae
Superfamily Gastrochaenoidea
Family Gastrochaenidae
Superfamily Arcticoidea
Family Arcticidae
Family Trapeziidae
Superfamily Glossoidea
Family Glossidae
Family Kelliellidae
Family Vesicomyidae
Superfamily Cyamioidea
Family Cyamiidae
Family Sportellidae
◆Superfamily Sphaerioidea
  ◆Family Corbiculidae
  ◆Family Sphaeriidae
Superfamily Cardioidea
Family Cardiidae
Family Hemidonacidae
Superfamily Veneroidea
Family Veneridae
Family Glauconomidae
Family Neoleptonidae
Superfamily Tellinoidea
Family Tellinidae
Family Donacidae
Family Psammobiidae
Family Semelidae
Family Solecurtidae
Superfamily Solenoidea
Family Solenidae
Family Pharidae
Superfamily Mactroidea
Family Mactridae
Family Anatinellidae
Family Cardiliidae
Family Mesodesmatidae
Superfamily Dreissenoidea
Family Dreissenidae
Order Myoida
Superfamily Myoidea
Family Myidae
Family Corbulidae
Family Erodonidae
Superfamily Pholadoidea
Family Pholadidae
Family Teredinidae

CLASS SCAPHOPODA
Order Gadilida
Suborder Entalimorpha
Family Entalinidae
Suborder Gadilimorpha
Family Pulsellidae

Family Wemersoniellidae
Family Gadilidae
Order Dentaliida
Family Dentaliidae
Family Fustiariidae
Family Rhabdidae
Family Laevidentaliidae
Family Gadilinidae
Family Omniglyptidae

CLASS GASTROPODA
Subclass Eogastropoda
Order Patellogastropoda
Suborder Patellina
Superfamily Patelloidea
Family Patellidae
Suborder Nacellina
Superfamily Nacelloidea
Family Nacellidae
Superfamily Acmaeoidea
Family Acmaeidae
Family Lepetidae
Family Lottiidae
Subclass Orthogastropoda
Superorder Cocculiniformia
Superfamily Cocculinoidea
Family Cocculinidae
Family Bathysciadiidae
Superfamily Lepetelloidea
Family Lepetellidae
Family Addisoniidae
Family Bathyphytophilidae
Family Caymanabyssiidae
Family Pseudococculinidae
Family Osteopeltidae
Family Cocculinellidae
Family Choristellidae
Family Peltospiridae
Superorder Vetigastropoda
Superfamily Pleurotomarioidea
Family Pleurotomariidae
Family Scissurellidae
Family Haliotidae
Superfamily Fissurelloidea
Family Fissurellidae
Superfamily Turbinoidea
Family Turbinidae
Family Liotiidae
Family Phasianellidae
Superfamily Trochoidea
Family Trochidae
Family Calliostomatidae
Family Skeneidae
Family Pendromidae
Superfamily Seguenzioidea
Family Seguenziidae
Superorder Neritopsina
Superfamily Neritoidea
Family Neritopsidae
Family Neritidae
Family Phenacolepadidae
Family Titiscaniidae
Family Hydrocenidae
■Family Helicinidae
Superorder Caenogastropoda

■Order Architaenioglossa
  ■Superfamily Cyclophoroidea
    ■Family Cyclophoridae
    ■Family Pupinidae
    ■Family Diplommatinidae
  ◆Superfamily Ampullarioidea
    ◆Family Viviparidae
    ◆Family Ampullariidae
Order Sorbeoconcha
Family Abyssochrysidae
Superfamily Cerithioidea
Family Cerithiidae
Family Dialidae
Family Litiopidae
Family Turritellidae
Family Siliquariidae
Family Planaxidae
Family Potamididae
◆Family Thiaridae
Family Diastomatidae
Family Modulidae
Family Scaliolidae
Superfamily Campaniloidea
Family Campanilidae
Family Plesiotrochidae
Suborder Hypsogastropoda
Infraorder Littorinimorpha
Superfamily Littorinoidea
Family Littorinidae
Family Pickworthiidae
Family Skeneopsidae
Superfamily Cingulopsoidea
Family Cingulopsidae
Family Eatoniellidae
Family Rastodentidae
Superfamily Rissooidea
Family Barleeiidae
Family Anabathridae
Family Emblandidae
Family Rissoidae
Family Epigridae
Family Iravadiidae
Family Hydrobiidae
Family Pomatiopsidae
Family Assimineidae
Family Truncatellidae
Family Elachisinidae
Family Bithyniidae
Family Caecidae
Family Hydrococcidae
Family Tornidae
Family Stenothyridae
Superfamily Stromboidea
Family Strombidae
Family Aporrhaidae
Family Seraphsidae
Family Struthiolariidae
Superfamily Vanikoroidea
Family Hipponicidae
Family Vanikoridae
Family Haloceratidae
Superfamily Calyptraeoidea
Family Calyptraeidae
Superfamily Capuloidea
Family Capulidae

647

Superfamily Xenophoroidea
  Family Xenophoridae
Superfamily Vermetoidea
  Family Vermetidae
Superfamily Cypraeoidea
  Family Cypraeidae
  Family Ovulidae
Superfamily Vellutinoidea
  Family Triviidae
  Family Velutinidae
Superfamily Naticoidea
  Family Naticidae
Superfamily Tonnoidea
  Family Bursidae
  Family Cassidae
  Family Ficidae
  Family Laubierinidae
  Family Personidae
  Family Pisanianuridae
  Family Ranellidae
  Family Tonnidae
Superfamily Carinoidea
  Family Atlantidae
  Family Carinariidae
  Family Pterotracheidae
Infraorder Ptenoglossa
Superfamily Triphoroidea
  Family Triphoridae
  Family Cerithiopsidae
Superfamily Janthinoidea
  Family Janthinidae
  Family Epitoniidae
  Family Aclididae
Superfamily Eulimidea
  Family Eulimidae
Infraorder Neogastropoda
Superfamily Muricoidea
  Family Buccinidae
  Family Colubrariidae
  Family Columbellidae
  Family Nassariidae
  Family Melongenidae
  Family Fasciolariidae
  Family Muricidae
  Family Coralliophilidae
  Family Turbinellidae
  Family Ptychatractidae
  Family Volutidae
  Family Olividae
  Family Olivellidae
  Family Pseudolividae
  Family Strepsiduridae
  Family Babyloniidae
  Family Harpidae
  Family Cystiscidae
  Family Marginellidae
  Family Mitridae
  Family Pleioptygmatidae
  Family Volutomitridae
  Family Costellariidae
Superfamily Cancellarioidea
  Family Cancellariidae
Superfamily Conoidea
  Family Clavatulidae
  Family Drillidae

  Family Pseudomelatomidae
  Family Turridae
  Family Conidae
  Family Terebridae
Superorder Heterobranchia
  Superfamily Valvatoidea
    Family Cornirostridae
    Family Orbitestellidae
    Family Xylodisculidae
  Superfamily Architectonicoidea
    Family Mathildidae
    Family Architectonicidae
  Superfamily Rissoelloidea
    Family Rissoellidae
  Superfamily Omalogyroidea
    Family Omalogyridae
  Superfamily Pyramidelloidea
    Family Pyramidellidae
    Family Amathinidae
    Family Cimidae
    Family Donaldinidae
    Family Ebalidae
Opistobranchia
  Order Cephalaspidea
    Superfamily Acteonoidea
      Family Acteonidae
      Family Bullinidae
      Family Aplustridae
    Superfamily Ringiculoidea
      Family Ringiculidae
    Superfamily Cylindrobulloidea
      Family Cylindrobullidae
    Superfamily Diaphanoidea
      Family Notodiaphanidae
      Family Diaphanidae
    Superfamily Philinoidea
      Family Cylichnidae
      Family Retusidae
      Family Philinidae
      Family Philinoglossidae
      Family Aglajidae
      Family Gastropteridae
    Superfamily Haminoeoidea
      Family Haminoeidae
      Family Bullactidae
      Family Smaragdinellidae
    Superfamily Bulloidea
      Family Bullidae
    Superfamily Runcinoidea
      Family Runcinidae
      Family Ilbiidae
  Order Acochlidea
    Superfamily Achochlidioidea
      Family Acochlidiidae
      Family Hedylopsidae
    Superfamily Microhedyloidea
      Family Asperspinidae
      Family Microhedylidae
      Family Ganatidae
  Order Rhodopemorpha
    Family Rhodopidae
  Order Sacoglossa
    Superfamily Oxynooidea
      Family Volvatellidae
      Family Oxynoidae

      Family Juliidae
    Superfamily Elysioidea
      Family Placobranchidae
      Family Elysiidae
      Family Boselliidae
      Family Gascoignellidae
      Family Platyhedylidae
    Superfamily Limapontioidea
      Family Caliphyllidae
      Family Costasiellidae
      Family Hermaeidae
      Family Limapontiidae
  Order Anaspidea
    Superfamily Akeroidea
      Family Akeridae
    Superfamily Aplysioidea
      Family Aplysiidae
  Order Notaspidea
    Superfamily Tylodinoidea
      Family Tylodinidae
      Family Umbraculidae
    Superfamily Pleurobranchoidea
      Family Pleurobranchidae
  Order Thecosomata
    Family Limacinidae
    Family Cavoliniidae
    Family Peraclidae
    Family Cymbuliidae
    Family Desmopteridae
  Order Gymnosomata
    Suborder Gymnosomata
      Family Pneumodermatidae
      Family Notobranchaeidae
      Family Cliopsidae
      Family Clionidae
  Order Gymnoptera
    Family Hydromylidae
  ○Order Nudibranchia
    ○Suborder Doridina
    ○Superfamily Anadoridoidea
      ○Family Corambidae
      ○Family Goniodorididae
      ○Family Onchidoridae
      ○Family Polyceridae
      ○Family Gymnodorididae
      ○Family Aegiretidae
      ○Family Vayssiereidae
    ○Superfamily Eudoridoidea
      ○Family Hexabranchidae
      ○Family Dorididae
      ○Family Chromodorididae
      ○Family Dendrodorididae
      ○Family Phyllidiidae
    ○Suborder Dendronotina
      ○Family Tritoniidae
      ○Family Bornellidae
      ○Family Marianinidae
      ○Family Hancockiidae
      ○Family Dotidae
      ○Family Scyllaeidae
      ○Family Phylliroidae
      ○Family Lomanotidae
    ○Suborder Arminina
      ○Family Arminidae
      ○Family Doridomorphidae

- ○ Family Charcotiidae
- ○ Family Madrellidae
- ○ Family Zephyrinidae
- ○ Family Pinufiidae
- ○ Suborder Aeolidina
  - ○ Family Flabellinidae
  - ○ Family Eubranchidae
  - ○ Family Aeolidiidae
  - ○ Family Glaucidae
  - ○ Family Embletoniidae
  - ○ Family Tergipedidae
  - ○ Family Fionidae

Pulmonata
- ■ Order Systellommatophora
  - ■ Superfamily Otinoidea
    - ■ Family Smeagolidae
  - ■ Superfamily Onchidioidea
    - ■ Family Onchidiidae
  - ■ Superfamily Rathousioidea
    - ■ Family Rathousiidae
    - ■ Family Veronicellidae
- Order Basommatophora
  - ■ Superfamily Amphiboloidea
    - ■ Family Amphibolidae
  - Superfamily Siphonarioidea
    - Family Siphonariidae
  - ■ Superfamily Lymnaeoidea
    - ■ Family Lymnaeidae
    - ■ Family Ancylidae
    - ■ Family Planorbidae
    - ■ Family Physidae
  - ■ Superfamily Glaucidorboidea
    - ■ Family Glacidorbidae
- Order Eupulmonata
  - Suborder Actophila
  - Superfamily Ellobioidea
    - Family Ellobiidae
  - ■ Suborder Trimusculiformes
  - ■ Superfamily Trimusculoidea
    - ■ Family Trimusculidae
  - ■ Suborder Stylommatophora
  - ■ Infraorder Orthurethra
  - ■ Superfamily Achatinelloidea
    - ■ Family Achatinellidae
  - ■ Superfamily Cionelloidea
    - ■ Family Cionellidae
  - ■ Superfamily Pupilloidea
    - ■ Family Pupillidae
    - ■ Family Pleurodiscidae
    - ■ Family Vallonidae
  - ■ Superfamily Partuloidea
    - ■ Family Enidae
    - Family Partulidae
  - ■ Infraorder Sigmurethra
  - ■ Superfamily Achatinoidea
    - ■ Family Ferussaciidae
    - ■ Family Subulinidae
    - ■ Family Megaspiridae
    - ■ Family Achatinidae
  - ■ Superfamily Streptaxoidea
    - ■ Family Streptaxidae
  - ■ Superfamily Rhytidoidea
    - ■ Family Rhytididae
  - ■ Superfamily Acavoidea
    - ■ Family Caryodidae

- ■ Superfamily Bulimuloidea
  - ■ Family Bulimulidae
- ■ Superfamily Arionoidea
  - ■ Family Punctidae
  - ■ Family Charopidae
  - ■ Family Helicodiscidae
  - ■ Family Arionidae
- ■ Superfamily Limacoidea
  - ■ Family Limacidae
  - ■ Family Milacidae
  - ■ Family Zonitidae
  - ■ Family Trochomorphidae
  - ■ Family Helicarionidae
  - ■ Family Cystopeltidae
  - ■ Family Testacellidae
- ■ Superfamily Succineoidea
  - ■ Family Succineidae
  - ■ Family Athoracophoridae
- ■ Superfamily Polygyroidea
  - ■ Family Coriidae
- ■ Superfamily Camaenoidea
  - ■ Family Camaenidae
- ■ Superfamily Helicoidea
  - ■ Family Helicidae
  - ■ Family Bradybaenidae

CLASS CEPHALOPODA
Subclass Nautiloidea
  Superfamily Nautiloidea
    Family Nautilidae
Subclass Coleoidea
  Order Sepioidea
    Family Spirulidae
    Family Sepiidae
    - ○ Family Sepiadariidae
    - ○ Family Sepiolidae
    - ○ Family Idiosepiidae
  - ○ Order Teuthoidea
    - ○ Suborder Myopsida
      - ○ Family Pickfordiateuthidae
      - ○ Family Loliginidae
    - ○ Suborder Oegopsida
      - ○ Family Lycoteuthidae
      - ○ Family Enoploteuthidae
      - ○ Family Octopoteuthidae
      - ○ Family Onychoteuthidae
      - ○ Family Walvisteuthidae
      - ○ Family Cycloteuthidae
      - ○ Family Gonatidae
      - ○ Family Psychoteuthidae
      - ○ Family Lepidoteuthidae
      - ○ Family Architeuthidae
      - ○ Family Histioteuthidae
      - ○ Family Neoteuthidae
      - ○ Family Ctenopterygidae
      - ○ Family Brachioteuthidae
      - ○ Family Batoteutidae
      - ○ Family Ommastrephidae
      - ○ Family Thysanoteuthidae
      - ○ Family Chiroteuthidae
      - ○ Family Promachoteuthidae
      - ○ Family Grimalditeuthidae
      - ○ Family Joubiniteuthidae
      - ○ Family Cranchiidae
  - ○ Order Vampyromorpha

- ○ Family Vampyroteuthidae
- ○ Order Octopoda
  - ○ Suborder Cirrata
    - ○ Family Cirroteuthidae
    - ○ Family Stauroteuthidae
    - ○ Family Opisthoteuthidae
  - ○ Suborder Incirrata
    - ○ Family Bolitaenidae
    - ○ Family Amphitretidae
    - ○ Family Idioctopodidae
    - ○ Family Vitreledonellidae
    - ○ Family Octopodidae
    - ○ Family Tremoctopodidae
    - ○ Family Ocythoidae
    - **Family Argonautidae**
    - ○ Family Alloposidae

649

INDEX *of* SPECIES BY COMMON NAME

A

Adams' miniature cerith 388
Admete viridula *see* Greenish admete
Admiral cone 590, 592
Alabaster murex 464
Alaska miter-volute 549, 550
Almond ark 45
American crown conch *see* Common crown conch
American pelican's foot 304, 305
American thorny oyster 86, 87
Amethyst gem clam 131
Ampulle bubble 625
Anatoma crispata *see* Crispate scissurelle
Angel wing 134, 135, 164, 165
Angled olive 499, 509, 511
Angular triton 379, 381
Angular volute 494, 496
Antillean crassatella 92
Arctic saxicave 114, 115
Arctic wedge clam 161
Arcuate pectinodont 185, 186
Armored hammer vase 471, 472
Asian cup-and-saucer 311
Asian flame volute 491
Atlantic carrier shell 318, 319, 320
Atlantic deepsea scallop 81, 84
Atlantic distorsio 374, 375
Atlantic kitten's paw 88
Atlantic nut clam 38, 39
Atlantic ribbed mussel 51, 53, 55
Atlantic Spengler clam 116
Atlantic surf clam 157, 158, 160
Atlantic turkey wing 43, 44, 46
auger-like miter *see* tessellate miter
Australian awning clam 40
Australian brooch clam 91
Australian cardita 94
Australian trumpet 479

B

Baby's ear moon 357
Baggy pen shell 65
Baird's bathybembix 225
Ballot's moon scallop 80, 81
Banded creeper/vertagus *see* Striped cerith
Banded tun 382, 383
Barnacle rock shell 189, 451, 452
Barren carrier shell 318, 321
Barthelow's latiromitra 481
Bartsch's false tun 368
Basket lucina *see* Common basket lucina
Bat volute 486, 490, 497
Beaded periwinkle 271, 274
Bear ancilla 503, 511
Bear paw clam 127, 130
Beautiful caecum 285
Beck's volute 496
Bednall's volute 490
Belcher's latirus 439, 441
Bell clapper 270
Bifrons scallop *see* Queen scallop
Black abalone 206
Black atlantic planaxis 262
Black coriocella 347
Black-lined limpet 8, 181, 182
Black mouth moon 355
Bleeding-tooth nerite 239, 242
Blistered marginella *see* Bubble marginella
Bloodstained dove shell 419
Blotchy strombina 421
Blotchy turrid 574
Blue mussel *see* Common blue mussel

Boat ear moon 356
Bonelike miter 539
Bonin Island limpet 181, 183, 184
Broad yoldia 42
Brown-lined ammonicera 612, 613
Bubble akera 621, 626
Bubble marginella 537
Bull conch 295, 296
Bullmouth helmet 368, 369, 371
Butterfly cone 588, 598
Butterfly miter 544
Butterfly moon 350
Button top *see* Common button top

C

Cadulus simillimus 170
Cake nassa 427
Camp pitar venus 133
Cancellate nutmeg 563
Cancellate vanikoro 310
Carinate false cerith 251
Carnelian olive 498, 501
Carribean carrier shell 319, 320
Carribean piddock clam 95
Caribbean risso 611
Caribbean vase 475, 476, 479
Casket nassa *see* Cake nassa
Cat's ear pyram 614
Caymanabyssia spina 193
Chambered nautilus 13, 632, 633
Channeled duck clam 157, 159
Chestnut dwarf triton 415
Chestnut frog shell 361, 362, 363
Chick-pea cowrie 325
Chiragra spider conch 300, 302
Cingulate ancilla 502, 508, 510
Circular slipper 314
Clasping stenochiton 31
Clench's helmet 371
Cock's comb oyster 61, 62
Coffee bean trivia 344, 345, 346
Colbeck's scallop 77, 79
Collar auger 602, 605
Colorful Atlantic moon 359
Commerical top 230
Common basket lucina 108, 109, 110
Common blue mussel 52, 54
Common button top 221
Common crown conch 432
Common distorsio 374, 375
Common dove shell 420
Common egg cowrie 342, 343
Common European limpet 176, 179
Common fig shell 372, 373
Common janthina 389, 390
Common music volute 484, 494
Common northern buccinum 407, 410
Common pelican's foot 304, 306
Common periwinkle 277
Common prickly winkle *see* Beaded periwinkle
Common screw shell *see* Great screw shell
Common turtle limpet 183, 184
Common watering pot 98
Common wentletrap 391, 392, 393, 396
Cone miter 541
Conical moon 352
Cooper's nutmeg 561, 563, 565
Coquina donax 145, 146
Costate miter 554
Costate tuskshell 172
Covered modulus *see* Tectum modulus
Crested turrid 576
Crispate scissurelle 198, 199

Crown conch *see* Common crown conch
Crusty nautilus 633
Crystalline thyca 398, 399
Cuming's afer 408, 409
Cyclopecten pernomus *see* Pernomus glass scallop
Cyrtulus spindle 440, 442

D

Deepsea volute 489
Deer antler murex 456, 459
Deer cowrie 338
Dennison's morum 522
Dentate top 231
Diadem latiaxis 467, 468
Diala albugo *see* White spotted diala
Dimidiate auger *see* Divided auger
Diphos sanguin 147, 148
Distorsio *see* Common distorsio
Distorted eulima 398, 400
Distorted triumphis 515
Divaricate nut clam 38, 39
Divided auger 603, 606
Dog conch 288
Dove shell *see* Common dove shell
Dubius spindle shell 436
Duplicate auger 600, 603
Du Petit's spindle 446
Dwarf Atlantic planaxis *see* Black Atlantic planaxis
Dwarf Pacific planaxis *see* Ribbed planaxis

E

Eastern American oyster 62, 63
Eastern white slipper 315
Edgar's pleurotomella 587
Egg cowrie *see* Common egg cowrie
Elated onion shell 516
Elegant abalone 203
Elegant marginella 535
Elegant star turrid 577
Elephant's snout volute 493
Elephant tusk 171, 172, 173
Elongate giant clam 129, 130
Elongate janthina 389, 390
Elongate tusk 173
Emerald nerite 237
Episcopal miter 547
Ethiopian macron *see* Ribbed macron
European giant lima 70
European limpet *see* Common European limpet
European panopea 114, 115
Eyed cowrie 331

F

False angel wing 134, 135
False cup-and-saucer 309
Fig cone 589, 590
Fig shell *see* Common fig shell
Fine-sculptured frog shell 361
First pagoda shell 473, 474
Fischer's gaza 224, 225
Flaming turrid 572
Flamingo tongue 340
Flat skeneopsis 280
Flattended stomatella 221, 223
Florida crown conch *see* Common crown conch
Florida horse conch 444, 447
Florida miter 545
Fluted giant clam 128, 129
Fly marginella 531

Foliated tellin 141, 144
Fool's cap 317
Four-handed turrid 575, 576
Fragile barleysnail 281, 283
Friend's cowrie 334
Frilled frog shell 362, 363
Frond oyster 61
Fulton's cowrie 330

G

Gaping ancilla 502, 506
Garter cone 583, 588, 598
Geography cone 597
German's volute 487
Gevers's trophon 448, 453
Ghastly miter 552
Giant American bittersweet 48, 49
Giant baler 497
Giant clam 68, 124, 128, 130
Giant cockle 125, 126
Giant eastern murex 463, 465
Giant forerria 448, 460
Giant hairy melongena 433
Giant knobbed cerith 255
Giant Mexican limpet 178, 180
Giant owl limpet 188, 190
Giant razor shell 155, 156
Giant sundial 610
Gilchrist's volute 489, 495
Girdled horn shell 264
Girdled marginella 527, 528, 529
Girgylla star shell 213, 217
Gladiator nutmeg 562, 564
Glans nassa 428
Globose nassa 424
Glory-of-the-sea cone 590
Glossy dove shell 417
Goliath conch 301, 303
Golden amauropsis 349
Golden cowrie 331, 335, 336
Golden moon 354
Goodall's cowrie 324, 329
Graceful fig shell 372, 373
Grammatus whelk 413
Grand ark 45, 46
Grand typhis 450
Granular spindle 438
Gray bonnet 365, 367
Great keyhole limpet 209, 211
Great ribbed cockle 122, 125
Great screw shell 258, 259
Great spotted cowrie 333
Green abalone 201, 206, 207
Green jewel top 226
Green mussel 52, 54
Greenish admete 557
Greenland wentletrap 391
Green-lined paperbubble 619, 620
Grinning tun 383, 384
Grooved razor clam 154
Guilding's lyria 482
Gumboot chiton 35

H

Heavy bonnet 365
Helen's miter 548
Helmet vase 475, 476
Hieroglyphic cone 583
Honeycomb oyster 64
Hooded ark 47
Hooked mussel 51, 53
Horned helmet 369, 370, 371
Horned paper nautilus 636
Horny nerite 239
Humpback conch 287, 290
Humped turrid 573

I

Imperial cone 592
Imperial delphinula 213, 214
Imperial harp 525, 526
Imperial turrid 566
Incised moon 358
Indian babylon 517, 519
Indian shank 477, 478, 479
Indian turrid 581

Inflated olive 498, 505
Italian keyhole limpet 210, 212
Ittibittium parcum see Poor ittibittium

J

Jacna abalone 200
Janthina see Common janthina
Japanese wonder shell 593
Java turrid 568, 569
Jenner's false cowrie 339, 342
Jourdan's turban 216, 218, 219
Julia's pagoda shell 473, 474
Junonia 492

K

Kelp limpet 179
Kelp scallop 71
Kerguelen Island eatoniella 281
King's crown conch see Common crown
    conch
Kiosk rock shell 451
Knobbed nutmeg 559, 562, 565
Knobby scallop 72, 74
Knobby snail see Tectum modulus

L

Laciniate conch 292, 295
Lamarck's glassy carinaria 386
Lance auger 603
Large ostrich-foot 308
Large perverse turrid 578
Laurent's moon scallop 76, 80, 82
Lazarus jewel box 112, 113
Leather donax 145, 146
Lesser girdled triton 377, 378
Lettered olive 507
Lewis' moon 360
Lienard's ancilla 502, 503, 506
Lightning whelk 412, 414
Linda's miter-volute 549
Lion's paw 75, 83
Lister's conch 293, 297
Lister's keyhole limpet 209, 211
Little dog turrid 570, 571
Long-ribbed limpet 177, 178, 180
Lumpy morum 521
Lyre-formed lyria 485, 488

M

Maculated dwarf triton 416
Magnificent calypto clam 120
Magnificent wentletrap 394, 397
Magpie shell see West Indian top
Mangrove jingle shell 89
Mangrove periwinkle 275
Map cowrie 332, 337
Maple leaf triton 376
Marble cone 595, 596
Marlinspike auger 601, 607
Matchless cone 585, 586
Mawe's latiaxis 469
Maximum nerite 246
Mediterranean pelican's foot 304, 305, 306
Melanioid abyssal snail 248
Melanioid diastoma 268
Miller's nutmeg 558, 560
Miniature moon snail 348
Miniature triton trumpet 403
Minor harp 523, 524
Miraculous torellia 316
Miter-shaped nutmeg 557, 559
Modest triphora 387
Modest worm snail 260, 261
Moltke's heart clam 118, 119
Money cowrie 326, 327
Morse code cone see Garter cone
Moskalev's macleaniella 191
Mud creeper 265, 266, 267
Mud tube clam 68, 166, 167
Music volute see Common music volute
Mytiline limpet 182

N

Naval shipworm 166, 167
Navicula paperbubble 622, 623

Nerite mud snail 423
Netted peristernia 437
Netted tibia 286, 297, 299
network beak shell
New England nassa 426
New England neptune 18, 407
Nicobar spindle 443
Nifat turrid 567
Noble pen shell 66, 68
Noble wentletrap 395
Norris's top 227, 228
Northern buccinum see Common
    northern buccinum
Northern quahog 117, 136, 139
North's long whelk 405
Nucleus cowrie 326
Nuttallochiton mirandus 31, 32

O

Ocean quahog 117
Orange auger see Divided auger
Orange marginella 532
Orange-banded marginella 530
Orbigny's cone 591, 595
Oxheart clam 118, 119

P

pachydermia laevis
Pacific asaphis 148, 149
Pacific half cockle 123
Pacific partridge tun 382, 383, 384
Pacific sugar limpet 188, 189
Pacific tiger lucine 106, 108, 110
Pagoda cerith 258
Pagoda prickly winkle 278
Panama harp 524, 526
Paper argonaut see Paper nautilus
Paper nautilus 636, 637
Paper piddock 164
Paradoxical blind limpet 192
Paz's murex 449, 454
Pea strigilla 141
Pear melongena 431
Pear triton 378, 380
Pearl oyster 56, 57
Pearly lyonsia 97
Pedum oyster 73, 77
Pelican's foot see Common pelican's foot
Penguin wing oyster 56, 57
Pennsylvania lucina 107
Pepper conch see West Indian chank
Perforate babylon 520
Periwinkle see Common periwinkle
Pernomus glass scallop 85
Peron's sea buttorfly 385
Portusus cone 585, 591
Peruvian hat 312
Peruvian olive 501, 505, 509
Petit's marginella 534
Pfeffer's whelk 404
Pharaoh cuttlefish 635
Pharaoh's horn see Striped cerith
Pimpled nassa 430
Pink-mouth murex 463
Plesiotriton vivus 561, 565
Pleurotomella edgariana see Edgar's
    pleurotomella
Plicate conch 288, 289
Plicate scallop 75, 78
Plum marginella see Orange marginella
Pointed ancilla 508, 510
Polished nerite 243
Polished nut clam 41
Ponderous dosinia 138, 140
Poor ittibittium 249
Porcupine castor bean see Strawberry
    drupe
Precious wentletrap 396
Pringle's marginella 538
Propellor ark 43, 44
prostitute venus
Pubescent thracia 101
Pulley iravadia 284
Punctate cerith see Striped cerith
Purple dwarf olive 513
Purple raphitoma 584, 593
Purplish American semele 150, 151

651

Q

Queen conch 7, 298, 301, 303
Queen marginella 533, 534
Queen scallop 74, 78
Queen tegula 227
Queen vexillum 554, 556
Quoy's slit shell 196, 197

R

Radial sundial see Sharp-edged sundial
Radish murex 457, 458
Radula nerite 236
Ramose murex 456, 466
Rapa snail 470
Raphitoma purpurea see Purple
    raphitoma
Recluse vitrinella 282
Recurved strombina 422
Reddish callista 138, 140
Red-ringed frog shell 364
Red-toothed corbula 163
Regal thorny oyster 86, 87
Reticulate miter 541, 542
Reticulate venus 132, 136
Rhino cerith 256
Ribbed macron 406
Ribbed nerite 240
Ribbed planaxis 263
Ribbed spindle 445
Ribbon bullia 429
Ribbon cerith 253
Rice worm snail 322, 323
Ridged turrid 579
Ringed blind limpet 187
Ringed miter 541
Rooster-tail conch 296, 298
Rose drupe see Strawberry drupe
Rose-branch murex 459, 464
Rostrate cuspidaria 105
Rosy diplodon 111
Rota scissurelle 198, 199
Rough lima 69, 70
Rough-ribbed nerite 245
Royal paper bubble 619, 620
Rufula olive 500
Rugged vitularia 450, 453
Rugose miter 552, 555
Rugose pen shell 9, 65, 66
Rumphius' slit shell 196, 197

S

Saddle tree oyster 58
Saffron miter 551, 553
Saint James scallop 82, 84
Sallé's auger 599
Samar conch 287
Sandbeach auger 599, 601
sansonia tuberculata
Saw-toothed pen shell 67, 68
Say's top 232
Scaly Pacific scallop 72, 74
Scissurella rota see Rota scissurelle
Scorched auger 600
Scorpion murex 454
Scraper solecurtus 153
Seaweed limpet 188, 189
Seba's spider conch 302
Select maurea 234
Serpulorbis oryzata see Rice worm snail
Sharp-edged sundial 609
Sharp-ribbed verticord 103
Shinbone tibia 286, 297, 299
Shiny marginella 536
Short coral shell 468
Short shield limpet 208, 212
Shoulderblade sea cat 627
Shuttlecock volva 341, 343
Sieve cowrie 328
Silver conch 292
Sinum incisum see Incised moon
Sitka periwinkle 273
Slender miter 543
Slit worm snail 260, 261
Small false donax 147, 149
Smaragdinella calyculata 624
Smooth European chiton 30, 32

Smooth exilia 480
Smooth pachyderm shell 195
Snake miter see Reticulate miter
Snipe's bill 461, 462
Snout otter clam 159
Snowflake marginella 527, 529
Soft shell clam 162
Solander's trivia 345, 346
Solid pupa 618
Solid semele 150, 151
South African turban 218
Southern zemira 514
Specious scallop 73
Spengler's lantern clam 99
Spiny bonnet 366, 367
Spiny cup-and-saucer 313
Spiny hammer vase 471, 472
Spiny paper cockle 121, 126
Spiny venus 132, 133, 137
Spiral babylon 518, 519
Spiral melongena 434, 435
Spiral tudicla 408, 409
Spirula 25, 634
Splendid olive 500, 504
Sponge finger oyster 59
Spoon limpet 176, 177, 178
Spotted babylon 517
Spotted marginella 529
Staircase abalone 202, 204
Stocky cerith 250
Stolid cowrie 324, 329
Stout American tagelus 152, 153
Strawberry conch 290, 291
Strawberry drupe 455
Strawberry top 222, 223
Striate cone 594
Striate myadora 102
Striated turrid 570
Striped acteon 617
Striped cerith 253, 254
Striped conch 289, 291, 293, 297
Striped engina 401
Sulcate astarte 93
Sulcate swamp cerith 264, 265, 267
Sunburst carrier shell 319, 321
Sunburst star shell 217
Sunrise tellin 142, 143
Sunset siliqua 155, 156
Supreme turrid 580, 582
Swollen nassa 425
Swollen pheasant 220

T

Tabled neptune 411
Tajima's limopsis 50
Tapestry turban 216, 219
Tectum modulus 269
Telescope snail 266
Telescoped dove shell 418
Tent olive 504, 511
Terebellum conch 307
Terebra pyram 615
Tessellate miter 546
Tessellate olive 498, 499, 507
Tessellate pyram 615, 616
Textile nerite 244
Textile venus 134
Thick-tail false fusus 435
Thorn latirus 440, 441
Three-rowed carenzia 235
Three-toothed cavoline 629
Tiger cowrie 332, 337
Tiger maurea 232, 233
Tiger moon 353
Toothed cross-hatched lucina 106, 107
Toothless lucina 109
Tower latirus 439
Tower screw shell see Great screw shell
Triangular nutmeg 560
Tricolor niso 399, 400
Trifle peasiella 271
Triseriate auger 602, 605
Triumphis distorta see Distorted
    triumphis
True heart cockle 121, 123, 124
True spiny jewel box 112, 113
True tulip 444, 447

Trumpet triton 11, 379, 381
Trunculus murex 449, 457
Tryon's scallop 76, 79
Tuberculate sansonia 279
Tuberculate emarginula 208, 210
Turned poromya 104
Turned turrid 568, 569
Turnip whelk 412, 414
Turtle limpet see Common turtle limpet
Turton's wentletrap 392, 393

U

Umbilical ovula 339, 341
Umbrella shell 628
Unequal bittersweet 48, 49
Unequal pandora 96

V

Variable abalone 201, 203, 205
Variable worm snail 322, 323
Varicose wentletrap 394, 397
Variegated sundial 608
Ventral harp 523, 525
Venus comb murex 461, 462
Venus sugar limpet 247
Vexillate volute 485
Violet mactra 158, 160
Violet moon 351
Violet spider conch 294
Volute erato 344
Volute-shaped dwarf olive 512, 513

W

Wandering triton 377, 380
Watering pot see Common watering pot
Waved goblet 402, 515
Wedding cake venus 131, 137
Wentletrap see Common wentletrap
West Indian chank 477, 478
West Indian chiton 34
West Indian top 226, 229
Western spoon clam 100
Western strawberry cockle 122
Whirling abalone 202, 204
White-cap limpet 185, 186
White cerith see Striped cerith
White crested tellin 142, 144
White giant turrid 577, 580
White hammer oyster 59, 60
White Pacific atys 622, 623
White spotted diala 257
White-toothed cowrie 331, 335, 336
Widest neritina 241
Winding stair shell see Spiral melongena
Windowpane oyster 89, 90
Woody canoebubble 621
Woody chiton 33

Y

Yellow cerith 252, 255
Yoka star turban 9, 215, 217
Yoldia 42
Young's false cocculina 194

Z

Zebra nerite 238
Zebra periwinkle 276
Zebra volute 483
Zebrina omalogyra 612
Ziervogel's miter 551
Zigzag periwinkle 272

INDEX *of* SPECIES
BY SCIENTIFIC NAME

Page numbers of related species
references are in *italics*

A

Abyssochrysos melanioides 248
Acanthosepion pharaonis see Sepia
   pharaonis
Acesta excavata 70
Acila divaricata 38, 39
Acmaea mitra 185, 186
Acteon virgatus 617
Adamussium colbecki 77, 79
Addisonia excentrica 192
Adelomelon beckii 496
Admete viridula 557
Aequipecten glyptus 76, 79
Afer cumingii 408, 409
Aforia circinata 579
Afrivoluta pringlei 538
Agaronia acuminata 508, 510
Akera bullata 621, 626
Amaea magnifica 394, 397
Amauropsis aureolutea 349
Americardia biangulata 122
Ammonicera lineofuscata 612, 613
Amoria zebra 483
Amusium balloti 80, 81
 Amusium laurenti see Euvola laurenti
Anadara grandis 45, 46
Anatoma crispata 198, 199
Ancilla alba see Bullia livida
Ancilla aperta 502, 506
Ancilla lienardi 502, 503, 506
Ancillista cingulata 502, 508, 510
Ancistrolepis grammatus 413
Angaria delphinus melanacantha 213, 214
Anodontia edentula 109
Antalis longitrorsa 173
Antiplanes perversa 578
Aporrhais occidentalis 304, 305
Aporrhais pespelecani 304, 306
 Aporrhais serresianus 304, 305, 306
Arca zebra 43, 44, 46
Architectonica maxima 610
Arcinella arcinella 112, 113
Arctica islandica 109
Argonauta argo 636, 637
Argonauta cornuta 636
Asaphis violascens 148, 149
Astarte sulcata 93
Astraea heliotropium 217
Atlanta peroni 385
Atrina serrata 67, 68
Atys naucum 622, 623
Austroginella muscaria 531

B

Babelomurex diadema 467, 468
Babylonia papillaris 517
Babylonia perforata 520
Babylonia spirata 518, 519
Babylonia zeylanica 517, 519
Barbatia amygdalumtostum 45
Barleeia subtenuis 281, 283
Bassina disjecta 131, 137
Bathybembix bairdii 225
Biplex perca 376
Bolma girgylla 213, 217
Brechites penis 98
Buccinum undatum 407, 410
Bufonaria bufo 361, 362, 363
Bufonaria foliata 362, 363
Bulla ampulla 625

Bullata bullata 537
Bullia livida 429
Bursa ranelloides tenuisculpta 361
Busycon coarctatum 412, 414
Busycon perversum 412, 414
Busycotypus spiratus 372

C

Cadulus simillimus 170
Caecum pulchellum 285
Calliostoma sayanum 232
Callista erycina 138, 140
Callochiton septemvalvis 30, 32
Calpurnus verrucosus 339, 341
Calyptogena magnifica 120
Calyptraea extinctorium 311
Campanile symbolicum 270
Cancellaria cancellata 563
Cancellaria cooperi 561, 563, 565
Cancellaria mitriformis 557, 559
Cancellaria nodulifera 559, 562, 565
Cantharus undosus 402, 515
Capulus ungaricus 316, 317
Cardita crassicosta 94
Cardium costatum 122, 125
Carenzia trispinosa 235
Caribachlamys imbricata see
   Caribachlamys pellucens
Caribachlamys pellucens 72, 74
Carinaria lamarcki 386
Casmaria ponderosa 365
Cassis cornuta 369, 370, 371
Cassis madagascariensis spinella 371
Cavolinia tridentata 629
Caymanabyssia spina 193
Cellana mazatlandica 181, 183, 184
Cellana nigrolineata 8, 181, 182
Cellana testudinaria 183, 184
Cenchritis muricatus 271, 274
Cerithidea cingulata 264
Cerithium citrinum 252, 255
Cerithium litteratum 260
Cerithium nodulosum 255
Chama lazarus 112, 113
Charonia tritonis 11, 379, 381
Cheilea equestris 189, 309
Chicoreus cervicornis 456, 459
Chicoreus palmarosae 459, 464
Chicoreus ramosus 456, 466
Chiton tuberculatus 34
Chlamys squamata 72, 74
Cirsotrema varicosum 394, 397
Cittarium pica 226, 229
Clanculus puniceus 222, 223
Clavatula imperialis 566
Clavocerithium taeniatum 253
Clavus canalicularis 570, 571
Clavus flammulatus 572
Cochlespira elegans 577
Codakia tigerina 106, 108, 110
Colubraria castanea 415
Colubraria muricata 416
Columbarium pagoda 473, 474
Columbella haemastoma 419
Columbella mercatoria 420
Coluzea juliae 473, 474
Concholepas concholepas 189, 451, 452
Conopleura striata 570
Conuber conicus 352
Conus ammiralis 590, 592
Conus cedonulli 585, 586
Conus figulinus 589, 590
Conus genuanus 583, 588, 598

Conus geographus 589, 597
Conus gloriamaris 596
Conus hieroglyphus 583
Conus imperialis 592
Conus marmoreus 595, 596
Conus orbignyi 591, 595
Conus pertusus 585, 591
Conus pulcher 588, 598
Conus striatus 594
Coralliophila abbreviata 468
Corbula erythrodon 163
Corculum cardissa 121, 123, 124
Coriocella nigra 347
Crassostraea virginica 62, 63
Crepidula plana 315
Crucibulum spinosum 313
Cryptobranchia concentrica 187
Cryptochiton stelleri 35
Cryptospira elegans 535
Cucullaea labiata 47
Cuspidaria rostrata 105
Cyclope neritea 423
Cyclopecten pernomus 85
Cymatium succinctum 377, 378
Cymatium femorale 379, 381
Cymbiola vespertilio 486, 490, 497
Cymbium glans 493
Cyphoma gibbosum 340
Cypraea argus 331
Cypraea aurantium 331, 335, 336
Cypraea corvus 338
Cypraea cicercula (takahashii) 325
Cypraea cribraria 328
Cypraea friendii 334
Cypraea fultoni 330
Cypraea goodalli 324, 329
Cypraea guttata 333
Cypraea leucodon 331, 335, 336
Cypraea mappa 332, 337
Cypraea moneta 326, 327
Cypraea nucleus 326
Cypraea stolida 324, 329
Cypraea tigris 332, 337
Cypraeacassis rufa 368, 369, 371
Cyrtopleura costata 134, 135, 164, 165
Cyrtulus serotinus 440, 442

D

Daffymitra lindae 549
Decatopecten plica 75, 78
Dendostrea frons see Lopha frons
Dentalium elephantinum 171, 172, 173
Diala albugo 257
Diastoma melanioides 268
Diodora italica 210, 212
Diodora listeri 209, 211
Discotectonica acutissima 609
Distorsio anus 374, 375
Distorsio clathrata 374, 375
Divaricella dentata 106, 107
Dolabella auricularia 627
Donax variabilis 145, 146
Dosinia ponderosa 138, 140
Drillia gibbosa 573
Drupa rubusidaeus 455
Duplicaria duplicata 600, 603
Duplicaria ustulata 600

E

Eatoniella kerguelensis 281
Eburna monilis see Bullia livida
Echinolittorina ziczac 272
Emarginula tuberculosa 208, 210

653

Enaeta guildingii 482
Engina medicaria 401
Enignomia aenigmatica 89
Ensis directus see *Ensis siliqua*
Ensis siliqua 155, 156
Entemnotrochus rumphii 196, 197
Entodesma beana 97
Epitonium clathrum 391, 392, 393, 396
Epitonium greenlandicum 391
Epitonium scalare 396
Epitonium turtonis 392, 393
Equichlamys bifrons 74, 78
Erato voluta 344
Eucrassatella antillarum 92
Euspira lewisii 360
Euvola laurenti 76, 80, 82
Exili blanda 480

F

Fasciolaria tulipa 444, 447
Fastigiella carinata 251
Ficus communis 372, 373
Ficus gracilis 372, 373
Fimbria fimbriata 108, 109, 110
Fissidentalium megathyris 172
Fissurella costaria see *Diodora italica*
Forerria belcheri 448, 460
Fulgoraria rupestris 491
Fusinus crassiplicatus 445
Fusinus dupetitthouarsi 446
Fusinus nicobaricus 443

G

Galeodea echinophora 366, 367
Gaza fischeri 224, 225
Gemma gemma 131
Geukensia demissa 51, 53, 55
Gibberula lavalleana 527, 529
Glabella pseudofaba 533, 534
Gloriapallium speciosum 73
Glossus humanus 118, 119
Glycyimeris americana 48, 49
Glycymeris inaequalis 48, 49
Glyphepithema alapapilionis 350
Granulifusus niponicus 438
Guildfordia yoka 9, 215, 217

H

Haliotis asinina 200, 205
Haliotis cracherodii 206
Haliotis cyclobates 202, 204
Haliotis elegans 203
Haliotis fulgens 201, 206, 207
Haliotis jacnensis 200
Haliotis scalaris 202, 204
Haliotis varia 201, 203, 205
Haminoea navicula 622, 623
Harpa amouretta 523, 524
Harpa costata 525, 526
Harpa crenata 524, 526
Harpa ventricosa 523, 525
Harpulina arausiaca 485
Hastula hectica see *Impages hectica*
Hastula lanceata 603
Hastula salleana 599
Haustellum haustellum 461, 462
 Hecuba scortum 145, 146
Helicacus areola 608
Hemifusus crassicaudus 435
Heterodonax bimaculatus 147, 149
Hexaplex fulvescens 463, 465
Hexaplex radix 457, 458
Hexaplex trunculus 449, 457
Hiatella arctica 114, 115
Hiatula diphos see *Soletellina diphos*
Hippopus hippopus 127, 130
Homalocantha scorpio 454
Hormospira maculosa 574
Hydatina amplustre 619, 620
Hydatina physis 619, 620
Hyotissa hyotis 64

I

Impages hectica 599, 601
Imbricaria conularis 540
Imbricaria punctata 539
Iravadia trochlearis 283, 284
Ischadium recurvum 51, 53
Isognomon ephippium 58
Ittibittium parcum 249

J

Janthina globosa 389, 390
Janthina janthina 389, 390
Jenneria pustulata 339, 342

K

Kuphus polythalamia 68, 166, 167

L

Lambis chiragra 300, 302
Lambis truncata sebae/truncata 302
Lambis violacea 294
Laternula spengleri 99
Latiaxis mawae 469
Latiromitra barthelowi 481
Latirus belcheri 439, 441
Leptopecten latiauratus 71
Lima scabra 69, 70
Limopsis tajimae 50
Lioconcha castrensis 133
Lithophaga teres 55
Littorina littorea 277
Littorina scabra (angulifera) 275
Littorina sitkana 273
Littorina zebra 276
Lopha cristagalli 61, 62, 64
Lopha frons 61
Lophiotoma indica 581
Lottia gigantea 188, 190
Lottia insessa 188, 189
Lucina pensylvanica 107
Lunulicardia auricula 123
Lutraria rhynchaena 159
Lyria lyraeformis 485, 488
Lyropecten nodosus 75, 83

M

Macleaniella moskalevi 191
Macron aethiops 406
Mactra violacea 158, 160
Malea ringens 383, 384
Malleus albus 59, 60
Mamilla melanostoma 355
Marginella glabella 536
Marginella petitii 534
Marshallora modesta 387
Maurea selecta 234
Maurea tigris 232, 233
Megathura crenulata 209, 211
Meiocardia moltkeana 118, 119
Melapium elatum 516
Melo amphora 497
Melongena corona 432
Mercenaria mercenaria 117, 136, 139
Mesodesma arctatum 161
Mitra florida 545
Mitra incompta 546
Mitra mitra 547
Modulus tectum 269
Monilispira quadrifasciata 575, 576
Mopalia lignosa 33
Morum dennisoni 522
Morum tuberculosum 521
Murex pecten 461, 462
Murex pele see *Homalocantha anatomica*
Mya arenaria 162
Myadora striata 102
Mytilus edulis 52, 54

N

Nacella mytilina 182
Nassarius arcularius 427
Nassarius gibbosulus 425
Nassarius glans 428

Nassarius globosus 424
Nassarius papillosus 430
Nassarius trivittatus 426
Natica aurantia 354
Naticarius canrena 359
Nautilus pompilius 13, 632, 633
Nautilus scrobiculatus 633
Neocancilla papilio 544
Neotrigonia margaritacea 91
Neptunea lyrata decemcostata 407
Neptunea tabulata 411
Neptuneopsis gilchristi 489, 495
Nerita costata 240
Nerita maxima 246
Nerita peloronta 239, 242
Nerita polita (antiquata) 243
Nerita scabricosta (ornata) 245
Nerita textilis 244
Neritina communis 239, 243
Neritina latissima 241
Neritodryas cornea 239
Neritopsis radula 236
Niso tricolor 399, 400
Nitidella nitida 417
Norrisia norrisii 227, 228
Northia pristis 405
Notocochlis tigrina 353
Notocrater youngi 194
Nucula proxima 38, 39
Nuculana polita 41
Nuttallochiton mirandus 31, 32

O

Oliva bulbosa 498, 505
Oliva carneola 498, 501
Oliva incrassata 499, 509, 511
Oliva peruviana 501, 505, 509
Oliva porphyria 504, 511
Oliva rufula 500
Oliva sayana 507
Oliva splendidula 500, 504
Oliva tessellata 498, 499, 507
Olivancillaria urceus 503, 511
Olivella biplicata 513
Olivella volutella 512, 513
Omalogyra zebrina 612
Onustus caribaeus 319, 320
Onustus exutus 318, 321
Oocorys bartschi 368
Opeatostoma pseudodon 440, 441
Orectospira tectiformis 258
Otopleura auriscati 614
Ovula ovum 342, 343

P

Pachydermia laevis 195
Pandora inaequivalvis 96
Panopea glycymeris 114, 115
Paphia textile 134
Paphia undulata see *Paphia textile*
Papyridea soleniformis 121, 126
Patella cochlear 176, 177, 178
Patella compressa 179
Patella longicosta 177, 178, 180
Patella mexicana 178, 180
Patella vulgata 176, 179
Patelloida saccharina 188, 189
Paziella pazi 449, 454
Peasiella tantilla 271
Pecten imbricatus see *Caribachlamys
     pellucens*
Pecten maximus jacobaeus 82, 84
Pectinodonta arcuata 185, 186
Pedum spondyloideum 73, 77
Periglypta reticulata 132, 136
Perotrochus quoyanus 196, 197
Pedum spondyloideum 73, 77
Periglypta reticulata 132, 136
Periploma planiusculum 100
Peristernia nassatula 437
Perna viridis 52, 54
Perotrochus quoyanus 196, 197
Persicula cingulata 527, 528, 529
Persicula persicula 529

Pervicacia ustulata see Duplicaria ustulata
Petaloconchus varians 322, 323
Petricola pholadiformis 134, 135
Phalium glaucum 305, 307
Phasianella ventricosa 220
Phasianotrochus eximius 226
Pholadidea loscombiana 164
Pholadomya candida 95
Phylloda foliacea 141, 144
Phyllonotus erythrostomus 463
Pinctada margaritifera 56, 57
Pinna nobilis 66, 68
Pinna rugosa 9, 65, 66
Pisania pusio 403
Pitar lupanaria 132, 133, 137
Placopecten magellanicus 81, 84
Placuna placenta 89, 90
Plagiocardium pseudolima 125, 126
Planaxis nucleus 262
Planaxis sulcatus 263
Pleioptygma helenae 548
Plesiothyreus cytherae 247
Plesiotriton vivus 561, 565
Pleurotomella edgariana 587
Plicatula gibbosa 88
Polystira albida 577, 580
Poromya microdonta see Poromya tornata
Poromya tornata 104
Prunum carneum 532
Pteria penguin 56, 57
Pugilina cochlidium 434, 435
Pugilina morio 433
Pupa solidula 618
Puperita pupa 238
Pusionella nifat 567
Pyramidella terebellum 615
Pyramidella tessellata 615, 616
Pyrene punctata 418

R

Raeta plicatella 157, 159
Ranella olearium 377, 380
Ranularia pyrum 378, 380
Rapa rapa 470
Raphitoma purpurea 584, 593
Rhinoclavis fasciata 253, 254
Rhinocoryne humboldti 256
Rhynocoryne humboldti see Rhinocoryne
   humboldti
Rissoella caribaea 611

S

Sansonia tuberculata 279
Scabricola fissurata 541, 542
Scalenostoma subulata 398, 400
Scaphander lignarius 621
Scaphella junonia 492
Scissurella rota 198, 199
Scutus antipodes 208, 212
Seila adamsii 388
Semele purpurascens 150, 151
Semele solida 150, 151
Sepia pharaonis 635
Serpulorbis oryzata 322, 323
Sigapatella novaezelandiae 314
Siliqua radiata 155, 156
Sinum cymba 356
Sinum incisum 358
Sinum perspectivum 357
Siphonalia pfefferi 404
Siratus alabaster 464
Skeneopsis planorbis 280
Smaragdia viridis 237
Smaragdinella calyculata 624
Solecurtus strigilatus 153
Solemya australis 40
Solen marginatus 154
Soletellina diphos 147, 148
Spengleria rostrata 116
Spinosipella acuticostata 103
Spinosipella agnes see Spinosipella
   acuticostata
Spirula spirula 25, 634
Spisula solidissima 157, 158, 160

Spondylus americanus 86, 87
Spondylus regius 86, 87
Stellaria solaris 319, 321
Stenochiton longicymba 30, 31
Sthenorytis pernobilis 395
Stomatella planulata 221, 223
Stramonita kiosquiformis 451
Streptopinna saccata 65
Strigilla pisiformis 141
Strombina maculosa 421
Strombina recurva 422
Strombus canarium 288
Strombus dentatus 287
Strombus gallus 296, 298
Strombus gibberulus 287, 290
Strombus gigas 7, 298, 301, 303
Strombus goliath 301, 303
Strombus lentiginosus 292
Strombus listeri 293, 297
Strombus luhuanus 290, 291
Strombus plicatus (sibbaldi/columba)
   288, 289
Strombus sinuatus 292, 295
Strombus taurus 295, 296
Strombus vittatus 289, 291, 293, 297
Struthiolaria papulosa 308
Subcancilla attenuata 543
Sveltia gladiator 562, 564
Syrinx aruanus 479

T

Tagelus plebeius 152, 153
Taron dubius 436
Tectarius pagodus 278
Tectonatica pusilla 348
Tectonatica violacea 351
Tectus dentatus 231
Tectus niloticus 230
Tegula regina 227
Teinostoma reclusum 282
Telescopium telescopium 266
Tellidora cristata 142, 144
Tellina radiata 142, 143
Tenagodus modestus 260, 261
Tenagodus squamatus 260, 261
Terebellum terebellum 307
Terebra armillata 602, 605
Terebra buccinoidea see Bullia livida
Terebra dimidiata 603, 606
Terebra maculata 601, 607
Terebra triseriata 602, 605
Terebralia palustris 265, 266, 267
Terebralia sulcata 264, 265, 267
Teredo navalis 166, 167
Thatcheria mirabilis 593
Thracia pubescens 101
Thyca crystallina 398, 399
Tibia fusus 286, 297, 299
Tonna perdix 382, 383, 384
Tonna sulcosa 382, 383
Torellia mirabilis 316
Tractolira germonae 487
Tridacna gigas 68, 124, 128, 130
Tridacna maxima 129, 130
Tridacna squamosa 128, 129
Trigonostoma milleri 558, 560
Trigonostoma scalare 560
Triplofusus giganteus 444, 447
Trisidos tortuosa 43, 44
Triumphis distorta 515
Trivia pediculus 344, 345, 346
Trivia solandri 345, 346
Trochita trochiformis 312
Trophon geversianus 448, 453
Tudicla spirillus 408, 409
Tudivasum armigera 471, 472
Tudivasum spinosum 471, 472
Turbinella angulata 477, 478
Turbinella pyrum 477, 478, 479
Turbo jourdani 216, 218, 219
Turbo petholatus 216, 219
Turbo sarmaticus 218
Turricula javana 568, 569
Turricula tornata 568, 569

Turridrupa bijubata 576
Turrilatirus turritus 439
Turris crispa (intricata/variegata/
   yoddoencia) 580, 582
Turritella terebra 258, 259
Tutufa bufo 364
Typhisala grandis 450

U

Umbonium vestiarium 221
Umbraculum umbraculum 628
Ungulina cuneata 111

V

Vanikoro cancellata 310
Varicospira crispata 286, 299
Vasum cassiforme 475, 476
Vasum muricatum 475, 476, 479
Vexillum cadaverosum 552
Vexillum citrinum 554, 556
Vexillum costatum 554
Vexillum crocatum 551, 553
Vexillum regina see Vexillum citrinum
Vexillum rugosum 552, 555
Vitularia salebrosa 450, 453
Volema paradisiaca 431
Voluta musica 484, 494
Volutoconus bednalli 490
Volutocorbis abyssicola 489
Volutomitra alaskana 549, 550
Volva volva 341, 343
Volvarina avena 530
Vulsella vulsella 59

X

Xenophora conchyliophora 318, 319, 320

Y

Yoldia thraciaeformis 42

Z

Zemira australis 514
Ziba annulata 541
Zidona dufresnei 494, 496
Zierliana ziervogeli 551

655

ACKNOWLEDGMENTS

M. G. HARASEWYCH

I would like to thank the many shell collectors and dealers who have, over the years, brought to my attention and allowed me to photograph numerous extraordinary specimens of mollusks. Al and Bev Deynzer of Showcase Shells, Sanibel, Florida and Sue Hobbs Specimen Shells, Cape May, New Jersey have been especially helpful in providing specimens for photography for this book. Many of the illustrated shells, including specimens from the William D. Bledsoe, Roberta Cramner, and Richard M. Kurz collections, are in the collection of the National Museum of Natural History, Smithsonian Institution. Ms. Yolanda Villacampa provided invaluable assistance in the production of the scanning electron micrographs of the minute shells. Special thanks are due to Kate Shanahan, Jason Hook, Caroline Earle, Michael Whitehead, and Kim Davis of Ivy Press as well as to reviewers and readers for their many contributions to the concept, design, organization, and content of *The Book of Shells*. I especially thank my wife and daughters for their patience and support during the production of this work.

FABIO MORETZSOHN

I am grateful to Colin Slater and Steve Luck, as well as Allison and Justin Knight, of Corpus Christi, Texas, for their assistance with research. Several people contributed information on queries about certain species; in particular I thank Marcus Coltro of Femorale.com, Brazil; Marta deMaintenon of the University of Hawaii at Hilo, Hawaii; Brian Hayes of Algoa Bay Specimen Shells, Raleigh, North Carolina; Richard Petit of North Mirtle Beach, South Carolina; and, Fabio Wiggers of Florianópolis, Brazil. Robert and Juying Janowsky of MdM Shell Books, Wellington, Florida, and Klaus and Christina Groh of ConckBooks, Hackenheim, Germany, provided most of the books used in this project and were instrumental in fetching some hard-to-find books. I thank all of the staff at Ivy Press, in particular Caroline Earle and Kate Shanahan, for their help and guidance, and the reviewers and readers for their constructive suggestions. And last, but not least, I thank my wife and daughters for their constant support and encouragement.